VERTICAL SEISMIC PROFILING AND ITS EXPLORATION POTENTIAL

SEISMOLOGY AND EXPLORATION GEOPHYSICS

E. I. GALPERIN
Academy of Sciences of the USSR, Institute of Physics of the Earth, Moscow, U.S.S.R.

VERTICAL SEISMIC PROFILING AND ITS EXPLORATION POTENTIAL

Editorial Supervision by

PETER KENNETT
Seismograph Service (England) Ltd.

D. REIDEL PUBLISHING COMPANY

A MEMBER OF THE KLUWER ACADEMIC PUBLISHERS GROUP

DORDRECHT / BOSTON / LANCASTER / TOKYO

Library of Congress Cataloging in Publication Data

CIP

Galperin, E. I. (Evseĭ Iosifovich), 1920-
Vertical seismic profiling and its exploration potential.

Translation of: Vertikal'noe seĭsmicheskoe profilirovanie.
Includes bibliographies.
1. Seismic reflection method. 2. Seismology. 3. Seismic prospecting. I. Title.
QE539.G3313 1985 551'.0?8 85-11863
ISBN-13: 978-94-010-8797-1 e-ISBN-13: 978-94-009-5195-2
DOI: 10.1007/978-94-009-5195-2

Translated by M. M. Samokhvalov.

Editorial supervision by Peter Kennett.

Published by D. Reidel Publishing Company
P.O. Box 17, 3300 AA Dordrecht, Holland

Sold and distributed in the U.S.A. and Canada
by Kluwer Academic Publishers,
190 Old Derby Street, Hingham, MA 02043, U.S.A.

In all other countries, sold and distributed by
Kluwer Academic Publishers Group,
P.O. Box 322, 3300 AH Dordrecht, Holland.

To my teacher – academician
Grigori Aleksandrovich
Gamburtzev

TABLE OF CONTENTS

FOREWORD

The present book is the author's third on the subject of vertical seismic profiling (VSP). Ten years have elapsed since the publication of the first book. During this period, VSP has become the principal method of seismic observations in boreholes and the chief method of experimental studies of seismic waves in the real earth. VSP combines borehole studies in the seismic frequency band, well velocity surveys, proximity or aplanatic surveys, all of which previously existed as separate methods.

The high effectiveness of VSP, its great practical value, the express nature and clarity of the results obtained have all contributed towards a very rapid acceptance of the method. In the USSR VSP has been used in an overwhelming majority of areas and is being used increasingly in many foreign countries as well. This has been greatly facilitated by the translation into English and the publication in the U.S.A. by the Society of Exploration Geophysicists of the book *Vertical Seismic Profiling* (Tulsa, Oklahoma, 1974). As the method has become more familiar, it has attracted growing interest outside the USSR This has been substantiated by the special seminar on VSP (Oklahoma, 1979) which was organized for 22 U.S. companies and universities and presented by the author.

Extensive experience in the use of VSP has confirmed that the effect of the transition from observing the first arrivals only to the inclusion of the later portion of the record can be compared only to the effect of a similar transition in the surface observations carried out in the thirties that resulted in the discovery for seismic exploration of the reflection method.

The high efficiency of VSP has attracted the attention of many specialists, and the method has made further progress. VSP studies have been made on continental shields of crystalline rocks; on plateaus with sedimentary beds of various thickness, both terrigeneous and carbonate; in conditions of folding consisting of sedimentary deposits of great thickness. A great number of surveys have been carried out in the course of studies of complex structures with high-dip interfaces such as salt domes, diapirs, reefs, tectonic faults, intrusions, faulted and fractured formations, etc.

VSP is carried out both on land and at sea in a wide range of distances from the source (from several metres to several kilometres) and in a wide frequency band (from several hertz to kilohertz), depending on the objectives of the observations.

VSP is also employed for solving a very wide range of problems that influence the effectiveness of all the seismic methods including reflected and refracted longitudinal, transverse and mode-converted waves, in solving the problems encountered in oil and mineral geology, in problems encountered in research and in foundation engineering, and those encountered in the study of thick beds of sedimentary deposits at comparatively great depths, and in the study of sea-bed placers at depths of 10–20 m.

However, despite the extensive application of VSP, the principal potentialities of the method have by no means been fully realized during the first stage of its development. The introduction of the general polarization method for seismic studies marked the start of the second stage of VSP development and not only enabled its potentialities to be fully utilized, but also substantially enhanced the importance of seismic studies in the general field of geologic exploration methods. The generality and the effectiveness of the polarization method stem from the vector analysis of the wave-field that makes use of all the field's parameters, including polarization, and of all wave types (monotype longitudinal, transverse and mode-converted) to obtain information about the Earth. The author, by invitation of the US Geophysical society, delivered a lecture entitled 'The Polarization Method – a General Method of Seismic Studies' at the 49th Assembly of the Society (New Orleans, 1979).

During the first stage of development, when only the vertical component of motion and the kinematic treatment of waves were considered, VSP was called upon chiefly to improve surface observations and to solve structural problems. Extensive combination of VSP with surface observations at that stage increased the importance of seismic studies for the selection of structures for deep drilling.

The Polarization method, PM–VSP, greatly improved the prospects for solving lithologic and stratigraphic problems and for studying the borehole locality. This created the conditions necessary for the expansion of the field of application of seismic studies, and made it possible to extend the limits of traditional exploration and selection of structures for deep drilling, and to be used during production and exploitation of the fields. PM–VSP formed the corner-stone for the development of a new trend – production seismics, the term implying seismic studies at the time of exploration and exploitation of fields. Production seismics, or seismics of the borehole locality, forms the basis for a substantial improvement in the effectiveness of the entire field of geologic exploration methods and exploitation work, and for a reduction in the amount of extremely costly deep drilling. When appraising the value of production seismics for national economy, it should be remembered that, despite the progress in seismic studies achieved mainly as the resuflt of digital recording and data processing, the effectiveness of geologic exploration methods for oil and gas drops as the depth of interest increases, and that more than half of the budget allocated for geologic exploration is spent on drilling dry wells.

In such conditions the chief factor for a substantial improvement in the effectiveness of geologic exploration work is a sharp reduction in the amount of costly unproductive deep drilling, both during the search and the exploration stages.

The list of references included in this book contains over 200 titles produced since the publication of the first book in 1974. In the course of preparation of the second book on VSP the author, in addition to describing the specific properties of the method, the equipment used and the procedures employed, has also endeavoured to digest the results of the research carried out and to generalize the experience gained in the application of VSP. In addition the exploration potential of the method is discussed, and the scope of problems solved by it at the various stages of geologic exploration, e.g. during the initial search for structures, their selection for drilling, and during exploration and exploitation drilling. However, owing to the rapid development of the method and to the growing number of publications, and to the limited extent of the book, the author could hardly have completely succeeded in his purpose. Special attention has been focused therefore on production seismics.

In 1982 appeared in the USSR the author's second book on VSP (*Vertikal'noye*

Seismicheskogo Profilirovania, Nedra, Moscow). The present book differs from it in that it is much more complete and contains more new experimental data, in particular, the results of studies of the effect of an oil and gas deposit on the wave field parameters, as well as the description of the automated VSP and PM–VSP data processing system.

The book contains a large amount of work carried out since the publication of the authors first book 'Vertical Seismic Profiling'. The work was directed by the author at the Institute of Physics of the Earth at the Academy of Sciences of the USSR (IPE) in collaboration with the 'Krasnodarneftegeofizika' trust, with the Kazakhstan branch of the All-Union Institute for Exploration Geology (VIRG) and with Volgograd expedition of the Lower Volga Scientific Research Institute for Geology and Geophysics. It also contains materials published by numerous research and industrial institutions, and the author is grateful to all those specialists who have made this material available to him.

In the preparation of the book the author has been greatly aided by R. M. Gal'perina, A. V. Frolova, and Yu. D. Mirzoyan to whom the author is very indebted. The author would also like to thank Professor I. I. Gurvich, who read the manuscript and made specific suggestions and remarks, and G. I. Petrashen' for providing a constant opportunity to discuss the problems concerning the development of the method.

The author will carefully consider the opinions and comments that readers would care to make concerning both the essence of the method and the book as a whole, and is grateful to them in advance. In the two decades during which the method has been developing, the author did not meet a single specialist who, having started work on VSP, subsequently rejected it. The method arouses enthusiasm on account of the unambiguity and clarity of its results. The number of specialists who feel enthusiastic about VSP grows every year, and the author wishes them all possible successes in their endeavours.

The Russian manuscript has been translated into English by M. M. Samokhvalov.

Author's note: The last proof of the present book was received by the author in July 1985, i.e., almost a quarter of a century after his first papers on VSP had been published (Galperin E. I., Irolova A. V., 1961 [15]. These papers had initially formulated the physical and geological foundations of the method and discussed the exploration potential of its two modifications (single- and three-component-PM VSP) that distinguished it from all preceding methods of seismic borehole observations. It is those papers which mark the beginning of VSP. Since that time, the author has published over 60 papers and several books on VSP.

The translation into English done in Moscow, the editing of the English translation of the manuscript carried out in England, along with several rounds of discussions on the text and proofs with the inevitable mailings, all combined to hold up the book's publication for three years. In addition, two years were spent preparing the Russian text.

During these five years the VSP method developed rapidly in the USSR in its PM VSP form. In the West, PM VSP has been developed during the last two years. This has been considerably facilitated by the publication of the English translation of the author's book *The Polarization Method of Seismic Exploration* (Reidel, 1983). Digital processing of VSP data and VSP modelling have been extensively developed. Many of the originally-formulated advantages of VSP have been put into practice, and this has substantially expanded the field of problems solved with the aid of VSP.

Unfortunately, for purely technical reasons, the author has been unable to reflect these developments in the present book, or even to include them in the list of references, and is, accordingly, obliged at this point to offer his apologies to the authors of numerous papers and books on VSP published recently.

I regard it as my pleasant duty to express special gratitude to Peter Kennett who undertook the task of editing the book and has done a great amount of work improving the English text.

The author is also grateful to the D. Reidel Publishing Company who realized the potential of the methods we have developed and overcame a lot of technical problems connected with the publication of our books.

INTRODUCTION

The effectiveness of the entire complex of geologic exploration methods for oil and gas depends greatly on the effectiveness of seismic exploration. In recent years over 90% of new oil and gas reserves have been discovered in areas selected for drilling by seismic methods. The volume of seismic exploration for oil and gas grows continuously. The importance of seismic studies in mining geology, in engineering studies and research grows from year to year. In mining geology, the seismic method has largely replaced geologic mapping in the study of ore bodies, of the beds containing them and of the structures that control them.

During the last 10–15 years, mainly due to the introduction of multiple subsurface coverage, digital recording and data processing, substantial improvement in the effectiveness of seismic exploration has been achieved, and principally in the increase in the depths studied. This has been accompanied by a sharp rise in the costs of seismic exploration. However, the effectiveness of seismic studies grows at a much slower rate than the complexity of the problems posed. A new problem formulated in seismic exploration for oil, in addition to the problems of increasing depths of interest, improved resolution and precision of the results, as well as that of studying complex formations, is one of predicting the geologic cross-section. This includes predicting the presence of reservoirs and their productivity, and establishing the genesis of the fields and mapping anticlinal traps. In addition, the physical and mechanical properties of the geologic cross-section, and the presence of zones of abnormally high pressures, etc. need to be determined for the design of boreholes.

In mining geology the difficulties arise out of the need to study weakly differentiated high-velocity formations with high-dip interfaces of complex shapes and formations with continuously changing parameters, as well as to make direct forecasts of ore deposits.

The effectiveness of seismic exploration depends upon numerous factors. The decisive ones are: knowledge of the process of propagation of seismic waves in the real earth, methods of analyzing complex wave fields and the types and parameters of the waves being used. Consider these factors in detail.

(1) Because of the complexity of the real earth in which the seismic waves propagate, the wave fields observed on the seismograms feature a large number of waves of various types and extensive interference patterns. As the studies become more detailed, the real earth appears increasingly heterogeneous, and the wave fields become increasingly complex. In such conditions, the methods of experimental study of seismic waves in the real earth assume special importance, over and above the theoretical study of the propagation of seismic waves, based on the application of modern computer techniques. The physics of the generation and propagation of seismic waves in the real earth has been studied far less extensively. This can be explained to a large extent by the fact that for a long time experimental studies of seismic waves have been limited mainly to surface observations. The data obtained on wave kinematics and dynamics in such observations

are inadequate for the understanding of the physics of their formation and propagation. For this reason valid criteria for the identification of the nature of recorded waves are not always available, and from the physical viewpoint sound methods for interpreting the observed data are often lacking. As a result, there are still no absolutely clear ideas about the effect of various inhomogeneities in the geologic section on the wave field, and on how these affect the characteristics of waves of various types.

In the search for ways of perfecting studies of seismic waves in the real earth, it seemed expedient to stop studying particular waves in wave-fields already formed, and instead to study the process of the formation of individual waves, and of the wave field as a whole. This encouraged us to develop down-hole observations in boreholes. It was imperative first of all to perfect the methods of analysing seismic waves, and the best prospect for this was to study not only the first arrivals but the entire later portion of the seismic record.

(2) Current methods of analysis of complex wave fields are founded on the discrimination of waves according to two parameters that characterize the wave field. In the early stages the discrimination was based mainly on the signal frequency. Subsequently, methods of discrimination based on the direction of propagation and the apparent velocity of the waves were developed extensively. The latter parameter is used in various types of source and detector arrays, in controlled directional reception, and, most importantly at present, the common depth point technique. Discrimination according to the two above-mentioned parameters is determined on the basis of recordings of a fixed component in space, i.e., the Z-component in the case of longitudinal waves. Hence, seismic exploration exhibits a scalar approach to the wave field, which in essence has a vector nature.

(3) Of all the various wave types, it is mainly longitudinal waves that have been used until now. Research is in progress, which is based on the use of transverse and mode-converted waves. Each wave type requires its own special methodology, e.g. the method of longitudinal reflected waves, the method of transverse reflected waves, the method of mode-converted transmitted waves, etc. Each method studies the wave field recorded by one particular geophone fixed in space (a Z component geophone for the longitudinal and an X or Y component geophone for the transverse waves). However, these methods until now have not been widely used in industry, and some of them have not advanced beyond the experimental stage.

Fundamentally, new results were to be expected not so much from the improvement of each of the methods enumerated above as from the development of a general method based on the simultaneous use of various wave types generated by a single source, and conditioned by the heterogeneities of the geologic cross-section. This has proved possible due to the results obtained in VSP studies of the conditions of excitation and propagation of waves in various media. Research in this direction carried out at IPE resulted in the creation of the polarization method of seismic studies. The polarization method involves the simultaneous discrimination and tracking of a large number of regular waves of different types, and this in turn requires the perfection of methods for analyzing complex wave fields, and the nature of waves, and for improving the accuracy of stratigraphic correlation.

Work on developing VSP was begun by IPE in 1959. From the viewpoint of its physical approach and its governing ideas, VSP is a development of the experimental principles for the study of seismic waves in the real earth advocated by G. A. Gamburtzev for many years. In the development of the method, prototypes of equipment were constructed and techniques were devised, and plans drawn up for the interpretation of results. Experimental studies of wave fields were also carried out in regions with a variety of geologic structures.

In the USSR surveys have been carried out in plateau areas consisting of sedimentary beds of thin high velocity layers and in formations with low velocity differentiation (Kuibyshev and Krasnodar provinces); in regions of thick high-velocity carbonate beds with either deep or comparatively shallow basement (Volgograd province); in regions with well-developed terrigeneous deposits and shallow basements (Kuschev area, Krasnodar province). Observations have been carried out in fold areas with thick beds of coarsely fragmented terrigeneous deposits (Fergana depression, Northern Tyan-Shan) and with thin-layered sandy-clay beds (Afip area, Krasnodar province). The results of this research formed the subject of the first book on VSP.

The effectiveness of the method resulted in its rapid introduction into industrial practice. An enormous volume of VSP work has been completed in the USSR. It was established as a result of these studies that the wave fields observed on surface seismic records are characterised by a high degree of complexity, and that the upper part of the geologic section plays a major role in the nature of the seismogram. The studies not only enabled the variations in the propagation of seismic waves in formations with different structures to be studied, but also made it possible to find out and appraise the potential of the method. It was established that in comparison with surface observations VSP makes it possible to determine the nature of waves and to establish their stratigraphic correlation more accurately, and to evaluate exploration potentialities of seismic techniques, and to choose in each specific geologic situation the best field procedure. All such factors exercise a substantial effect on the effectiveness of seismic studies, and are of major importance in the practice of seismic exploration. The best prospects are offered by a combination of surface observations and of VSP. The results obtained have led to a critical re-appraisal of certain fundamental and practically important concepts concerning the part played by different types of waves in the formation of the total wave field, as well as drawing attention to the exploration potentialities of seismic methods. The main amount of work has been carried out in the seismic frequency band. Many of the results obtained may also be of interest for the understanding of the specific properties of the wave fields observed in deep seismic measurements and in earthquake seismology.

Despite the importance of the results obtained, the fundamental advantages of VSP have been realized only since the polarization method has been developed. This has substantially broadened the scope of problems being solved, primarily, in the study of the borehole locality.

This is due to the fact that, despite the improvement in seismic methods and the use of more sophisticated recording systems (with a substantial rise in costs), the increase in the depths of interest and work in areas with more complex structures have brought about a reduction in the effectiveness of seismic exploration, and thereby of the whole field of geologic exploration for oil and gas. The result is a rise in the percentage of dry wells and soaring costs of the geologic exploration process. In such conditions we thought it expedient, in order to cut the amount of costly deep drilling of dry wells, at the stage of search and of exploration, to exploit more fully the potentialities of VSP in the study of the borehole locality and to simultaneously improve methods of surface observations. VSP, in contrast to all other geophysical methods, enables the geologic section to be studied not only along the axis of the borehole, but also in the borehole locality at distances up to 2.0–2.5 km from it. Such studies are possible not only down to the depth penetrated by the borehole, but also beneath total depth. The use of PM–VSP for the study of the borehole locality not only improves the effectiveness of seismic studies during the traditional seismic exploration stage and selection of structures, but also becomes the basis of production seismics – i.e. seismic studies at the exploration and exploitation stages. Typical of production seismics is the extensive use of the combination of surface seismic observations and other forms of geophysical observations (electric, sonic, radioactive logs) in boreholes in close association with the drilling activity. Production seismics provides an opportunity to draw up drilling plans with open eyes and hence to reduce the number of dry wells. Naturally, production seismics envisages a substantial growth in the volume of seismic studies in boreholes and consequently in their costs. However, since the only alternative to seismics is drilling, where costs are many times higher than those for seismic observations, there can be no doubt that additional structural information obtained with the aid of production seismics will enable a reduction in the costs of drilling.

The problems being solved at present with the aid of VSP can be subdivided into two main groups, according to their nature: those connected with the interpretation of surface seismic data, and those dealing with direct studies of the borehole locality.

Problems in the first group include: the study of the velocity distribution (the determination of average, and interval velocities of longitudinal and transverse waves); the study and selection of optimum conditions for excitation of various wave types; the study of wave fields (determination of the nature and parameters of unwanted waves); the discrimination of various types of reflected waves, their nature and their stratigraphic correlation; the identification of various wave types associated with the same elements of the geologic section; the determination of the intensity of various wave types (especially multiple reflections) and providing data to determine the optimum methods and recording systems for each specific geologic situation etc. Such problems can be solved with both the methods of reflected and refracted waves.

The task of studying the borehole locality involves two groups of problems: the structural and the lithologic and stratigraphic. Problems in the former group include: mapping interfaces intersected by the borehole and located beneath total depth; the study of high-dip interfaces from which reflected waves are not observed on the surface; discovery and tracing of tectonic faults etc.

The solution of lithologic and stratigraphic problems is based on the study of the physical properties of the rock. Using the different velocities of the various wave types together with the absorption and reflection properties of the geologic layers, the sub-surface can be sectioned in accordance with the elastic parameters, e.g. Young's modulus and Poisson's ratio. With a simultaneous improvement in resolution, such information makes it possible to predict the geology of the borehole locality.

Among the geologic problems, in addition to structure, location of the deposits is the main one during the field exploration stage. This may include: the discovery of a deposit penetrated by the borehole, but not detected earlier; predicting a deposit not penetrated by the borehole, but located in its vicinity; the study of the deposit itself, and its dimensions, volume, boundaries, geometry, internal structure, and gas-oil contacts; predicting zones with abnormal pressures and the initial evaluation of the reserves.

In the course of field exploitation VSP makes it possible to monitor the characteristics of the deposit enumerated above, to regulate the injection of water into the reservoir, to evaluate the effectiveness of production methods as well as solving several technological problems associated with the drilling of the borehole itself.

Part I of the book (Chapters 1–4) deals with general methodological aspects of VSP. In Part II (Chapters 5–8) the attention is focussed on the methods used to improve the effectiveness of surface seismic observations. It describes the experience gained and the results obtained in the application of VSP in the study of various wave types that form the basis of the major seismic exploration methods.

Because of their importance, topics dealing with the use of PM–VSP in production seismics and with the study of the borehole locality are presented in Part III (Chapters 9–13). Chapter 13 – the last one – discusses the exploration potential of the method and ways of perfecting it.

Not all the potentialities of VSP are being put to work at present. However, there can be no doubt that the development of the method will open up new possibilities, which are at present difficult to foresee. Any progress in the method should be valuable, and one can unreservedly spend both money and effort trying to achieve it.

PART ONE

VERTICAL SEISMIC PROFILING (VSP)

CHAPTER 1

PHYSICAL AND GEOLOGICAL PRINCIPLES OF VSP

The principal difficulties involved in the interpretation of data from all existing seismic methods are governed by the complexity of the wave fields, by the absence of reliable criteria for the analysis and classification of recorded waves. This is to a large extent due to the inadequacy of available information about the process of wave propagation in the real earth.

The main trend in the development of seismic exploration in the last 50 years has been the continuous improvement of methods to increase the signal-to-noise ratio, with the aim of discriminating and tracing progressively weaker regular waves [1]. But the discrimination of regular waves is only the first step in the analysis of complex wave fields, and does not constitute the sole cause of difficulty. It is also necessary to determine the nature of the discriminated waves, and to establish their stratigraphic correlation. However, since the effect of the geology on the formation of different wave types has not been adequately studied for all models of the real earth, the nature of recorded waves is not completely understood. This is true primarily for methods utilizing the recording of waves from comparatively great source offset distances.

In the most common method of reflected longitudinal waves, the limitation of the effectiveness and the depth of investigation in the majority of regions is due to the presence of multiple reflections. In many regions, there are no regular waves on the seismograms at all, because of intensive interference of unwanted waves connected with the upper portion of the geologic section, and whose origin is not always clear*.

To improve the effectiveness of seismic studies, it appeared necessary to extend the potential for analysing complex wave fields, and to use waves of different types to obtain information about the earth. The polarization method (PM) [2] provided the means by which both these aims could be attained. This method, in addition to the conventional parameters used in the analysis of wave fields (frequency and apparent velocity), utilizes the discrimination of waves at the detector in terms of the direction of particle motion, i.e. the wave polarisation. This facilitated both discrimination and interpretation of different wave types, thereby greatly improving the effectiveness of the method. However, this required the determination of the nature of a large number of different wave types discriminated in PM, as well as the identification of different wave types connected with the same inhomogeneities of the earth.

To overcome the difficulties cited above, experimental methods for a more reliable analysis of the wave pattern on the seismograms had to be developed, and first of all methods of studying the process of generation and propagation of individual waves and the formation of the wave field as a whole had to be evolved. This made a transition from predominantly surface to down-hole observations necessary.

* Here and below the term upper portion of the geologic section implies the layer between the surface and the first seismic marker reflector.

1. Down-Hole Seismic Observations

Seismic observations in boreholes have been carried out for almost 40 years and form part of the standard range of seismic surveying. They may be classified according to two criteria: the problems to be solved and the frequency band used. For a long time they have been used mainly to measure the propagation velocity of seismic waves, and to solve certain structural problems. Accordingly, two different modifications of the down-hole method were developed: well velocity surveys and proximity, or aplanatic, surveys.

Waves from various sources in different frequency bands are used to measure velocities in boreholes. According to the frequency content of the waves, which determines the resolution and the accuracy of measurements, well velocity surveys can be subdivided into two types – one employing the seismic and the other the sonic frequency bands (sonic logging).

Well velocity surveys

The overwhelming mass of well velocity surveys is performed using direct *longitudinal* waves. This method is the most reliable source of information about the average velocity of seismic waves [3, 4].

Surveys based on *refracted* waves involve laborious observations at great distances from the source, and have not found extensive application.

Surveys based on direct *transverse* waves have been used in recent years thanks to the development of the shear wave method [5]. In this method direct transverse SH waves excited by a directional source are studied.

There are very limited possibilities for studying the detailed velocity distribution in boreholes and for constructing velocity models of the real earth in the seismic frequency band. It is an established fact that, whereas it is possible to obtain an average velocity curve with the aid of a longitudinal vertical locus, the determination of interval velocities, especially in the case of media with poor velocity differentiation, or of thin high-velocity layers, presents great difficulties. The conventional methods of velocity logging are practically worthless for determining the nature of seismic interfaces and for the study of transition layers and continuously-changing media.

To improve the accuracy and the resolution of the method, efforts have been underway to perform seismic observations using short source-receiver bases. Electric discharges, piezoelectric transducers and electromagnetic vibrators have been used as sources. But unhappily, much of this work has not gone beyond the experimental stage, although no doubt the prospects for it are good.

Sonic logging

This method of high-frequency observations is based on the study of the propagation of sonic (50-5000 Hz) and ultrasonic (20-50 kHz) pulses in rock with short source-receiver spacings, enabling the geologic column to be sectioned into layers down to fractions of a meter thick. Sonic logging has found extensive applications: (a) in mining geophysics for lithologic sampling, for studying porosity, the presence of caves and the determination of gas- and water-oil contacts, as well as for assessing the quality of borehole casing cementation; (b) in engineering geology for studying the mechanical and strength properties of soil; (c) in seismic exploration, where sonic logging serves as a source of detailed information on propagation velocities in the real earth. Note that velocities obtained from sonic logging may differ from the velocities obtained from observations made in the seismic frequency band. Inhomogeneities of the earth, velocity dispersion, changes of rock properties in the vicinity of the borehole due to fluid invasion, saturation of dry terrigeneous rock etc. may also serve to create such a difference. There have been numerous papers dealing with sonic logging, and this subject is treated at length in this book.

Proximity (or aplanatic) surveys

They are employed for the solution of structural geological problems, e.g., to determine the stratigraphic details of high-dip interfaces in the vicinity of the well, and particularly in the study of the flanks of salt domes. As a rule, only the first arrivals of longitudinal waves are used.

Seismic up-hole surveys

Another method conventionaly classified under seismic well shooting is up-hole surveys, in which the observations are carried out on the surface and charges are exploded at various depths in the borehole. Such a method is generally used to study the upper portion of the geologic section, and the extent of its application grows from year to year.

The study of waves in the later parts of the well-geophone record

In spite of the large amount of well velocity surveys carried out world-wide, it has failed to exert a radical influence on the effectiveness of seismic studies, because, in the majority of cases, the major difficulty is not the result of the lack or inadequacy of average velocity data but the result of limited potential for the analysis and interpretation of complex wave fields. Down-hole observations can provide much important information about the nature of the recorded waves, not only from the first arrivals, but also from waves observed in the later portion of the record.

Many authors have pointed out the advantages connected with the study of waves in the later portion of the record and the scope for such studies [3, 6, 7, 8].

The ability to obtain undistorted down-hole records permits the study of the details of wave propagation, as well as the dynamics of direct longitudinal and transverse waves with the object of determining the attenuation properties (absorption and dispersion) of the real earth [6]. The evolution of the method of transverse waves was responsible for significant progress in borehole observations. The results of this work served as a basis for studies begun at the Institute of Physics of the Earth of the Academy of Sciences of the USSR (IPE) in 1959, which led to the development of the method of Vertical Seismic Profiling [9–15]. Since the first book was published, which encouraged the application of the method, a large amount of experimentation aimed at further development of the method has been carried out. Great attention has been paid to methods of processing the recorded data. Here, the main tendency has been the application of techniques and procedures widely tested and used in surface seismic methods adapted for use along the vertical profile. Software for processing VSP data on various types of computers has been programmed, and digital processing became part of production technology. Methods of studying the dynamic characteristics of waves, the relative intensity of multiple reflected waves, and the physical properties of the geology from VSP data have evolved. A noteworthy contribution in this field has been the work carried out at the Leningrad State University and at the Leningrad Institute for Optical Mechanics (T. I. Vavilova, B. Ya. Gal'chinsky, N. I. Berdennikova, G. G. Pogonyailo [16–22]), at the Perm'University (Yu. A. Byakov, V. A. Bereznev, A. A. Malovichko, V. A. Silaev). Methods and instrumentation for measurements have been perfected (IPE – A.G. Gamburtzev, V. V. Kuznetzov, A. V. Seval'nev [23, 24], VNII Geofizika – Shekhtman [25–27]). Several modifications of the method applicable to the solution of diverse problems have been developed.

For the solution of structurals problems, of major importance the work has been carried out at VNIIYa GG (E. V. Karus, L. L. Khudzinskiy, G. E. Rudenko [28–32]) and at VNIIGNI (V. A. Teplitzkiy [33–37]).

Vertical seismic profiling using sonic frequencies up to 2000 Hz has been developed, to study coastal seabed sediments 20–30 m thick, in the search for mineral placers, and for the solution of problems in engineering geology.

The development of the polarization method [2], born at the IPE, created new opportunities for VSP. The polarization method of VSP (PM–VSP) substantially enhanced the effectiveness of VSP studies of complex media. Of great value in the evolution of the PM–VSP for oil geology has been the work carried out by the 'Krasnodarneftegeofizika' trust (Yu. D. Mirzoyan) and for mining geology – the work at KazVIRG (R. N. Khairutdinov).

2. Principal Features of VSP

For the solution of a variety of problems, the effectiveness of seismic measurements depends to a substantial degree on the orientation of the line of observations with respect to the inhomogeneities of the geologic section being studied. Observations along a line in a plane parallel to that of the interface enables the maximum length of interface to be mapped. From this point of view surface observations are most effective when horizontal or low-dip interfaces are being studied. However, observations along a line that intersects the inhomogeneities of the geologic section are most conducive to the study of the propagation of seismic waves, the determination of their origin and their stratigraphic correlation, the study of the reflection and absorption properties, as well as other characteristics of seismic waves which are of great importance for the improvement of the studies. For horizontal and low-dip strata such a line is the vertical one.

Let us compare the principal features of observations made along a vertical profile (VSP) with observations along a horizontal (surface) profile.

The potentialities of VSP are primarily determined by the following differences:

(1) In contrast to surface observations in which only the emergent wave pattern can be studied, VSP enables the very process of individual seismic wave formation and propagation in the real earth as well as the formation of the total wave field to be observed and studied. To observe the evolution of the wave process, we, so to speak, split up the entire wave field into its individual wave components. This makes it possible to determine the origins of waves of different types, to study their kinematic and dynamic properties and, thus, to gain more comprehensive information about the Earth. We are

then able to find out how a comparatively simple and short duration source impulse is able to generate a complex wave field, where and when this process takes place, and what part is played in it by various wave types, and by the inhomogeneities of the Earth. It is an established fact that the propagation of an impulse in a homogeneous absorbing medium is accompanied only by changes in the shape and the amplitude of the impulse. The presence of interfaces and heterogeneities in an inhomogeneous (layered) earth results in the appearance of secondary waves, and the energy is partitioned between them.

VSP enables the processes of reflection and refraction at each interface, as well as the entire generation of secondary waves, to be studied. Down-hole observations permit the study of the distribution of energy between the secondary waves each of which may in turn, become the source of new waves. Waves reaching the zone of low velocities or the surface, are reflected back into the earth. The number of waves grows like an avalanche. The result is a superposition of all the waves, so that waves observed in the real earth are for the most part composite waves. By studying the wave fields corresponding to different geologic sections, it is possible to evaluate the part played by different wave types in the formation of the wave field.

(2) Generally, the characteristics of waves measured by surface seismic observations in a horizontally layered earth, vary very slowly and smoothly along the seismic profile. In VSP the line of the profile cuts various elements of the geologic-section with which the waves are connected. In this case, a sharp difference is observed between the kinematic and dynamic characteristics of the different wave types, with the appropriate characteristics of each wave changing along the vertical profile.

Different waves may exist in different parts of the sub-surface. For this reason, the kinematic and dynamic characteristics of the waves may often be of limited value in surface observations, while in VSP they may provide reliable criteria for the identification of wave types, even without preliminary velocity determinations.

Of special importance in VSP is the feasibility to study waves just above or below an interface before the newly-formed waves are distorted by the superposition of other waves, whereby the origin of the waves and their stratigraphic association can be determined with the utmost certainty. This makes detailed and precise studies possible, in particular the study of low-amplitude distortions and small scale structures.

(3) In contrast to surface observations, the detectors used in VSP are located both below and above reflecting interfaces, so that not only waves arriving from below can be recorded by the geophones, but also direct and reflected waves arriving from above; these may be downward travelling reflections from the underside of the surface or from shallow layers in the upper part of the geologic section. Generally these are conspicuous by high amplitudes and can be reliably tracked at great depths beneath the surface. For all practically accessible depths (down to about 7000 m) the surface and the upper portion of the geologic section will be seen to have a marked influence on the wave field. VSP enables the waves to be tracked continuously along their entire paths, this being of essential importance for understanding the mechanics of formation, and the origin of the many waves recorded in the later portions of surface seismograms. Specifically, downward travelling waves on VSP seismograms may provide a more reliable determination of the relative intensity of multiple reflections than the data obtained in surface observations, which records only waves arriving from below. It should be specially

emphasized at this point that VSP can discriminate waves which create interference patterns on surface seismic profiles.

(4) Unlike vertical seismic profiles, surface seismic profiles are located on the Earth's surface which is itself an interface and a most pronounced one at that.

Pronounced heterogeneity and large velocity gradients typical in the upper portion of the geologic section, cause a substantial part of the energy of an explosion in the near-surface to be propagated close to the surface in the form of low-velocity waves of various types (surface, channel and reverberations, etc), with comparatively little depth of penetration. Many such waves have high amplitudes and persistence on surface seismograms. In addition, scattered and mode-converted waves of various types are generated at the heterogeneities in the upper portion of the geologic section. In conditions of abundant heterogeneities, the interference of multiply scattered waves creates a persistent unwanted wave-field which obscures and may totally prevent the tracking of deep waves on surface seismograms. However, whereas surface waves can be effectively suppressed with the aid of source and geophone arrays, the interference pattern of scattered waves can be diminished only by means of statistical averaging, this being a less effective method [38]. The intensity of such unwanted waves rapidly decreases with depth with the resulting improvement in the signal-to-noise ratio for VSP that makes possible the discrimination of waves reflected from deep interfaces, and hence the application of VSP for solving structural problems.

(5) The level of seismic background noise in down-hole observations is perceptibly less than at the surface, so that the useful sensitivity of the instrumentation, limited mainly by its inherent electrical noise, is much higher than in the case of surface observations. This facilitates the study of waves in a wider frequency band than in the case of surface observations and the study of waves distinguished by their extremely low amplitudes (e.g. head-waves associated with thin layers).

(6) The pronounced filtering effect of the upper portion of the geologic section, responsible for major variations in the signal's spectrum at interior points of the earth, is less severe for VSP and allows a broader signal frequency band to be detected, thus improving the resolution of recording; this constitutes the basis for high-resolution seismics. This is especially important in the study of the dynamic characteristics of waves and the physical properties of a geologic section.

Hence, the principal advantages of VSP over surface observations are, on the one hand, the capability of VSP to move away from the surface and the shallow sub-surface, and, on the other hand, the capability to conduct observations close to geologic interfaces. In the latter case, this permits the improvement in resolution and in the stratigraphic correlation of the waves.

(7) The direction of particle motion in simple body-waves, observed at interior points in homogeneous or thick layered isotropic formations, corresponds to the direction of wave propagation, especially near the wave front. The trajectories of motion can be used in geologic interpretation of VSP data. Of special interest is the polarization of seismic waves, which is a very sensitive space-time characteristic of a wave. Its study enables additional data on the structure of the real earth and on the individual characteristics of wave propagation in it to be obtained.

(8) In VSP the direction of approach of various waves from a single shot-point, may be sharply differentiated and may vary markedly not only with an increase in the

distance from the source, but also with the detector position along the vertical profile. The laws governing these variations may differ for different wave types, thus greatly complicating the discrimination and tracking of waves using a Z-component geophone. At the same time, this creates an opportunity to discriminate waves in accordance with the polarization of particle motion. This is a major contributing factor in the effectiveness of PM–VSP.

The features listed above substantially extend the potentialities of VSP for studying the velocity distribution and the borehole locality. At present, in many areas, conventional well velocity surveys have been completely replaced by VSP, and soon this will be generally true everywhere.

It is worth noting that most of the advantages, enumerated above, of vertical seismic profiling in media with horizontal or low-dip interfaces, are gained when the line of the profile intersects the interfaces. Some of the advantages listed above are lost in complex media, with high-dip or even vertical interfaces, when the line of the profile does not intersect the interfaces, and may even be parallel to them. In oil geology, this is the case for regions with salt-domes or diapir tectonics, but is more frequent in mining applications, where studies of faulted metamorphised strata with extremely high-dip interfaces are required. In such conditions, a vertical seismic profile is mainly used for mapping interfaces that cannot be located by surface observations. The nature of the waves may be studied effectively using horizontal (surface) profiles when these intersect the interfaces. However, there are always specific problems involved in surface observations due to the surface itself being a very sharply defined interface.

3. Polarization Method of VSP (PM–VSP)

In the early stages of development of the VSP method, studies were limited to two groups of parameters that characterize the wave field: those associated with the propagation of the wave in space, and those with particle motion at any point through which the seismic wave passes. VSP observations were carried out using three-component geophone arrays, to study the directions of particle motion (15). Subsequently, because of specific difficulties involved in three-component observations in deep boreholes, the trend towards the study of the propagation of the waves in space and recording only the vertical component became prevalent. This stage of VSP development was characterized chiefly by the employment of kinematic properties of the waves. However, the 'kinematic' stage was fraught with difficulties that made the full potential of VSP impossible to realise. This served as an incentive to continue research aimed at the study of wave dynamics, and the improvement of methods to discriminate different wave types and their processing. Such difficulties included, the complexity of the wave fields observed in VSP; variations in the direction of approach of the waves along the vertical profile, which may influence observations from a single shot-point, even in homogeneous and horizontal layered strata and may lead to erroneous conclusions concerning the relative intensities of different waves; ignorance of the specific nature of the wave field, which in complex media may result not only in substantial errors, but may also make the results of observations practically worthless in many cases. Studies of wave polarization, [2, 39–40], initiated in conjunction with VSP, resulted in the creation of the polarization method (PM) – a general method of seismic investigation, [2, 41–44].

VSP results obtained in a variety of seismological conditions provided the incentive for the development of the polarization method. The results proved two important points: (a) not only longitudinal, but also transverse waves are excited in normal explosions; and (b) waves of various types, both monotype and converted (longitudinal, transverse) are formed at the same geologic interfaces. This provided the corner-stone for the creation of a general method based on the vector analysis of the wave field and on the combined discrimination, tracking and interpretation of different wave types. At present, different versions of the method applicable to surface (PM reflected wave, PM common-depth-point), and to borehole (PM–VSP) observations are being developed. A special monograph deals with the polarization method [2]. At this point we shall present only a brief description of the main properties of PM–VSP. Traditionally methods employing longitudinal and transverse waves made use of separate optimum excitation conditions for each type of wave. This substantially simplifies the wave fields and improves chances for analysis, discrimination and tracking of useful waves. However, this limits the exploration capabilities of each method to waves of a single type. The polarization method greatly increases the potential for the analysis of complex wave fields by enabling the discrimination of wave types (by reference to their polarization) at the point of detection, which frequently obviates the necessity to use separately the longitudinal and the transverse wave methods. This special property of PM–VSP requires the simultaneous excitation of longitudinal and transverse waves of adequate intensity at the energy source.

PM–VSP makes use of the polarization of waves both for the analysis of the wave field and to gain direct information about the earth. Consider these two aspects separately. Firstly, the polarization of the waves is used as a parameter, which, in addition to such traditional parameters as frequency and apparent velocity, is used for discrimination of the waves. This improves the discrimination and tracking of the waves with a corresponding increase in the reliability of identification of the wave types and their stratigraphic correlation, and allows the reflecting properties of the interfaces of interest and other problems associated with the study of wave dynamics to be solved. This trend is at present widespread and guarantees the effectiveness of VSP.

Secondly, the study of polarization parameters (direction of particle motion) is aimed at gaining additional information about the nature of the earth. This trend has, as yet, not become widespread, because the real earth, as a rule, has a thin-layered structure and displays anisotropy. In anisotropic media, the direction of particle displacement does not generally coincide with that of wave propagation, and the theory of wave propagation in such media is still in its infancy. Another factor in favour of VSP, is that, at interior points of the earth, the polarization parameters are less distorted by the effects of the low-velocity layers in the upper portion of the geologic section, thus making it possible to gain additional information about the earth following quantitative processing of particle trajectories.

PM–VSP results [45], accumulated in recent years both in mines and oil wells, are proof of its effectiveness. It should be specially emphasized that not only does PM improve VSP, but that VSP itself assumes special importance in the general PM field. This is so as it becomes necessary to determine the nature of a large number of different wave types, recorded and tracked with the aid of the polarization-positional correlation on the surface, and to establish their stratigraphic correlation. Accordingly, PM–VSP can be recommended for all-round use, not only in conditions of complex media, but also in comparatively simple areas with a layered structure, especially for the solution of problems that cannot be tackled by the traditional VSP method, and in particular for the simultaneous study of *P*- and *S*-wave velocities that are required for geological analysis. It is expected that part of VSP field work wil be carried out using the polarization method exclusively when mass-produced borehole equipment becomes available. PM–VSP assumes paramount importance in connection with the problem of studying the area around the borehole and with the application of seismic studies in the exploration and exploitation stages of oil fields.

VSP IN COMBINATION WITH OTHER TECHNIQUES

VSP is, as a rule, combined with surface observations, such a combination being of particular importance for the determination of the nature of waves observed at the surface. The essence of such a combination is that waves observed at the surface can be tracked back along the vertical profile down to the interfaces, with which they are associated. Using combined observations, it is possible to make use of the variation of the wave parameters with the separation of source and receiver (in the case of surface measurement) and with the distance between the receiver and the interface (in the case of borehole observations). VSP is part of the entire range of seismic exploration methods which study the nature of the earth and its corresponding wave fields, and which also includes conventional sonic velocity logs and theoretical calculations [46]. VSP is also part of the combination of methods used in the construction of an effective seismic model, which includes other non-seismic geophysical methods [47].

The combination of VSP and sonic logging is a very effective one, because synthetic seismograms, calculated on the basis of sonic logs enable a complex wave field to be split up, and the effects on it of various geological heterogeneities to be studied [48, 29]. The current use of computers facilitates the calculation of wave fields involving a large number of layers, and with refraction and absorption effects taken into account. The shape of the source impulse necessary for these calculations may be obtained from VSP. The comparison of the calculated synthetic wave patterns with the wave field actually observed in the real earth by VSP sheds light on the mechanism of wave formation, and makes possible a rigorous stratigraphic correlaction of the wave, as well as the evaluation of the influence of different geologic elements on the wave field. This is especially important considering the thin-layered structure typical of the real earth. Considerable experience accumulated with the combined use of VSP and sonic logging agrees well for such a combination. It would be expedient to work with lower sonic frequencies more suitable for observations in cased boreholes. With adequate resolution, this would improve the prospects of combining VSP with sonic logging. The combination of VSP with seismic acoustic-impedance sections, which are now being widely used, and which enable a more detailed splitting of the velocity cross-section required for stratigraphic work [49, 50] should also prove to be effective.

VSP has also been responsible for a substantial increase in the application of seismic studies in the general geologic exploration process. In particular, it has become feasible to use VSP for both wildcat and appraisal drilling, and also at such important later stages as the development and exploitation of oil-fields. In such studies, VSP is part of a broad complex of surface and borehole observations which constitute the subject of production seismics [51].

Production seismics requires a considerable increase in the amount of work performed in each borehole, and this involves putting the borehole out of operation for a prolonged period. Accordingly, VSP work in the context of production seismics should be allowed for in the timetable for drilling, and integrated with the drilling operation.

In addition, great importance should be attached to the possibility of operating in cased boreholes. Not only are such operations much safer and cheaper than the operation in uncased boreholes, but they also enable the extent of the method to be broadened if basic production, injection, tensometric and other cased boreholes are used for VSP.

4. Wave Correlation in VSP

The correlation of waves in VSP can be determined because of the reproducibility of records obtained at different points in the earth under conditions of constant source excitation. The correlation of waves recorded along vertical profiles became possible as a result of the suppression of unwanted signals such as mechanical noise from the borehole itself (see Chapter II). In the description of wave correlation the seismograms are presumed to be free of noise. The special properties of wave correlation in vertical seismic profiling are determined by the kinematics and dynamics of the waves along the profile, which can be substantially different from similar properties along a horizontal profile. This is governed by the variation of travel times, apparent velocities, amplitudes, and frequencies, and even by the zones where various waves exist.

The main features facilitating wave correlation in VSP are as follows:

(1) Observations carried out just above or below geologic interfaces enable the waves associated with them to be studied in the early portion of the seismogram, which, as a rule, is less distorted by regular and irregular unwanted signals. Most waves in this case originate from above, and can be relatively easily attenuated, because of the difference in the sign of the velocity.

(2) The broader frequency spectrum of the waves in VSP, compared to surface observations provides better resolution of recorded waves having small changes of apparent velocity.

(3) The direction of particle displacement is not distorted by the effect of the low-velocity layer, thus contributing to better discrimination of the waves in terms of their polarization.

(4) The records are not subject to the attenuating effect of the low-velocity layer and the upper part of the geologic section, or to the effect of low-velocity unwanted signals such as surface waves.

However, wave correlation in VSP is, as a rule, more intricate than in the case of surface observations. This is the result of:

(1) The presence of a much larger number of waves than in the case of surface observations. Waves recorded along the vertical profile include multiples reflected downwards above from detector points, and downward travelling converted waves that are not observed on the surface.

(2) A more pronounced variation in the direction of incidence of the waves at different

points of the vertical profile. This may create a false impression about the relative intensities of different waves, if only one directional component in the wave is studied.

(3) Kinks on the travel-time curves, drastic changes in the dynamic characteristics of the waves, and the superposition of secondary waves in parts of the profile close to geologic interfaces.

The difficulties involved in the correlation of waves in the upper part of the geologic section, and especially close to the surface, deserve special mention. In addition to a great diversity of wave types, this region is characterized by the presence of noise of mechanical origin.

(4) The dynamic properties of the records are greatly influenced by absorption of the energy as well as by spherical divergence of wave fronts, by interference and by the variation of elastic properties of the Earth along the vertical profile. As the interfaces are traversed, wave shape and amplitude may experience a sharp change. When correlating waves along the vertical profile, one should be aware of possible changes in amplitude and wave shape as well as phase distortion associated mainly with interfaces.

(5) The coupling between the well geophone and the borehole wall exercises a much greater effect than in surface observations, because the variation of rock type and structure along the vertical profile may be greater than along the horizontal profile.

Wave correlation in PM–VSP

Of great importance in the discrimination and tracking of waves in VSP is the discrimination according to wave polarization upon which the polarization method (PM–VSP) is based.

Wave field analysis in PM–VSP is based on the discrimination of waves at a point according to their polarization, combined with the discrimination of waves in the body of the earth according to their direction of propagation.

Polar correlation and polarization-positional correlation methods have been specially devised for discriminating and tracking waves in PM–VSP [39, 2].

The polar correlation* method is based on the analysis of recorded data as a function of the direction of motion of the wave components in space. PC is executed with the aid of polar seismograms, which consist of a set of traces of wave components each with a specified orientation in space. PC enables the nature of wave polarization and the trajectory of particle motion to be determined visually. The principal equations in polar correlation the – line-up equation and the equations for amplitudes on the polar seismogram – determine laws governing the variations of phase and amplitude constants for waves having different polarizations. Analyzing those equations, one can formulate criteria for the discrimination of linearly-polarized and space-polarized oscillations. Particular features of the former include: (a) vertical line-ups, i.e., the phase shift on the polar seismogram is independent of the orientation of the wave components in space, however, a phase reversal of 180° is possible for some components and (b) the identical shape of all traces on every seismogram, and a cosine law of amplitude variation. Violation of even one of the above criteria is proof of violation of linear polarization.

* The term 'azimuthal correlation' formerly used refers to a particular case of polar correlation, in which the only variable is the azimuth of a component.

Curved or oblique line-ups are typical for elliptically polarized waves. Analyzing the shape of the line-up, one can determine the orientation in space of the polarization plane and the shape of the ellipse. Controlled directional reception of the first kind (CDR–I) is also based on polar correlation [52].

Controlled directional reception of the first kind discriminates waves according to the polarization at a point. It is based on the directional sensitivity of the detector array, which in polar coordinates is represented by two tangent spheres. In the region close to the axis the sensitivity of the system is very low, increasing significantly in directions orthogonal to the axis. The improvement in the signal-to-noise ratio in CDR–I is achieved by means of changing the orientation of the directivity, in such a way as to discriminate in favour of a component of the useful wave for which the signal-to-noise ratio is at its maximum. Such a component serves as a tracking component. In some particular cases, it is possible to discriminate the useful wave in the plane of zero displacement of unwanted waves. It should be pointed out that there may be a substantial difference between amplitudes of the components recorded near the zero displacement plane, even if they make only a small angle with each other.

In contrast to the traditional positional correlation, the component being tracked in PPC is not a fixed component (e.g. Z, X, or Y), but the tracking component whose orientation in space may vary for each wave. The tracking component for each wave is determined on the one hand by the direction of the total wave vector, and on the other hand by the superimposed unwanted waves. In different situations these factors have a different effect. For dominant waves, and for waves recorded as first arrivals against a quiet background, the tracking component coincides in direction with that of the total wave vector. In the presence of noise, the tracking component deviates from the direction of the total wave vector. Not only does polarization-positional correlation substantially improve the discrimination of longitudinal waves, but – this is still more important – it also enables regular waves, irrespective of their type, velocity and propagation direction, to be tracked.

Figure 5b depicts the seismograms of components along the axes of two coordinate systems: a XYZ and a local PRT (Y). The seismograms have been obtained in comparatively simple conditions. The thickness of sediments is 2600 m, with horizontal layering. Even in such simple conditions the wave field will be seen from the seismograms to be made up of a set of waves of different origin and type (P, S, PP, PS). The first longitudinal wave on the Z component is the most intense and is multiphase. On the R component it is greatly attenuated, and the secondary waves (reflected and refracted) are trackable to the region of the first arrivals, thus greatly improving the accuracy of their stratigraphic correlation. Longitudinal multiples are formed on the sharpest velocity boundary (with a velocity increase of 900 $\mathrm{m\,s^{-1}}$). This interface acts as the conversion interface. Converted (reflected and refracted) waves are trackable on the seismogram.

Quite reliably trackable along the vertical profile are the direct transverse S waves distinguished from waves of other types by a low-frequency spectrum. The visible frequencies of the S waves lie in the 10 to 17 Hz interval.

The transverse wave has been recorded on the R components in the entire interval of observations (Figure 5). However, in the region of the traveltime curve's minimum they are more reliably trackable on the Y component, this being evidence that tangential components predominate in this part of the cross-section.

Only digital data processing techniques can assure full realization of the potential of polarization-positional correlation. The procedure for finding tracking components has been formalized, and algorithms for discriminating tracking components in the PM converted wave and the PM common-depth-point methods have been devised. These algorithms are currently being evaluated.

When establishing the correlation of waves along a vertical profile, one observes the same types of wave interference patterns as in the case of surface observations [53, 54]: intersection of line-ups with the waves being tracked on both sides of the interference zone; the coincidence of line-ups; the superposition of wavelets in the case of parallel line-ups. The interference zones most frequently encountered in VSP are associated with the intersection of line-ups resulting from the interference of primary and multiple reflections propagating in different directions. The coincidence of line-ups is typical where the direct and secondary waves intersect and is observed mainly in sections of the profile close to geologic interfaces. The superposition of wavelets, where there are parallel line-ups, is the result of the interference of a direct wave with its ghost reflection, of reflected waves with multiples propagating in the same direction, and of waves reflected from different but closely spaces interfaces. In this case the superposition creates stable and extensive interference zones and severely hinders the analysis of the wave pattern. Such zones are most extensive when the difference between the apparent velocities is small, and when the duration of the wavelets is large. VSP seismograms usually display a combination of various types of interference.

On the transpositional correlation of waves in VSP

It is sometimes necessary to identify waves recorded by the vertical profile from different source offset positions. In contrast to observations on horizontal profiles, the transpositional correlation in VSP cannot be based on the reciprocity principle. Waves can be identified with the aid of a system of combined horizontal and vertical observations. In this case the waves recorded on vertical profiles from different source locations are identified with the aid of horizontal traveltime curves. For correlation methods in the vertical plane, when the shotpoints are sufficiently close to each other, the waves can be identified with the aid of 'fixed geophone level' profiles. Waves can be reliably identified by combining VSP with the inverse VSP method. Frequently the identification of waves on vertical profiles arriving from shotpoints with different offsets may be based on the special characteristics of vertical travel-time curves that can be effectively used to determine the nature of the waves arriving from individual shotpoints. For reflected waves, such characteristics are the points of intersection of the first arrival curves of direct waves with the time-depth curves of reflected waves; for head waves the points of intersection of the travel time curves of the direct wave and the head wave and points at which the head wave changes to the transmitted wave; for refracted waves – points of minimum on the time-depth curves of reflected waves etc. General regularities in the disposition of such characteristics at various distances from the source may be used to identify the waves recorded on the same vertical profile, but arriving from different shot-points.

CHAPTER 2

VSP INSTRUMENTATION AND TECHNIQUES

Since its inception, VSP has made increasing use of the experience gained from surface observations in seismic studies. This applies to both the instrumentation and the procedures for recording and processing the data: the number of recording channels has been increased, depending on the problems being solved, sophisticated recording systems with multiple subsurface coverage and signal stacking are being employed. This is largely determined by the extensive use of the combination of surface and down-hole methods fundamental for production seismics.

However, evolution in this direction is limited by the special nature of seismic observations in deep boreholes including the difficulties associated with the large number of separate observations needed, and, as a consequence, the large number of repeatable, consistent shots required to complete a profile. It is also complicated by the difficulty of devising complete observation techniques to record both sub-parallel and overlapping time-depth curves, and by the presence of a much greater number of waves propagating in different directions compared to surface observations. In addition, the task of obtaining high-quality data is greatly complicated by numerous unwanted waves of different types. Accordingly, we shall discuss the available facilities for obtaining high-quality data.

1. Unwanted Waves of Mechanical Origin in VSP

The data quality and the ability to discriminate and track waves in VSP are primarily determined by the intensity of unwanted waves recorded in down-hole observations. The study of the origin of such waves and the mechanism governing their formation, their characteristics and the methods of suppressing them is one of the prerequisites for improving the quality of data obtained in VSP. In many cases, the results depend on the extent to which the unwanted waves can be suppressed.

The unwanted waves observed in VSP can be conveniently subdivided into two groups: mechanical noise due to the presence of the borehole itself and the recording conditions in it, and elastic waves generated in the earth e.g. Stoneley, side-waves, etc. Only the first type of noise will be discussed in this chapter.

In the first group are hydraulic waves which propagate along the fluid column filling the borehole, waves determined by the structure of the borehole, e.g., casing and cable waves, as well as resonant oscillations set up at the contact between the well-geophone and the borehole wall. The velocities of most of the unwanted waves listed above, with the exception of casing waves, are lower than that of the first arrivals propagating in rock, and such waves are recorded as later arrivals. Accordingly, although such waves have been recognised for a long time, they did not affect conventional velocity surveys, and their formation mechanism and characteristics were not studied adequately.

With the inception of VSP those waves became the main source of noise obscuring the tracking of useful waves propagating in the rock. The quality of VSP data is most seriously affected by hydraulic waves and vibrational casing waves. In recent years an extensive amount of theoretical and experimental work aimed at studying such waves has been carried out, and the results have been systematically presented in [25].

HYDRAULIC WAVES*

These waves (P_{n}) propagate along the column of fluid filling the borehole being studied. They have a high intensity and are one of the major sources of noise in VSP. Accordingly, a study of their characteristics and their formation mechanism is of great interest.

The velocity of hydraulic waves in an uncased borehole is determined by the elastic constants of the drilling fluid and of the rock surrounding the borehole as expressed by the formula:

$$v = v_0 \sqrt{\frac{\mu}{\mu + K_0}},$$

where μ is the shear modulus of the rock making up the borehole walls, v_0 is the velocity in the fluid or mud filling the borehole, K_0 is the bulk modulus of elasticity of the mud.

The presence of metal casing in the borehole increases the velocity of the hydraulic wave, which is then expressed as [55]:

$$v = v_0 \left(1 + \frac{K_0 D}{Ed}\right)^{-1/2},$$

where D and d are the diameter and the thickness of the casing respectively and E is Young's modulus of elasticity of the casing material. The increase in the velocity may be as high as 40%. Calculations have shown that in a borehole filled with fluid and having good bonding between the casing and the rock, the velocity of the hydraulic waves differs from the velocity in the fluid, its value being dependent on the thickness of the casing and of the surrounding rock. A value of 1430 m s^{-1} has been obtained for the velocity in a cased borehole [55]. Phase velocity dispersion is not observable for ratios of wave length to borehole diameter typical for seismic exploration.

The mechanism of excitation of hydraulic waves in boreholes has not been adequately studied. It may be as a result of a sharp change in pressure, which takes place in some sections of the borehole. As a rule, hydraulic waves are excited in the upper portion of the borehole where there are usually several casing strings. However, they may also be excited at large depths at the boundaries of zones of poor cement bonding. Hydraulic waves are excited by the incidence of direct waves on such sections, but they can also be excited as the result of the incidence of high-intensity waves of other types. Once excited, the hydraulic wave propagates along the borehole shaft and can be reflected from different parts of the borehole, where physical parameters such as the density of the drilling fluid, the bulk elasticity modulus, the elastic parameters of rock surrounding the borehole, the thickness of the walls or the diameters of casing strings experience a change.

* Generally referred to in the West as tube waves.

The reflection coefficient is determined by the equation [55]:

$$K = \frac{\rho_2 v_{fl_2}/d_2^2 - \rho_1 v_{fl_1}/d_1^2}{\rho_2 v_{fl_2}/d_2^2 + \rho_1 v_{fl_1}/d_1^2},$$

where ρ, v_{fl}, and d are the density of the drilling fluid, the velocity of the hydraulic wave, and the internal diameter of the borehole, respectively, and the wave arrives at the interface from medium 1.

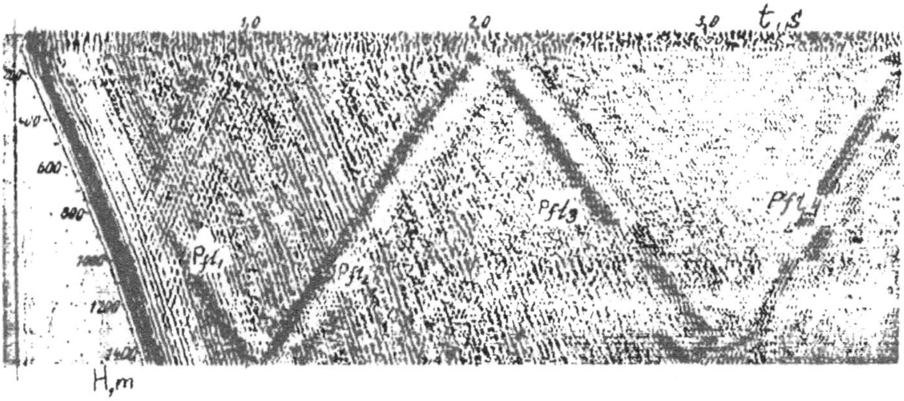

Fig. 1a.

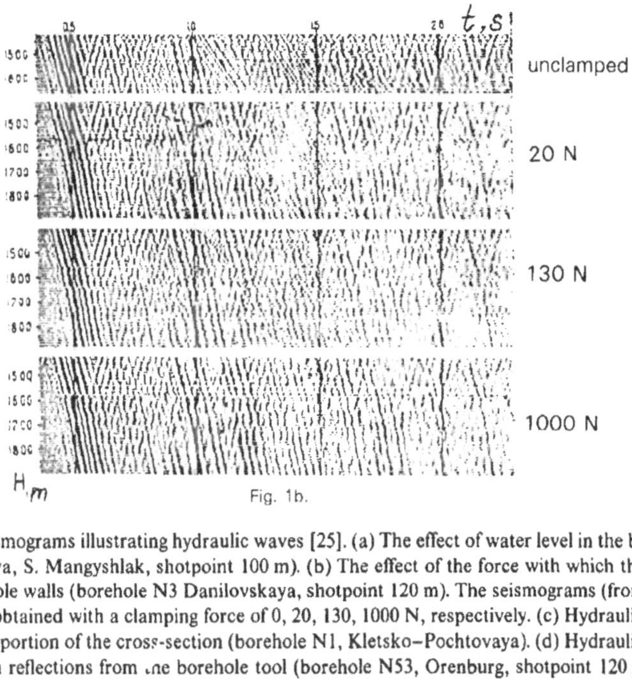

Fig. 1b.

Fig. 1a-d. VSP seismograms illustrating hydraulic waves [25]. (a) The effect of water level in the borehole (borehole Uzen'skaya, S. Mangyshlak, shotpoint 100 m). (b) The effect of the force with which the probe is clamped to borehole walls (borehole N3 Danilovskaya, shotpoint 120 m). The seismograms (from top to bottom) have been obtained with a clamping force of 0, 20, 130, 1000 N, respectively. (c) Hydraulic waves formed in the upper portion of the cross-section (borehole N1, Kletsko–Pochtovaya). (d) Hydraulic waves associated with reflections from the borehole tool (borehole N53, Orenburg, shotpoint 120 m).

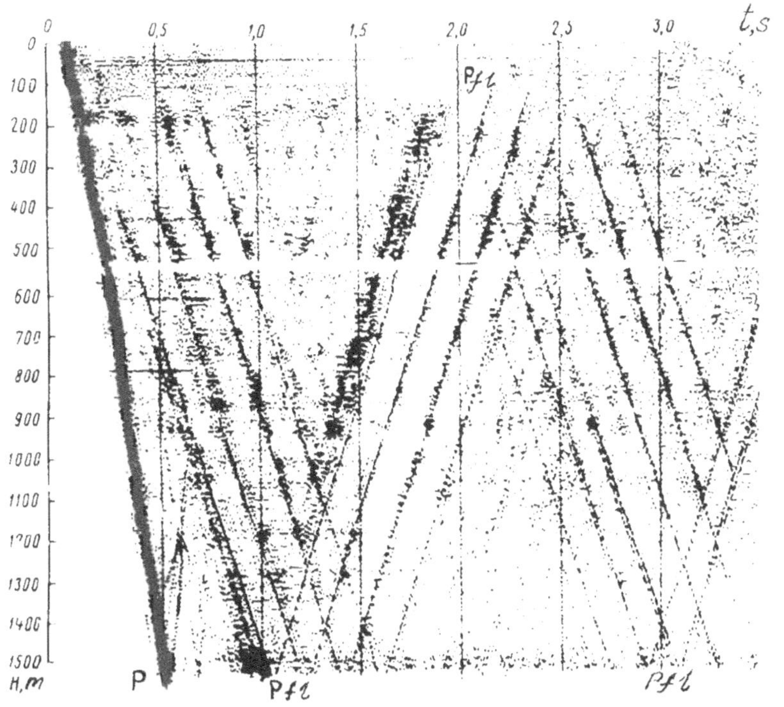

Fig. 1c.

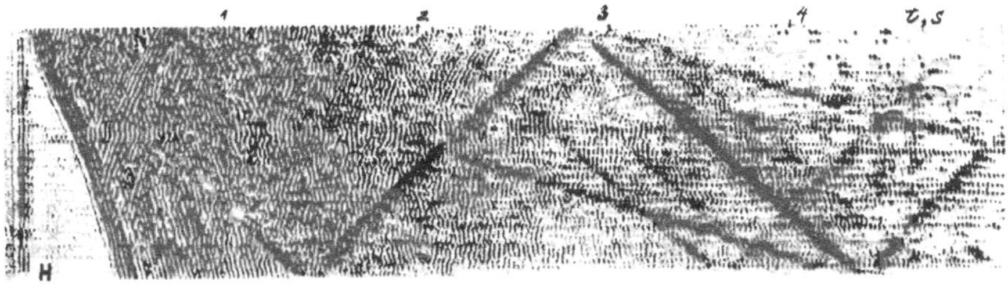

Fig. 1d

Hydraulic waves that have been reflected six or seven times from the surface of the fluid in the borehole have been seen on VSP seismograms. It is noteworthy that the point of excitation of hydraulic waves may not coincide with the point of subsequent reflection. A convincing experiment is described in [25]. The level in the borehole was deliberately changed. After 12 geophone points starting at the top of the borehole were recorded, the level of water was raised from 200 m to 100 m. Another 33 geophone positions were measured, and the level was raised a further 60–70 m. The change of the water level had no effect whatsoever on the downward wave $P_{\text{п}_1}$ or on the wave $P_{\text{п}_2}$ reflected from the lower end of the casing (Figure 1d), i.e., the conditions of formation of hydraulic waves experienced practically no change. However, the change in the level had a dramatic effect on the multiple P_{fl_3} reflected in the upper portion of the borehole by the surface of water. The displacements of the line-ups of the downward wave P_{fl_3} and the wave P_{fl_4} reflected from below, corresponding to the changes in the water level in the borehole, are visible on the seismogram.

Because the formation of hydraulic waves depends heavily on the contact between the casing and the surrounding rock, and principally on the quality of the cementation to the borehole wall which cannot always be adequately controlled, the intensity of hydraulic waves may vary considerably between similar boreholes drilled in identical geological conditions, even when close to one another. It should be noted however that the contact is not always governed by the quality of cementation, particularly in boreholes of complex design, when one casing string is cemented inside another. On the other hand, with a single string of casing in the well and without special cementation, the contact can sometimes be quite good. This is mostly true of boreholes drilled in porous rocks that have been standing for some time with casing in them. In such conditions the casing can be firmly held by flowing rock. This is substantiated by the large volume of observations carried out in old cased boreholes, in which records free of hydraulic waves have been obtained.

The intensity of the hydraulic waves recorded is influenced by the density of the drilling fluid (the intensity decreases with lower density) and by the force with which the borehole tool is clamped to the borehole wall. The intensity of the hydraulic wave may be substantially reduced by as much as an order of magnitude, if the tool is clamped with a force 2 or 3 times that of its own weight. This is clearly evident from Figure 1.b. The hydraulic waves are reliably trackable on the upper seismogram recorded with an unclamped tool. They are still trackable on the middle seismograms recorded with a clamping force of 20 and 130 N, but it is impossible to identify them on the lowest seismogram (f = 1000 N). As a rule, hydraulic waves on a seismogram form a long train of pulses, which can sometimes be resolved as in Figure 1c, but at other times may form a continuous complex wave train. Such differences are determined by the formation conditions in the upper portion of the well where the hydraulic waves originate.

The shape and the spectral content of hydraulic waves recorded by unclamped geophones are distinguishable from the first direct wave by both higher frequency and shorter duration. As the hydraulic wave propagates, its spectral band-width is reduced, and the peak frequency is shifted towards the low-frequency end. As the number of cycles in the hydraulic wavetrain increases, the frequency of the later cycles may be sometimes lower than the first cycle, in the case of a long wavetrain. The first direct wave propagating through rock has a lower frequency content than the hydraulic wave and is more stable along the line of observations.

It should be pointed out that the hydraulic waves are generally recorded only on the vertical geophone components. On the horizontal components, they are either non-existent or recorded with very low amplitudes.

Hence the following features are typical for hydraulic waves: their velocity is independent of the distance from their source; largely independent of the rock's lithology; their interval on the record is determined exclusively by the time of arrival of the wave that excited (mainly at the boundaries of uncemented sections of the casing string) them. Their intensity decreases sharply with increasing depth of water level below the top of the borehole and with increasing offset. It also varies with the intensity of the waves that excite them. These features may be used to distinguish hydraulic waves from other low-velocity and converted waves in the upper portion of the geologic section.

Hydraulic waves propagating with a velocity about three times lower than normal have been observed. They are recorded only in the form of downgoing waves, and their traveltime curves intersect with the traveltime curves of the hydraulic waves described above, where they are reflected from the surface of water in the upper portion of the borehole, and from the bottom of the hole (Figure 1d). Paper [25] states that there is an unambiguous correlation between these waves and a poor coupling of the casing string to the surrounding rock, and, in our opinion, erroneously interprets these waves as flexural casing waves. Analysis of such waves shows that they are formed, when the borehole tool itself provides a good reflecting interface. This happens when the tool's diameter is large enough and its head is flat. In such cases, the borehole tool serves, so to speak, as an artificial hole bottom whose position along the vertical profile changes continuously. The hydraulic wave, is reflected by the surface of the drilling fluid in the borehole, and detected again by the geophone. The inability of the borehole geophone to record the reflection of this wave from the casing shoe or from the hole bottom, as pointed out in [25] is explained simply by the fact that the wave is not associated with the lower end of the casing string or with the hole bottom, and is reflected from such depths only when the tool is located at total depth or at the casing shoe. Waves of this type observed in the upper part of the section, may be mistaken for converted or transverse waves (because of their low velocities).

The waves listed above are not the only types of hydraulic waves. However, waves of other types occur less frequently, do not display such high intensity, and do not constitute a major source of noise in VSP observations. It is interesting to note that, if the casing string has not been cemented, hydraulic waves can propagate in the mud-filled annular space around the casing, even if the borehole itself is empty of mud. The transition to higher frequencies adds to the problem of suppressing the hydraulic waves.

The most effective way to combat hydraulic waves is to move the source away from the top ot the borehole. Substantial attenuation of hydraulic waves on VSP seismograms can be achieved by increasing the clamping force of the borehole tool to the borehole in uncased boreholes or to the casing when there is cement bonding. Lowering the level of the drilling fluid is also an effective method, provided the level is lowered beyond the section where the hydraulic wave is first excited. However, this is not always technically possible, and so the method is generally of little value. It is worth noting that hydraulic waves have high intensities on VSP seismograms recorded using pressure sensitive geophones, and this is one (but not the only) factor that makes velocity sensitive detectors preferable for borehole work.

Unwanted waves associated with the mechanical design of the borehole

The quality of data obtained in uncased boreholes is, as a rule, superior to that obtained in cased boreholes, for which the presence of unwanted waves associated with the mechanical design of the borehole, such as casing strings, jackets etc., and with the level and the quality of cementation, is a typical feature. Here the major role is played by the so-called casing signal. There are two types of casing signals (a) longitudinal waves

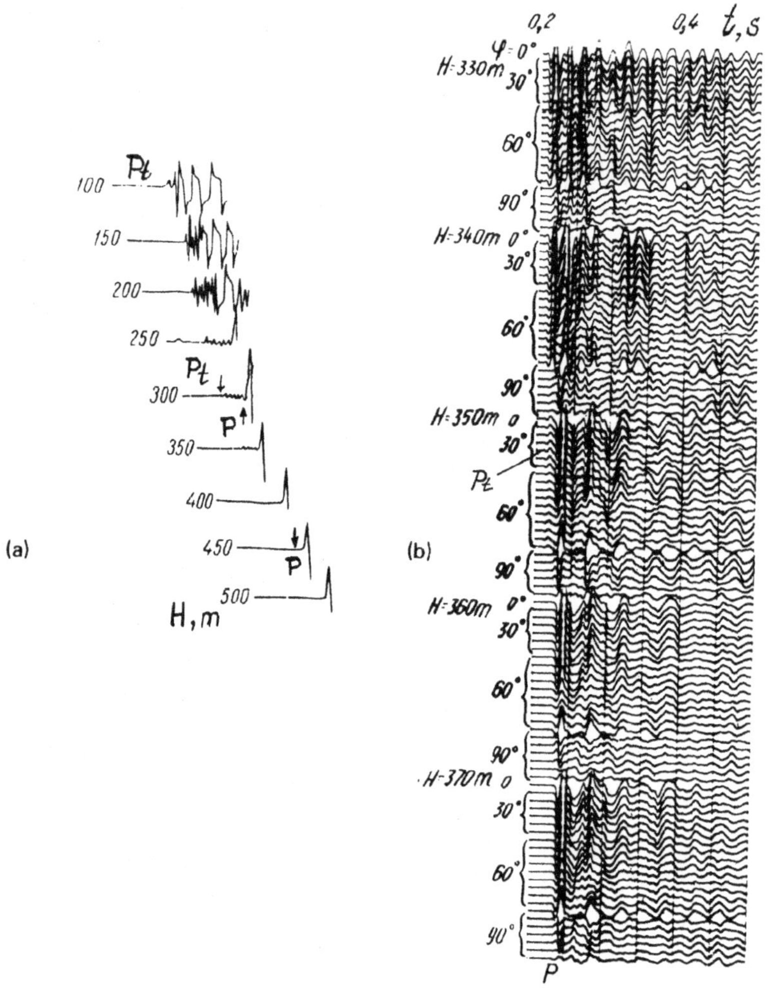

Fig. 2. A tube wave (a) on a Z-seismogram; (b) on polar seismograms (at $\varphi = 90°$ the tube wave does not register).

propagating along the walls of the casing; (b) casing signal associated with resonant vibrations of poorly bonded casing strings. Following [25], we shall term the waves of the latter type vibrational casing signal.

Casing signals

A longitudinal wave appears as the result of elastic waves impinging on the casing column. This creates a high frequency wave which propagates with the velocity of longitudinal waves in steel (4600–4900 m s^{-1}) and is recorded as the first arrivals. The phase velocity of the casing signal largely is independent of the wave velocities in the surrounding rock and drilling fluid, and the density and thickness of the casing. The casing signals P_T are, as a rule, irregular, have a high frequency and distort the first formation arrivals. A high damping coefficient is typical for them, the experimental value being $2.5 \times 10^{-2}\ m^{-1}$ at a frequency of 250 Hz. In the shallow part of the borehole their intensity may be comparatively high (on the vertical geophone component), with the intensity rapidly decreasing with depth (Figure 3) and with the distance from the shotpoint to the borehole. At a depth of about 800–1000 m their intensity is so low that for all practical purposes they do not interfere with the recording of waves propagating in the rock. In the case of poor cementation of the casing column, these signals may

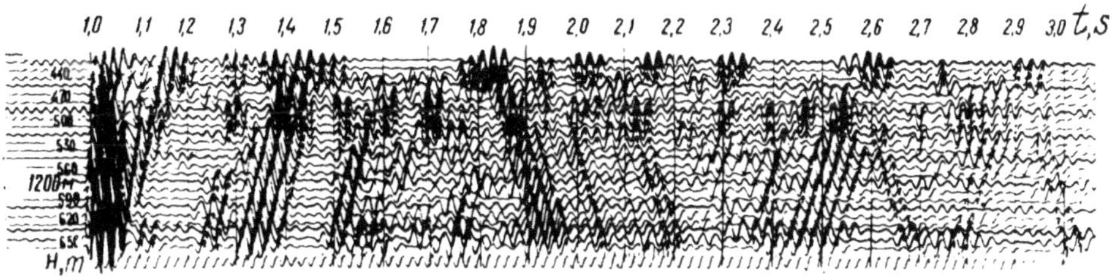

Fig. 3. The effect of the quality of cementation of the casing string. The upper part of the seismogram has been obtained in an uncemented string.

display a high intensity and propagate to great depths. We would like to point out that PM–VSP is capable of suppressing casing arrivals whose oscillations are linearly polarized in directions approaching the vertical. At the same time direct waves in the upper portion of the geologic section, generally arrive from directions other than the vertical (Figure 2b). In such cases polar positional correlation using observations of the tracking components of the direct waves, may be very effective. This is most important, since it allows seismic velocity from well velocity surveys to be extended up to the surface. At the same time, when only the vertical component is being recorded, no information is, generally, obtained about the upper portion of the geologic section. When studying casing arrivals, one can excite them by impacts on one end of the casing sting. Casing signals can be effectively suppresed by frequency discrimination.

Vibrational casing signals

In the case of good cementation, the presence of the casing column in the borehole does not introduce noticeable distortions into the recorded waveform in the seismic frequency band. However, in sections of the borehole where the casing column has been poorly cemented, parasitic vibrations of the column occur, and unwanted waves of high amplitude associated with them are recorded in this depth range at all distances between shotpoint and borehole. Such waves are represented by oscillations with vertical alignments. The number and intensity of unwanted waves grows rapidly where several casing strings are present, and when the cementation of each column does not reach back to the surface. The frequency content of the waves is very irregular and depends on the length of the loose sections of casing. Unwanted waves with the highest intensity are generally observed in the upper portion of the borehole (Figure 3). Primary and multiple reflected waves can be reliably traced in the lower part of the figure corresponding to the uncased section of the profile. The correlation between almost all waves observed above 540 m is poor. There are as yet no effective means for suppressing such unwanted waves during recording. For this reason in order to obtain undistorted records, the upper portion of the profile should be observed in a shallow uncased borehole specially drilled for the purpose.

CABLE WAVES

Elastic waves, excited by explosions and propagating in the near surface between shot-point and borehole, act on the logging cable suspending the geophone when under tension, or on the well-head, causing oscillations in the cable-geophone system that propagate along the cable and become superimposed on the recorded seismic waves.

The recorded arrival time of cable waves on the seismograms depends on the ratio of the velocities in the cable and the rock. In the upper-most portion of the section where low velocities prevail, the cable waves may be recorded as the first arrivals and distort the seismic signals [3]. In such cases, by moving the source further away from the borehole, it is usually possible to obtain records of formation first arrivals free of the cable wave. The decrease in arrival times, sometimes observed on surface geophones situated closest to the shot-hole in uphole and low-velocity-layer surveys, may also be explained by cable waves propagating towards the surface along the firing line suspending the charge [25].

However, in the majority of cases the arrival of cable waves occurs in the later portion of the records. Such waves are one of the major sourses of noise that degrade the quality of VSP data. For this reason theoretical and experimental studies of cable waves have been carried out in recent years.

Cable waves constitute a superposition of waves of various types that propagate along a cable under tension. Of major importance are the longitudinal and torsional waves. The calculation of velocities of such waves for armoured multi-conductor cables produced by the industry is shown in the following Table I [25].

Cable wave velocities can also be easily determined experimentally by recording impacts made on the cable at the top of the borehole. It should, however, be kept in mind that the calculated and the epxerimental values may disagree, since the drilling fluid in the borehole will reduce the velocity as its density increases. Thus, the experimental velocities of longitudinal waves measured for the KSB–6 and the KB 9–8 cables lie in the range 2500 to 3500 m s^{-1}. Even for measurements in the same borehole, a decrease in the velocity of cable waves with depth, from 3750 m s^{-1} in the upper portion down to 2280 m s^{-1}, as the density of drilling fluid increased, has been observed [25]. Torsional wave velocities may exceed those of longitudinal waves by 30%.

The effects of cable waves can be assessed in any individual case by recording impacts made on the cable when the tension of the cable is reduced by clamping the geophone to the borehole wall. Good coupling of the geophone to the borehole wall, and mechanical isolation by slackening the cable by several metres after the tool has been clamped enables records free of cable waves over the entire range of depth measurements to be obtained; these records are then suitable both for kinematic and dynamic process-

TABLE I

Cable	Velocity of longitudinal Wave (m s^{-1})	Velocity of torsional wave (m s^{-1})
KOBD-6	4220	3730
KTB-6	3940	3300
KSB-6	3830	5060
KBG-8	3690	4880

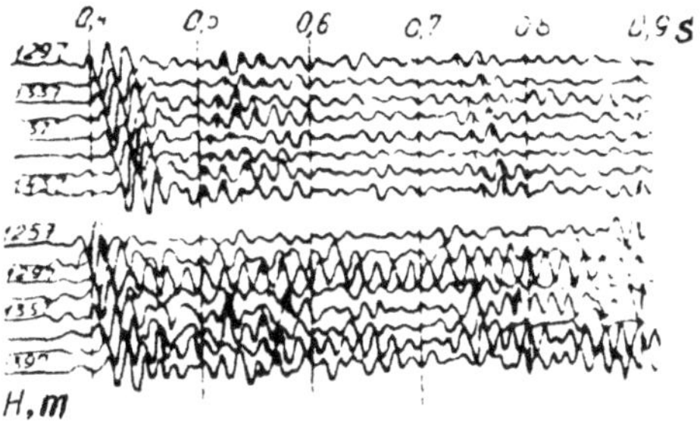

Fig. 4. Records obtained with strongly (upper seismogram) and weakly (lower seismogram) clamped tools [24].

ing [6, 15, 57, 58]. However, to be effective the geophone clamping must support the full weight of the borehole tool (Figure 4). When this is not the case, additional isolation of the geophone from the cable is needed in order to attenuate the cable wave. Mechanical filters in the form of a shock-absorbing mass or a separate clamping device located above the geophones may be used for this purpose. The shock absorbers can conveniently be made detachable which enables the degree of isolation to be adjusted, and in particular, to be more effective for observations in the upper portion of the borehole, where the level of cable waves is especially high. However, the effectiveness of such damping depends greatly on the design of the filters, and the damping itself cannot, as shown by experience, be regarded as adequate for quantitative studies of the dynamics of seismic waves. The full potential of isolation by means of mechanical filters has, as yet, not been exhaustively studied. Borehole tools whose clamping device is a bowspring or sliding types are also sometimes used. Seismograms recorded with the aid of such probes can only be used for kinematic processing.

Resonance

In the case of poor contact between the geophone and the borehole wall, resonant vibrations may appear both in the cable to tool and the geophone-clamp to borehole wall systems. The frequencies of such vibrations are determined by the design of the tools and, as experience shows, may often lie inside the seismic frequency band. Being of long duration, such vibrations may completely dominate the seismogram. Reduction of the tool's weight is the most effective method of shifting the parasitic resonances (associated with the contact and the clamp) towards higher frequencies. A rigid contact with the borehole wall enables the parasitic resonances to be suppressed almost completely, and an undistorted record to be obtained (Figure 4). It should be pointed out that many of the unwanted waves referred to above may provide information about the structure of the surrounding rock and the mechanical characteristic of the borehole, which may be useful for production seismics and the application of seismic methods to the solution of technological drilling problems.

Stray electrical pick-up on the cable should also be classified as noise. A long cable hanging in the borehole acts like an aerial that picks up noise from industrial electricity and ground currents. Of major importance for the suppression of such waves is good cable insulation and comparatively low resistance of the cable. The signal-to-noise ratio, can be improved by increasing the sensitivity of the borehole assembly by using groups or stacks of geophones. However, this results in an appreciable increase in the length of the tool and makes multiple detector arrays quite unwieldy. With the same goal in view, geophones can be provided with preamplifiers mounted directly in the borehole tool. In addition to these methods of damping unwanted waves during field recording there are other methods of attenuating unwanted waves during data processing. Such methods will be discussed in Chapter III.

In addition to unwanted regular waves, one sometimes comes up against a background of irregular waves, especially at long travel times when the intensity of useful waves decreases rapidly. However, the problem of background irregular oscillations has not been adequately studied. There is no clear idea as to the origin of these oscillations and to their properties.

2. VSP Instrumentation

The instrumentation used for VSP differs little from that used for surface observations either in its seismic or its electrical characteristics. The main distinctions are due to the requirement to make observations in deep boreholes in conditions of high temperatures and pressures. In such conditions, the reliability and good technical specification of the recording system assumes paramount importance. The main trend in the development of VSP instrumentation during the last decade has been the increase in the number of recording channels from one, which was typical in the initial stages of VSP development, to as many as six or twelve in recent years. VSP instrumentation can be subdivided into the down-hole and surface. The down-hole instrumentation is the most essential and specific for VSP.

Down-hole instrumentation

The main characteristics influencing the quality of recorded data are the number of detectors and the method of clamping the borehole tool to the borehole wall. With single detector instrumentation, one has to make a large number of explosions (up to 100 and more) from each shotpoint to complete a vertical profile. In such conditions, it is very difficult to ensure constant source conditions. The use of multiple detectors enables the cost of VSP, the amount of drilling and shooting, and the time required to complete a vertical profile to be reduced, and this is of major importance for drilling boreholes.

In 1963, a three detector probe was first developed and employed at IPE in which every geophone was individually clamped to the borehole wall [14]. At present, VSP is being executed with three- and six-detector probes. Despite the fact that further increase in the number of recording channels involves great technical difficulties, several assemblies of multiple detector equipment have been developed in recent years. Such equipment can be subdivided into two groups depending on the method of signal transmission from the borehole geophones to the surface: one utilizing direct recording with separate cable conductors for all the channels, and the other utilizing multiplexing of the different signals.

Instrumentation utilizing multiple cable conductors

Here one should first of all mention the VSP–I station for six-detector single-component observations developed and mass-produced by the 'Neftepribor' plant [59]. The borehole equipment VSP–I consists of the VSP–6/7 probe made up of six geophones and two dampers or shock absorbers.

The probes are provided with nonadjustable leaf-spring clamps and are designed chiefly for work in cased boreholes. The technical specification of the borehole equipment of this system is presented below:

Number of measuring channels	
with a seven-conductor cable	6
with a three-conductor cable	2
Operating temperature range	from −20 to 160 °C
Maximum operating pressure	98 MPa
Phase duplication between seismic recording channels	less than 6%
Noise referred to input	less than 0.5 μV
Frequency response of the seismic recording channels,	9–125 Hz
Crossfeed between seismic recording channels,	less than 1%
Dynamic range of the signal recording system on an oscillograph,	not less than 68 dB
on magnetic tape,	not less than 80 dB
Suppression of 50 Hz Hi-line	not less than 40 dB
Range of borehole diameter,	from 100 to 300 mm
Dimensions (without leaf springs) of the tool and the shock-absorber	length 534 mm diameter 75 mm
Number of shots recorded on magnetic film with a six detector probe	4

To improve sensitivity and resistance to interference, the VSP–1M system was developed, which, in addition to the VSP–6/7 probe, is equipped with the VSP–2 probe with downhole amplifiers enabling the signal at the logging cable input to be amplified, and thus the signal-to-noise ratio to be increased.

Power to the borehole tools in the VSP–1M system is supplied from the surface via the conductors that carry the signals from the probe. To amplify signals without disturbing the crossfeed compensation, the downhole amplifier is connected to the cable as a current generator. A single conductor serves to transmit the signal to the surface instruments, and to supply power to the drain of the field-effect transistor.

The probes described above utilise sliding clamp devices, and this makes it practically impossible to use data obtained with them for dynamic processing.

Other variants of multichannel equipment utilizing direct data transmission have also been developed. Among these, the three-detector probe with controlled clamping devices (SPPU–2M), and the six-detector probe (ASPU–6–48) developed at VNIIGAS, and designed for VSP in cased and uncased boreholes [60], are worthy of note. Independent control of clamping devices enables each tool to be clamped in succession, thereby enabling them to be isolated from one another, while facilitating observations during both the lowering and the raising of the probe. Observations in any desired succession can also be carried out in the most interesting geologic zones. The last point is very important for production seismics. Seven-conductor cables are used to supply power to the electric circuits and to transmit seismic signals to the surface equipment. Technical specification of the equipment is listed in Table II. A general view of one section of the probe is shown in Figure 5.

Both ends of the shaft that drives the clamping arm are mounted in hermetically sealed chambers closed to outside pressure. This enables the effect of external pressure on the clamping force to be excluded*. In the ASPU–6–48 probe,** the arm of the clamping device moves rapidly until its end comes into contact with the borehole wall, after which its speed is reduced, but the clamping force is increased.

Signal preamplification is provided in the probe. The amplifying stage utilizes a metal ceramic tube (6S 51N) whose plate resistor is connected via one of the cable's conductors to the surface control panel. This makes it possible to use the same conductor for supplying power to the tube and for the transmission of signals. The motion of the probe in the borehole during its descent is controlled by the signals of a horizontal geophone connected to a cable conductor.

Table II shows the specification of VSP instrumentation produced by some US companies.*** The equipment listed in Table II will be seen to belong to the single-detector type. The borehole tools are as a rule equipped with electromechanical clamps and for the most part have large dimensions and weight. The tool L–10.3 DSWC (Mark products) has a mechanical clamp released by a heavily loaded spring upon contact of the tool with the bottom of the borehole. The majority of borehole tools house either only a vertical component or a three-component *XYZ* geophone array.

A six-detector probe employing amplitude modulation (with a separate conductor for each channel) has also been developed. The probes are provided with a telemetering system for programmed amplification control. Symmetrical matching transformers are often used to minimize crossfeed.

* USSR patent No. 254803.

** USSR patent No. 654796.

*** Data are reprinted from the maker's catalogues.

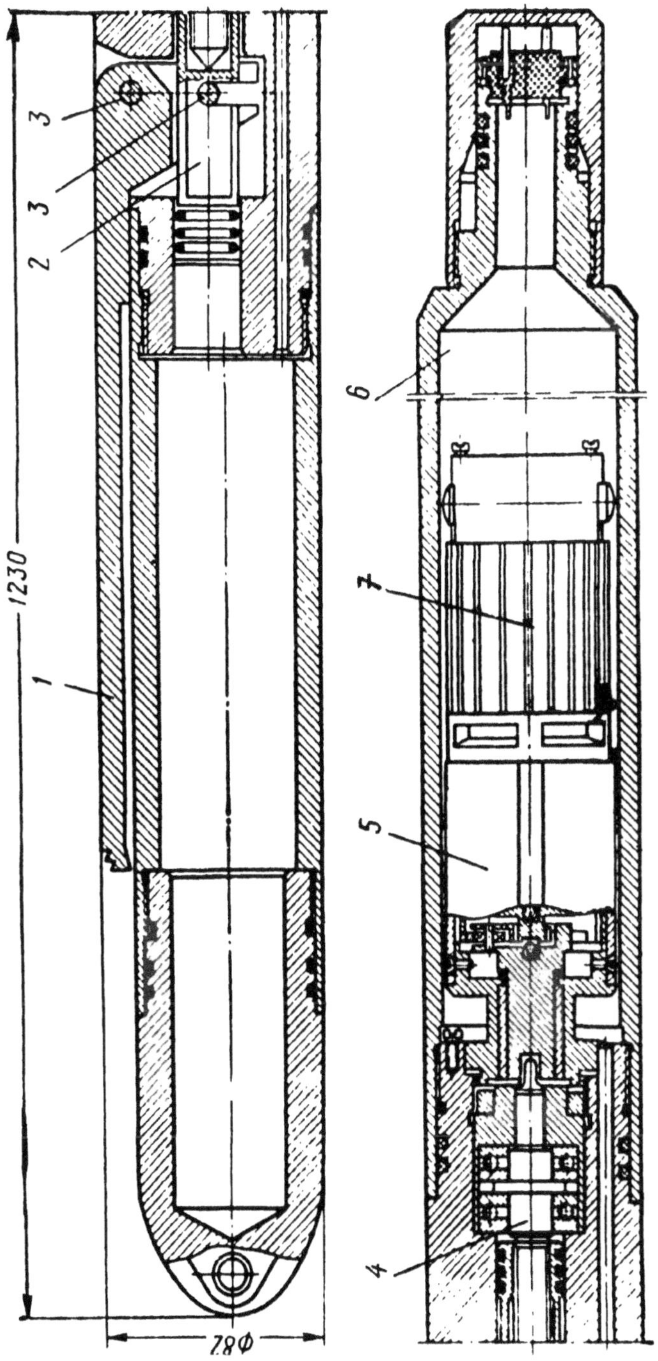

Fig. 5a. A general view of the SPPUU borehole tool [60]: 1 – clamping lever; 2 – rod; 3, 4 – pin; 5 – screw; 6 – step-down gear; 7 – electric motor.

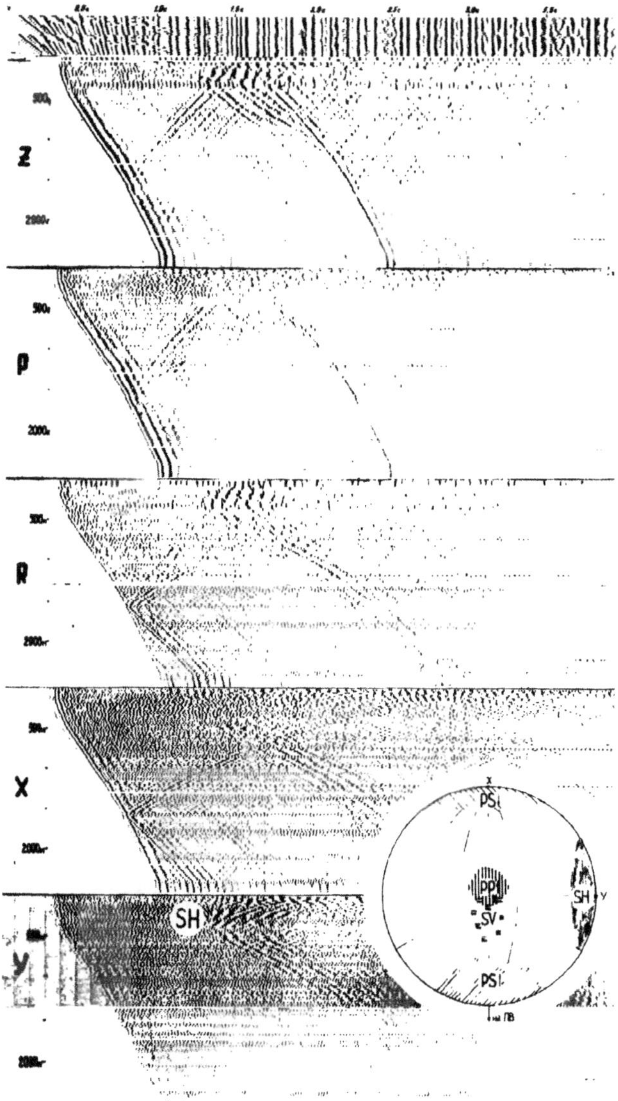

Fig. 5b. PM–VSP seismograms of components of the original XYZ and the local PRT (Y) coordinate systems. Borehole 1 Vostochno–Molodezhnaya SP 600 m. The seismograms have not been subjected to any processing procedures. Various wave types are trackable on the seismograms. The regions of their tracking components are shown on the stereographic projection.

EQUIPMENT WITH ELECTRONIC MULTIPLEXING OF SIGNAL CHANNELS

Various types of multiplexing are in use: frequency modulation, frequency division with amplitude modulation and pulse-position modulation. At present, a multichannel system of transmitting data over a single conductor cable and utilizing pulse-position modulation and time sharing, seems to be of most interest. The first version of such a system was developed by L.L. Khydzinskiy [23] and claims high reliability and compact dimensions. The system may be used as a 24-channel tool device.

The main specification of the system is shown below:

TABLE II

Frequency range	10–125 Hz
Dynamic range,	60 dB
Non-linear distortion in the operating frequency band	2%
Inter-channel crossfeed	2%
Maximum cable length	5000 m
Maximum operating temperature	120 °C
Maximum operating pressure	60.6 MPa

This system served as the basis for the development of 6- and 12-channel VSP probes of 42 and 69 mm diameter, respectively. There are different versions of such equipment in existence, the distinction between them being chiefly in the arrangement of their component parts. It should be pointed out that the depth at which measurements can be carried out is limited by the temperature range.

With modulation systems, the equipment has to satisfy more rigorous requirements. In many cases the noise generated in such systems greatly distorts the signals of weak reflected events. This applies principally in the study of wave dynamics.

For VSP work in mining applications, small-diameter probes are employed. At present probes developed by VIRG have a diameter of 42 mm. Such probes have occasionally been employed in uncased boreholes of 59 mm diameter drilled with diamond drills. The borehole tool uses a mechanical clamping device.

For VSP high-frequency measurement in coal fields, a three-detector probe of 60 mm diameter with a controlled clamping device is used, with a frequency band of 50–400 Hz and a dynamic range of 90 dB [61].

For VSP work in engineering geology, multichannel probes with controlled azimuthal orientation of geophones are in use [62]. Two-component geophones are mounted on a rigid support fixed to a flexible band of sufficient rigidity to prevent spontaneous rotation of individual geophones in the probe. Each geophone cluster is clamped with the aid of a pneumatic device. The probe provides for simultaneous observations at 11 points at 1 m intervals and is designed for observations in boreholes up to 20–30 m deep and 100–127 mm diameter. In VSP studies of porous sediments in marine wells, carried out in the search for placer fields and for the solution of problems in engineering geology, a single-detector probe with a piezoelectric pressure-sensitive geophone is used. Frequencies up to 2000 Hz are recorded.

Clamping devices

In modern borehole tools the clamping devices should solve several problems simultaneously: ensure a reliable contact between the tool and the borehole wall, exclude parasitic resonances associated with tool to borehole contact, suppress cable waves by taking the weight off the cable, and effect a substantial attenuation of hydraulic waves on the record.

Recent studies aimed at evaluating the optimum clamping force of multi-detector probes [25, 59] have substantiated the results [14, 15] previously obtained to the effect that, in order to obtain a high-quality record free of cable waves and of resonant vibrations in the seismic frequency band, the borehole tool must be clamped with a force several times that of its own weight. As to the conditions of contact, the frequencies of parasitic oscillations induced at the contact should lie outside the operating spectrum of the instrumentation. This makes it necessary to increase the clamping force (for a tool of the same weight) if the frequency band is extended. The quantitative dynamic data processing aimed chiefly at studying the physical properties of rocks, especially where the polarization method is used, makes very exacting demands on the quality of recorded data. This depends to a large extent on the quality of clamping. Thus, in order to solve a variety of problems, the clamping force as a function of the frequency band, the dynamic range and sensitivity of the instruments, the design of the probe and the type of cable to be used, must be evaluated by experiment. This is of special importance, when multi-detector probes are being employed. Figure 4 shows seismograms obtained from a multi-detector probe for both weak and strong clamping forces.

Under favourable circumstances the use of sliding clamp devices enables the dominant waves to be tracked, and their kinematics to be studied. However, the use of such records for rigourous dynamic processing may lead to considerable errors, and can hardly be recommended. Sliding-clamp probes used in VSP are clamping devices little different from similar devices widely used in down-hole geophysical studies, e.g. resistivity-logs.

The clamping devices themselves may be subdivided into three types – uncontrolled, controlled and combined. *A controlled device* makes it possible, by sending commands from the surface, to clamp the tool to the borehole wall and release it at any depth, and thus enables observations to be carried out both running in and pulling out of the borehole. Controlled clamping devices are mainly of the hydraulic and the electromechanical types [64]. Self-clamping devices having a force proportional to the tool's weight are also in use. Probes have been developed in which each geophone is clamped when the weight of the detector beneath it is taken off; after the deepest detector has been clamped the rest lock one after the other. Such a successful controlled clamping device is the SPPU–2 in which both ends of the actuating rod are housed in hermetically sealed chambers to exclude the effect of external pressure on the clamping force*.

* USSR patent No. 254803.

The technical specification of SPPU-2 is shown below:

Diameter of boreholes	127–300 mm
Clamping force	
with a long clamping arm	1600 N
with a short clamping arm	33560 N
Current with no load on	
clamping arm	0.35 A
Maximum ambient pressure	80 MPa
Maximum ambient temperature	125 °C
Dimensions	80 × 1310 mm

Single-action clamping devices that enable the tool to be clamped only once at the maximum depth reached, can also be classified as controlled clamping devices. Subsequent obervations can be carried out only by raising the tool in the clamped state. Such clamps can be recommended only for use in cased boreholes. A mechanical actuator is generally used with such single-action devices. The spring is released mechanically as the tool makes contact with the bottom of the borehole, or immediately the tool begins to be lifted. Other devices used for this purpose include sundry cam and hinged-lever devices with various springs and latches. It should be pointed out that in cased boreholes, cam-type clamps that ensure a constant contact of the tool with the casing string have proved worth while. *Uncontrolled clamping devices* are lowered and withdrawn from the borehole with clamps already actuated. In such cases leaf-spring clamps, ensuring a sliding contact of the tool pressed against the borehole wall, are used (they are sometimes termed sliding clamps). As previously stated, with sliding clamps, it is not possible to disengage the tool from the surface, or suppress cable waves or to obtain undistorted records necessary for the qualitative processing of the waves' dynamic properties.

Combined devices are frequently used with multi-detector probes. In such devices one of the detectors (usually, the uppermost) is equipped with a controlled clamp, while the lower tools are equipped with uncontrolled sliding clamps. Naturally, the probes with independent controlled clamping claim the maximum potential with substantial advantages over probes of other types.

It should be pointed out that when performing VSP with multi-detector probes, one should emloy special devices to ensure the safety of the lowering and lifting operations. To this end a standard pulley is usually equipped with a lever system, which enables successive tools to pass over the pulley. It is moreover necessary to observe the probe's motion in the borehole to prevent paying out excessive lengths of cable. Several methods can be used for this purpose. One of them is to monitor the noise level of the geophone signals. When the tool stops, background noise, especially on the horizontal components, drops drastically.

GEOPHONES

VSP usually employs mass-produced electrodynamic geophones widely used in surface observations. The characteristics of the geophone are determined by the object of the study and by the design of the borehole tools. Geophones operating at low frequencies (with a natural frequency in the range 3 to 10 Hz) have been used for studying the dynamic properties of the wave field, when a wide frequency band is required. This includes the study of the dynamics of direct waves and of the mechanism of absorption in the real earth. Geophones normally used for the study of reflected waves, as well as in mineral seismic exploration, have been of the electrodynamic type with a natural frequency in the range of 10 to 30 Hz. Grouping of several geophones per channel is being used to increase sensitivity.

In regional VSP studies, when vertical seismic profiles have been made from shotpoints located up to distances of 60 km, geophones with a natural frequency of 3 to 5 Hz have been used. Borehole geophones with a natural frequency of 1 Hz have been used for some special problems, in particular, for the study of the wave field excited in the process of deep crustal seismics. As a rule, low-frequency geophones have to be specially adjusted and mechanically damped during lifting and lowering operations. The need to improve the resolution of recordings makes the use of high-frequency geophones with a natural frequency of 80 to 200 Hz imperative. We think it expedient to use accelerometers to record oscillations in the 100 to 400 Hz frequency band.

To improve the signal-to-noise ratio, most borehole tools are provided with pre-amplifiers with a gain of 30 to 50 dB.

In addition to displacement or velocity sensitive geophones, pressure detectors, freely suspended by the logging cable without being rigidly clamped to the borehole wall, have also been tested for VSP work. This greatly simplifies the design of the borehole tools and the measurement procedures. However, several deficiencies are common to pressure sensitive devices. Whereas it is possible to trace undistorted first arrivals on records obtained with pressure geophones both in cased and uncased boreholes, all subsequent portions of the record are greatly distorted by the presence of a large number of high amplitude hydraulic waves propagating along the fluid column; this greatly degrades the seismogram and limits any possibility of wave field analysis. The situation is made worse by the fact that pressure detectors, being omni-directional in response, record a variety of wave types, thus greatly complicating the seismograms. In addition, pressure detectors cannot discriminate waves according to their polarization. All this limits the potential for pressure detectors in VSP applications. Nevertheless, pressure devices may be used to solve certain problems involving the study of primary waves observed in uncased boreholes.

SURFACE EQUIPMENT

The recording of VSP seismograms is performed on the surface using standard equipment. For multichannel seismograms, magnetic tape recording synchronized to shot instants is used. In order to obtain records suitable for dynamic processing by analogue computers, magnetic tape recording is duplicated by oscillograph records without filtering. Frequently, discrimination is executed during the course of subsequent processing.

Special surface equipment used with multichannel probes has provision for (i) controlling the probes, (ii) recording demodulated and direct signals on magnetic tape, (iii) recording frequency modulated signals, and (iv) producing oscillograph records.

When wave dynamics is studied in VSP, the dynamic range of the recording instrumentation is of special importance. The ratio of the amplitude of the first arrival (at a shallow detector position) to that of waves reflected from great depths can be as high as 100 to 110 dB. In order to obtain undistorted records, equipment with a dynamic range of 120 dB must be available.

The use of digital recording with high dynamic range enables the full potential of VSP, based on the application of wave dynamics, to be realized. Digital recording, aimed at obtaining undistorted signals of reflected waves with travel times of 4 to 5 seconds, is normally carried out with three, or sometimes five, different amplification levels, separated by amplification factors of 4 to 5 times. Various types of programmed gain control have been used for this purpose followed by reproduction of the signals over the entire dynamic range.

Low-frequency recording equipment employing correlation methods has been used to study refracted waves. When observations are carried out at great distances between source and borehole, the dynamic range of recorded signals is much smaller than in the reflection method, and, as a rule, two amplification levels are sufficient.

The gain of the recording channel is controlled with the aid of the magneto-electric constant-amplitude generator or with a conventional audio-frequency signal generator. In the latter case the generator is monitored by a special galvanometer.

The recording equipment used for VSP has to satisfy some specific requirements, as distinct from that used for surface observations. These are mainly a lower level of internal noise and a wider frequency band. The level of seismic noise in boreholes drops rapidly with depth, and at some specific depth, the sensitivity will be limited by the internal noise of the instrumentation, thus preventing the recording of weak high-frequency waves [65]. One has substantially to increase the sampling rate in order to record weak undistorted signals in the frequency band up to 400 to 450 Hz. Special design and development is required to reduce equipment noise for operations at great depths where high temperatures and pressures prevail.

Instrumentation for PM–VSP

The PM–VSP method specifically requires three-component observations and the recording of the components of wave motion oriented in space.

A four-detector three-component probe with pulse-position modulation, has been developed for PM–VSP [28]. The main technical features of the equipment have been listed above. It is, however, worth pointing out that it would be more sensible to use a three-detector four-component probe as described below; this enables continuous seismic control of channel identity for each of the three-component sets to be performed. To study seismic wave dynamics at comparatively shallow depths (down to 700 m), Wuenschel [66] developed a special tool comprising two sections. The lower section houses a three-component geophone, two independent hydraulic clamps that lock the tool to the borehole wall with a force of about 80 kg, and two mutually perpendicular vibrators used to determine the quality of geophone coupling at different points of the lithology. The upper section houses signal amplifiers and reversible electric motors that drive hydraulic pumps to clamp and release the tool from the borehole wall. The two sections are separated by about one metre and are connected by means of an electric cable and vacuum hoses.

PM–VSP instrumentation is described in detail in [2] and here we shall point out only its principal features.

Two types of mutually orthogonal geophone assemblies are currently in use for borehole tools – the *XYZ* and the symmetrical. The former is traditional and has been in use for many years. The symmetrical type is also made up of three mutually orthogonal geophones each of whose axes is inclined at the same angle of about 35 °C to the horizontal, and with 120° of azimuth between adjacent geophones. In surface observations when the angle of emergence of the waves is close to the vertical, the *XYZ* arrangement provides a visual record of the three components because of the separation of the waves. The longitudinal waves are recorded mainly by the *Z* component, and the transverse waves by the *X* and *Y* components. Such visual records are only of any advantage in visual data processing. In the case of digital processing, this visual advantage of the *XYZ* arrangement becomes irrelevant. For VSP observations, when the direction of arrival of the waves does not generally coincide with the vertical, the advantages of the *XYZ* arrangement disappear. This configuration has several drawbacks, including particularly the difficulty of monitoring the frequency and phase characteristics of the horizontal and vertical geophones, and the polarity of the channels in the field. The need to use geophones of different types (horizontal and vertical) in the same assembly is an additional disadvantage.

In contrast to the *XYZ* arrangement, the symmetrical arrangement enables seismometer channel, duplication to be tested on the surface and also in the borehole. On the surface this can be done by recording signals with all the geophones having the same azimuthal direction, or by recording signals from depth with geophones arranged in their normal azimuthal directions. In the former case, the first longitudinal wave arrives in a near vertical direction, and all the geophones should produce identical records [2]. In the borehole, the control is performed by recording the direct longitudinal wave from a source near the top of the hole.

In the process of studying the trajectories of particle displacement, a very convenient method of continuous seismic control is achieved by recording an additional (the fourth) component of motion. With the symmetrical arrangement, the fourth component recorded is the *Z*-component. The output is compared with the sum of signals from the other three components in order to monitor the duplication of the geophone elements.

This method not only enables the control procedure to be formalized, but also provides facilities for the necessary corrections. In the course of the control procedure, the amplitude correction and data selection are performed automatically by a computer.

In recent years IPE, VNIIGIS and 'Krymmorgeologiya' have jointly developed a three-detector four-component probe (APM VSP–3–4). The type SPPU–2 borehole tool described above, has been used to incorporate a 12-channel electronic pulse-position multiplexer. Each borehole tool is equipped with a container housing a three-component symmetrical cluster, a fourth vertical geophone and a multiplexer. The circuitry of the APM VSP–3–4 equipment allows the non-linear distortion of signals, the internal noise level and cross-talk between the channels to be reduced, as compared with the equipment described above. Two types of control are provided: the seismic and the electric. The seismic control is executed continuously by comparing the output of the *Z*-component obtained by direct recording and that obtained by summation of the signals from the other three components. Such comparison continuously and auto-

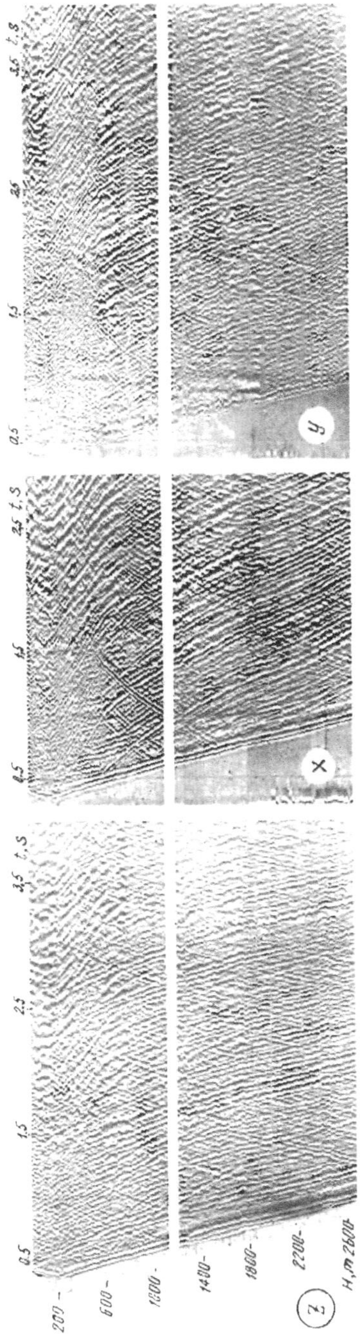

Fig. 6. **PM–VSP seismograms obtained with a forced-orientation tool (borehole N2 Skrychalovskaya, by courtesy of M. N. Isayenko).**

matically performed releases the operator from the obligation to carry out a constant watch over the channel duplication. A special circuit transmits a calibration signal when duplication is in doubt. This electric control is executed by supplying calibrated d.c. pulses, and is actuated automatically when the seismic control system gives out an unbalance signal. The equipment has been tested in boreholes in the Crimea at depths down to 3000 m and at temperatures up to 120°.

It should be pointed out that in PM–VSP work involving the recording of different wave types, including waves with horizontal particle displacement, a pre-requisite for high-quality data is the suppression of possible parasitic rotational vibrations of the borehole tool. This requires stronger sidewall clamping than in the case of conventional single-component observations. This is conveniently done by locking the tool in two mutually orthogonal directions. To reduce torsional vibrations of the cable, the latter and the tool are sometimes joined by means of a slipping assembly. However, this greatly complicates the probe's design, reduces the reliability of the equipment and increases the inherent noise. The expediency of employing such connection is open to doubt.

Specifically in PM–VSP, it is expedient to obtain components of motion oriented in space. Such components substantially enhance the effectiveness of PM–VSP studies, especially in the case of complex formations. Oriented records are very important, when the polarization parameter is used for quantitative interpretation. To orient borehole tools in cased boreholes, gyroscopes are being used. Other devices in use include borehole tools with controlled orientation of the geophone assembly, and equipped with orientation sensors that enable the actual orientation of the assembly to be determined [2]. Figure 6 depicts three-component seismograms obtained using a borehole tool with gyroscopic geophone orientation designed by M. N. Isayenko. It is worth noting that, despite single-detector observations, there is a sufficiently reliable correlation between different wave components on the records, especially high-frequency waves. However, borehole tools with controlled orientation are intricate mechanisms difficult to produce, and for this reason, in our opinion, tools equipped with orientation sensors offer substantial advantages. With such sensors it is not the geophone assembly, but the records themselves that have to be oriented. The orientation is performed on the surface by transforming the coordinate system either in an analogue form directly in the field, or in the course of subsequent computerized data processing.

In uncased boreholes the Earth's magnetic field may be used for determining the geophone orientation. In cased boreholes, one has to make use of such gyroscopic orientation devices as DOU-1. In contrast to the inclinometer, which determines the attitude of the borehole in space by measuring the two angles it makes with the vertical (ϕ_0) and the horizontal (ω_0), DOU–1 additionally measures the angle of rotation of the tool about its axis (γ).

To determine the geophone orientation from the angles ϕ_0, ω_0 and γ, one marks the position of the geophone axis on the stereo net (point O in Figure 7a). Next one constructs the plane Q at right angles to the axis. The vertical plane Q_a passing through the zero axis of the geophone intersects the plane Q in the direction 'a'. Then the angle γ is measured in the plane Q from this direction, and the direction 'b' is obtained. The first direction (point 'a') corresponds to the line of intersection of the plane Q and the vertical plane Q_a that passes through the borehole axis. The second direction (point 'b') is the line of intersection of the plane Q and the plane Q_l passing through the axis of

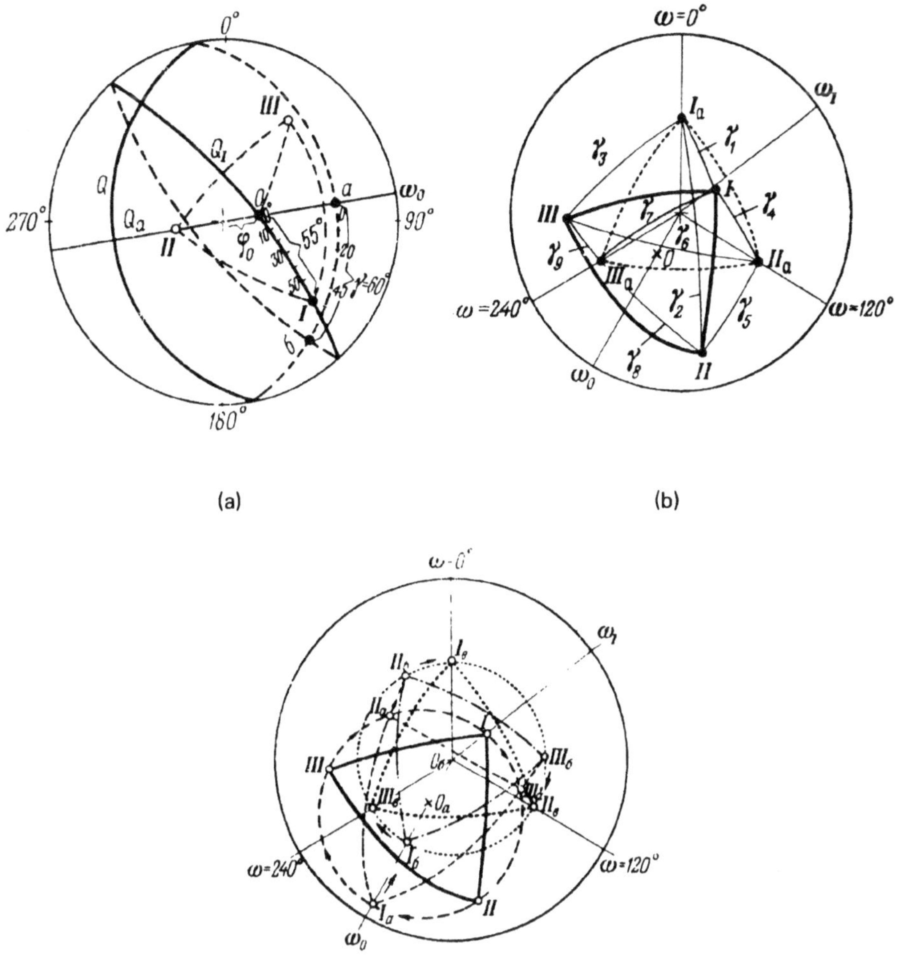

Fig. 7. Orientation of seismograms and determination of the geophone's orientation: (a) direct orientation; (b) stage-by-stage orientation; (c) stereo net.

the geophone (I) and the borehole axis (O). To find the axial position of the geophone (I) in the plane Q_I, one has simply to mark on the net the angleof 55° from the point O in the direction of point 'b'. To obtain the position of the two remaining geophones, one should draw a plane Q_{II-III} at right angles to the geophone (I) axis. This plane will pass through point K. The direction of the axes of geophones (II) and (III) are symmetrical about point K and make an angle of 45° with the direction K. The example cited refers to the case when $\phi_0 = 2°$ and $\gamma = 60°$. The actual position of the geophone assembly thus determined enables oriented wave components to be obtained. Two orientation procedures are feasible: the direct and the step-by-step. *Direct orientation* consists of the direct transition from the signals of a geophone assembly arbitrarily oriented in a specified direction by means of a weighted summation of the former. The weighting is determined by correction coefficients proportional to the cosines of the angles between the axial directions of the actual (I, II, II) and of the oriented (Ia–IIIa) geophones. Nine angles are measured on the stereo net (Figure 7b): Ia–I = γ_1; Ia–II = γ_2; Ia–III = γ_3; IIa–I = γ_4 IIa–II = γ_5; IIa–III = γ_6; IIIa–I = γ_7; IIIa–II = γ_8; IIIa–III = γ_9.

The values of cos γ_i correspond to the values of the correction coefficients sought.

Step-by-step orientation, in contrast to the direct, where each imaginary geophone is oriented independently, enables the orientation of the assembly as a whole to be executed (Figure 7c). Let the actual assembly I, II, III be oriented in space so that the direction of its axis (the borehole axis) is determined by the angle of tilt ϕ_o and the azimuth ω_o (point O_a), and let the azimuth of the first geophone of the assembly be ω_I. In our case $\omega_I = 53°$. The first step (operation (a)) consists of adjusting the azimuth of the first geophone ω_I to coincide with the azimuth of the assembly axis (ω_o). As a result, the geophone axes of the actual assembly I, II, III will take up a position coinciding with that of the oriented assembly (a): Ia, IIa, IIIa. The second step (operation (b)) consists of placing the axis of the oriented assembly (a) in a vertical position (point O_a moves to the position O_b). This will orient the axes of all three components of the assembly (a) in the directions of the axes of the imaginary assembly (b): Ib, IIb, IIIb, so that they will take up positions on the surface of a cone whose generatrix makes an angle of 35° with the horizontal. The third step (operation (c)) involves the rotation of the assembly (b) clock-wise about the vertical axis until the azimuth of the first geophone Ib coincides with the northward direction ($\omega = 0°$) or any other specified direction. Thus the imaginary assembly (c) is now oriented. In our case the directions of the axes of the geophones Ic, IIc, IIIc are described by the following coordinates: $\omega_{\mathrm{I,c}} = 0°$, $\omega_{\mathrm{II,c}} = 120°$, $\omega_{\mathrm{III,c}} = 240°$ and $\varphi_{\mathrm{I,c}} = \gamma_{\mathrm{II,c}} = \gamma_{\mathrm{III,c}} = 53°$.

Step-by-step orientation is technically convenient since it enables the data obtained from DOU (φ_o, ω_o, ω_{I}) to be used directly and obviates the necessity for executing the operations of angle measurement and coefficient determination in the field. The step-by-step orientation procedure may form the basis for the design of an instrument, convenient to work with, in which orientation corrections would be introduced automatically from parameters read off DOU. Special orientation devices have been evolved to obtain oriented records in analogue form. Operational models of such devices as well as orientation procedures have been discussed in detail in [2]. In the absence of orientation devices it is possible to orient records by the direction displacement in the primary longitudinal wave, if certain simplifying assumptions are made about the earth.

To realize the full potential of VSP for solving various geological problems associated with the space surrounding the borehole, in particular for locating and investigating deposits and when a quantitative evaluation of the fine detail of the dynamics of the record is required, borehole equipment must be available that is capable of recording undistorted signals in a wide frequency range (approximately 10 to 450 Hz) and with a high dynamic range (up to 120 dB).

The potential effectiveness of combining VSP and sonic logging data, makes it expedient to design instrumentation that would enable both types of recording to be carried out in a single logging operation. In addition to saving time, this would improve the accuracy of data correlation with the geologic section and with each another. The latter is required, most of all, to exploit the potential of production seismics (see Chapter 13).

CHAPTER 3

RECORDING PROCEDURES AND DATA PROCESSING

VSP is being used to solve a wide range of problems, and each specific case determines its own procedures for recording and data processing. However, there are certain general features which are typical of VSP as a whole.

1. Source Excitation Conditions

There is little difference in principle between the exictation conditions in VSP and in surface observations. Thanks to VSP, it is possible to study objectively the pronounced influence that the source conditions have on the wave field and on the effectiveness of seismic studies. This influence manifests itself first of all on the direct *P*- and *S*-waves that initiate the entire wave generation process, and through them – on the shape of each particular wave and the seismogram as a whole*. In many cases, the excitation conditions limit the potential of seismic studies, and for this reason we shall have to deal with them at length.

The main point about source excitation in VSP work is that a large number of seismic disturbances are required to complete one profile. Accordingly, to obtain high-quality VSP data, the most important thing is to guarantee constant source excitation conditions. This makes it necessary to study the effect of a number of factors relating to excitation conditions, primarily, the depth and nature of the source, and the lithology at the source location. It has also been necessary to work out procedures for controlling excitation conditions.

VSP, as well as surface observations, can be executed using separate excitations of longitudinal and transverse waves. In conjunction with the development of the polarization method, simultaneous excitation of *P*- and *S*-waves from a single source is being increasingly used. In addition to excitation by explosions, various methods using non-explosive sources are gaining importance in VSP.

Excitation of *P*-waves

A large amount of special experimental work has been carried out to study excitation conditions and to evaluate the effects of various factors. From this, the following conclusions may be drawn:

Charge depth

The strong influence that charge depth has on the excitation conditions is the result of lithology changing with charge depth, and the relative position of the charge to any lithological heterogeneities. With homogeneous near-surface conditions, a comparative-

* Some aspects of the effects of source excitation conditions on the shape of the direct wave are discussed also in Chapter 5.

ly simple pulse is obtained from shot hole charges. The spectrum of the pulse is inhomogeneous. The high frequencies are concentrated in the initial portion of the record, with the subsequent phases having progressively longer periods. An increase in the depth of the shot usually results in higher frequencies being generated, which in turn simplifies the record and improves its resolution.

In inhomogeneous near-surface conditions, the form of the first wave and its spectrum are determined chiefly by superposition of various waves formed at the heterogeneities. The number of such waves, especially the ghost reflections, and the way they are superpositioned depends largely on the relative position of the charge and the heterogeneities, which exercise a decisive influence on the shape of the downward wave. In

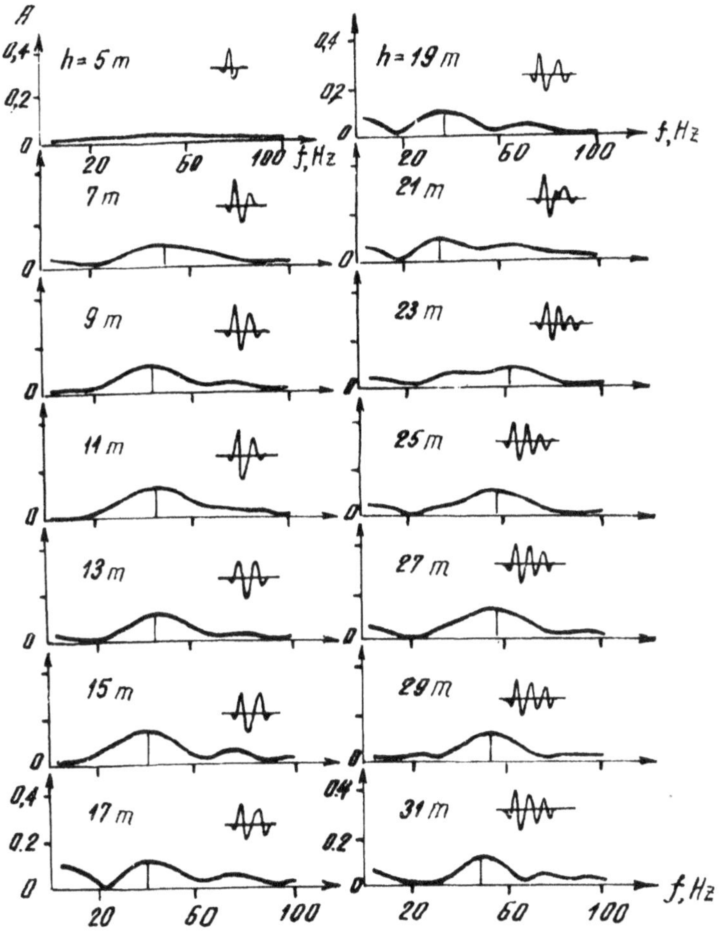

Fig. 8. Shape and spectra of combined pulses excited by explosions at various depths (borehole N3 Danilovskaya, l = 100 m, H = 1060 m [27]).

the case of a low-velocity zone of comparatively simple structure, it is possible to get rid of the ghosts almost completely by placing the charge directly at the base of the low-velocity zone. In such a case, increasing the charge depth means inferior excitation condi'ions and a more complicated seismogram. This is clearly evident from Figure 8,

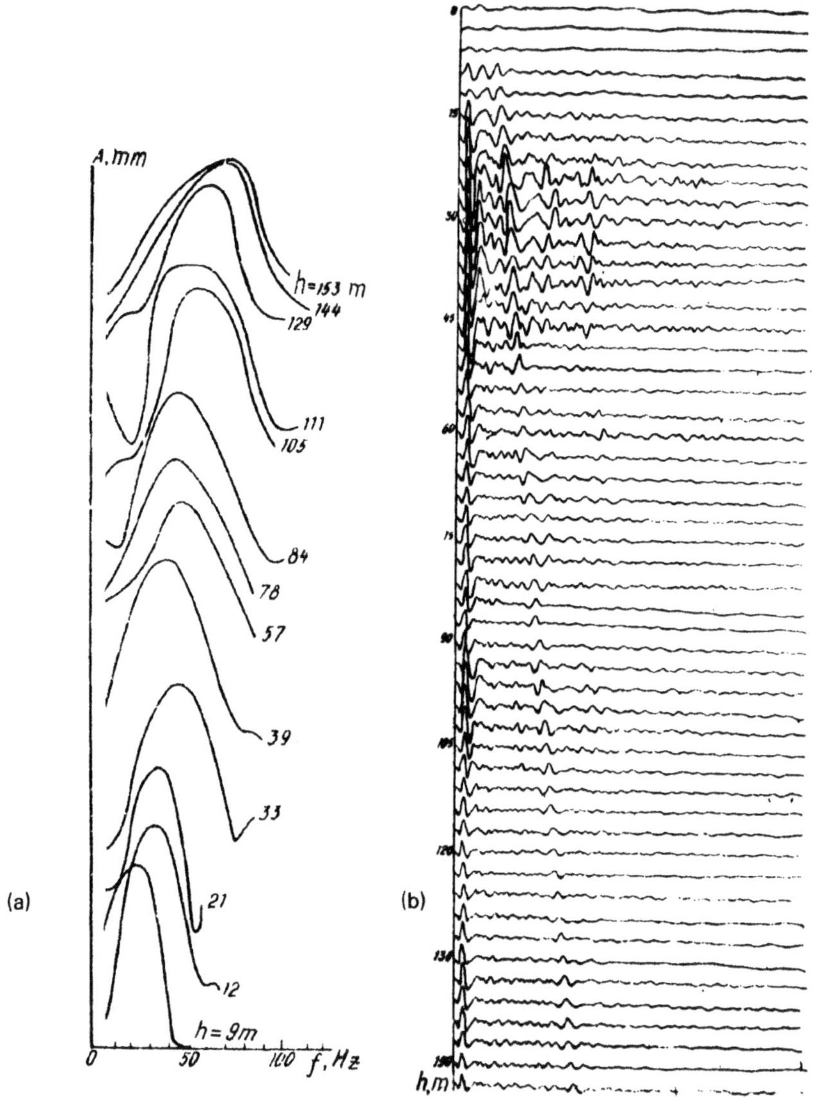

Fig. 9a-b.

Fig. 9a-c. Illustrating the influence of charge depth on the shape of the primary wave and on the structure of the initial portion of the seismogram [67]: (a) amplitude spectra; (b) direct wave and ghost; (c) initial portion of the seismogram for charge depths h = 24 m (left) and h = 121 m (right).

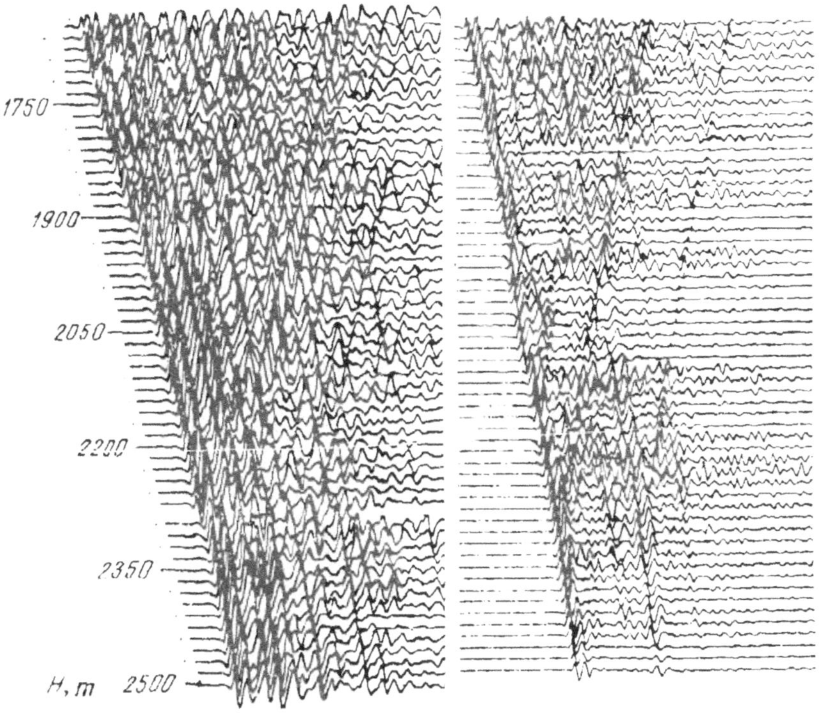

Fig. 9c.

where in conditions of a very thin low-velocity zone (2–6 m), a simple and very short duration pulse was recorded at a depth of 1060 m, only for shots fired immediatly below the low-velocity zone. Even a slight change in the charge depth may result in an appreciable change in the spectrum. Thus, for example, a change in charge depth from 21 to 23 m shifts the dominant frequency from 34 to 63 Hz. In the case of a complex zone containing several layers, it is usually impossible to get rid of the ghosts completely. To obtain a direct wave of simple shape and short duration, it is necessary to separate the ghosts from the direct wave; this can be effectively achieved, although it is money- and labour-consuming, by increasing the charge depth.

We shall use recorded data obtained in Uzbekistan [67] to illustrate the dependence of the direct wave, and of the wave field as a whole, on the charge depth. Increasing the charge depth, h, from 9 to 153 m (Figure 9a) changes the frequency spectrum of the first arrival pulse of the direct wave shifting the peak frequency systematically from 20 to 85 Hz. The structure of the initial part of the seismogram changes as well. When the depth of charge is between 25 and 45 m, the heterogeneities of the earth in the vicinity of the charge cause a complex, long train of waves with amplitudes comparable to that of the direct wave to be formed (Figure 9b). At charge depths greater than 45 m on the seismogram, in addition to the first arrival there is only one ghost whose time lag with respect to the first wave increases monotonously with h. The particular processes of formation of the direct wave at the source (Figure 9a, b) are responsible for a great difference between the wave fields recorded in the initial part of the seismogram (Figure 9c). The strong effect of charge depth on the spectrum of the first arrival can be used to monitor the spectrum of the downward wave, and this can be of great importance in exploration. The effect of charge depth on the shape of the direct wave can clearly be seen from Figure 10a, which shows seismograms obtained at the depths of $H = 1010$ m and $H = 100$ m using 0.4 kg charges fired at various depths. Only when the charge depth exceeds 55 to 60 m does the ghost move far enough away, i.e. 50 to 60 m s, from the direct wave for the latter to assume a simple shape. At shallower depths, the ghost interferes with the direct wave, and its shape becomes complex and changes with depth. The shape of the direct wave changes most radically when the charge is located in the low-velocity zone (the two upper traces). In the course of studying the spectrum and the shape of direct wave pulses produced by various sources (see Figure 10b) it has been demonstrated that shots using electric detonators only, produce high-frequency pulses of a simple shape (Figure 10b–3). Similar results were also obtained using a hydrogen-oxygen mixture (Figure 10b-1). The peak frequency of signals from the seismic vibrator, GSK–10, located on the surface was much lower (Figure 10b-2). As the charge was increased to 0.4 kg, the spectral peak decreased to 34 Hz (Figure 10b-4).

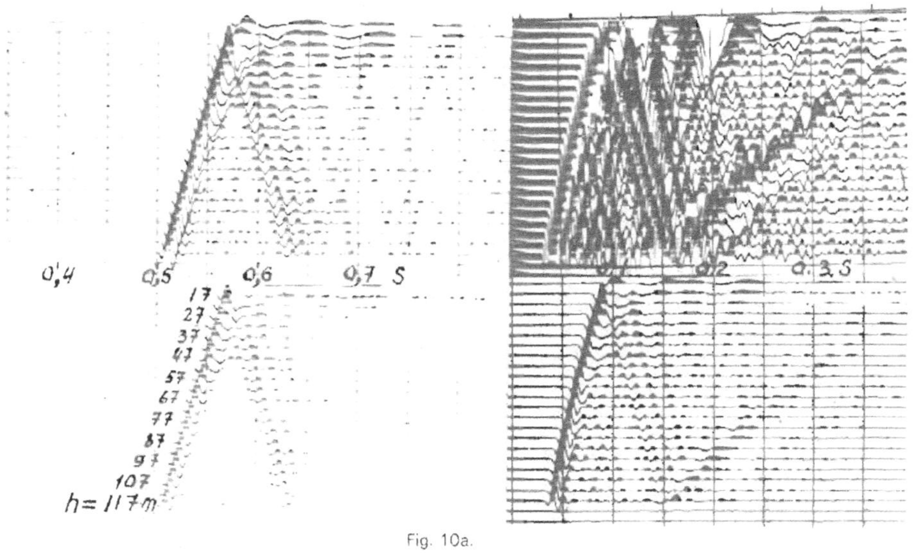

Fig. 10a.

Fig. 10. Illustrating selection of source excitation conditions: (a) effect of charge depth on the shape of the direct wave. $H = 1010$ m (left) and $H = 100$ m (right), $Q = 0.4$ kg, $h = 0$–117 m. Recorded on two amplification levels; (b) spectra of the direct wave with various sources. $H = 1500$ m. Borehole B210 Kaluzhskaya: 1 – gas seismic source (IGS). Depth $h = 12$ m; 2 – seismic vibration (GSK–10), surface excitation; 3 – explosions of ten electric detonators; 4 – explosions of 0.4 kg charges; (c) shape of the direct wave with the same sources, $H = 80$ m.

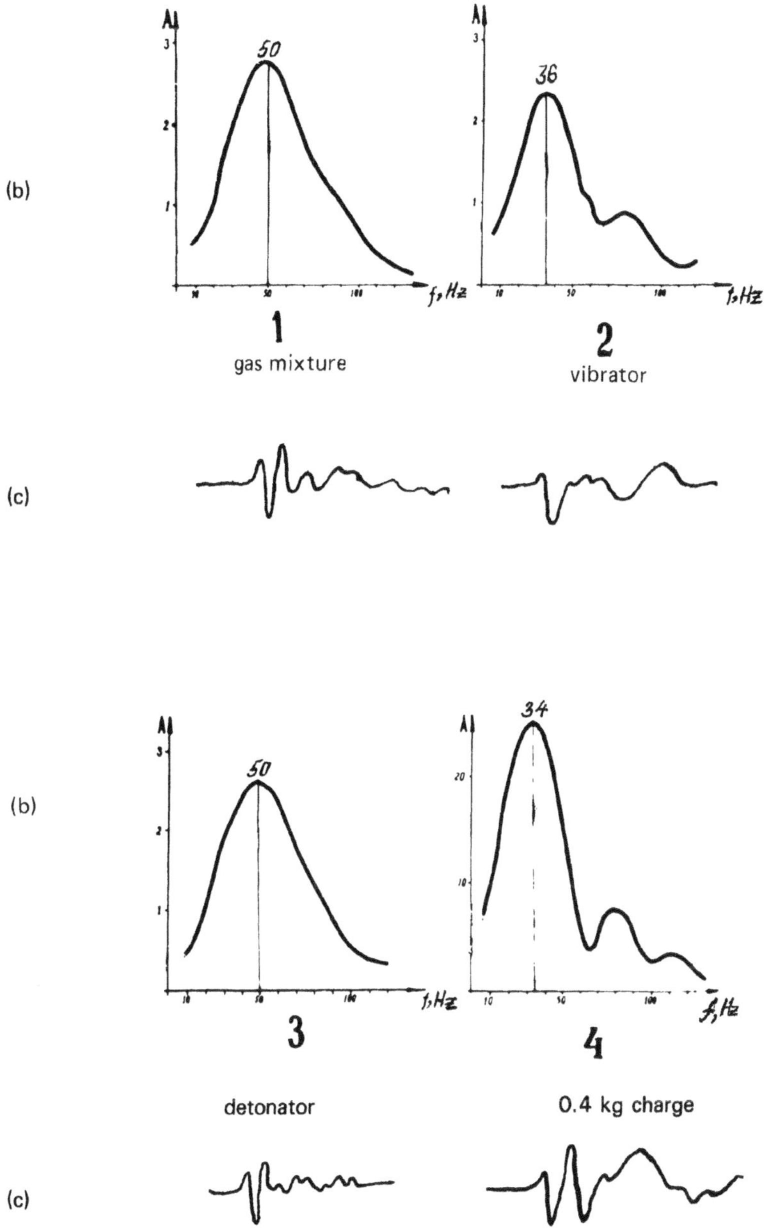
(b)
A
3
2
1
50
f, Hz
1
gas mixture
36
2
vibrator
(c)
(b)
3
50
34
20
10
4
detonator
0.4 kg charge
(c)

Fig. 10b-c.

Lithology

Lithology exercises an influence primarily on the repeatability of source excitation conditions. In explosions in sandy-clay rock, the form of the record generally stabilizes after the first three or four shots and from then on remains constant. It has sometimes been possible to fire up to 30 or 40 shots in one shot-hole. The reproducibility of the record with explosions in hard rock (limestones, marls) is considerably poorer, and the constancy of source excitation conditions deteriorates rapidly as the charge size is increased. Usually subsequent explosions in such a hole produce divergence in the ratios of the amplitudes and the periods of individual phases.

Thus, the effect of charge depth on the frequency content of the record is determined by two factors: the superposition of waves and the lithology. The usual effect of lithology is to produce higher frequencies in the direct wave from shots at greater depths. Shots in the low-velocity zone, in porous media, excite waves of lower frequencies than those excited by explosions beneath the zone.

Charge weight

In VSP, the minimum size of charge providing sufficient energy is chosen by direct observations at great depths. Experience shows that the shape of the direct wave at great depths is largely independent of the charge. At the same time an increase in charge size greatly complicates the wave field observed at shallow depths, mainly because of surface waves of various types and unwanted waves associated with heterogeneities in the near surface. Besides, an increase in charge sizes makes reproducibility of the record worse. Accordingly, it is advisable to work with minimum charge sizes. When comparing data obtained from charges of varying size one has to take into account possible differences in the wave field. The reproducibility of records establishes another criteria for borehole measurements.

Small diameter boreholes have poor stability and, as a rule, withstand relatively few explosions. At the same time repeated explosions in large diameter holes (250–500 mm) leave the spectral content of the direct waves unchanged even when the number of explosions is large. Such boreholes have withstood 200 to 300 explosions of 0.2 kg charges, and made it possible to shoot a vertical profile from one hole. Devices have been developed that make possible the repeated use of conventional diameter holes [66, 68]. With comparatively small charges, a special centraliser that keeps the charge aligned with the hole's axis can be used to prevent damage to the casing string. With such a device in one case, a cemented 8-inch steel casing string withstood two hundred explosions of charges up to 1.0 kg having a linear charge density of 300 g m^{-1} [2].

Experience gained to date shows that a repeatable source having a constant source signature spectrum for use in shotholes is very difficult to construct and not always possible. In this respect demolition cord is of great interest. When such charges are exploded in uncased boreholes, the wave field generated by them is much less dependent on the charge depth than in the case of a compact TNT charge, and an uncased hole withstands a greater number of such explosions (up to 300) The charge length was 5 m, and its magnitude was in each case specially selected, but did not exceed five sections placed approximately 5 to 7 m below the low-velocity zone. In trials of detonating cord charges, over thirty thousand meters of VSP observation have been carried out, and they confirmed the effectiveness of the said method.

Much experience has also been accumulated on the use of horizontal detonating cord buried in the ground to a depth of 0.5 to 1 m [69]. By detonating the line at the end remote from the borehole (i.e. in the direction of the hole), one obtains a downward wave with a front dipping towards the borehole. The wave field from such explosions is simpler than that from conventional explosions in boreholes, chiefly because of the much lower amplitude of the ghost. Consequently, the relative intensity of downward multiples in later arrivals is approximately 25% lower than in the case of explosions in boreholes.

A lot of experience has been accumulated with VSP in marine boreholes. VSP observations offshore are characterised by more stable source excitation conditions. The sources used include air and gas gun equipment of various capacities. They enable recording depths of 4.0 to 4.5 km to be studied. In marine work, where it is possible to move the source rapidly along the surface, it is comparatively easy to ensure constant geophone depth observations. The chief difficulties are associated with navigation. When it is required to make a large number of observations aimed at studying the region surrounding the borehole, a network of sensors is installed on the sea-bed in the vicinity of the borehole whose signals are used to orient the ship carrying the energy source.

The use of grouping of explosions

Vertical arrays of sources enable the direction diagram and the spectral content of the source to be varied, and hence the ghosts to be suppressed, permitting the relative intensity of deep waves to be enhanced.

Source arrays may be of a special interest in the studies of the upper portion of the geological sequence, where the wave field can be very complex. Here it is possible to set up a borehole version of the common-depth-point method used in surface seismics, utilizing both vertical and horizontal summation of signals. Vertical summation does not suppress multiple reflections associated with deep interfaces. This is best done by horizontal summation of recored waves from shallow holes arranged along a horizontal profile, with the same depth of charge in each shothole. When necessary, it is possible to combine both types of summation, such a combination being capable of suppressing both down- and upward multiples. It should be pointed out, however, that work with such systems is very labour-consuming. The use of vertical source arrays require intricate technology and detailed knowledge of the velocities in the near surface, but it is more economical than horizontal grouping. VSP has demonstrated that frequency selection can be achieved more effectually at the source than at the detector. The pulse spectrum can also be varied by introducing time lags between shots or by changing the array dimensions in case of air-shots. In delayed explosions, the interference of waves excited by indivicual explosions causes the energy to be re-distributed between the spectral components. This makes it possible to shift the peak frequencies of the downward wave towards the lower end, which is often of practical importance, particularly when mapping deep interfaces in conditions of high signal absorption.

Non-explosive sources

Difficulties and deficiencies associated with explosives make it expedient to use non-explosive sources for VSP. These have several essential advantages over explosives, the most important being good reproducibility and shorter operating times required for the survey. Various types of non-explosive sources have been tested.

Seismic Vibrators

These guarantee high record reproducibility coupled with frequency and amplitude control of the source. Directivity can also be controlled enabling *P*- and *S*-waves to be excited separately [70]. High productivity (it takes several minutes per shot point) enables complex recording systems using multiple source positions to be developed. With vibrators, it is usually possible to complete a verical profile without moving the source. The main drawback of vibrators is the difficulty of obtaining high-frequency records.

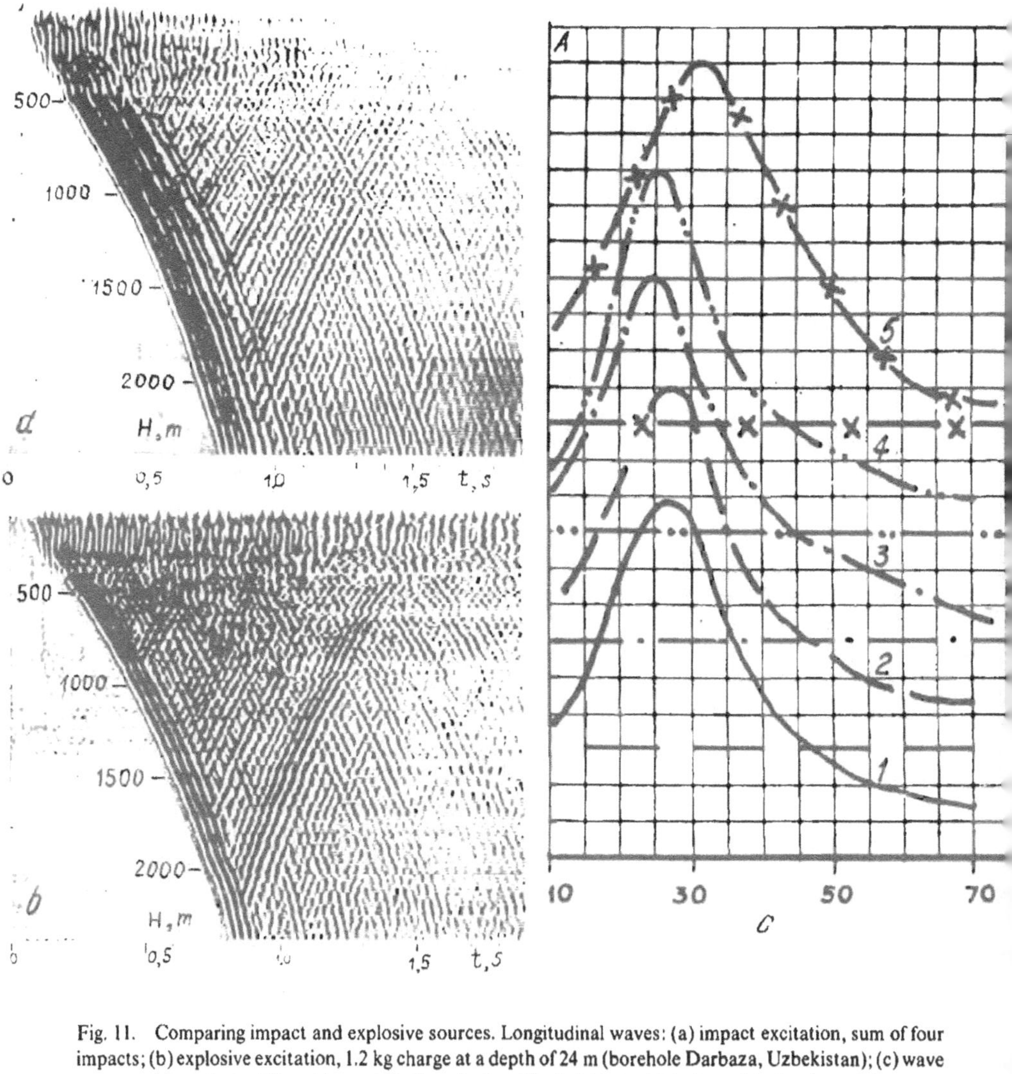

Fig. 11. Comparing impact and explosive sources. Longitudinal waves: (a) impact excitation, sum of four impacts; (b) explosive excitation, 1.2 kg charge at a depth of 24 m (borehole Darbaza, Uzbekistan); (c) wave spectra with explosive (5) and impact (1–4) excitation. The height at which the weight is released decreases from 4 m to 1 m [35].

Impulsive sources utilizing the detonation of gas charges are also in use. The direct wave from this generator is less compact than that from an explosive source fired at an optimum depth, but its spectrum is simpler and narrower, with a single maximum at the frequency of 35 to 40 Hz. A particular feature of seismic vibrabors is the generation of a long unresolved train of low-frequency surface waves. In addition to the fundamental Rayleigh wave mode, higher harmonics are also recorded. Good results have been obtained with a source employing a free-falling weight. Wave fields obtained with such sources generally display greater simplicity and regularity (Figure 11a), than in the case of excitation by explosions (Figure 11b). Modern impact sources enable information about the Earth to be obtained down to depths of 3000 to 4000 m. It is, however, worth mentioning that in the case of the impact source, the spectrum of the signal has a lower frequency content than in the case of explosives (Figures 11c), and it shifts towards lower frequencies as the height from which the weight is released diminishes. Also, intense surface waves that obscure the tracking of deep waves back to the surface on VSP seismograms are registered on all the shallowest levels in the borehole.

In addition to surface sources, downhole devices are also employed, such as a pneumatic pulse generator designed to operate in three-inch holes.

Among the non-explosive sources, the electric discharge source may be of special interest for VSP. Experience gained with a source having a discharge energy of 70 000 J, a voltage of 20 000 V, and a storage capacitance of 350 μF [71] looks promising. The electric generator produces a high-frequency pulse, and thus makes high-resolution VSP observations possible which is of decisive importance for solving many VSP problems. Moreover with electric pulse sources the noise level is lower, and there are no unwanted waves propagating along the fluid column. By controlling the shape of the direct wave, it is possible to influence the wave field by enhancing useful waves and suppressing unwanted ones. In this connection, it is most important to study the low-velocity layer and the depth of weathering so as to maintain the charge depth with respect to the base of the low-velocity zone. VSP allows the study of the process of seismic wave formation, excited by a variety of sources.

Generation of *S*-waves

S-waves have been excited by different methods: by directional action (impacts and explosions) and by grouping asymmetrical sources. VSP permits the directivity of the source, its intensity and stability to be studied. In this respect, it is most convenient to study direct transverse waves, which have been excited by both directional impacts and by explosions.

Excitation of *SH*-waves has been carried out by means of impact devices mounted on a mobile crane. A 1300 kg weight suspended at a height of 4 m is allowed to strike a vertical plate as a horizontal pendulum. It has been established by experience that a horizontal impact has good directivity. Such impacts normally enable direct transverse waves to be excited and recorded down to depths of 1200 to 1400 m, depending on the geology. Some companies outside the USSR also make use of similar equipment. Impact installations using a falling weight produce intense regular transverse waves which can be seen on VSP seismograms (Figure 12a).

In a recent development, groups of shotholes arranged in three parallel rows, have been used as an explosive source of directional waves. Shots fired both sides of a shattered zone, produced beforehand by firing the middle row of shotholes, enable transverse wave polarised in opposite directions to be obtained.

A comparison of impact and explosive sources shows that an explosive charge excites transverse waves of higher intensity, but of longer duration and more complex form. As

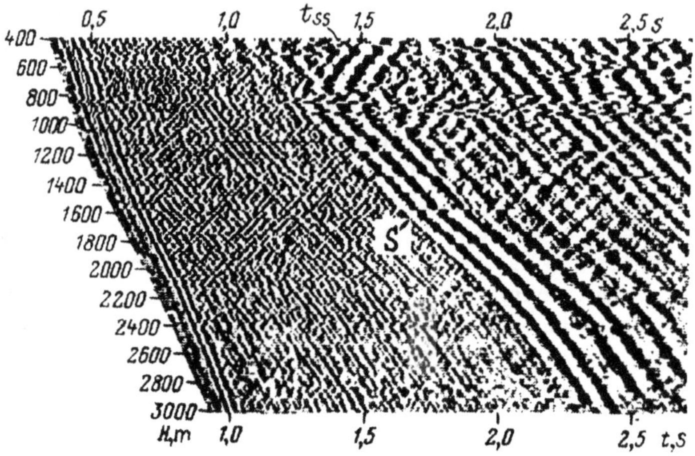

Fig. 12a.

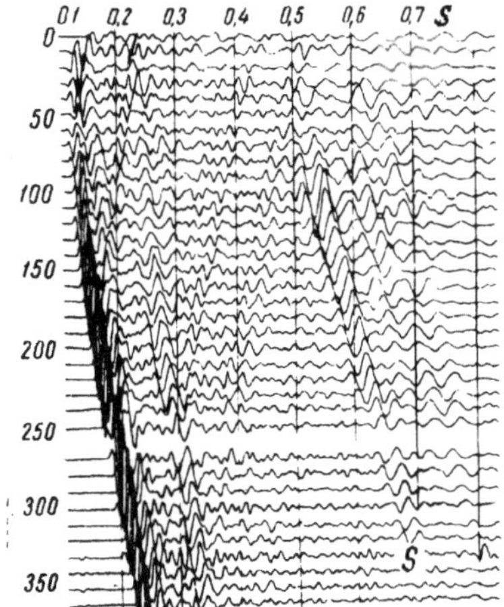

Fig. 12b.

Fig. 12a-c. Illustrating the excitation of transverse waves: (a) seismogram of the X-component in the case of horizontal impact excitation (X-component, borehole N2, Chernin area; Pripyat' depression, shotpoint 670 m [72]; (b) seismogram of Z-component in the case of a symmetrical explosion (borehole N90–a Kudinovskaya, shotpoint 100 m, Q = 600 g); (c) seismogram of Z-component (top) in the case of directional excitation and a difference seismogram of Y-component (bottom), shotpoint 110 m, Q = 400 g, h = 3 m, borehole N90 Kudinovskaya).

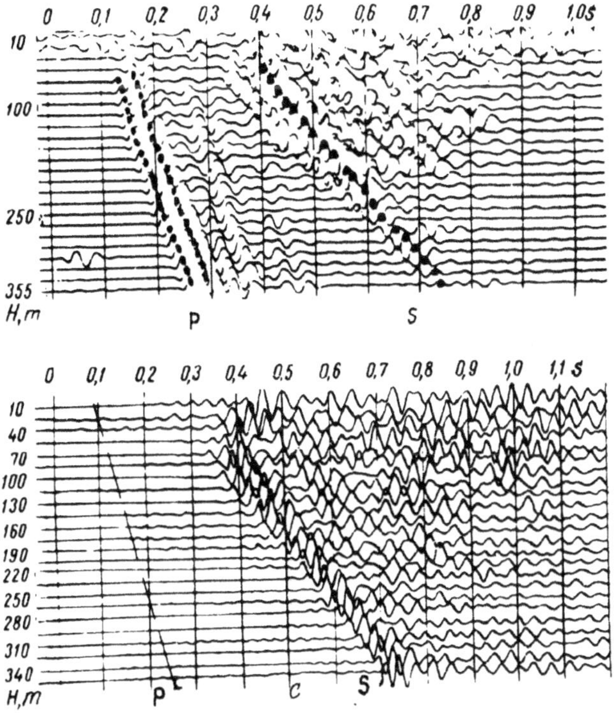

Fig. 12c.

a rule, substantially fewer unwanted waves are recorded in down-hole observations from impact sources than from explosives. The directivity of an explosive source depends to a large extent on the lithology in which the explosions are fired. In clay for instance, the directivity of an explosive source is similar to that of an impact one, but is worse in sands.

The technique of using shotholes in three parallel rows is useless when the level of ground water is high, and it is labour consuming and harmful to the environment [69].

It has been established as a result of much VSP work carried out in a variety of sub-surface conditions that *S*-waves of sufficient intensity are generated using concentrated explosive charges, such as is used in conventional surface seismic exploration [14]. This is the result of the so-called natural directivity caused chiefly by high gradients of elasticity of the rock in the vicinity of the explosion. A vertical velocity gradient attributes to the explosion the properties of a cylindrical radiator.

The excitation of *SH*-waves may be associated with horizontal heterogeneities in the soil structure, with a horizontal gradient of elasticity, and with anisotropy in the horizontal plane.

Natural directivity manifests itself most sharply when explosions are carried out in dry soil. However, as seen by both surface and down-hole observations, transverse vibrations are excited in wet soil as well.

It would be interesting to compare records obtained from directional and from conventional explosives. In the case of conventional explosive charges (Figure 12b), numerous regular downward travelling and reflected longitudinal waves are recorded between the first arrival curve and the *S*-wave, and the background noise between them is much higher than in the case of the directional explosion in Figure 12 (top). Explosions of distributed charges placed at the opposite walls of a pit were used as a pair of opposite-sign sources. With good source excitation conditions, by reversing the polarity of the transverse *SH*-wave, it is possible to discriminate waves according to their polarization at the source, by using explosive sources of opposite directivity, and at the same time suppressing waves that have not been inverted, such as the longitudinal, converted and transverse *SV*-type waves. We shall illustrate the effectiveness of such discrimination by means of the seismogram in Figure 12c (bottom) on which all waves prior to the arrival of the *S*-wave are almost totally suppressed (the travel times of the first arrival longitudinal wave are shown by the dashed line). When choosing source conditions, one should pay attention not only to the intensity, but also to the shape of the vibrations being excited. The dependence of the shape of transverse vibrations on source conditions (charge depth, charge weight etc.) has not been studied sufficiently. The recorded shape of the transverse wave-train is determined by the degree of superposition of both the *SV*- and *SH*-waves, as well as the *SS*-waves reflected from the surface, and other waves. Accordingly, the shape of surface recordings of the *S*-wave is more stable than that observed below the low-velocity zone. The direction of propagation and charge depth also affect the shape of the record. Generally, in single-cover surface observations the shape of the *S*-wave changes more drastically with a change of charge depth than in the case of observations at the same depth but at different azimuths with respect to the shotpoint. In observations below the low-velocity zone, the shape of the record can change drastically both with a change in charge depth and in the azimuth of the detector position.

SIMULTANEOUS EXCITATION OF *P*- AND *S*-WAVES

It has been established as a result of numerous VSP observations that a single source can be used for the simultaneous excitation of *P*- and *S*-waves. Specifically, using symmetrical explosive sources, natural directivity causes transverse waves to be excited in addition to longitudinal ones. In order to determine the optimum conditions for the generation of *P*- and *S*-waves, special research was carried out in a variety of near-surface and low-velocity zone conditions.

Observations were carried out along both a circular and several linear profiles. Shots were taken at various depths in a borehole located in the centre of the circular profile of radius 100–150 m, and observations were carried out at six points arranged uniformly around the profile. At each point, a three-component symmetrical cluster with one vertical geophone was installed on the surface and a similar one under the low-velocity zone ($H = 0$ m) and oriented in a uniform manner: the azimuth of the first geophone was opposite to the azimuth to the source. Shots were taken at intervals of 3 m ranging from 10 to 60 m. The charges at each point varied in the range from 0.4 kg to 7.9 kg. Recording was carried out using equipment capable of producing undistorted data in the frequency band 7 Hz to 125 Hz with continuous seismic control of channel duplication of the symmetrical clusters.

We present here the results of observations in the Novo-Dmitriyevskaya area in

Krasnodar province. The first point to be noted is the dependence of the velocity on the direction of propagation of the wave. Thus, for *P*-waves, the maximum velocity recorded at the surface (1340 m s^{-1}) was observed in the azimuthal direction 30° and the minimum (1210 m s^{-1}) in the direction 180°. Beneath the low-velocity zone these directions reduce by about 60°. The ratio V^P/V^S in the low-velocity zone varies between 3.35 and 3.5 increasing in the rock underlying the LVL to 4.4 at a depth of 35 m.

Shots taken beneath the low-velocity zone (25–60 m) produce *P*-waves with much greater intensities than the *S* waves both on the surface and beneath the low-velocity zone. Towards the bottom of the low-velocity zone, the intensity of *P*-waves diminishes, and the amplitude ratio A_S/A_P attains its maximum (7 to 10 on the surface and 2.5 to 3.5 beneath the low-velocity zone) (Figure 13). Observations from a number of different regions have demonstrated that shots taken near the bottom of the low-velocity zone produce the best simultaneous excitation of *P*- and *S*-waves. The ratio A_S/A_P varies between 1.5 and 3, depending on the waves' direction of propagation.

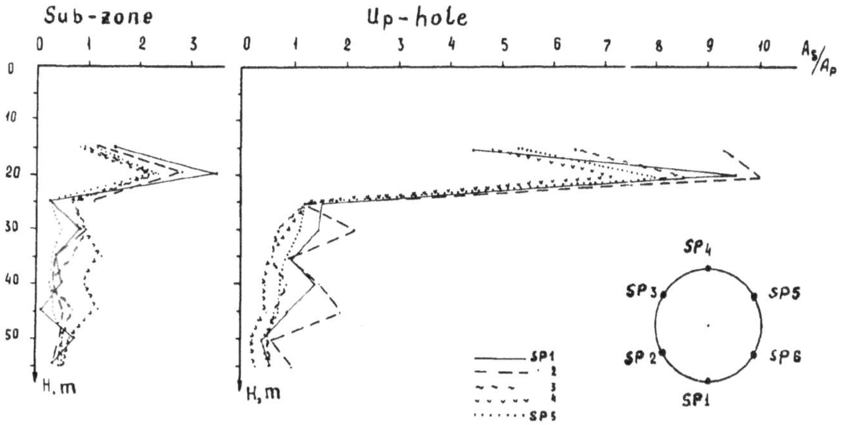

Fig. 13. Dependence of amplitude ratio of direct *P*- and *S*-waves on charge depth for different points of a circular profile below the low-velocity zone and on the surface.

DEMOLITION CORD AS AN ENERGY SOURCE

Good results have been obtained in the simultaneous excitation of *P*- and *S*-waves using horizontal linear *DSh* charges. The advantages of line-cord used to generate longitudinal waves have been demonstrated by experience. They include easy handling, comparative safety, little environmental damage and technological convenience. Their coefficient of artificial directivity is 1.3 to 1.5 times higher than that of the three rows of shotholes. *S*-waves can be excited in conditions of high ground-water level when conventional explosives are not effective enough. Good simultaneous excitation of *P*- and *S*-waves has been obtained with 1 m, separation between *DSh* lines, at charge depths between 40 cm and 80 cm, with lines 50 to 60 m long. The *DSh* lines may be regarded as a universal source. Hence, sources previously used for exciting *P*-waves only can be employed for the simultaneous excitation of *P*- and *S*-waves.

Controlling excitation conditions

Excitation conditions can be selected and controlled using the direct wave and the reflected waves. In the former case, the criterion for optimum conditions is a short and simple pulse-form for the first event, and in the latter – maximum signal-to-noise ratio. The choice of method is conditioned by the problems to be solved. When studying the wave field as a whole, it is more convenient to use the direct wave, but for waves of geological interest at depth it is advisable to make use of the signal-to-noise ratio.

To record the direct wave, a control geophone is placed beneath the shot-point in a special shot-hole at a depth where the form of the direct wave has already stabilised (see Chapter 5). Recording is performed using a low-amplification channel (sometimes even without an amplifier) and with a wide frequency band (10–300 Hz). Usually, this makes it possible to obtain records similar in shape to those of deep geophones. The consistency of excitation conditions is monitored by recording the output of a group of geophones located on the surface around the circumference of a circle of 70 to 100 m radius, with its centre at the shotpoint. Source conditions can also be monitored by overlapping observation points on the vertical profile.

In observations at large distances from the source, a special station is used to monitor excitation conditions, which compares the shapes of the downward waves produced at different places by different charges. To determine the relative amplitude curves of waves recorded from different shotpoints, the dependence of the pulse amplitude on the charge should be assessed experimentally. When a large number of shots are fired from one shotpoint, a spread in the arrival times results and kinematic corrections need to be applied. These corrections take into account the difference in the up-hole times at the top of the borehole being studied. However, discrepancies may arise due to the difference in the wave paths of waves arriving at the surface and at the down-hole geophones. The corrections are often determined empirically. Because excitation conditions can be monitored, VSP can be employed to examine and select the optimum excitation conditions for solving various problems by means of surface observations; for example, to evaluate delayed detonations in order to control the shape of the pulse in the downgoing wave.

2. Observation Procedure

The range of problems that can be solved by VSP is very wide, and, as in surface observations, each problem requires its own appropriate VSP procedure.

Observation methods

VSP methods utilizing moving geophones and fixed source positions (one or several) have found the widest application. For some special problems and those involving complex wave fields, methods utilizing moveable sources with geophones at fixed predetermined depths are being used (we call such systems level systems, because they can track waves along a horizontal profile at some specified depth of observation).

Methods combining observations along the vertical, the level and the surface profiles are also in use. These have found wide application in studies of the area surrounding the borehole. In VSP, as in surface observations, the term 'in-line profile' is understood to mean one shot from a source position on the same line as the geophone profile, that is, at the top of the borehole, where $l = 0$. The term 'off-line profile' is understood to mean one shot from a position away from the line of the profile. Observation methods used in VSP may be linear (one-dimensional), planar (two-dimensional), or spatial (three-dimensional).

A linear system consists of an in-line or off-line vertical profile shot from one source position.

A planar system consists of a vertical profile shot from a series of shotpoints arranged on the surface along a line passing through the top of the borehole (the source profile), with the separation between shotpoints permitting the tracking of waves from adjacent shotpoints, both in the vertical and in the horizontal directions at various levels* by correlation.

One method of observation in the vertical plane is to shoot from one point and record in a series of holes arranged along a line (a profile). However, such an order of shooting a vertical system is impractical.

At present it is technically easier and more economical to use a number of shotpoints, and to make recordings in one borehole.

A spatial (volumetric) system consists of a vertical profile shot from a series of source positions arranged at different azimuths with respect to the borehole including one that does not pass through the borehole.

Various modifications of the VSP method are employed for a variety of specific problems, the main difference between them being the relative disposition of source and detector positions. For instance, the method used in level observations is a particular case of the planar method with a fixed observation point. In the study of various wave types, source offset distances from 0 to 60 km have been used.

The resolution of the method is determined by the distance separating the detector points along the line of the profile, and between the shotpoints on the surface. Depending on the nature of the problem, VSP observations may be carried out along the entire profile or in selected sections, e.g. when solving structural problems.

The criteria determining the separation between geophone positions along the vertical profile are the same as in surface observations. For reliable wave correlation, the phase shift between signals recorded at adjacent points should not exceed one half of a wavelength. Because of the difference in velocities along the vertical profile, phase correlation is possible using variable distances between the observation points. Such intervals can be selected in advance with the aid of electric logs and on the basis of a general knowledge of the cross-section. However, when operating with multi-detector probes, technical difficulties make the variation of detector intervals practically impossible.

To correlate the vertical and the horizontal observations, it is often necessary to reduce the detector interval in the upper portion of the borehole, where the wave field displays additional complexity on account of the presence of various unwanted waves as well as the interference pattern of the up- and the downward travelling waves reflected

* Actually, a family of overlapping off-line traveltime curves is obtained in VSP, when the profile is shot from a series of shotpoints at different distances from the top of the borehole. However, for the sake of convenience and simplicity of presentation, let us agree to term such a set of observations a series of off-line vertical profiles.

from the surface and from the low-velocity zone. Let us consider the different VSP methods used to study reflected and refracted waves. Those used to study the upper portion of the cross-section will be described separately.

Recording technique for the reflection method

In the majority of cases VSP problems in the reflection-method are associated with the wave pattern recorded on surface seismograms and can be subdivided into two classes; (a) those where stable regular waves are recorded on surface seismograms at appropriate time intervals when the VSP problems becomes the determination of their nature and stratigraphic correlation and (b) when regular waves cannot be tracked on surface seismograms, or their tracking is not sufficiently stable, and VSP is then carried out chiefly to study the reflection properties of the interfaces of interest from the point of view of the geology. In the case where reflected waves have been detected at geophone postions close to the interfaces, the problem becomes a study of the reasons for the poor correlation on surface seismograms.

Except for the last of these, linear methods are most commonly used for the above problems, particularly the *in-line* vertical profile system. Because of the specific wave kinematics on an in-line profile, the zone of interference between the reflected waves and the direct wave is the shortest, and the waves are best discriminated in time. In addition, observations along an in-line profile in either a horizontally-layered medium or in the presence of low-dip interfaces enables normally incident waves to be studied, in which case the wave pattern takes its simplest form and is easily calculable. In this respect, data obtained on an in-line vertical profile may be of special interest for dynamic processing. Normally in VSP operations, the in-line vertical profile is completed first. This, in addition to providing information about the wave pattern, enables data concerning the velocity distribution to be obtained, specifically, interval velocities (from first arrivals) required for further clarification of the observations.

Off-line profiles are very important for the study of reflected waves. They are needed to study high-amplitude reflected and transcritically-reflected waves at variable angles of incidence (or distances). This is of special importance, because it is rather difficult to identify transcriticaly-reflected waves from surface observations at great distances from the source. Off-line observations are used to determine the kinematic (velocities, trajectories of propagation of multiple waves, delay functions of the multiples) and dynamic characteristics of seismic waves, as well as to select optimum parameters for a common-depth-point system (tracking multiplicity, length of observation base and the inverval between detector positions).

Observations along off-line vertical profiles are of special importance for VSP reflection studies undertaken to solve structural problems. In such cases observations are performed both along the vertical and the horizontal profiles. Operating procedures and the methods employed will be discussed in detail in Chapter 9.

Recording technique for the correlation method of refracted waves

VSP is usually carried out along off-line profiles to establish the origin of the waves and to identify refracting horizons in the geologic section. When designing a VSP field technique for the purpose of studying refracted waves, one should very carefully analyse the spectral content of the wave field on surface profiles on which to base the selection of source offset position.

To study the wave pattern associated with the entire mass of sedimentary deposits, VSP has been executed along both in-line and a family of off-line profiles at distances up to 30 km, and sometimes up to much greater distances (up to 60 km). This allows an examination of the characteristics of all waves formed in the earth, at angles of incidence from 0 to 180°, including the wave field variation with distance.

Recording technique for studying the upper portion of the geologic section and the wave field associated with it

The factors that make it expedient to use spatial surveys are the heterogeneity of the layers in the upper portion of the sub-surface with its extremely complex wave field, as well as the comparative ease of setting up. Strongly differentiated interfaces, low velocities and an abundance of waves with different origins all contribute to a complex wave-field interference pattern, which makes the problem a difficult one requiring very detailed observation procedures.

Combination planar systems consisting of one or more vertical profiles and a family of horizontal profiles have been used in these studies. Using horizontal intervals of 5 or even 3 m enables the main wave groups to be tracked both at the surface and at various depths. These surveys allow the presence of wave groups to be established by contouring, and a time field constructed for each wave. In the majority of regions studied, the vertical array extends over a zone about 200 m in depth and horizontally 1500 to 2000 m in length. With such a system, 400 to 500 shots have been fired. In rough field conditions, one can set up spatial systems for detailed studies using different arrangements of geophones in shallow boreholes (along a line or in a plane) and sources. Special systems are set up to study complex media with steeply dipping interfaces and no axial symmetry. Such systems have been used to examine diapir, reef and salt-dome tectonics, and highly fractured media, and to solve structural problems with the VSP method.

Note that a spatial VSP system is usually augmented by surface observations in the vicinity of the borehole, which together constitute a space element. Spatial elements set up around different boreholes can be combined into a general spatial system in which all the vertical and the horizontal profiles are shot from common source positions. Such systems provide the maximum information about the wave field and are very effective. However, they involve massive amounts of data acquisition, and for this reason are used only to study complex media.

An example of such a system employed to study the wave field associated with the diapir tectonics of the Taman peninsula, is depicted in Figure 14a. The system consists of a network of surface profiles, with vertical profiles located at the intersections. All the profiles in the system, both horizontal and the vertical, were shot from common source positions.

In planning VSP operations, the choice of field technique is no less important and perhaps even more so than in surface observations. This is due on the one hand to the high labour and financial costs of deep borehole observation, which make an excessively detailed field technique undesirable, and, on the other hand to the difficulty of carrying out repeated observations in a borehole. This is a frequent occurence and requires that the system should be sufficient only for the solution of the particular problem. Accordingly, for the sound planning of VSP operations it is advisable to avail oneself of a synthetic or modelled wave field calculated from sonic logging data, which

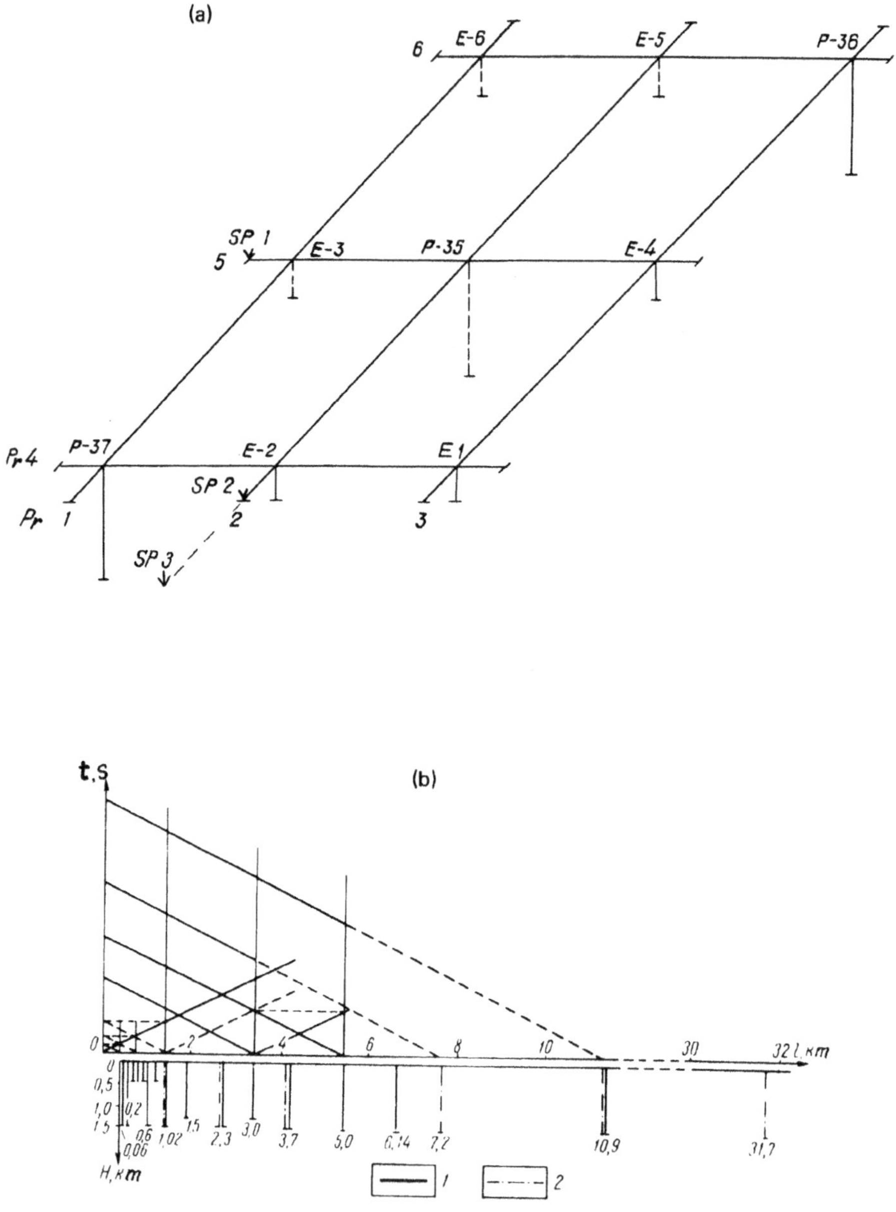

Fig. 14. Combined observation systems: (a) a PM–VSP spatial observation system in region of diapir tectonics of the Taman' peninsula; (b) a two-dimensional system in the vertical plane shot in regional VSP-correlational-method-of-refracted-waves studies (Kletsko–Pochtovaya area of the Volgograd province): 1 – Z-registration, 2 – three-component registration.

can be used to assess the presence and relative intensities of both useful and unwanted waves.

Combining VSP with surface observations

The most comprehensive information about the wave field, and most importantly about the origin of the waves recorded on surface seismograms, and about the earth itself, can be obtained by combining observations from both the vertical and horizontal profiles. Waves recorded in surface observations must be tracked along the vertical profile down to their associated interfaces if intersected by the borehole, or to the bottom of the borehole. The relative amounts of observations between vertical and horizontal profiles is determined by the object of the study and by the complexity of the wave pattern. It is advisable to set up such a combined observation system where the vertical profiles are located at positions where on the surface records the wave pattern is most indistinct or difficult to analyse.

Because of the need for multiple shots in VSP work, spatial surface observations in the vicinity of the borehole can be carried out without any increase in the amount of drilling and shooting. The combined processing of surface and VSP data makes possible a more objective analysis of the wave field and better structural interpretation.

An example of a combined two-dimensional system is shown schematically in Figure 14b, designed to study the wave field associated with the entire section of sedimentary deposits and the crystalline basement beneath it. It consists of eleven shotpoints. The series of shotpoints between 0 m and 2300 m are used to study reflected and refractred waves from the carbonate beds ($H = 1450$ m) The PM–VSP technique with a three-component geophone enables converted reflected and transmitted waves associated with the interfaces in the upper portion of the section to be studied at shotpoint offsets of 400, 800, 1000, 2300 m. Remote shotpoints between 3000 and 7000 m were used to record refracted waves from deep interfaces as well as the carbonate deposits located at shallow depths. From the shotpoints at 10 900 and 31 700 m, waves propagating in the crystalline basement were recorded as first arrivals along the entire vertical profile. This system was used in the Russian platform where the geology consists of axially-symmetrical beds of sedimentary terrigeneous deposits. In complex media with dipping interfaces three-dimensional combined systems must be employed.

Field Operating Conditions

When observations are carried out along a vertical profile intersecting layers with different elastic properties, the borehole conditions for the geophone may, in principle, not be identical at all points. However, a factor of decisive importance in VSP is the contact between the geophone and the borehole wall, which is mainly determined by the clamping force. In cased boreholes, the clamping force remains unchanged. VSP experience shows that much more constant planting conditions can be achieved with clamping devices than in surface observations.

Consider the spectral properties of recordings employing multi-detector probes. Comparison between the spectra of the direct wave recorded in single- and multi-detector observations shows that the mechanical connection between individual detectors of the probe is irrelevant and of no importance in conventional observations. However, when studying wave dynamics, it is necessary to examine the effect of the connecting cable between the individual tools especially when under stress. It would be expedient to provide in multi-detector systems facilities for the sequential clamping of each detector individually, so that the cable between the tools could be slackened.

When the probe is lowered on a heavy multi-conductor cable whose weight greatly exceeds that of the tool, it becomes necessary to control the tool's descent along the borehole shaft. If this is not done, an excessive length of cable may be paid out into the borehole, and this may result in knotting and unavoidable damage to the cable, or still worse render the borehole unuseable. The descent of a tool is conveniently controlled with the aid of a microammeter connected into the cricuit of the lower-most geophone of the probe, or by recording one of the parameters usually studied in production geophysics (e.g., apparent resistivity). In the latter case, such control makes possible a rigorous correlation between the observation points and the lithology, this being necessary to study the wave pattern. This is especially important, since the depth of the geophone is not always accurately measured.

The duplication characteristics (amplitude and phase) on each channel of a multi-detector probe can be monitored in the field by methods similar to those used in surface seismic exploration. However, in this case, it is important to exclude the effects of the interconnections between the separate detectors of the probe; to this end, all the detectors should be mounted rigidly on a special rack, which is subsequently lowered into a shallow hole or simply burried. To monitor channel characteristics during the survey, a method has been devised based on the processing of direct waves recorded by the different detectors of the probe [24]. The systematic deviation of the amplitude of the direct wave from a mean value serves as an indication that the channels are not identical. However, it is worth pointing out that to detect such deviations, rather laborious statistical processing of large amounts of data is required, and this is possible only with a computer under laboratory conditions.

VSP in the studies of porous marine deposits

VSP methods, when used in continental shelf studies of water-saturated porous deposits, which have experienced a boom in recent years in connection with the search for placer deposits, as well as problems in engineering geology associated with the construction of hydrotechnical projects [73], are characterized by the following:

- the thickness of sea-bed deposits being studied is comparatively small – up to 40 or 50 m;
- recordings are carries out in the 200 to 2000 Hz frequency band;
- the sea-bed deposits (the object of the studies) and the source are separated by a layer of water up to 20 or 30 m deep;
- the low stability of the unconsolidated marine sediments penetrated by the borehole;
- an electric pulse generator is used as a source.

Recording difficulties stem chiefly from high noise levels typical of high-frequency operations, and associated mainly with waves propagating along the walls of steel casing. This makes it necessary to substitute the steel casing with a type made from polyethylene whose acoustic properties are similar to those of the porous water-saturated sediments (the density of polyethylene is 0.91 to 0.99×10^3 kg m^{-3}, and the velocity of longitudinal waves in it is 1300 to 2000 m s^{-1}). The amplitude and phase

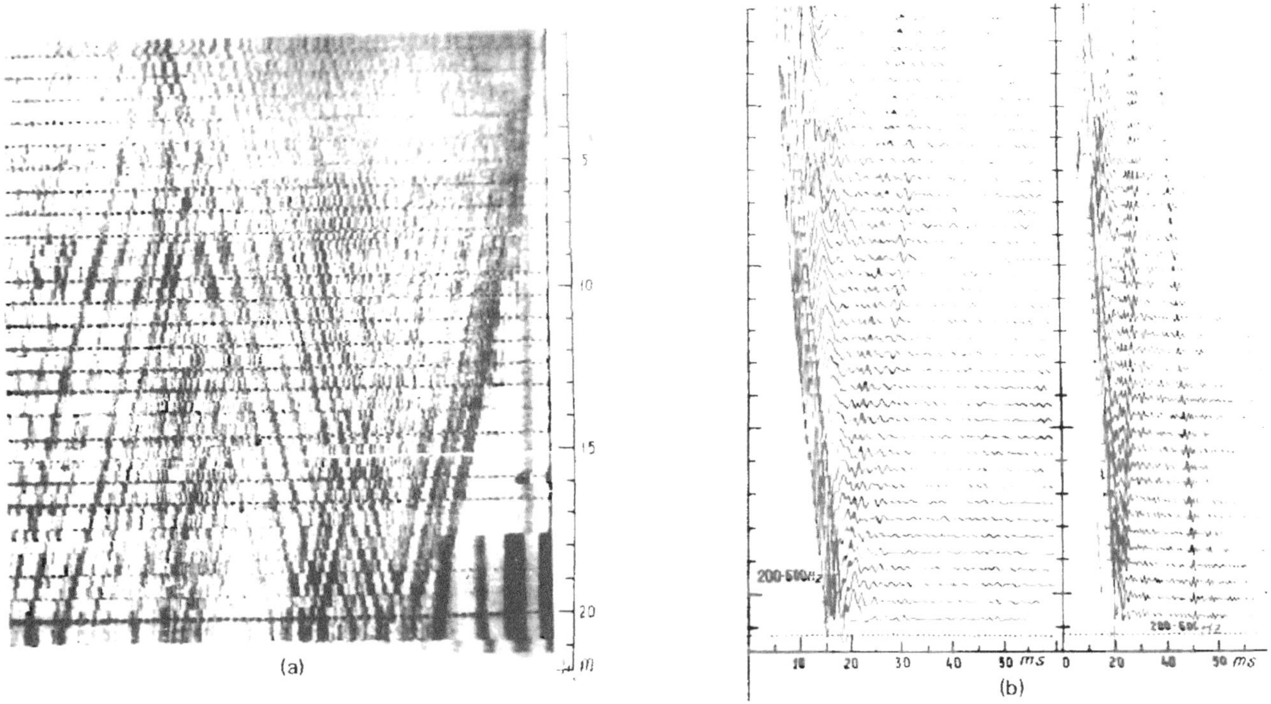

Fig. 15. VSP seismograms obtained in the course of the study of porous water-saturated marine deposits: (a) seismograms obtained in a steel-tube-cased hole illustrate numerous unwanted waves; (b) seismograms obtained in a polyethylene-tube-cased hole are free from unwanted waves. The time scales and intensities on both seismograms are not identical.

characteristics of wave transmission through polyethylene tubes in the frequency band used (up to 2000 Hz) guarantee an almost undistorted reception of the wave fields throughout the entire range of variation of acoustic properties. The records obtained with and without the polyethylene tube are practically identical. Figure 15 depicts for comparison purposes seismograms obtained in boreholes with both steel and polyethylene casing strings. To make such a casing, first a steel casing tube of 144 mm diameter is sunk into the borehole. Next a string of polyethylene tubes of 100 mm diameter is sunk into the steel string by welding together sections each three meters long. After the polyethylene casing is installed the steel casing is withdrawn. When work has been completed, the polyethylene tubes are withdrawn and can be used again.

The main source of interference when recording in a polyethylene casing is the hydraulic wave. However, it decays rapidly at frequencies of about 1000 Hz. The fact that hydraulic waves make their appearance below the sea bed interface, and that their propagation velocity in water-saturated deposits is always substantially less than that of longitudinal waves in the surrounding rock, creates favourable conditions for recording longitudinal downward and reflected waves. A 100 to 200 J electric pulse generator used as a source enables a complex cross-section 50 m thick to be studied. Single-channel equipment with a pressure sensitive detector is used for such observations. Recordings are made at depth intervals of 0.5 m. The recorded signal period from visual observation is 1.25 ms. The pulse is repeated after 15, 30, and 60 s. The total amplitude variation is 5–7%. The procedure for processing these VSP data is practically identical to the traditional methods.

VSP AND UP-HOLE SHOOTING (INVERTED VSP)

In contrast to conventional VSP, in up-hole shooting the shots are fired in the borehole at different depths, i.e., along the vertical profile (hence the name 'up-hole'), with detectors on the surface*. Up-hole surveys may be of interest when VSP cannot be carried out in studies of the upper portion of the geologic section where, because of heavy interference, it is usually impossible to obtain good quality data; also for studies at great depths where poorly cemented casing presents a problem.

The advantages of up-hole surveys stem from the fact that they are very flexible, and that there is no longer any need to use heavy equipment such as logging cable trucks, and this enables observations to be carried out in inaccessible regions, as well as for surface area observations. The main drawback of uphole shooting is the destruction of boreholes by the explosion of concentrated charges. However, the use of a length of demolition cord held in place along the borehole axis by a special device prevents the destruction of the borehole or of the casing string in a cased borehole. In the course of processing up-hole VSP data one can identify waves tracked in two directions – the horizontal at various depths and the vertical at various distances from the top of the borehole.

With the use of up-hole surveys, it is often cheaper and technically easier to set up a detailed observation system than with conventional VSP. The main difficulty of up-hole shooting is to find a deep borehole in which shots could be fired. In addition,

* Such observations are sometimes termed inverse VSP, this term being more appropriate to describe the essence of the procedure.

up-hole shooting almost totally precludes utilizing particle displacement directions in the analysis of the wave pattern to obtain information about the earth.

In terms of kinematics, uphole shooting is in principle – capable of providing an observation system similar to that obtained with VSP. In such cases, all source excitation points are assumed to be coincident with the top of the borehole, and all transit times of waves recorded on the surface are determined between the depth point of the explosion and the appropriate at distances from the borehole of the surface geophones. Taking into account the ease of operation, as well as the ease with which very detailed systems can be obtained, it would be deemed expedient to use up-hole surveys for the study of the upper 100 or 200 m portion of the geologic section, which is easily drilled. However, by perfecting the technology of repeated explosions in deep boreholes without the destruction of casing strings, for example by employing specially aligned charges, it will be possible to use up-hole surveys extensively to study the deep portions of the geologic section. Such studies can be performed already during drilling, with the charges being lowered through the drill-pipe and exploded at the bottom of boreholes being drilled. Up-hole shooting of deep boreholes can augment VSP observation systems and guarantee reliable correlation between surface and down-hole observations. Such a combination can be of special significance in production seismics in studying the region surrounding the borehole.

Since explosions of 5 to 8 kg charges in an open borehole do not obstruct continued drilling, it becomes possible to combine VSP and inverse VSP by performing stage-by-stage observations during drilling. Additional reasons in favour of this procedure are that it allows sonic logging of the entire borehole and prediction of the section below the bottom of the borehole. This latter point is of particular importance, and accordingly such a technique, which provides the most comprehensive information, should be recommended for wide-scale trials.

3. Particular Aspects of VSP Data Processing

In VSP, just as in surface observations, the accuracy, validity and the entire results depend on the study of individual waves and on a knowledge of their origin. The latter end cannot be achieved without reliable discrimination and tracking of the waves on VSP seismograms. For this reason, and despite the fact that different data processing may be required for different problems, the discrimination and tracking of regular waves remains the main and indispensable basis of the procedure. The complexity of the VSP wave field, resulting from the interference of down- and upward travelling waves, as well as the long duration of the signals, makes discrimination and tracking of individual waves difficult. Acoordingly, the first stage of processing aims at improving the signal-to-noise ratio, to aid wave correlation along the vertical profile, and resolution.

To improve the signal-to-noise ratio, VSP utilizes the same methods of wave discrimination according to the wave field characteristics, the frequency content and the apparent velocity, that have been used for the surface data. Note that the opportunities for discriminating waves according to such parameters and to polarization are better in VSP than in surface observations. This is the result of reduced filtering by the low-velocity zone, as well as the fact that many of the waves are distinguishable not only by their magnitude, but also by the sign of their velocity. Consider each of the three methods of wave discrimination used in VSP:

Discrimination by Frequency

The wider band of frequencies recorded in VSP necessitates the use of filters with continuously and widely variable parameters. Of great significance for improving data resolution, is the inverse filter. The seismic pulse excited at the source may be short and simple. However, at the point of detection, due to the superposition of waves both in the regions of excitation and reception, as well as the filtering properties of the earth and the equipment being used, it is recorded as a more or less complex, poorly differentiated wave train. This greatly reduces the data resolution and the exploration potential of the observations. *Inverse filtering*, which enables the input signal to be reproduced from a known output signal and a known impulse response of the filter, is one of the most effective methods of extending the frequency spectrum, of improving data resolution and the wave correlation from the interference pattern. The pulse duration can be minimised by eliminating frequency distortions in the recording equipment and those due to unfavourable conditions of excitation and reception.

The effectiveness of inverse filtering depends essentially on the correct choice of the seismic signal for the computation of the inverse filter operator (both the amplitude and the phase characteristics should be known). In VSP, the direct wave pulse can be used for this. Experience shows that the use of the direct downward wave as the basic signal for the computation of the inverse filter operator greatly enhances the effectiveness of deconvolution. Sometimes a complex wave train, which also includes ghosts and reflected waves associated with them, is chosen as the basic pulse [74, 75]. The results of processing by inverse filtering facilitates the effective use of procedures aimed at subsequently improving the signal-to-noise ratio. The extension of the frequency spectrum from 5 to 100 Hz as a result of inverse filtering makes the application of frequency discrimination possible.

While inverse filtering reconstitutes the entire frequency band, *selective filtering*, a modification of inverse filtering, preserves the data in a specified frequency band. Such filtering requires the consistent application of an inverse and a band-pass filter within the specified band. Outside the filter pass band, the signal is drastically attenuated, and inside the inverse filter reconstructs the spectrum. *Matching filters* are applied to detect weak signals against the background of irregular unwanted waves. All these methods are based on the exploitation of the frequency characteristics of the record. When the upward reflected waves are several times weaker than the downward, inverse filtering will be more effective, if carried out after wave discrimination according to the direction of propagation.

It is the ability to discriminate down- and upward waves in VSP that ensures the higher effectiveness of inverse filtering. Apart from that, it is possible, by eliminating the multiples that interfere with the initial pulse and make it more complex, to obtain information about single reflections, that are required to determine earth characteristics.

Discrimination according to apparent velocity

Discrimination according to v_a enables the downward travelling and the reflected waves to be separated in the early stages of processing. Such discrimination is very effective in VSP, since the wave velocities differ in sign. At subsequent processing stages, discrimination according to v_a is used to separate upward waves of various types, and waves associated with particular interfaces (especially in off-line profiles). This discrimination is carried out with the aid of various stacking and array patterns and utilizes the kinematic parameters of the waves. Such systems are quite familiar and known to be effective [1, 4].

The main difference in the application of methods extensively tested in surface observations, is that the wave field changes more rapidly along a vertical profile than along a horizontal one, and this makes it difficult to use long bases. However, in cases when there is a substantial difference between the parameters of the useful (upward) and of the unwanted (downward) waves, and they are generally known, it is possible to achieve adequate selectivity of the system with a comparatively small number of channels to be summed. A more simple and handy method of wave discrimination is directional detection (the first modification of CDR–II), which can be easily executed in analogue form. However, there are certain deficiencies inherent in the method, the main one being the possible appearance of false alignments.

In-phase summation

In-phase summation of traces, when each recorded signal has identical amplitude and phase characteristics, will result in an n-fold multiplication of its amplitude. The amplitudes of the waves propagating in directions other than the summation direction increase much less than this. However, the summation of random noise and of regular unwanted waves with fluctuating amplitudes may result in the appearance of false alignments. Typical of false alignments is the variation of apparent velocity with that of the summation direction, whereas for genuine waves the apparent velocities are independent of the summation direction. This criterion has been tested, and the results proved its effectiveness when used for analyzing VSP data [27]. The main drawback of the method of in-phase summation, as well as of other interference systems, is distortion of the records due to the suppression of high frequencies when the summation direction is inaccurately specified – a frequent case in VSP – when the average velocity along the profile varies as the result of changing velocities in the heterogeneities intersected by the profile.

Summation of single shot records

To discriminate waves having almost equal apparent velocities, and to detect weak waves against the background of high interference, we have employed methods of summing records from sources located at different offset distances from the detector positions. Such summations may be vertical (shots fired at different depths) and horizontal (shots fired along a horizontal profile at different distances from the borehole). In the case of vertical summation, we suppress ghost reflections associated with the interfaces above the source, by introducing static time corrections and reducing them to one depth. Horizontal summation reduces the level of multiples whose kinematics differ from that of the useful waves. However, such methods have not found extensive use.

Discrimination by fan filtering

When the spectra and kinematic parameters of the useful and unwanted waves are similar or overlap, it is expedient to employ fan filtering for wave discrimination. The increase in the signal-to-random noise ratio (statistical effect) in the case of fan filtering depends on the spectrum of the random noise falling with increasing frequency. Fan filtering is usually employed as a preliminary procedure prior to inverse filtering, and reduces interference.

Subtraction

Subtraction increases the amplitude of the useful signal. However, the effectiveness of the method drops when the traces are not identical. The use of subtraction to discriminate useful waves in VSP does not produce false alignments [76]. If it is still necessary to improve the signal-to-noise ratio after subtraction, it is advisable to employ in-phase summation followed by an appraisal of the alignments thus obtained. At present, various methods of subtracting noise are in use, including one based on fan filtering, or simultaneous subtraction of several unwanted waves whose apparent velocities differ from those of the useful wave. This method is noise-free and enables waves of specified direction to be discriminated during further data processing in the presence of variable noise conditions. In this method the unwanted waves are enhanced by the suppression of useful waves on the original record. This is achieved by summing the original records without time delays, and is in effect an equivalent fan filter. The simultaneous subtraction of several unwanted waves consists of enhancing the unwanted waves, reconstructing the original record, and performing the subtraction itself. This can be carried out on both analogue and digital computers.

To discriminate regular waves, it is sometimes advisable to transform VSP seismograms beforehand into time cross-sections in which the apparent velocities vary much less than on the original seismograms. The subtraction of entire wave packets, while ensuring high-speed operation, raises the effectiveness of subtraction and proves an efficient method of discriminating the down- and upward wavefields. With the latter method, the downward waves can be suppressed at much greater depths than with fan filtering. At the same time the form of the useful signal is retained and makes the discrimination of weak signals possible. Knowledge of the velocities of regular unwanted waves, which is a special property of VSP, greatly enhances the effectiveness of the subtracion method. Substraction of a wave packet is used to discriminate useful waves in their pure form and in particular to obtain the direct wave.

Optical filtering

In the course of processing VSP data in analogue form, much experience has been gained in the use of optical filtering [77]. Optical filtering is a flexible high-speed system that enables longitudinal and transverse waves to be discriminated over a wide velocity range, the dominant frequency band of seismic events to be determined, the variation of frequency with depth to be studied, mechanical noise to be suppressed, up- and downward waves to be separated and the conditions for tracking useful waves to be improved. For all these properties, optical filters are more advantageous than other methods of analogue processing. Moreover, optical filtering does not produce false alignments. Because results can be obtained rapidly in optical filtering, it is useful as an express method for preliminary processing, followed by other processing methods, and can be combined with digital processing.

Straightening alignments

The method of straightening alignments or traveltime curves [78] is used to improve visual wave correlation and to make subsequent processing more convenient. The method boils down to the application of static, positional and kinematic corrections for the variation of distance from the reflecting interfaces as the detector moves along the profile.

In the case of an in-line profile, straightening can be achieved by introducing a correction to the observed arrival times of reflected waves that takes account of the changes in their paths as the point of observation moves along the vertical profile. The arrival times of waves at every point of the profile may conveniently be reduced to the surface. The following expression effects such a transformation for upward waves:

$$t_{\text{str}} = t_{\text{obs}} + \Delta t_{SP} + \Delta t_{gh} + t_H$$

and for downward waves

$$t_{\text{str}} = t_{\text{obs}} + \Delta t_{SP} + \Delta t_{gh} - t_H$$

where t_{str} are reduced times, t_{obs} are observed times, Δt_{gh} are static corrections for the positions of the shotpoint and the reception point; t_H is the transit time of the direct wave along the in-line vertical profile from the surface to the observation point in the borehole at depth H.

The transformed alignments of upward reflected waves have no extrema, they intersect the traveltime curve of the direct wave (that has been reduced to the surface) at a point corresponding to the depth of the reflecting interface, and for comparatively small dips they have an almost infinite apparent velocity. The alignments of downward reflected waves that obscure the tracking of upward waves have almost twice as great a slope as compared with the field recorded data. This facilitates visual correlation of upward waves as well as the interpolation and extrapolation of the traveltime curves in sections where correlation is uncertain. VSP data with vertical alignments are convenient for processing as well as for computer input.

Reflected waves on a vertical seismic profile with straightened alignments of upward reflected waves (see Figure 16) can be reliably identified with the waves on a surface seismogram and stratigraphically correlated with the geologic section. It will be seen from the VSP display that the last singly-reflected wave recorded on the surface seismogram at the time 2.1 s, corresponds to a strong reflecting interface at the depth 3450 m. Despite the fact that observations have been carried out down to the depth of 5460 m, reflections from deeper interfaces could not be detected; all arrivals later than 2.10 s on the VSP display in the depth range from 0 to 3400 m are multiples. Straightened alignments also improve conditions for the correlation of downward waves.

Computerized VSP data processing complexes

Some of the processing discussed above can be performed on analogue computers. However, the most comprehensive processing is possible only with a digital computer. Special systems have been devised for automatic computer processing of VSP data

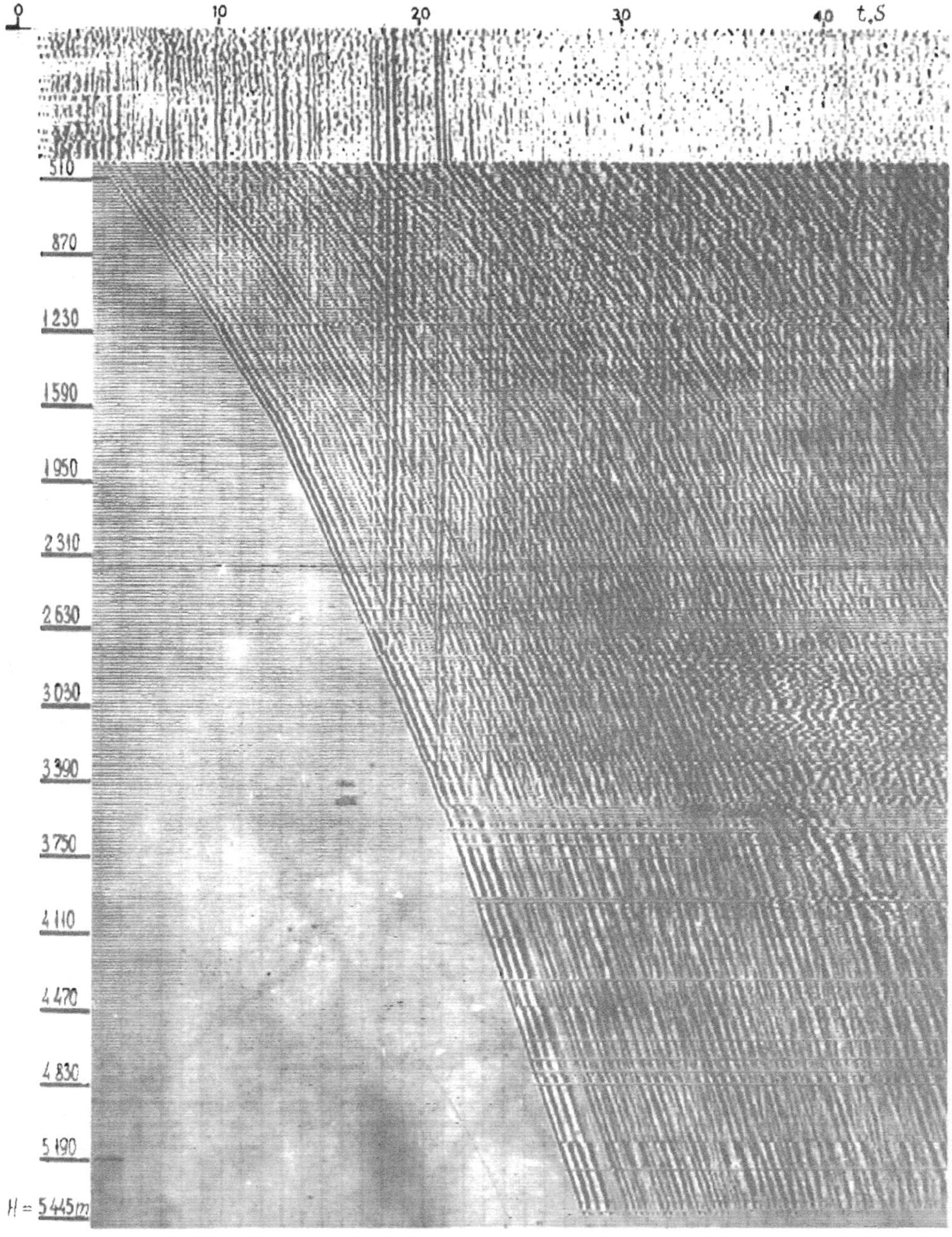

Fig. 16. A VSP seismogram (Z-component) with straightened line-ups of reflected waves (courtesy G. N. Gogonenkov).

appropriate to the various computer types and to the solution of various problems associated with wave field analysis [60, 79, 80] and with structural mapping [32]. They consist of a set of programs which serve to plan, to control and to execute the processing.

The automated VSP data processing system VNIIX/GIS [60] is universal, flexible and capable of using the same procedures for different processing lay-outs. The functions of planning and solution have been separated, and the processing programs are in the form of autonomous blocks. The system consists of a control program (a monitor), a library of processing programs, sub-programs and auxilliary programs. The system can be updated by the inclusion of new programs.

Without going into technical details, let us cite just a few from the list of almost thirty processing programs contained in the library, which enables all the chief procedures of preliminary processing of VSP data to be executed: straightening of alignments of up and downward waves; static correction to eliminate time lags; amplitude correction of waves; computation and automatic determination of static corrections from cross-correlation functions; quality control of wave correlation: automatic amplitude adjustment; computation of digital filters (LF, HF, reject); digital filtering – both fixed and

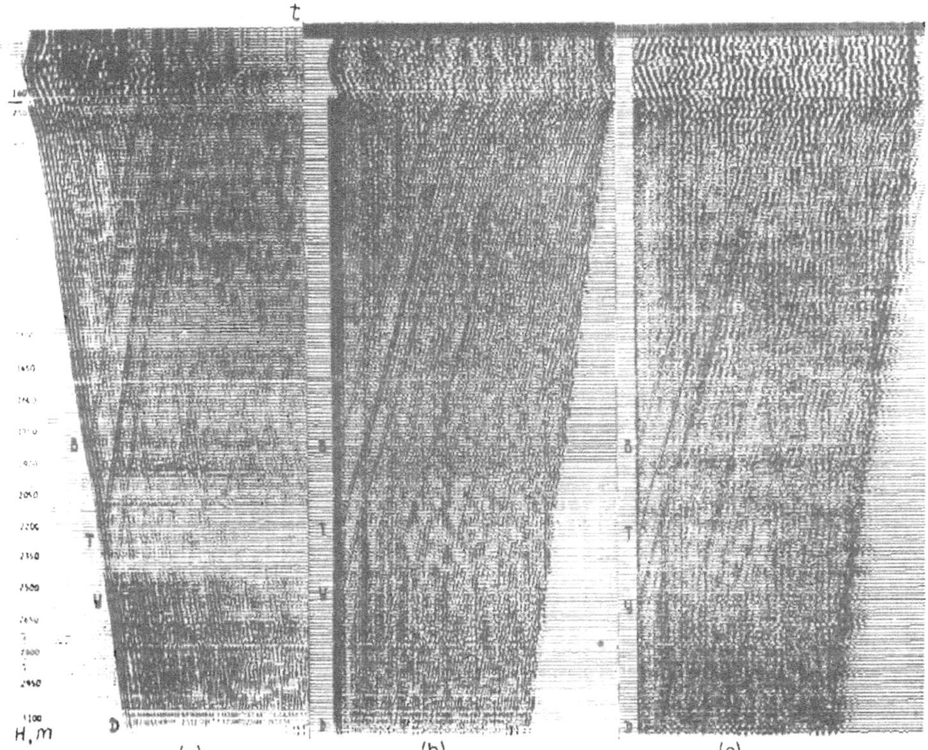

(a) (b) (c)

Fig. 17a-c.

Fig. 17a-f. Digital processing of VSP data (borehole N23 Dergunovskaya [60]).

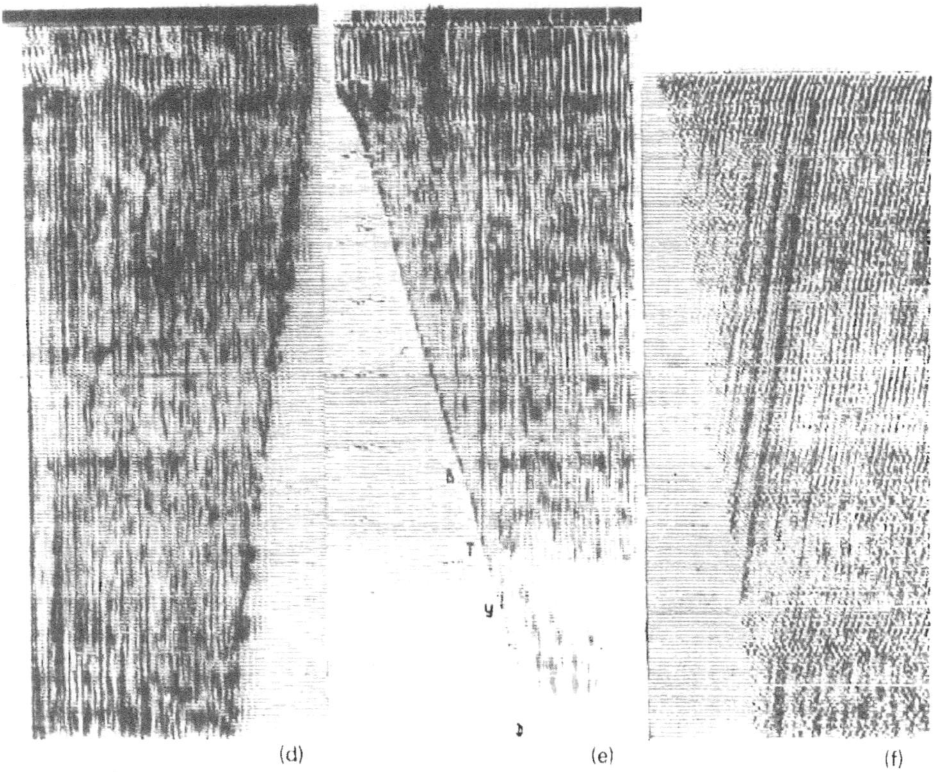

Fig. 17d-f.

time-variant; reject and recursive filtering; matched filtering – to detect useful signal against the background of random noise; inverse filtering; weighed summation; wave subtraction to simplify the form of the record; fan filtering.

Adherence to the block principle in devising processing programs enables them to be used for the solution of a variety of problems. The experience gained with this system demonstrated its effectiveness in the solution of the following problems: discrimination and tracking of useful waves and their stratigraphic correlation, determination of absorption and the reflecting properties of a geologic section etc.

We shall illustrate the operation of this system with the example of VSP data processing (borehole No. 23, Dergunovskaya) executed by Yu. G. Antipin (Figure 17). Several reflected waves can be seen and reliably tracked on the initial seismogram (Figure 17a) down to about 2000 m. However, at greater depths downward waves are the dominant ones and it is practically impossible to track the reflected waves, although they can still be discriminated against this background. After the downward wave alignments have been straightened, static corrections applied and digital amplitude adjustments made, the correlation of the downward waves is improved (Figure 17b).

Subsequent selective inverse filtering reconstitutes the 5 to 90 Hz frequency band and substantially reduces the wavelet breadth and improves the resolution (Figure 17c). Fan filtering after the inverse filtering removes all the upward waves (Figure 17d). Similar procedures (fan filtering after inverse filtering) enables the downward wavefield to be practically suppressed (Figure 17e). Then substracting the downward wavefield after inverse filtering allows reflected waves to be tracked at great depths (Figure 17f).

The digital processing system designed by A. A. Tabakov [81], based on the use of VSP data in a reproducible dynamic mode in a digital computer, has great potentialities for the application of modern procedures for wave field analysis. The system is capable of reproducing the initial dynamic relationship by means of amplification adjustment of; discriminating the downward and reflected wave fields; of prediction and shaping filtering; of discriminating the filtered reflected wavefields for the solution of the reciprocal dynamic problem yielding the acoustic impedance section below the deepest level of observations (see Chapter 13).

We shall illustrate the operation of this system with the example of VSP data (borehole No. 7, Khauzak) processed by A. A. Tabakov [81]. VSP observations were carried out down to 2460 m (See Figure 18). It is evident from the initial seismogram with straightened downward wave alignments (Figure 18a) that the entire seismogram is made up chiefly of downward waves (amplitude compensation has been introduced into the seismograms).Upward reflected waves are discernible, though obscured by the downward waves, but only at appreciable intervals after the first arrvals. It is practically impossible to track them and, still less, to determine their parameters or establish their stratigraphic correlation. Filtering, according to apparent velocity, enabled downward (Figure 18b) and the reflected (Figure 18c) wave fields to be separated. The downward waves are easily tracked along the entire profile (the correlation coefficient between adjacent traces is 0.985 ± 0.01). The removal of downward waves substantially improved the trackability of reflected waves, especially the later arrivals. Close to the first arrivals, the correlation of waves is much worse because of the superposition of high-intensity downward waves. The downward waves have been used as input signals to compute the deconvolution operator with a length of 0.070 s, a gate of 0.4 s, and a coefficient of noise resistance of 0.4. The operator obtained is convolved with the original traces. After repeated application of discrimination procedures, filter computation and convolution, the intensity of signals in the downward wave became much less than the signal intensity in the zone of the first arrivals (Figure 18d, left). The downward wave thus obtained has been used to compute an optimum shaping filter to be applied twice to the original traces. The pulse obtained as a result is less than 0.040 s duration (with the main maximum more than three times greater than the secondary maximum, Figure 18d (right)).

For effective attenuation of the downward wavefield and discrimination of the reflected wave field, the original traces have been subjected to prediction deconvolution. The use of a noise resistance parameter of 0.4, and an operator length of 0.070 s, results in low noise amplification. The cycle 'wave discrimination-deconvolution' when applied twice almost completely suppresses the downward wave field (Figure 18e).

In the computation of an optimum shaping filter, a symmetrical signal of required duration, with a spectrum as close to that of the original signal as possible, is chosen as the desirable output. This improves resolution and ensures maximum resistance to

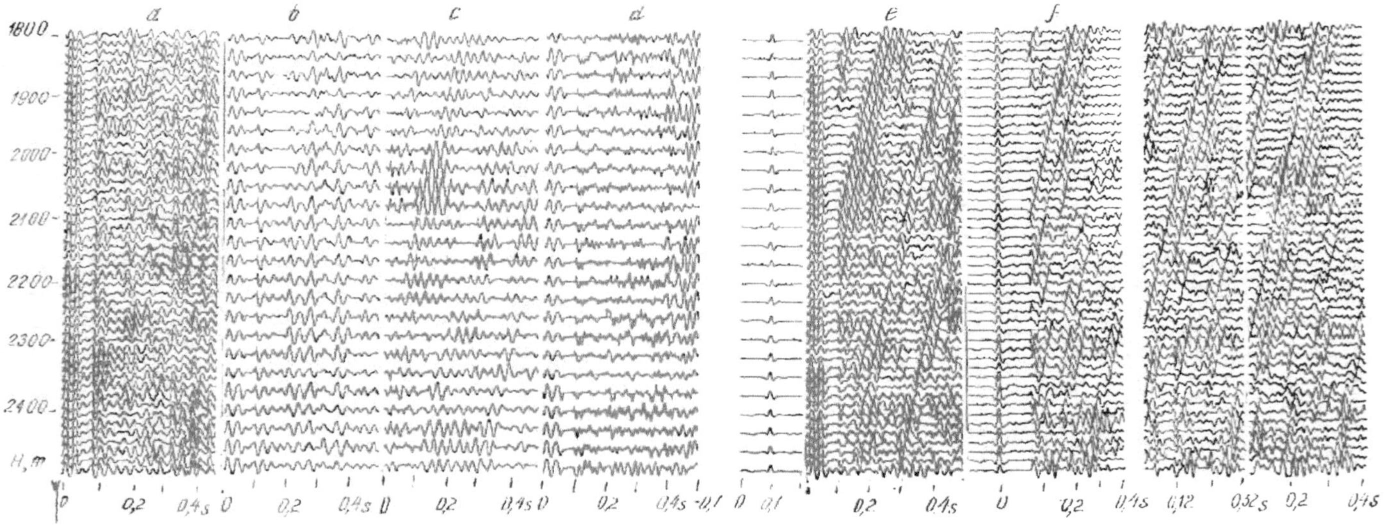

Fig. 18. Digital processing of VSP data (borehole N7 Khauzak [81]).

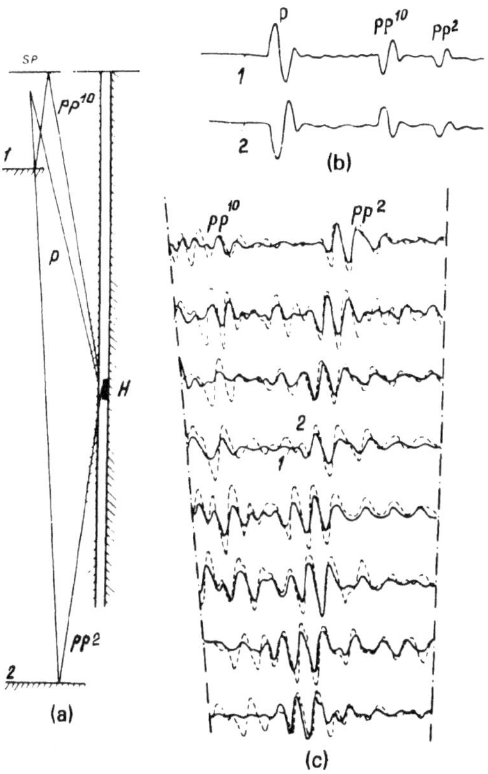

Fig. 19. Discrimination with combined detectors: (a) ray-path schematics; (b) records of velocity (1) and pressure (2) (schematic representation); (c) comparison between a section of the observed seismograms obtained with combined (2) and velocity (1) detectors.

noise. After the optimum shaping filtering the source pulse will be made up of three half-periods, the normal negative half-periods constituting not more than 30% of the main positive half-period. The reflected wave field can be reliably correlated from the time of 0.020 s. Simultaneously, good suppresion of high-frequency noise is achieved. Figure 18f (left) shows the output data from zero time. In the centre are shown the output data from the time 0.020 s and on the right from 0.1 s. The use of wave discrimination enables the source pulse and the reflected wave field to be separated from the first arrivals. Effective wave discrimination is ensured by the use of a short source pulse.

To sum up, the procedures for wave discrimination and inverse filtering enable the reflected waves that cannot be tracked and are hardly visible on the original seismogram to be separated over the entire record length from the very first arrivals.

One possible method of wave discrimination, which, it is true, has not yet found wide application, is the two-dimensional filter using a computerized method of natural orthogonal functions.

Discrimination by means of combined detectors

It would be possible to simplify the wavefield appreciably, if the useful waves only, e.g. upward waves, could be recorded at every individual point of the vertical profile. This would simplify the wave pattern on the seismogram without detriment to the resolution and improve appreciably the accuracy and the validity of the results of research. A seismic receiver with such an exclusive directivity diagram can in principle be designed by combining a pressure detector with a displacement/velocity sensitive detector.

An electrodynamic geophone in its operational frequency band is a displacement/velocity detector, and the directivity diagram of the first type (according to displacement) is represented in polar coordinates by two tangent spheres. For this reason waves of one type (e.g. pressure waves), up- and downward, will record opposite in phases. A pressure detector whose directivity diagram is a sphere records all waves of the *P*-type in-phase regardless of their direction of arrival. This difference in the directivity diagrams forms the basis for a combined detector having the necessary directivity, i.e., that does not record downward waves.

The most intense downward waves experience multiple intermediate reflection at the interfaces in the upper portion of the geologic section and propagate deep into the Earth. By choosing the polarity of the recording channel so that the pressure and the velocity detectors would record useful (upward reflected) waves in-phase irrespective of the sign of the acoustic impedance change at the interface, and the downward multiple reflected waves in opposite phases, we should be able, by summing the signals from both detectors, to suppress the downward multiples. The suppression efficiency depends on the similarity of the records (amplitudes and shapes) obtained with the pressure and the velocity detectors. The results of experimental observations show that in the real earth the shape of the pressure pulse and the displacement/velocity pulse are similar, and when observations are made in the comparatively narrow frequency band employed in seismic exploration – almost identical.

Here is an example to illustrate the point. Let a direct *P*- and a reflected PP^2-wave be recorded at point *H* in the borehole (Figure 19) together with a multiple PP^{10} reflection associated with the interface 1 in the upper portion of the geologic section. In the example cited the dipping interfaces 1 and 2 have a positive reflection coefficient while that of the surface is as always negative.

The direct *P*- and the multiple PP^{10}-waves arrive at the point of observation *H* from above with positive apparent velocities. On the seismogram recorded with the velocity detector the reflected pressure wave PP^2 records in the opposite phase to that of the direct wave *P* (Figure 19b), and the multiple PP^{10}-wave although it coincides with the direct wave in direction of arrival, has the opposite polarity on the seismogram on account of phase reversal during intermediate reflection. On the other hand, on the record of the pressure detector the reflected PP^2-wave will be in-phase with the direct *P*-wave, and the PP^{10}-wave will have a phase reversal of 180° with respect to the direct wave. By choosing the polarities of the channels so as to make the direct wave have the opposite phase on the records obtained with the velocity and the pressure detectors we shall be able, by summing both records, to suppress the multiples. Examples of seismograms obtained with the aid of combined detectors are shown in Figure 19c and are proof of a substantial suppression of the PP^{10}-wave. The summation of the signal from both detectors can be performed either in the borehole tool, or on the surrface if they have been separately recorded on a magnetic tape. This allows rigorous control of the channel sensitivities of both detectors, and ensures high-quality data.

The design of a combined detector instrument utilizes two compartments: an upper hermetically sealed one housing the velocity detector and a lower oil-filled one housing the pressure detector. In addition to the geophone, the upper compartment contains a electronic matching device (we have used transformers and emitter followers).

The method described above has not as yet gone beyond the experimental stage. The results obtained so far encourage hopes that such a discrimination method may prove an effective means of suppressing the downward multiples directly in the course of observations at each point in the borehole.

4. Processing PM–VSP Data

The methods of processing VSP data described above dealt with single-component seismograms, and their effectiveness has been repeatedly proved by surface observation experience. There are several specific aspects peculiar to the processing of PM–VSP data stemming from the three-component recordings and from the discrimination of waves according to their polarization. A substantial part of such procedures can be executed only with the aid of a computer. At present several software packages for processing PM–VSP data have been devised and are being used. We shall cite here only the algorithms of specific PM–VSP data processing procedures devised by I. A. Bykov [82], S. I. Alexandrov [83], and Kh. Agayev [84].

Amplitude Correction of a Three-Component Record

By continuously recording the signal from a fourth control geophone oriented in the direction of the axis of symmetry of the three-component cluster, control of channel duplication can be established, as well as the spread in the sensitivities of each of the three-components in the geophone cluster to be corrected.

If the three channels are identical, the following equality must hold:

$$\sum_{i=1}^{3} U_i(t) = \sqrt{3}\, U_4(t),$$

where $U_i(t)$ is a seismic signal of the ith channel. In practice this condition may be violated. The correction coefficients a_i in the form of amplitude factors are determined from the condition of a minimum for the function

$$\varepsilon = \int_t \left[\sum_{i=1}^{3} a_i U_i(t) - \sqrt{3}\, U_4(t) \right]^2 \mathrm{d}t$$

by solving the equation system

$$\sum_{i=1}^{3} a_i (U_i, U_4) = \sqrt{3}\,(U_i, U_4) \quad \text{where} \quad i = \overline{1, 3}.$$

In case the required duplication cannot be achieved, the algorithm can be used to reject recorded data which does not conform with quality standards.

Comparison of three-component records

Sometimes it may be of interest in PM–VSP to compare records obtained in different coordinate systems, when a comparison between the components themselves is not feasible.

If $U(t)$ and $V(t)$ are records obtained in different coordinate systems, then the orthogonal transformation $N = (N_{ik})$ connecting the coordinate systems is determined from the condition that the function

$$\varepsilon = \sum_{i=1}^{3} \int_{(t)} [U_i(t) - c \sum_{k}^{3} N_{ik} V_k(t)]^2 \mathrm{d}t$$

be minimum where

$$N = [(U_i V_p)]^T M.$$

The amplitude coefficient c is determined from the expression

$$c = \sum_i M_{ii} \Big/ \sum (V_i, V_i).$$

The correlation of the record, irrespective of the coordinate system being used, is determined by the generalized correlation coefficient

$$\rho = \sum_i M_{ii} \Big/ \sqrt{\sum_i (U_i, U_i) \sum_k (V_k, V_k)},$$

where M is the modulus of the matrix $[(U_i, V_p)]$ unlikely determined by the equality

$$M^2 = [(U_i, V_p)]^T [(U_i, V_p)].$$

The coefficient ρ is always positive and can vary from zero to unity; N is the orthogonal transformation that describes the rotation of an arbitrary vector. The rotation angle and the direction cosines l_1, l_2, l_3 of the rotation axis are determined by the expression

$$\cos\alpha = 0.5\left(\sum_i N_{ii} - 1\right); \qquad l_1 = (N_{3,2} - N_{2,3})/2 \sin\alpha;$$

$$l_2 = (N_{1,3} - N_{3,1})/2 \sin\alpha; \qquad l_3 = (N_{2,1} - N_{1,2})/2 \sin\alpha.$$

Orientation of records with the aid of the direct wave

In conditions of axial symmetry of the medium, the particle displacement direction coincides with the azimuth of the borehole-source direction. Within the framework of such assumptions, one can determine the orientation of the geophone in the borehole, and obtain oriented records.

Let X', Y', Z' be a left-handed orthogonal system in which the Z' axis points vertically upwards and the X' axis coincides with the horizontal projection in the direction of the shotpoint.

The directions of the three axes, I, II, III, of a symmetrical geophone cluster are described by the following matrix

$$N' = \begin{pmatrix} \cos\omega'\sin\theta & \sin\omega'\sin\theta & \cos\theta \\ \cos(\omega' + \frac{2}{3}\pi)\sin\theta & \sin(\omega' + \frac{2}{3}\pi)\sin\theta & \cos\theta \\ \cos(\omega' + \frac{4}{3}\pi)\sin\theta & \sin(\omega' + \frac{4}{3}\pi)\sin\theta & \cos\theta \end{pmatrix},$$

where ω' is the azimuth of geophone I, θ is the angle it makes with the vertical; $\cos\theta = 1/\sqrt{3}$; $\sin\theta = \sqrt{2}/\sqrt{3}$.

Should A_i be the projection of the full vector **A** on the direction of the geophone's axis, the projections of **A** on the X, Y, Z axes would be

$$A'_Z = \sum_{i=3}^{3} A_i \cos\theta = \sum_{i=1}^{3} A_i\sqrt{3},$$

$$A'_X = \sqrt{\sum_{i=1}^{3} A_i^2 - \frac{1}{3}\left(\sum_{i=1}^{3} i\right)^2},$$

$$A'_Y = 0.$$

The angle ω' is determined by the following expressions

$$\tan\omega' = \frac{\sqrt{3}(A_3 - A_2)}{3A_1 - \sum_{i=1}^{3} A_i}; \qquad \cos\omega' = \frac{3A_1 - \sum_{i=1}^{3} A_i}{\sqrt{2}\sqrt{3\sum_{i=1}^{3} A_i^2 - \left(\sum_{i=1}^{3} A_i\right)^2}},$$

$$\sin\omega' = \frac{\sqrt{3}(A_3 - A_2)}{\sqrt{2}\sqrt{3\sum_{i=1}^{3} A_i^2 - \left(\sum_{i=1}^{3} A_i\right)^2}}.$$

After substituting $\sin\omega'$ and $\cos\omega'$, the matrix N'' will assume the form:

$$N'' = \begin{bmatrix} \dfrac{3A_1 - \sum_{i=1}^{3} A_i}{\sqrt{3}R} & \dfrac{A_3 - A_2}{R} & \dfrac{1}{\sqrt{3}} \\[2ex] \dfrac{3A_2 - \sum_{i=1}^{3} A_i}{\sqrt{3}R} & \dfrac{A_1 - A_3}{R} & \dfrac{1}{\sqrt{3}} \\[2ex] \dfrac{3A_3 - \sum_{i=1}^{3} A_i}{\sqrt{3}R} & \dfrac{A_2 - A_1}{R} & \dfrac{1}{\sqrt{3}} \end{bmatrix},$$

where

$$R = \sqrt{3 \sum_{i=1}^{3} A_i^2 - \left(\sum_{i=1}^{3} A_i\right)^2}.$$

The matrix N and the angle $\Delta\omega$ to orient the cluster in the X, Y, Z coordinate system, where X is the northward, Y the eastward directions and Z points vertically upward, are determined from the formulae

$$\Delta\omega = \omega' - \omega; \qquad N = N'\Omega,$$

where ω is the azimuth of the borehole-source direction, and Ω is the matrix of rotation of the coordinate system around the vertical axis through the angle $\Delta\omega$:

$$\Omega = \begin{pmatrix} \cos\omega & \sin\omega & 0 \\ -\sin\omega & \cos\omega & 0 \\ 0 & 0 & 1 \end{pmatrix}.$$

Apart from the errors, which may stem from the basic assumption that the azimuth of the displacement direction coincides with the direction to the source, the chief cause of errors in the determination of the orientation matrix are the errors of determination of the projections of the full vector **A** on the coordinate axes. The accuracy of orientation falls sharply as the angle between the displacement direction and the borehole axis decreases, and orientation becomes impossible when those directions coincide. For this reason in the case of a vertical borehole, observations along off-line profiles, or of transverse waves along an in-line profile, are required to orient the record.

ORIENTATION OF RECORDS IN DEVIATED AND IN CURVILINEAR BOREHOLES

As demonstrated in Chapter 2, when carrying out observations in deviated boreholes, it is convenient to execute orientation in two stages. If the direction of the borehole axis at the point of observation is specified by means of the vector **l**, and the displacement direction of the direct wave is specified in the coordinate system whose axes coincide in direction with the geophone axes by the vector **A**′, then to orient a record it is first of all necessary to transform the vector **A**′ into a coordinate system symmetrical with respect to the Z axis, and then to rotate the geophone around the Z axis. The latter operation has already been described above.

The orthogonal transformation L transforms the vector **i** into one pointing in the direction of the Z axis. This transformation describes the rotation around an axis

$$\mathbf{l} \times i_3 = (l_2/\sqrt{l_1^2 + l_2^2})i_1 - (l_1\sqrt{l_2^2 + l_1^2})i_2$$

through the angle γ determined by the expression $\cos\gamma = l_3$. The rotation matrix is formed from the direction of the axis and the rotation angle:

$$\begin{pmatrix} l_3 + \dfrac{l_2^2}{1 + l_3} & -\dfrac{l_1 l_2}{1 + l_3} & -l_1 \\ -\dfrac{l_1 l_2}{1 + l_3} & l_3 + \dfrac{l_1^2}{1 + l_3} & -l_2 \\ l_1 & l_2 & l_3 \end{pmatrix}$$

The component coinciding in direction with that of the shotpoint is obtained by the substitution of the components of the vector $\mathbf{A} = L\mathbf{A}'$ into the orientation matrix. The composition of the N and the L transformations determines the position of the geophone system in space.

Calculation of components of a local coordinate system and arbitrary fixed components

A very effective method used in the analysis of the wave field in the early part of the seismogram is the transition from the coordinate system connected to the axes of a three-component geophone to the coordinate system connected to the displacement direction in the first arrival. Such is the orthogonal system PRT in which the P-axis coincides in direction with that of the particle displacement of the first arrival, and the two remaining axes lie in an orthogonal plane with the R-component lying in the same vertical plane as the P-component. Naturally, the orientation of a local system changes from point to point.

The transformation matrix (unnormalized) from a coordinate system connected to the three-component geophone to a local system is of the form

$$N1 = \begin{bmatrix} A_1 & A_2 & A_3 \\ A_1(A_2+A_3)-(A_2^2+A_3^2) & A_2(A_1+A_3)-(A_1^2+A_3^2) & A_3(A_1+A_2)-(A_1^2+A_2^2) \\ A_2-A_3 & A_3-A_1 & A_1-A_2 \end{bmatrix}$$

where A_i is the amplitude of the ith component of the three-component geophone. The transformation into the PRT system is performed with the aid of the formula:

$$\begin{pmatrix} U_P \\ U_R \\ U_T \end{pmatrix} = N1 \begin{pmatrix} U_1 \\ U_2 \\ U_3 \end{pmatrix}$$

The transformation of original three-component record data into a coordinate system XYZ fixed in space with its X axis pointing north is performed according to the formula:

$$\begin{pmatrix} U_x \\ U_y \\ U_z \end{pmatrix} = N^T \begin{pmatrix} U_1 \\ U_2 \\ U_3 \end{pmatrix},$$

where N^T is the transformed orientation matrix from formula (N'').

Arbitrary fixed components (e.g., I) are calculated as follows

$$U_{\mathrm{I}} = \sum_{i=1}^{3} K_i U_i(t),$$

where $K = N\mathrm{I}$.

By way of an example of primary PM–VSP data processing, Figure 20 depicts the seismograms in a fixed XYZ and a local PRT coordinate system obtained from original seismograms of a symmetrical I, II, III cluster. Waves of various types are detectable on the seismograms. By analyzing a wide range of components, it has been possible to establish the contours of the regions of tracking components of various waves. Such regions are shown on the stereogram. The tracking components of a wave type change their orientation along the line of the profile. On the entire profile they may

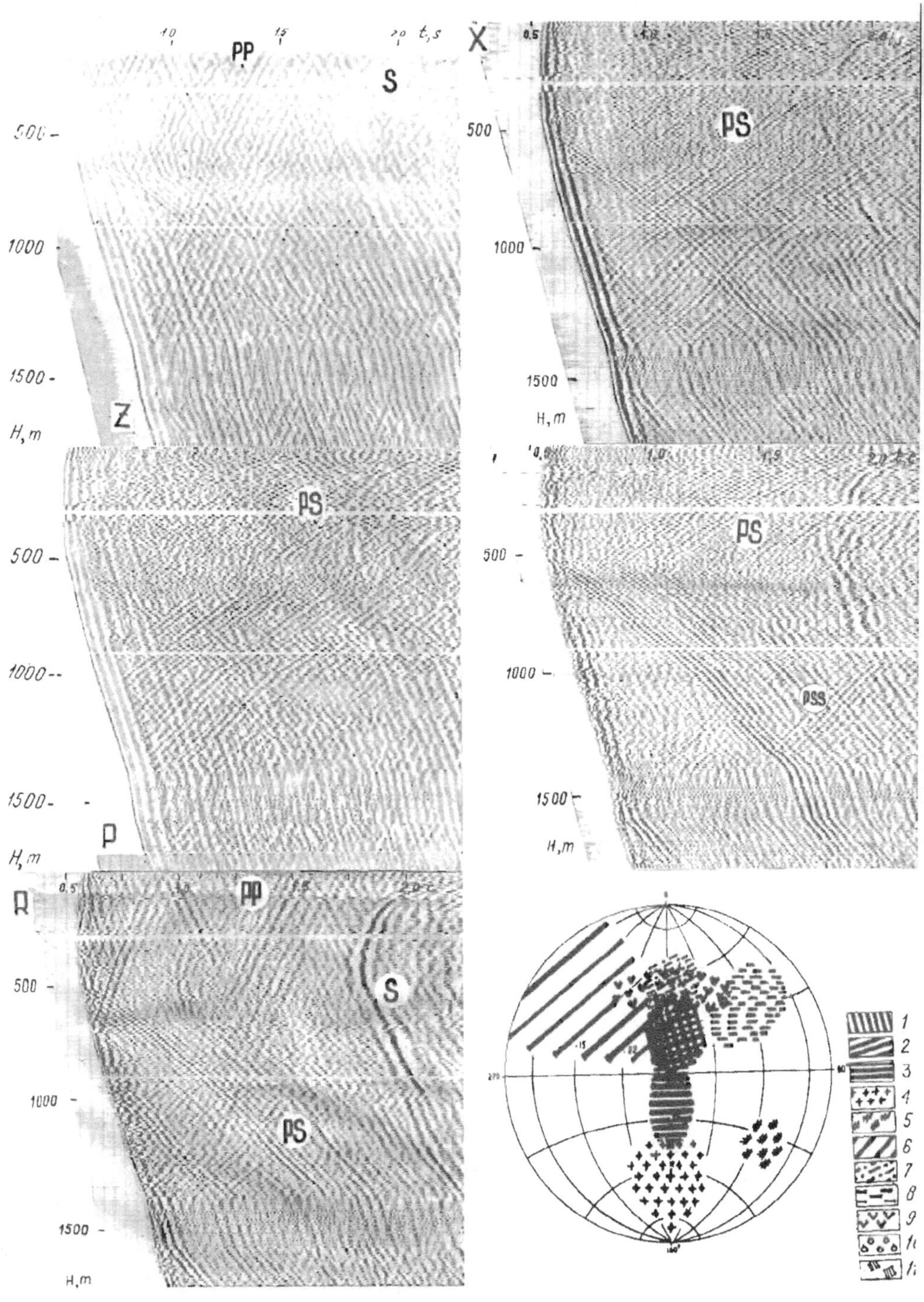

PP
S
PS
PSS
X
Z
P
R
t, s
H, m
500
1000
1500
1
2
3
4
5
6
7
8
9

be localized in a comparatively small sector, or such sectors may change drastically in various intervals of the profile, as in the case being considered. The regions of tracking components of waves of various types will be seen from the stereogram to occupy a large part of the space.

Discrimination of regular waves. Tracking component algorithm

The algorithms considered above have been designed for the processing of three-component records obtained at every point. The polarization method of discriminating regular waves envisages the polarization-positional correlation, which is a combination of discrimination along the line of observation of waves according to their directions of propagation with discrimination according to displacement direction at a point. For each wave, the components tracked in polarization-positional correlation are not components fixed in space, but variable tracking components for which the signal-to-noise ratio is at its maximum [2]. There are several algorithms in existence for the discrimination of regular waves. We shall cite the example of the algorithms for discriminating waves in a moving space-time interval.

The condition of discrimination of tracking components is that the expression $\rho = 1-(E_1/E)$ be at its maximum, where E is the total energy of seismic oscillations in k channels in the specified time interval, and E_1 is energy of the irregular component of the oscillations. To determine the tracking component, one will have to find the displacement direction corresponding to the maximum of ρ, which is the measure of regularity.

The value of ρ is determined from the expression

$$\rho = \left(\mathbf{a}, \left[\left(\sum_k U_{ki}, \sum_k U_{kj}\right)\right]\mathbf{a}\right) \Big/ N\left(\mathbf{a}, \sum_k \left[\left(U_{ki}, U_{kj}\right)\right]\mathbf{a}\right),$$

where U_{ki} is the ith component of the kth trace. $[(U_{ki}, U_{kj})]$ is the covariation matrix of the components of the kth trace.

The maximum value of ρ corresponds to the maximum eigenvalue of the matrix

$$\left[\left(\sum_k U_{ki}, \sum_k U_{kj}\right)\right]\left(\sum_k [(U_{ki}, U_{kj})]\right)^{-1}$$

and is attained when the direction of the tracking components sought coincides with that of the corresponding eigenvector.

Spectral polarization analysis

When processing PM–VSP data, one may be interested in analyzing the parameters of polarization as functions of the frequency content to study the effects of the Earth's

---

← Fig. 20. PM–VSP seismograms and the regions of tracking components (borehole N693 Anatasie-yevsko–Troitzkaya, l = 1150 m); longitudinal reflected PP-waves: $H_{ref} < 800$ m (1); $H_{ref} = 800$–1700 m (2); $H_{ref} > 1700$ m (3); converted reflected PS-waves (4); PS-waves from the 600 m-deep interface (5); converted transmitted waves: $H_{con} = 500$–700 m (6); $H_{con} = 800$–1100 m (7); $H_{con} = 1500$ m (8); transverse waves: S (9), SV (10), SH (11).

inhomogeneities on the polarization of various monochromatic components. We shall cite here the spectral polarization analysis algorithm '*P*-spectrum' worked out by S. I. Alexandrov. The *P*-spectrum of the signal $S(t)$ is a set of vectors [83]

$$\mathbf{P}_j(\omega) = \mu_j(\omega)\gamma_j(\omega), \quad j = \overline{1, 3},$$

where μ_j, γ_j are eigenvalues and normalized eigenvectors of the symmetrical matrix $G(\omega)$ with the elements

$$g_{ij}(\omega) = \int_{-\infty}^{+\infty} \mathbf{P}_{ij}(\tau) \cos \omega t \, dt, \qquad i = \overline{1, 3},\ j = \overline{1, 3},$$

where

$$\mathbf{P}_{ij}(\tau) = \int_{-\infty}^{+\infty} S_i(t)\, S_j(t + \tau)\, dt \quad \text{and} \quad S_i(t) = (S(t), X_i),$$

$S(t)$ being a vector function that specifies the seismic signal $\{X_j\}$, $j = \overline{1, 3}$ in a fixed Cartesian coordinate system.

The vectors $\mathbf{P}(\omega)$ coincide in direction with the corresponding axes of the polarization ellipses of individual monochromatic components of the signal $S(t)$, with their lengths $\mu_j(\omega)$ coinciding with the squares of the lengths of the corresponding semiaxes of the ellipses. The vectors $\mathbf{P}_j(\omega)$ satisfy the conditions

$$(\mathbf{P}_i(\omega), \mathbf{P}_j(\omega)) = \begin{cases} 0 & \text{for } i \neq j, \\ \mu_j(\omega) & \text{for } i = j, \end{cases}$$

in which it is assumed that

$$\mu_1(\omega) > \mu_2(\omega) \geqslant \mu_3(\omega) \equiv 0.$$

The parameter $\gamma_j(\omega)$ determines various orientations of the polarization parameters of individual monochromatic vector components. For an example of the application of the spectral polarizational analysis, see Chapter 12.

CHAPTER 4

SPECIFIC FEATURES OF VSP WAVE KINEMATICS

Studies of the kinematic characteristics of various wave types observed along horizontal profiles have been rather comprehensive. Equations have been derived and horizontal traveltime curves described for a wide range of seismic models. Kinematic characteristics of waves observed along vertical profiles have been studied much less so. In borehole seismics vertical traveltime curves of the first arrivals for both in-line and off-line profiles have been mainly considered. Until recently, the vertical traveltime curves of waves recorded in the later portion of the record have not been studied.

At the same time experience gained in VSP has demonstrated that the kinematic characteristics of waves on vertical profiles may serve as a source of information about the nature of the recorded waves and about the properties of the geologic section.

During the course of VSP development in recent years, equations have been derived for the vertical traveltime curves of various wave types and numerous earth structures.

In addition to describing wave traveltime curves, this chapter deals with certain specific problems relating to the determination of seismic wave velocities mainly in connection with VSP.

1. Kinematic Characteristics of Waves in Homogeneous Media

Along the entire profile, or in individual sections of the profile, the first arrivals to be recorded on VSP seismograms are the direct or transmitted, refracted and head waves in the same way as on surface seismograms, depending on the geologic structure and the distance between the shotpoint and the borehole. The waves recorded as subsequent arrivals are the reflected and the refracted waves (longitudinal, transverse and converted) and multiple reflections. The direct wave on a vertical profile in a homogeneous medium is taken to mean the first arrival with a positive apparent velocity. In layered media this wave is essentially a refracted (transmitted) wave. In continuously changing media, its wave path is curved as the result of refraction bending. The refracted wave on a vertical profile is understood to be the first arrival with a negative apparent velocity, which, depending on the law of velocity variation, may either be a bent (curved-ray refracted) wave or a head wave. The latter, in contrast to the curved-path wave, is understood to be a wave, whose mechanism of energy recuperation is connected with diffraction. Close to the point of emergence such a wave is an interferential curved path wave. It should be pointed out that in the presence of high-dip interfaces the signs of the apparent velocities of the direct and refracted waves indicated above may in individual cases be reversed. To gain a clearer insight into the properties of the traveltime curves of various waves, and the interrelationship between them, it is expedient to consider these for every model and for all wave types. The traveltime curves are studied in the plane of dip of the interface at right angles to the line of intersection of the interface with the vertical plane (Figure 21).

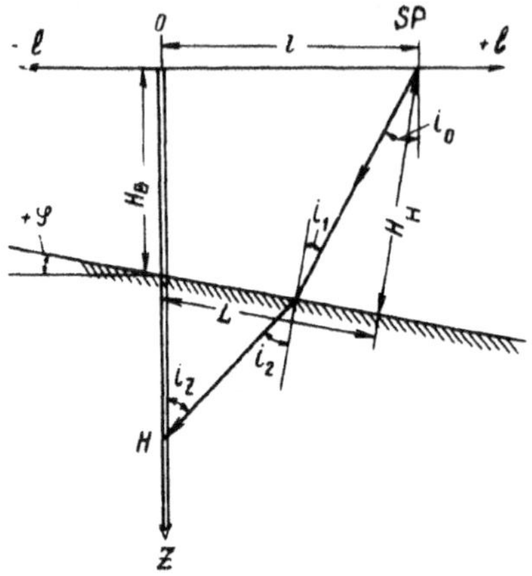

Fig. 21. Derivation of traveltime equations.

Consider the vertical traveltime curves of waves recorded in both constant-velocity, layered and continuously-changing media.

Constant-velocity medium

In a homogeneous medium only the direct longitudinal and transverse downward travelling waves are recorded as first arrivals. The traveltime curve of the longitudinal and transverse down-wave is determined by the equation

$$t = \frac{1}{v}\sqrt{H^2 + l^2} \tag{1}$$

and for an off-line vertical profile ($l \neq 0$) it takes the form of a hyperbola whose axis is symmetric with the abcissa. As H increases, the traveltime curve for an off-line profile approaches asymptotically the traveltime curve for an in-line vertical profile, which takes the form of a straight line $t = H/v$. In normal borehole seismics this expression serves to determine the average wave velocity v_{av}.

Continuously-changing medium

In the real earth, a continuous change in velocity with depth is usually observed even in the same lithologic unit. Abrupt changes in velocity are observed mainly at interfaces between different types of rock. For media with small velocity differentiation, the velocity gradient usually decreases with depth. For such media, the function

$$v(H) = v_0(1 + \beta H)^{1/a}, \quad 1 \leqslant a \leqslant 3 \tag{2}$$

is known [3, 85] to be a good approximation of the velocity distribution. For $a = 1$, the velocity versus depth function is linear with a constant gradient. In continuously-changing media, curved-path waves are recorded as first arrivals. Multiple curved path reflections from the surface are recorded as later arrivals.

The equation for the vertical traveltime curves of a family of rays in the case of a continuously-changing medium takes the form [5]

$$t = 2k \int_0^{H_r} \frac{pv(H)\,\mathrm{d}H}{\sqrt{1 - p^2v^2(H)}} \pm \int_0^{H} \frac{pv(H)\,\mathrm{d}H}{\sqrt{1 - p^2v^2(H)}} \tag{3}$$

$$t = 2k \int_0^{H_r} \frac{\mathrm{d}H}{v(H)\sqrt{1 - p^2v^2(H)}} \pm \int_0^{H} \frac{\mathrm{d}H}{v(H)\sqrt{1 - p^2v^2(H)}}, \tag{4}$$

where the parameter

$$p = \frac{\sin i}{v(H)} = \frac{1}{v(H_r)}.$$

The coordinates of the minima of the vertical traveltime curve are [5]:
for $a = 2$:

$$l = \frac{1}{\beta}(2k \pm 1)\sqrt{1 + \beta H_{\min}}\left(\frac{\sqrt{\beta H_{\min}}}{1 + \beta H_{\min}} + \arccos\frac{1}{\sqrt{1 + \beta H_{\min}}}\right), \tag{5}$$

$$t_{\min} = \frac{2}{\beta v_0}(2k \pm 1)\sqrt{1 + \beta H_{\min}}\arccos\frac{1}{\sqrt{1 + \beta H_{\min}}}, \tag{6}$$

for $a = 3$:

$$l = \frac{1}{\beta}(2k \pm 1)\,[1 + 2(1 + \beta H_{\min})^{2/3}]\,\sqrt{(1 + \beta H_{\min})^{2/3} - 1}, \tag{5'}$$

$$t_{\min} = \frac{3}{\beta v_0}(2k \pm 1)(1 + \beta H_{\min})^{1/3}\,\sqrt{(1 + \beta H_{\min})^{2/3} - 1}. \tag{6'}$$

The presence of both signs in the term $(2k \pm 1)$ takes into account any abrupt unit change in multiplicity at such points. From the velocity distribution given in Equation (2), the equation of the in-line ($l = 0$) vertical traveltime curve of the first arrivals is determined by the expression [3]:

$$t = \frac{a}{\beta v_0(a - 1)}[(1 + \beta H)^{(a-1)/a} - 1].$$

Differentiating Equation (4), one can obtain the expression for the apparent velocity of the first arrival ($k = 0$) such that

$$\frac{\mathrm{d}t}{\mathrm{d}H} = \frac{\sqrt{1 - p^2v^2(H)}}{v(H)},$$

where $H_{\min} = H$. Hence, the ordinate of the traveltime curve's minimum corresponds to the depth of maximum penetration of the ray (for $a \geqslant 1$).

Consider the kinematics of waves in the case of a linear ($a = 1$) law of increase in velocity with depth in more detail. Such a law is the one most frequently used to approximate a geologic section.

Linear law of increase in velocity

In this case the curved-path parametric traveltime curve equations take the form [5]:

$$H = \frac{1}{pv_0\beta}\left\{\sqrt{1-[(2k \pm 1)\sqrt{1-p^2v_0^2} - l\beta pv_0]^2} - pv_0\right\},$$

$$t^{\downarrow} = \frac{1}{\beta v_0}\ln\frac{(1+\beta H)(1+\sqrt{1-p^2v_0^2})^{2k+1}}{p^{2k}v_0^{2k}[1+\sqrt{1-p^2v_0^2(1+\beta H)^2}]},$$

$$t^{\uparrow} = \frac{1}{\beta v_0}\ln\frac{[1+\sqrt{1-p^2v_0^2(1+\beta H)^2}](1+\sqrt{1-p^2v_0^2})^{2k-1}}{p^{2k}v_0^{2k}(1+\beta H)}.$$

Traveltime curve equations of first arrivals of zero multiplicity can be deduced from the equations presented above, if the parameter P is excluded from them [3]:

$$t = \frac{1}{v_0\beta}\operatorname{arc\,cosh}\left[\frac{\beta^2(l^2+H^2)}{2(1+\beta H)} + 1\right].$$

In the particular case of an in-line profile

$$t = \frac{1}{v_0\beta}\ln(1+\beta H).$$

The latter equation has been derived by N. N. Pusyrev [3]. The off-line traveltime curve (for $l \neq 0$) has a minimum whose ordinate (depth) is determined by the expression:

$$H_{min} = \frac{1}{\beta}(\sqrt{1+\beta^2l^2} - 1).$$

An increase in the offset distance of the source brings about an increase in the minimum's ordinate. The locus of the minima is determined by the coefficient of increase in the velocity. The upper part of the traveltime curve corresponds to a negative apparent velocity, and the lower part of it – to a positive velocity.

The peculiarities of curved path vertical traveltime curves enable the presence of curved-ray refractions to be detected even in the case of media having small velocity gradients.

Consider the horizontal and vertical traveltime curves of the first arrivals calculated in the case of a linear velocity increase with depth, coefficient β being equal to 0, 0.0002, 0.0004, and 0.0010 m^{-1} (Figure 22). For the first three values of β, the horizontal traveltime curve is practically a straight line. A slight curvature is observed only for $\beta = 0.0010\ m^{-1}$. At the same time, the vertical traveltime curves (calculated for three values of l) are curvilinear whose minima ordinates increase rapidly along the abscissa with the increase in l and β. The family of lines through the minima of the traveltime curves, calculated for various values of the coefficient of velocity increase β, can be used

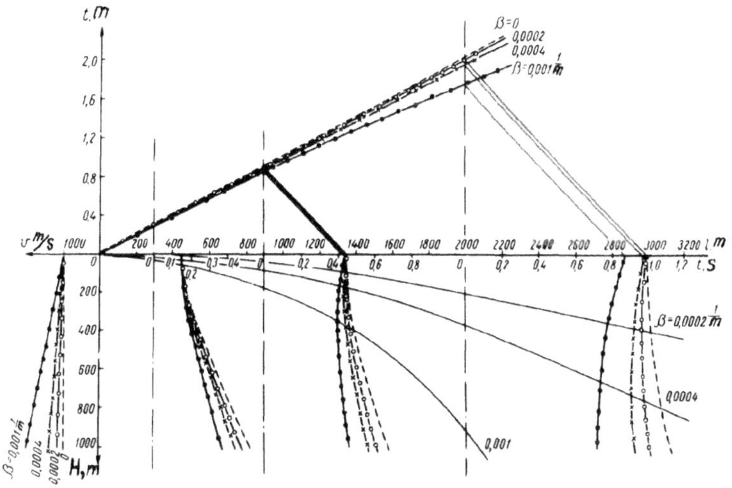

Fig. 22. Computed combined traveltime curves of the first curved-path waves.

as a chart for determining β from observated data. There is a fairly marked distinction between the vertical traveltime curves of waves in a homogeneous (dotted lines $\beta = 0$, Figure 22) and in a continuously-changing medium.

Figure 23 depicts observed traveltime curves of the first longitudinal curved path waves on a vertical profile shot from eight points. The locus of the minima (which defines the point of reversal of the direction of the wave's arrival) corresponds to a linear law of velocity variation with depth for $\beta = 0.00048\ \mathrm{m}^{-1}$.

Traveltime curves of curved-path multiples

In a continuously-changing medium in the absence of reflectors it is still possible to record multiple refractions.

The multiplicity of such a wave is determined from the number of points of maximum penetration of the ray. For each wave of any multiplicity, there is a section of the vertical profile where it propagates upwards with a negative apparent velocity and downwards with a positive apparent velocity. In the equations the signs '–' and '+' correspond to such sections.

The traveltime curves of multiple reflections have a typical acute point of inflection, corresponding to the reflection from the surface at $H = 0$, where the direction of propagation changes. Values of the arrival times of multiple reflections at the surface can be found from Equations (5) and (6) by substitution of the parameter P for $H = 0$

$$t = \frac{2k}{\beta v_0} \ln \frac{l\beta + \sqrt{4k^2 + l^2\beta^2}}{2k}.$$

The difference between the arrival times of waves of adjacent multiplicities decreases with the increase in the order of multiplicity. Multiples with curved paths have caustic surfaces in the vicinity of which the energy is focused, and where the normal cross-

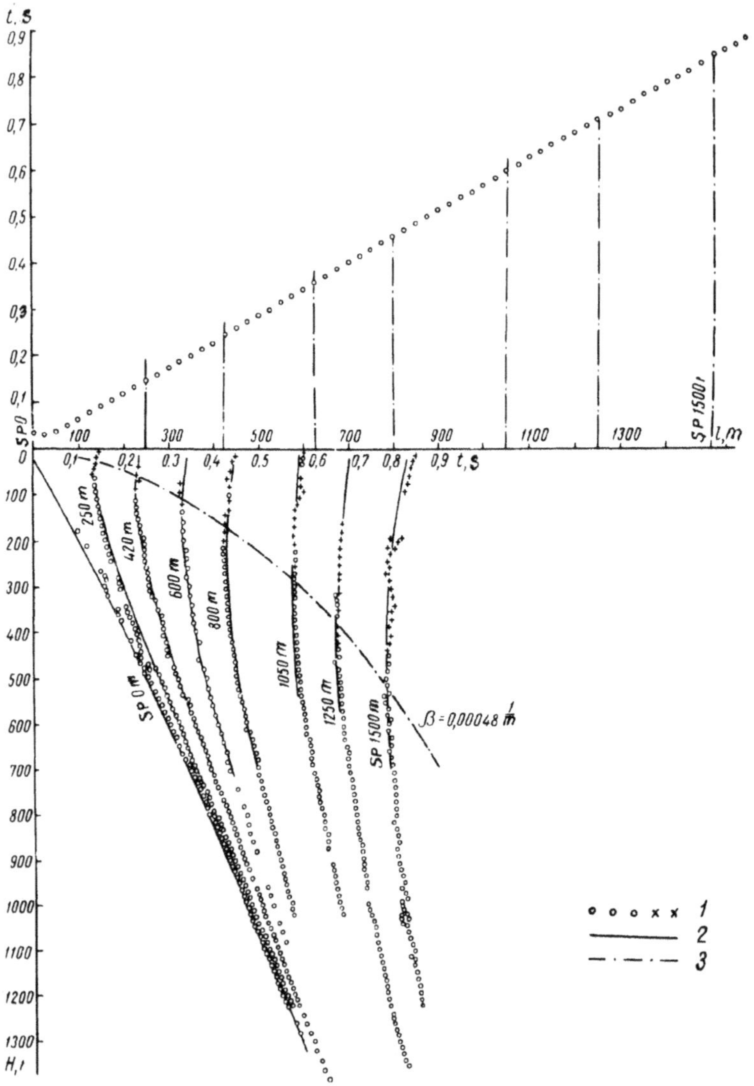

Fig. 23. Traveltime curves in a medium with a velocity gradient: (a) traveltime curves of the first bent wave: observed (1), theoretical (2) line of minima (1) (borehole N3 Afipskaya, Krasnodar province); (b) ray-path schematics of the singly-bent wave; (c) vertical traveltime curves of multiply-bent waves (5, 14) (Krasnodar province).

section area of ray packets is zero (Figure 23b). The envelope of the family of rays of multiple waves is a locus of caustic surfaces and is determined by the equation

$$H_c = \frac{1}{\beta}\left[\sqrt{1 + \frac{l^2\beta^2}{(2k+1)^2 - 1}} - 1\right].$$

It serves as the deepest boundary for the traveltime curves of multiple reflections. With the increase in multiplicity, the region of existence of multiples diminishes, and in the limit, $k \to \infty$, tends to zero. The coordinates of the minima on the traveltime curves of multiples are determined by the equation

$$H_{\min} = \frac{1}{\beta}\left[\sqrt{1 + \left(\frac{l\beta}{2k \pm 1}\right)^2} - 1\right],$$

$$t_{\min} = \frac{2k \pm 1}{\beta v_0} \ln \frac{l\beta + \sqrt{l^2\beta^2 + (2k \pm 1)^2}}{2k \pm 1},$$

and it follows that $H_{\min} < H_c$. Adjacent sections of traveltime curves are bent away in opposite directions from the depth axis.

The traveltime curve equation for an exponential law of velocity variation

In the case of a velocity gradient decreasing with depth, such that

$$v(H) = v_\infty - (v_\infty - v_0)e^{-aH}$$

the equation of an in-line vertical traveltime curve takes the form:

$$t = \frac{H}{v_\infty} + \frac{1}{av_\infty} \ln \frac{v_\infty - (v_\infty - v_0)e^{-aH}}{v_0}.$$

In the case of an increasing velocity gradient, when $v(H) = v_0 e^{aH}$, the equation for an in-line traveltime curve takes the form

$$t = \frac{1}{v_0 a}(1 - e^{aH})$$

and for an off-line one

$$t = \frac{\sqrt{2}}{v_0 a} \frac{\sqrt{\cosh aH - \cos al}}{e^{aH/2}}.$$

In this case the traveltime curve's minimum is determined by the expression

$$H_{\min} = \frac{1}{a} \ln (\cos al)$$

The traveltime curve equation in homogeneous anisotropic media

The determination of the parameters that characterise the anisotropy of rock is based on the analysis of the kinematic properties of waves in anisotropic media. The equations of vertical traveltime curves for the most common earth structures, and in which the wave fronts can be represented as rotational ellipsoids, have been obtained in [86].

In the case of a vertical axis of anisotropy, the traveltime curve equation for the direct wave takes the form:

$$t^2 = \frac{H^2}{v^2} + \frac{l^2}{A^2 v^2}, \tag{7}$$

where H is the depth of the detector, l is the distance of the source from the borehole, v is the velocity in the direction of the axis of anisotropy, and A is the coefficient of anisotropy.

In the case of an oblique axis of anisotropy, the vertical traveltime curve of the direct wave takes the form

$$t^2 = \frac{1}{v^2}(H \cos \varepsilon \pm l \sin \varepsilon)^2 + \frac{1}{A^2 v^2}(l \cos \varepsilon \pm H \sin \varepsilon)^2, \tag{8}$$

where ε is the angle between the axis of anisotropy and the vertical. The positive sign corresponds to shotpoints located in the direction opposite to the angle of inclination of the axis of anisotropy.

2. Kinematic Properties of Waves in a Two-Layered Medium

The primary longitudinal wave incident on a boundary at an angle less than the critical one results in the formation of secondary waves: two reflected longitudinal and converted waves and two transmitted waves – a longitudinal and a converted.* Normal incidence of a wave on an interface does not result in the formation of converted waves.

Consider vertical traveltimes curves of the first arrivals and secondary waves in a homogeneous medium with a single plane inclined interface.

Traveltime Curves of the First Arrivals above an Interface

For small offset distances the direct wave is recorded as a first arrival above an interface. The equation of its traveltime curve has been given in [1]. If the velocity in the second

* The terminology currently in use in the West is applicable chiefly to observations employing the reflection method and usually longitudinal waves. Now that a general polarization method has been devised and a new trend based on the extensive simultaneous use of waves of different types gains prominence in seismic exploration such terminology becomes very inconvenient. For example, why should a transverse wave emanating from a source be termed 'secondary' (with respect to what?), the term 'primary' being the priviledge of any longitudinal wave.

In this book the author uses the following terminology:

(1) Primary waves – waves emanating from the source. They may be both P and S waves. When boundary phenomena are being studied, the term applies to incident (both P and S) waves.

(2) Secondary waves – waves formed in the course of propagation on heterogeneities of the Earth (again they may be both P and S waves).

(3) Direct wave – a wave that has not experienced any reflections between the source and the seismometer. It may be recorded both as the first event and as a subsequent event.

(4) First arrival (first wave, first pulse). These concepts refer only to the first event on a seismogram and are used throughout in this sense. It is practically impossible to discriminate first arrivals for waves in the subsequent portion of a seismogram that are recorded as a rule against background noise.

We distinguish also other elements of a wave: first break, i.e. the instant of arrival of the first event, first phase of a wave. These concepts are applicable to any wave, both in the initial and in subsequent portions of a seismogram.

layer v_2 is greater than the first v_1, then, starting from some distance l the head (refracted) wave can be observed above the interface as the first arrival. Its traveltime curve is a straight line given by the equation [3]:

$$t = \frac{1}{v_1}[2H_v \cos i_{cr} \cos \varphi + l \sin (i_{cr} \pm \varphi) - H \cos (i_{cr} \pm \varphi)]. \tag{9}$$

In Equation (9) the positive sign corresponds to the case when the direction to the shotpoint is in the direction of the dip of the interface with respect to the borehole.

For a horizontal interface, the equation takes the form

$$t = \frac{1}{v_1}[(2H_v - H) \cos i_{cr} + l \sin i_{cr}].$$

The minimum value of l for which head waves can be formed at the interface is determined, by the point of emergence from the formula

$$l_{min} = H_v \tan (i_{cr} \pm \varphi).$$

For a horizontal interface ($\varphi = 0$)

$$l_{min} = H_v \tan i_{cr}.$$

For a value of l satisfying the inequality

$$H_v \tan i_{cr} < l < 2Hv \tan i_{cr}$$

the head wave will not be recorded along the entire vertical profile, but only in the section adjacent to the interrface (Figure 24). The length of this section ΔH_1 (from the refracting boundary) is determined by

$$\Delta H_1 = \frac{l - H_v \tan i_{cr}}{\tan i_{cr}} = \frac{l\sqrt{v_2^2 - v_1^2} - H_v v_1}{v_1}. \tag{10}$$

The head wave will only be recorded as a first arrival in the section between the interface and the point where it becomes coincident with the direct wave. The length of this section is

$$\Delta H_2 = \frac{1}{v_1}[H_v v_2^2 + H_v(v_2^2 - v_1^2) + lv_1\sqrt{v_2^2 - v_1^2} -$$

$$- 2v_2\sqrt{H_v}\sqrt{H_v(v_2^2 - v_1^2) + lv_1\sqrt{v_2^2 - v_1^2}}]. \tag{11}$$

Above this section the direct wave will be observed first, and the head wave will appear as a subsequent arrival. For $H_v \tan i_{cr} < l < 2H_v \tan i_{cr}$, the interval ΔH_3 of the vertical profile, where the head wave is recorded as a subsequent arrival, can be determined as the difference between the sections ΔH_1 and ΔH_2:

$$\Delta H_3 = \frac{2v_3\sqrt{H_v}(\sqrt{H_v(v_2^2 - v_1^2) + lv_1\sqrt{v_2^2 - v_1^2}} - v_2\sqrt{H_v})}{v_1^2}.$$

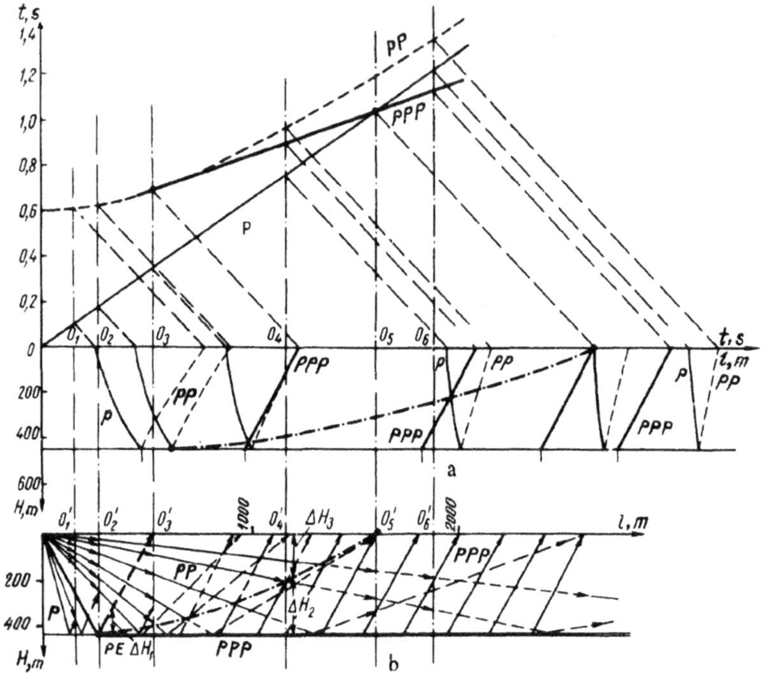

Fig. 24. Combined traveltime curves of the direct, the head and the reflected waves (a) and ray-path schematics (b).

The minimum distance, $l_{\min}$, at which the head wave is recorded as the first arrival at a point of the vertical profile located at a distance ΔH_2 from the interface, can be obtained by solving Equation (11) for l.

Starting with the distance $l = 2H_v \tan i_{cr}$, corresponding to the point of origin of the head wave, the head wave will be recorded along the entire vertical profile above the interface.

It should be kept in mind that for high-dip interfaces and for large differences in velocities, the head waves can be recorded as first arrivals along the entire in-line vertical profile ($l = 0$) above the interface or in individual sections of it [3].

Traveltime curves of secondary waves

The traveltime curves of these waves may be obtained from general equations written in a parametric form

$$H = \frac{1}{\sqrt{1 - p^2 v_1^2}\,(pv_2 \cos\varphi \pm \sin\varphi \sqrt{1 - p^2 v_2^2})} \times$$
$$\times \left[\sqrt{1 - p^2 v_1^2}\,(pv_2 H_v \pm l\sqrt{1 - p^2 v_2^2} \cos\varphi \mp pv_1 H_H \sqrt{1 - p^2 v_2^2})\right],$$

$$t = [H_H[p(v_2^2 - v_1^2)\cos\varphi \pm v_2 \sin\varphi\sqrt{1-p^2v_2^2}] + \\ + v_1 L \cos\varphi\sqrt{1-p^2v_1^2}]/[v_1v_2\sqrt{1-p^2v_1^2}(pv_2\cos\varphi \pm \sqrt{1-p^2v_2^2}\sin\varphi)], \quad (12)$$

where

$$p = \frac{\sin i_1}{v_1} = \frac{\sin i_2}{v_2}.$$

The upper sign corresponds to the transmitted waves, and the lower sign to the reflected waves. The arrival time of all the secondary waves at the depth of the interface H_v is determined by the expression

$$t = \frac{\sqrt{l^2 + H_v^2}}{v_1}$$

and is equal to the time it takes the direct wave to reach the interface. This reflects the fact that the incidence of the direct wave on the interface and the formation of secondary waves take place simultaneously.

Transmitted waves

The traveltime equation of a transmitted wave in parametric form and for horizontal layering is as follows

$$H = \frac{(l - H_v \tan i)\sqrt{\gamma^2 - \sin^2 i}}{\sin i} + H_v,$$
$$t = \frac{H_v}{v_1 \cos i} + \frac{(H - H_v)\gamma}{v_2\sqrt{\gamma^2 - \sin^2 i}}, \quad (13)$$

where $\gamma = v_1/v_2$.

The angle of incidence of the wave on the interface serves as the parameter in these equations. For $l = 0$ and $i = 0$ (in-line profile), we obtain

$$t = \frac{H_v}{v_1} + \frac{H - H_v}{v_2}.$$

The longitudinal transmitted waves constitute the first arrivals below the interface. The existence of two extrema is typical for the traveltime curves of first arrivals represented by direct, head and transmitted waves: the maximum value of t corresponds to the point where the direct wave changes to the head wave PPP, and the minimum value to the point where the head wave changes to the transmitted wave at the depth of the interface (Figure 24). At the former point, the apparent velocity changes from positive to negative. At a distance $l > H_v \tan i_{cr}$, there is a discontinuity in the arrival times on the traveltime curves of the direct and transmitted waves for $H = H_v$ equal to

$$\Delta t = \frac{v_2\sqrt{H_v^2 + l^2} - H_v\sqrt{v_2^2 - v_1^2} - lv_1}{v_1 v_2}.$$

In the case of a dipping interface ($\varphi = 0$) the traveltime curve, Equation (9), can also be represented in the form:

$$t = \frac{(H_v \cos \varphi + l \sin \varphi)\sqrt{v_{\text{in}}^2 - v^2}}{vv_{\text{in}}} + \frac{l \cos \varphi - H_v \sin \varphi}{v_{\text{in}}} + \frac{H(\cos \varphi \sqrt{v_{\text{in}}^2 - v^2} + v \sin \varphi)}{vv_{\text{in}}}.$$

It follows from this, that for shotpoints located downdip, a kink in the traveltime curves is produced at the transition from the head wave to the transmitted wave on the interface. However, there is no change of sign of the velocities. The traveltime curve of the transmitted wave is curvilinear, and the velocities change sign at a point corresponding to the horizontal direction of wave propagation, where $v_a = \infty$. The absence of a reversal in the sign of the velocities in this case, not only reduces the accuracy of stratigraphic correlation, but also makes the detection of a refracting interface much more difficult.

For shotpoints located up-dip, a kink is produced in the travel time curve at the transition from the head wave to the transmitted wave at the point where the borehole intersects the interface, and at the same time the sign of the velocities is reversed. In this case the interface can be detected with certainty, and stratigraphic correlation can be established much more accurately. With an increase in H, branches of the transmitted wave off-line traveltime curves approach that of the in-line vertical traveltime curve, which acts as their asymptote. For small values of φ (not exceeding 20 to 30°) and for $l \neq 0$, a clearly defined minimum can be observed on the traveltime curve at the depth of the interface (the curve is convex towards the depth axis).

Longitudinal reflected waves

The equation of the vertical traveltime curve of the reflected wave can be obtained from Equation (12), if the parameter p is excluded, and the conditions $v_1 = v_2$ are introduced:

$$t = \frac{1}{v_1}\sqrt{l^2 + H^2 + 4H_H(H_v - H)\cos \varphi}. \tag{14}$$

The vertical traveltime curve of a reflected wave is a hyperbola. When the shotpoint is located within the range $-H_v \cot \varphi < l < \infty$, the reflected waves are recorded above the interface. For $l < -H_v \cot \varphi$ the reflected waves will be recorded below the interface. The latter case is possible only in the case of large dips, and is a rare occurrence in normal seismic exploration. The traveltime curves of the reflected wave may have a minimum if $0 \leqslant 2H_H \cos \varphi < H_v$. The coordinates of the minimum are determined from the formulae:

$$H_{\text{min}} = 2H_H \cos \varphi.$$

$$t_{\text{min}} = \frac{H_v \sin 2\varphi - l \cos 2\varphi}{v_1} = \frac{2H_H \sin \varphi - l}{v_1}$$

With an increase in l, the ordinate of the minimum increases as well. In other cases the traveltime curves have no extrema, and their slopes correspond to either negative or positive apparent velocities, depending on the range of values of the angle φ.

For a fixed source offset distance, the vertical traveltime curves of reflected waves straighten out with an increase in the dip of the interface, and acquire a minimum for φ lying inside the interval $\arccos H_v/2H_H < \varphi < 90°$. For small values of φ, the apparent velocities of the traveltime curves are close to one another.

For shotpoints located up-dip, the effect of the dip is more pronounced than for shotpoints down-dip. The vertical traveltime curves of waves reflected downwards are confined to a narrow band for various values of φ, and a fixed value of the ratio l/H_v. With an increase in l/H_v this band broadens. The region of existence of reflected waves for l/H_v = a constant is limited on one side by the traveltime curve of the direct waves (the traveltime curves of direct and reflected waves are coincident at $\varphi = \pm 90°$) and on the other side by the reflected wave traveltime curves corresponding to the values $\varphi_1 = \frac{1}{2} \arctan (l/H_v)$ above the interface and $\varphi_2 = \frac{1}{2} \arctan (l/H_v) + (\pi/2)$ below the interface; φ_1 and φ_2 are the dip angles, for which the times on the traveltime curves are maximum.

In VSP practice one has very frequently to deal with overlapping traveltime curves obtained for different values of l's. For up-dip sources, the apparent velocities increase, and a minimum appears. Comparing the equations of the direct and the reflected waves, one can easily see them to be different sections of the same hyperbola. The section from the peak to the interface represents the traveltime curve of the direct wave, with the section from the interface to the surface representing that of the reflected wave. For this reason the traveltime curve of the reflected wave is closer to a straight line, and for a layer whose thickness is large compared with the distance l, will practically coincide with the hyperbola's asymptote. The distinction between the curvatures of the traveltime curves of the direct and the reflected wave is more pronounced for small values of l's.

At each point of a vertical profile in a homogeneous medium, the sum of the arrival times of the direct wave, reduced to the in-line profile, and the arrival time of the reflected wave is constant, and equal to $2H_v/v_1$. The difference between the traveltime curves of the reflected and direct waves is represented by a straight line $H = H_v - (v_1/2)\Delta t$, with a slope equal to $v_1/2$, which cuts off a segment $2H_v/v_1$ on the time axis and a segment H_v on the depth axis. This relation can be used to determine the depth of the interface in the case where the vertical profile does not intersect the interface, or in the case where the reflected wave has not been tracked along the entire profile down to the interface. For $\varphi = 0$ and for $l \ll H_v$, the reflected wave vertical traveltime curve is a mirror image of that of the direct wave, thus enabling individual sections of the reflected wave traveltime curve to be identified in places where interference or some other cause obscures the continuous tracking of the reflected wave.

TRAVELTIME CURVES OF REFLECTED WAVES IN CONTINUOUSLY-CHANGING MEDIA

For a linear law of velocity increase and for a plane horizontal interface, the equation of the vertical traveltime curve of the longitudinal reflected wave, in parametric form, is as follows:

$$H = \frac{\sqrt{1 - (2\cos i - \cos i_0 + \beta l \sin i_0)^2} - \sin i_0}{\beta \sin i}, \tag{15}$$

$$t = \frac{1}{v_0 \beta}\left(\ln \frac{\tan i/2}{\tan i_0/2} + \ln \frac{\tan i/2}{\tan i_H/2}\right),$$

where $\sin i = \sin i_0(1 + \beta H_v)$; $\sin i_H = \sin i_0(1 + \beta H)$. When the velocity in the overlying medium varies according to a more intricate law, one can write down the equation of the reflected wave traveltime curve making use of the method of averaging velocities [87]:

$$t = \frac{1}{v_{av}} \sqrt{(2H_H \cos \varphi - H)^2 + (2H_H \sin \varphi - l)^2} \times$$

$$\times \left[1 - \frac{(v_{av\,H} - v_{av})(H_v - H)}{v_{av\,H}(2H_H \cos \varphi - H + \tan \varphi(2H_H \sin \varphi - l))}\right]. \quad (16)$$

For $V_{av} = V_{av\,H}$, this equation turns into the equation of the reflected wave traveltime curve for a constant-velocity medium.

The presence of a horizontal velocity gradient, in addition to a vertical one, is typical for the upper portion of the geologic section in many areas. In the latter case the traveltime curves are displaced without a substantial change in their shape. In the direction $\beta > 0$ the traveltime curves are displaced to smaller times, and *vice-versa*. The effect of a horizontal velocity gradient on the kinematics of reflected waves in VSP is similar to that of a dipping interface, and this makes it very difficult to evaluate the effect of each of those factors separately from the observed traveltime curves.

Reflected wave traveltime curves in VSP in deviated boreholes

In practice one is often confronted with a necessity of executing VSP in deviated boreholes. This is primarily true for mining areas. However, in oil geology too there is a marked tendency towards an increase in the number of deviated boreholes. The equation of the traveltime curve of the reflected wave recorded along an inclined line of observations is of the form:

$$t = \frac{1}{v} \sqrt{(X - X_1)^2 + (Z - Z_1)^2}.$$

Here, X and Z are the coordinates of the borehole axis in a rectangular coordinate system, with the origin located at the shotpoint. The Z axis points downwards, and the X-axis points up-dip with respect to the interface. X_1, Z_1 are coordinates of the imaginary shotpoints. It should be pointed out that at all levels (except at the level of the interface) curvature in the borehole will have a more pronounced effect on the shape of the incident wave than or that of the reflected wave, the effect increasing with l. At the level of the interface, where the traveltime curves of the incident and the reflected waves have a common point, the effect of the borehole's curvature on the traveltime curves of both waves is similar.

Level traveltime curve of reflected waves

The term level traveltime curve applies to a traveltime curve obtained with down-hole detectors at a fixed depth, from shot-points located on the surface. In respect of kinematics, such observations correspond to observations along a horizontal profile laid out at a fixed depth level. The level traveltime curve of reflected waves obtained at a fixed geophone depth H can be used to identify waves registered from different shotpoints (transpositional correlation), to solve structural problems, to construct reflecting boundaries and to determine layer velocities.

In the case of a plane reflecting boundary, the level traveltime curve represents a hyperbola with the vertical traveltime curve as its asymptote. The equation of the level traveltime curve is of the form [88]:

$$t = \frac{1}{v_{av}} \sqrt{l^2 + H^2 + 4 \cos \varphi (H_v - H)(H_v \cos \varphi \pm l \sin \varphi)}, \tag{17}$$

where H_v is the depth to the reflecting boundary, H is the detector depth; φ is the dip; l is the distance of the shotpoint from the borehole; the signs '+' and '–' correspond to the down- and up-dip directions, respectively.

For a horizontal boundary

$$t = \frac{1}{v_{av}} \sqrt{l^2 + (2H_v - H)^2}.$$

For $l = 0$ (near the top of the borehole),

$$t_{ost} = \frac{1}{v_{av}} \sqrt{H^2 + 4H_v \cos^2 \varphi (H_v - H)}.$$

For $\varphi = 0$,

$$t_{ost} = \frac{1}{v_{av}} (2H_v - H),$$

where t_{ost}, in contrast to surface observations, represents the transit time of the reflected wave from the borehole entrance to the observation point along the vertical profile. This time is related to t_0 obtained in surface observations by means of the following expression

$$t_0 = t_H \cos \varphi + \sqrt{t_{ost}^2 - t_H^2 \sin^2 \varphi},$$

where t_H is the transit time of the direct wave along the line of the vertical profile from the borehole to the observation point.

In the case of curved boundaries, the level traveltime curves of reflected waves may have an intricate shape (with nodes, inflections, etc.). The curvature of interface excercises less influence on the shape of level traveltime curves than on the shape of surface curves.

Traveltime curves of waves in anisotropic media

The equations presented below have been obtained by S. I. Lapin [86] for elliptical wave fronts.

(a) When the axis of anisotropy is at right angles to the layering and the overlying medium is homogeneous, the traveltime curve of a wave reflected from a horizontal boundary is of the form:

$$t^2 = \frac{(2H_v - H)^2}{v^2} + \frac{l^2}{A^2v^2}, \tag{18}$$

where H_v is the boundary's depth; H is the depth of the detector point; v is the velocity in the direction of the anisotropy axis; A is the coefficient of anisotropy.

The equation of the vertical traveltime curve for the head wave in similar conditions is

$$t = \frac{(2H_v - H)(A_1^2 + \tan^2 i_{cr})^{1/2}}{A_1v_1} + \frac{l - (2H_v - H)\tan i_{cr}}{A_2v_2}, \tag{19}$$

where

$$\tan^2 i_{cr} = \frac{A_1^4v_1^2}{A_2^2v_2^2 - A_1^2v_1^2},$$

and A_2 and v_2 are the coefficient of anisotropy and the velocity in the direction of the axis of anisotropy of the second medium, respectively. For a wave surface in the second medium of an arbitrary shape, the velocity in the second layer along the interface should be substituted into the expression instead of A_2v_2.

(b) For an oblique axis of anisotropy, which is not normal to the interface, overlain by a homogeneous anisotropic layer, the traveltime curve equation for the reflected wave takes the form

$$t^2 = \frac{1}{v^2}[(2H_v - H)\cos\varphi \mp l\sin\varphi]^2 + \frac{1}{A^2v^2}(l\cos\varphi \pm H\sin\varphi)^2, \tag{20}$$

where φ is the dip of the interface. The upper sign corresponds to a shotpoint located up-dip with respect to the interface, and the lower sign to a shotpoint down-dip. The traveltime curve for the head wave is

$$t = \frac{[(2H_v - H)\cos\varphi \pm l\sin\varphi]\,(A_1^2 + \tan^2 i_{cr})^{1/2}}{A_1v_1} +$$

$$+ \frac{l\cos\varphi \mp H\sin\varphi - [(2H_v - H)\cos\varphi \pm l\sin\varphi]\tan i_{cr}}{A_2v_2}. \tag{21}$$

(c) If the anisotropy axis in the overlying medium is vertical and is not normal to the interface, the traveltime curve equation for the head wave is

$$t = \frac{\cos\varphi}{v'_{i_1}\cos i'_1}(H_v \mp l\tan\varphi) + \frac{\cos\varphi}{v''_{i_1}\cos i''_1}(H_v - H) + \frac{\cos\varphi}{A_2v_2}\times$$

$$\times [l\cos\varphi \pm H\tan\varphi - (H_v - H)\tan i''_1 - (H \mp l\tan\varphi)\tan i'_1], \tag{22}$$

where

$$v'_{i_1} = \frac{A_1v_1}{[A_1^2\cos^2(i'_1 \pm \varphi) + \sin^2(i'_1 \pm \varphi)]^{1/2}},$$

$$v''_{i_1} = \frac{A_1 v_1}{[A_1^2 \cos^2(i_1'' \mp \varphi) + \sin^2(i_1'' \mp \varphi)]^{1/2}}.$$

i_1' is the critical angle and i_1'' is the angle the emergent head wave makes with the normal to the interface as determined from the grazing wave condition

$$\frac{\tan(i_1' \mp \varphi) \mp A_1^2 \tan\varphi}{[A_1^2 + \tan^2(i_1' \pm \varphi)]^{1/2}} = \frac{\tan(i_1'' \mp \varphi) \pm A_1^2 \tan\varphi}{[A_1^2 + \tan^2(i_1'' \mp \varphi)]^{1/2}} = \frac{A_1 v_1}{A_2 v_2 \cos\varphi}.$$

The equation for the traveltime curve of the reflected wave is

$$t = \frac{(H_v \mp l \tan\varphi)\cos\varphi}{v_{\beta_1}\cos\beta_2} + \frac{(H_v - H)\cos\varphi}{v_{\beta_2}\cos\beta_2},$$

$$H = \frac{H_v(\tan\beta_1 + \tan\beta_2) - l(1 \pm \tan\varphi\tan\beta_1)}{\tan\beta_2 \pm \tan\varphi}, \qquad (23)$$

where β_1 and β_2 are determined from Snell's law for the wave reflected from the interface between two media having different axes of anisotropy, when β_1 and β_2 are substituted for the angles i_1' and i_1'' in Equation (22), and $v_{\beta 1}$ and $v_{\beta 2}$ are determined from expressions for v_i' and v_1'', in which β_1 and β_2 have been substituted for i_1' and i_1''.

TRAVELTIME CURVES OF CONVERTED REFLECTED AND TRANSMITTED WAVES

The same laws are valid for converted reflected waves as for converted transmitted waves. The vertical traveltime curve of a converted reflected or a transmitted wave in the case of horizontal layering can be represented by equations in a parametric form

$$H = H_v \pm \frac{(l - H_v \tan i)\sqrt{1 - \gamma^2 \sin^2 i}}{\gamma \sin i}$$

$$t = \frac{H_v}{v_{P_1}\cos i} \pm \frac{H_v - H}{v_S\sqrt{1 - \gamma^2 \sin^2 i}}. \qquad (24)$$

The upper sign corresponds to the converted reflected wave in which case $v_S = v_{S_1}$, and $\gamma = v_{S_1}/v_{P_1}$. The corresponding sign for the converted transmitted wave is the lower one in which case $v_S = v_{S_2}$, and $\gamma = v_{S_2}/v_{P_1}$. Substituting the value $H = H_v$ into Equation (24), we obtain the arrival time of the wave at the interface, $t = H_v/(V_{P_1}\cos i)$, which is equal to the time the direct wave is incident on the interface, i.e., the traveltime curves of these waves have a common point at the depth of the interface. In the case $v_{S_2} > v_{P_1}$, the converted transmitted wave front in the second medium, where $l > Hv \tan$ (arc sin V_{P1}/V_{S_2}) overlaps the longitudinal wave front, and there will no longer be a common point at the interface between the traveltime curves of the converted transmitted and the incident and reflected waves. The converted reflected wave vertical traveltime curves ($\gamma < 1$) for small values of φ have their own asymptotes and are almost parallel, especially for large values of l. The minimums on the traveltime curves appear only for large dips ($\sim 70°$) and vanish rapidly with an increase in l. The apparent velocity rises with an increase in the dip, and the traveltime curves for various values of l's remain almost parallel. In the case of converted transmitted waves ($\gamma < 1$, $\varphi = 0$), the vertical traveltime curves have no minima, and the traveltime curves for different values of l converge with an increase in the distance from the interface. There is no substantial change in the shape of the traveltime curves in the case of dipping interfaces. In the case $\gamma > 1$ and $\varphi = 0$, the traveltime curves of converted transmitted waves rapidly converge with an increase in H. As φ increases, the traveltime curves become almost parallel.

The theoretical traveltime curves of converted waves both reflected and transmitted, as well as longitudinal waves, may be conveniently constructed with the aid of the virtual shotpoint method. This method enables the trajectory of a particular wave type propagating from a virtual shotpoint to be considered, instead of the trajectory of the wave from the actual shotpoint. Such a method by construction may prove simpler and more visualizable than conventional computational methods [14].

MULTIPLY REFLECTED WAVES

Both upward and downward reflected waves are recorded on a vertical profile in a layered medium. The reflected waves enable detection of the principal interfaces in the geologic section between which multiple reflections are generated. Downward reflected waves enable the nature and the location of intermediate reflecting interfaces to be determined. The traveltime curves of waves reflected downward may be used to establish the dip of the principal and intermediate reflectors. In [15] we presented a qualitative comparison of the particular features of traveltime curves of multiple reflections obtained on both horizontal and vertical profiles. Consider the equations of vertical traveltime curves of multiple reflections.

For a single plane interface dipping at an angle φ to the horizontal, the vertical traveltime curves of reflected waves of any multiplicity can be expressed by the equation [5]:

$$t = \frac{1}{v_1}\sqrt{l^2 + H^2 + 4H_H \frac{\sin k\varphi}{\sin^2\varphi}(H_v \sin k\varphi \cos\varphi \pm H \sin\varphi \cos k\varphi)},$$

where k is the multiplicity. The positive sign corresponds to the incident wave on the interface, and the negative to the reflected wave. The equation of the traveltime curve of a wave reflected twice, in the case of a dipping interface, can be represented by [87]:

$$t = \frac{1}{v_1}\sqrt{(4H_H \cos\varphi \cos 2\varphi \pm H)^2 + (4H_H \cos\varphi \sin 2\varphi - l)^2}.$$

In the case of horizontal layering ($\varphi = 0$), the traveltime curves equation for waves of any multiciplicity is given by:

$$t = \frac{1}{v_1}\sqrt{(2H_v k \pm H)^2 + l^2}.$$

The traveltime curves for both upward and downward reflected waves can be written out separately

$$t = \frac{1}{v_1}\sqrt{[(m+1)H_v - H]^2 + l^2} \quad \text{for odd values of } m,$$

$$t = \frac{1}{v_1}\sqrt{(mH_v + H)^2 + l^2} \qquad \text{for even values of } m,$$

where m is the total number of reflections that the wave has experienced between the upper and lower interfaces. Putting $m = 0$, we obtain the traveltime equation of the direct wave

$$t^2 = \frac{l^2 + H^2}{v_1^{\,2}}.$$

For $m = 1$, we obtain the equation of a primary reflection

$$t^2 = \frac{l^2 + (2H_v - H)^2}{v_1^{\,2}}.$$

The vertical traveltime curves of waves of various multiplicity are represented by sections of a hyperbola. In case of an in-line vertical profile ($l = 0$), the traveltime curves

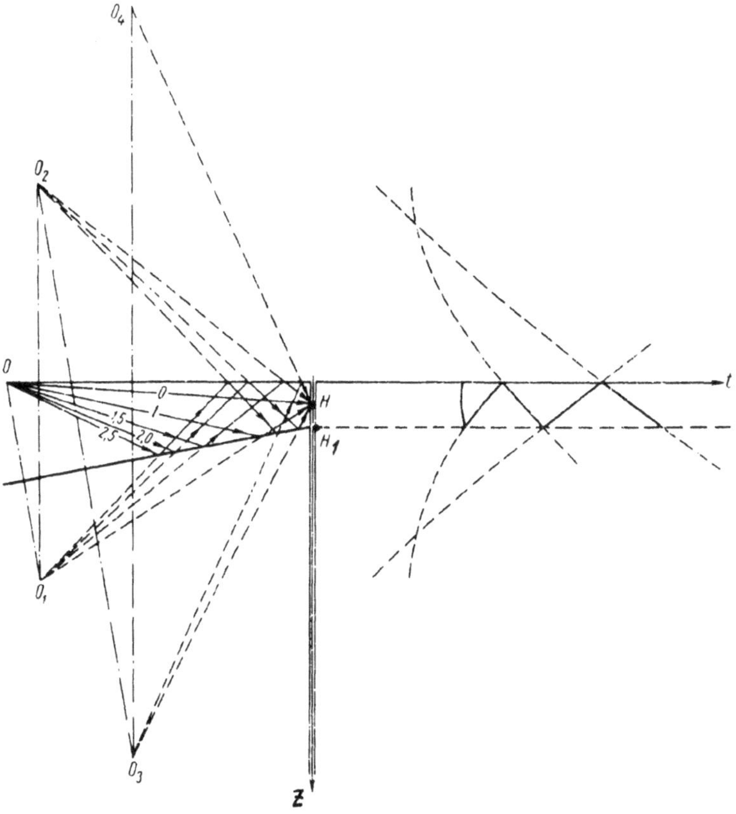

Fig. 25. Constructing vertical traveltime curves for multiply-reflected waves, $l \neq 0$.

are represented by an open polygon. The apparent velocities of primary and multiple reflections of different multiplicity for $\varphi = 0$ differ not only in magnitudes, but also in sign. At the depth corresponding to the interface there is a kink on the traveltime curve of a multiple. For off-line profiles ($l \neq 0$) at any fixed depth, as well as on the surface, the difference between the arrival times of waves of adjacent multiplicity increases with the multiplicity. In the particular case of an in-line vertical profile ($l = 0$), such differences are maximum, constant and independent of the multiplicity. If the vertical traveltime curve of a multiple reflection is interpreted as a section of the traveltime curve of a singly reflected wave from a deeper interface, the imaginary interface thus obtained will be located at a depth, which is a multiple of the depth of the primary interface.

The equation of the level traveltime curve of a multiple for $\varphi = 0$ is

$$t = \frac{1}{v_{av}} \sqrt{l^2 + (s \pm H)^2},$$

where s is the length of the travel path of the multiple: for a full multiple $s = nH$, and for a partial multiple $s = nH_1 + nH_2$. The positive sign corresponds to a downward wave, and the negative sign to an upward.

The vertical traveltime curves of reflected waves of any multiplicity can be constructed using the virtual shotpoint method. The traveltime curve of a wave reflected from the surface can be represented as that of a direct wave propagating from point O_2 (Figure 25), which is the mirror image of the point O in the plane of the surface. The vertical traveltime curve of a wave reflected twice is the mirror image of the direct wave propagating from the virtual point O_3. Vertical traveltime curves of waves of any multiplicity can be obtained by similar construction. The pattern of ray paths makes it possible to assess the regions of existence of waves of any multiplicity on a vertical profile, as well as the sections of the interface illuminated by such waves.

3. Kinematic Characteristics of Waves in Multilayered Media

The real earth is, as a rule, multilayered. Even the very simplest models of the Earth contain up to 10 and more layers. Consider the kinematic characteristics of waves in the simplest multilayered case.

Traveltime Curves of First Arrivals in Multilayered Media

In a medium with several interfaces between layers of great thickness ($d_1, d_2 \ldots d_n$) and constant velocities ($v_1 < v_2 < \ldots v_n$), the sections of the vertical traveltime curve of the first arrivals above reflecting interfaces, starting from definite distances, are made up of the vertical traveltime curves of head waves, and below the interfaces – of the traveltime curves of transmitted waves. Consider them separately. The traveltime curve equation of the head wave associated with the interface between the $(n-1)$th and the nth layers, observed in the mth layer (the layer's parameters are d_m, v_{pm}) is of the form

$$t = \sum_{k=1}^{n-1} \frac{d_k}{v_k} \cos i_{kn} + \sum_{k=m+1}^{n-1} \frac{d_k}{v_k} \cos i_{kn} + \frac{l}{v_n} - \left[\frac{H - \sum_{k=1}^{m} d_k}{v_m} \cos i_{mn} \right], \qquad (26)$$

where i_{kn} is the angle of the ray of the wave refracted in the layer at the place of its intersection with the boundary of the kth layer, i.e., $i_{kn} = \arcsin v_k/v_n$ is the wave's angle of incidence on the kth interface. For clarity rewrite the equation in the form

$$t = A\frac{\cos i_{mn}}{v_m} + C_m ,$$

where

$$A = \sum_{k=1}^{m} d_k - H; \qquad C_m = \sum_{k=1}^{n-1} \frac{d_k}{v_k} \cos i_{kn} + \sum_{k=m+1}^{n-1} \frac{d_k}{v_k} \cos i_{kn} + \frac{l}{v_n} .$$

Hence, inside the layer the traveltime curve of the refracted wave is a straight line with its slope equal to $\cos i_{mn}/v_m$ that cuts a segment C_m off the time axis. Since the changes resulting from the transition from one layer to another affect the slope and the segment cut off the time axis, the traveltime curve of a head wave in horizontal layered media is represented by an open polygon, the kinks on which coincide with the interfaces between the layers.

In a three-layer case the minimum value of l for which the formation of a head wave at the second interface becomes possible is determined by the expression

$$l = \frac{d_1 v_1}{\sqrt{v_3^2 - v_1^2}} + \frac{d_2 v_2}{\sqrt{v_3^2 - v_2^2}} .$$

An increase in l increases the subsurface coverage of the head wave in the second layer. The recording interval, ΔH, on the vertical profile in the second layer is equal to

$$\Delta H = \frac{l - d_1 \tan i_1 - d_2 \tan i_{2\,\mathrm{cr}}}{\tan i_{2\,\mathrm{cr}}} ,$$

where

$$l = d_1 \tan \arcsin \frac{v_4}{v_3} + 2d_2 \tan \arcsin \frac{v_2}{v_3} .$$

The head wave will be recorded over the entire interval of the second layer d_2, but only as the first arrivals in the section adjoining the lower interface. With further increase in l, the head wave will be recorded on the profile in the first layer as well. For $l = 2d_1 \tan i_1 + 2d_2 \tan i_{2\mathrm{cr}}$, the head wave from the deepest interface will be recorded over the entire profile, but not as first arrivals exclusively. Only when

$$l = \frac{2}{v_3 - v_1}\left[d_1 \frac{v_3^2 - 2v_1^2}{\sqrt{v_3^2 - v_1^2}} + \frac{d_2 v_1}{v_2} \frac{v_3^2 - 2v_2^2}{\sqrt{v_3^2 - v_2^2}}\right]$$

will the refracted wave at the lower interface be observed as first arrivals throughout the entire profile.

In multilayered media this distance is equal to

$$l = \frac{2}{v_n - v_1}\left[d_1 \frac{v_n^2 - 2v_1^2}{\sqrt{v_n^2 - v_1^2}} + v_1 \sum_{i=2}^{n-1} \frac{d_i(v_n^2 - 2v_i^2)}{v_i\sqrt{v_n^2 - v_i^2}}\right].$$

The first wave to be detected beneath the deepest refracting interface is the transmitted wave. The equation of the traveltime curve of a transmitted wave in parametric form is

$$t = \sum_{k=1}^{n-1} \frac{d_k}{v_k\sqrt{1 - p^2 v_k^2}} - \frac{1}{v_n^2}\sum_{k=1}^{n-1} \frac{d_k v_k}{\sqrt{1 - p^2 v_k^2}} + \frac{l}{v_n^2 p},$$

$$H = \sum_{k=1}^{n-1} d_k + \frac{l\sqrt{1 - p^2 v_n^2}}{p v_n} - \frac{\sqrt{1 - p^2 v_n^2}}{v_n}\sum_{k=1}^{n-1} \frac{d_k\, v_k}{\sqrt{1 - p^2 v_k^2}}, \tag{27}$$

where i_K is the angle of incidence on the kth interface, and

$$P = \frac{\sin i_1}{v_1} = \frac{\sin i_2}{v_2} = \ldots = \frac{\sin i_n}{v_n}.$$

The asymptote of the traveltime curve of the transmitted wave is a straight line

$$t = \frac{1}{v_n} H + \sum_{k=1}^{n-1} \left(\frac{d_k}{v_k} - \frac{d_k}{v_n}\right).$$

It can be demonstrated that the traveltime curve of the transmitted wave is a smooth curve convex towards the positive time axis, with its minimum for $p = 1/v_n$ having the coordinates

$$t = \sum_{k=1}^{n-1} \frac{d_k}{v_k} \cos i_k + \frac{l}{v_n}, \qquad H = \sum_{k=1}^{n-1} d_k .$$

The traveltime curves of the head and transmitted waves have a common point. The depth of the refracting interface can be determined from the reversal of polarities of the apparent velocities between the traveltime curves of the head and the transmitted wave.

Level traveltime curve of the reflected wave in a multilayered medium

For a horizontally layered medium (d_i, v_i) containing n layers penetrated by the borehole and m layers above the detector position, the equation of the level traveltime curve of the wave reflected from the bottom of the layer (d_j, v_j) can be represented by the parametric expression [65]:

$$l = \sum_{i=1}^{m} \frac{d_i v_i p}{\sqrt{1 - p^2 v_i^2}} + 2\sum_{i=m+1}^{n} \frac{d_i v_i p}{\sqrt{1 - p^2 v_i^2}} + 2\frac{d_j v_j p}{\sqrt{1 - p^2 v_j^2}} =$$

$$= l_i(p) + 2\frac{d_j v_j p}{\sqrt{1 - p^2 v_j^2}},$$

$$t = \sum_{i=1}^{m} \frac{d_i}{v_i\sqrt{1 - p^2 v_i^2}} + 2\sum_{i=m+l}^{n} \frac{d_i}{v_i\sqrt{1 - p^2 v_i^2}} +$$

$$+ 2\frac{d_j}{v_j\sqrt{1 - p^2 v_j^2}} = t_i(p) + 2\frac{d_j}{v_j\sqrt{1 - p^2 v_j^2}},$$

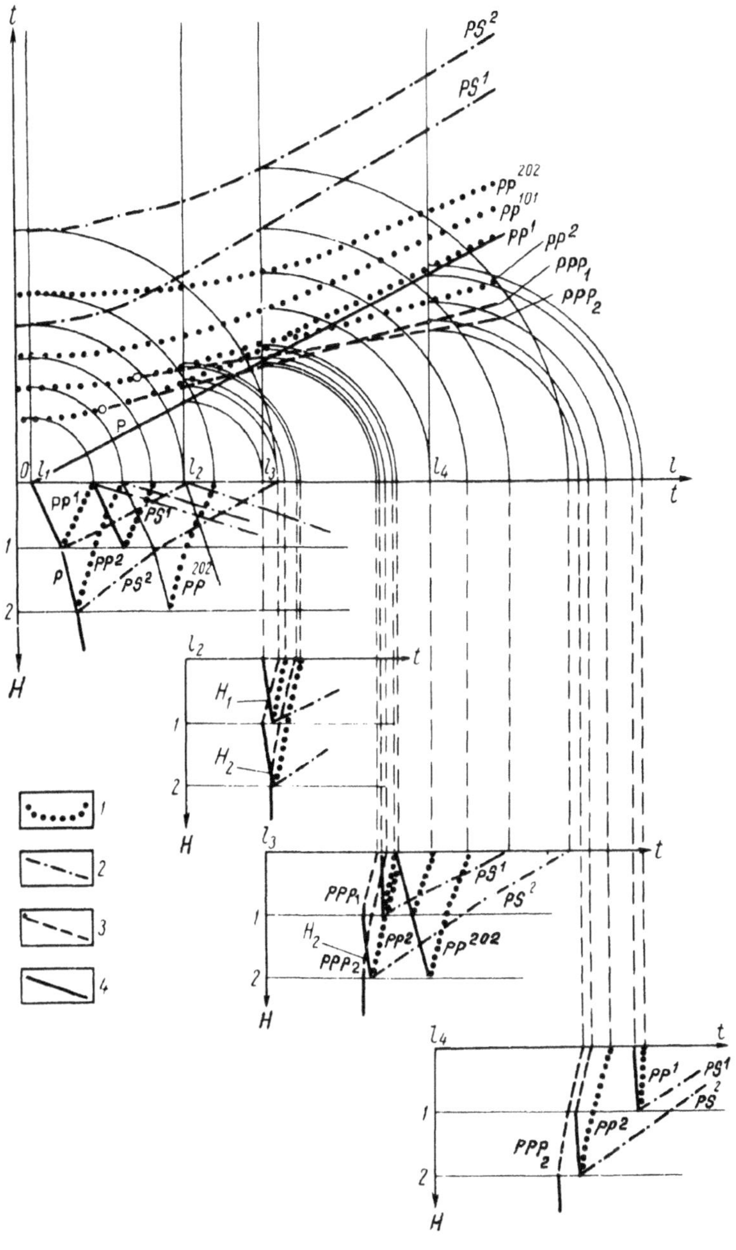

Fig. 26. Theoretical combined horizontal-vertical traveltime curves of various wave types: (1) waves propagating to the surface – the longitudinal reflected and the multiples; (2) same as (1) but for converted reflected waves; (3) same as (1), but for head waves; (4) waves propagating into the Earth. The horizontal and the vertical traveltime curves of the same waves are connected by arcs.

where t_H is the arrival time of the wave reflected in the borehole at the depth H; p is the ray parameter.

It is worth noting that these two equations may be used to determine the values d_j, v_j of the first layer not penetrated by the borehole that is below total depth.

Combined Horizontal and Vertical Traveltime Curves

Vertical seismic profiling is usually combined with surface observations. In such cases the horizontal and the vertical traveltime curves should be considered jointly [15]. Such consideration, in addition to providing a visualizable picture, facilitates a continuous transition from observations along a horizontal profile to those along a vertical profile.

Continuous tracking of waves along horizontal and vertical profiles shot at different values of l enables the nature of the waves recorded on the horizontal profile to be determined.

Consider the computed combined traveltime curves of the first and the subsequent waves in a medium with two interfaces (Figure 26) for four values of l. The definition and the notation of the waves is presented in Table III.

TABLE III

Wave type	Designation	
	First interface	Second interface
Reflected longitudinal	PP^1	PP^2
Reflected converted	PS^1	PS^2
Reflected multiple	PP^{101}	PP^{202}
Head longitudinal	PPP_1	PPP_2

On the vertical profile the first arrivals include: at the distance l_1 the direct transmitted wave, at the distance l_2 the conversion from the direct to the head wave propagating from the first interface at the distance l_3, the conversion from the direct to the head wave propagating from the second interface and at the distance l_4, the head wave propagating along the entire vertical profile. The traveltime curves of reflected and multiple longitudinal and converted waves, as well as of transmitted waves are shown as subsequent arrivals. The traveltime curves are discussed further in [14].

In combined horizontal and vertical observations, waves recorded on the surface can be tracked continuously along the vertical profile down to the interface with which they are associated.

An example of observed combined traveltime curves obtained in the process of the study of a sedimentary section is shown in Figure 68. The refraction from the shotpoints 600 m and 1000 m will be seen from the vertical traveltime curves to take place immediately below the Paleozoic carbonate deposits between 120 and 130 m. The wave recorded from the shotpoint at 1500 m has been refracted in the lower Carboniferous carbonate bed at 640 m and can be observed up to a distance of 4700 m. The refracted waves are recorded in turn along the horizontal profile: in the carbonate deposits of the upper part of the Paleozoic (up to a distance of 3 to 5 km), in the Lower Carboniferous carbonate bed (up to 8 to 5 km) and in the crystalline basement.

PART TWO

SEISMIC WAVES IN THE REAL EARTH

THE DIRECT WAVE

The direct wave formed in the vicinity of the energy source exerts a strong influence on the shape of each particular wave and on the structure of the entire seismogram. This, specifically, explains the strong dependence of the seismic record on the conditions of source excitation. The study of the conditions of formation of the direct wave and its propagation in the real earth is one of the most fundamental issues that to a great extent determines the potentialities of seismic studies. As the problems become more complex, and the methods of seismic exploration are improved, the direct wave, its shape, constancy and the means for regulating and controlling it, especially in methods based on the grouping of explosions from different shot points, specifically, the common depth point method, have to satisfy progressively more stringent requirements. This is primarily the outcome of the development in recent years of methods of interpreting the dynamic features of waves for the purpose of studying the physical properties of rock, their absorbtion and reflection characteristics and estimation of the geologic section. Moreover, the wavelet of the direct wave observed in VSP is being used to compute the filter operator for the deconvolution of surface seismic observations. The choice of the optimum observation procedure is of great importance for the study of the direct wave.

1. Procedure for Studying the Direct Wave

The direct wave has traditionally been studied with the aid of surface observations. However, a great volume of VSP work demonstrates that in many cases the effectiveness of such an approach is low, and it holds out little promise. The prime reason for this is that the direct wave observed on the surface may differ from the one recorded down-hole and which generates the entire wave field. The main causes of such differences are: (a) heaving phenomena associated with a free surface; (b) inhomogeneities in the upper portion of the geologic section, which exert a strong and unequal influence on the shape of the direct wave observed on the surface and down-hole; (c) the directivity of the source.

Ground heave phenomena

In the case of shallow explosions and surface observations close to the shothole where the direct wave is recorded first, its shape may vary substantially as a function of the distance to the shot hole and the charge. As the charge size is increased, the comparatively simple shape of the direct wave becomes more complex, and repetitive pulses with increasing arrival times appear in the tail portion. The visible shapes of the first arrival and the repetitive pulses are similar. The spectra of the repetitive pulses are narrower and are displaced towards the lower frequencies. At first, such pulses recorded in surface observations close to the source, were interpreted as repeated shocks from explosions in the shot holes, and they were even supposed to affect the shape of reflected waves. Subsequently, it has been demonstrated that these repetitive pulses are the result of the transformation by geophones of the dome-shaped displacement of the free surface and the low-velocity zone adjoining it. In the presence of porous soils, direct waves may be distorted by pulses associated with heaving phenomena even with explosions of comparatively small charges.

In the immediate proximity of the shot-hole, the shape of the direct wave may also be affected by the disruption of the continuity of the medium as the result of drilling.

Distortion of the direct wave becomes smaller when the observation points are somewhat removed from the shot-hole.

Inhomogeneities of the upper portion of the geologic section

Various types of waves are formed at the heterogeneities in the near-surface that may interfere with and substantially affect the shape of the direct wave. The greatest effect is that of the multiple reflected waves. The shape of the direct wave is influenced by superposition of the multiples. It should be pointed out however, that the superposition of multiples associated with the near-surface forms a stable wave pattern. An intricate shape of the direct wave may persist along the whole vertical profile. The presence of interfaces below the shot-point may result in enhancement of the direct wave observed on the surface and in energy losses caused by reflection in the case of the direct wave propagating downwards. And *vice-versa*, the presence of interfaces above the shotpoint may result in the attennuation of the direct wave observed on the surface, and in changes in the shape and intensity of the downward wave caused by ghost reflections. In the case of surface observations, interference is also possible with waves associated with the low-velocity zone, and with surface waves excited by Lamb waves propagating in the mud column filling the borehole as well as with transverse waves. The filtering properties of the near-surface may also affect the shape of the direct wave. This is primarily true for the high-frequency components of oscillation.

Source directivity

In conjunction with the development of PM–VSP, the source directivity becomes an important factor in addition to the traditional parameters and has not been adequately studied seismically. On the basis of numerous experimental data, a normal explosion in a borehole may be said to exhibit a natural directivity due chiefly to the elastic anisotropy of the rock in the vicinity of the explosion. This is confirmed by the fact that for normal symmetrical explosions used in seismic exploration, both longitudinal and transverse waves are excited simultaneously (see Chapter 3).

The factors discussed above, as well as others and in particular those associated with non-linear processes in the vicinity of the explosion, make it difficult and in certain cases even unadvisable to study the direct wave by means of surface observations. This concerns, first and foremost, the study of the conditions of excitation and propagation of the waves in the real earth.

For downhole observations, we shall understand the term direct wave to mean the first downward wave. In a layered medium, this wave experiences refraction at the interfaces and is, in effect, a transmitted wave. VSP enables the influence of source excitation conditions and the mechanism of their influence to be studied, as well as the depth zone in which the direct wave pulse is completely formed, to be determined. The shape of the direct wave can be determined after it has stabilised by observation at points far removed from the interfaces. And, finally, by continuous tracking of the direct wave along a vertical profile, it is possible to study the conditions of its propagation in the real earth. The latter point is extremely important, since it enables information about the absorption and the reflection properties of the geology to be obtained. The wave characteristics observed on surface seismograms naturally depend on the shape and the intensity of the direct wave impinging on reflecting interfaces. The information value of the direct wave in VSP can be said to be much higher that that of any wave in surface observations.

The direct wave has been studied with differing amounts of detail in all regions where wave fields have been studied by VSP observations. In some regions, special experiments have been carried out to this end. When describing the influence of the near-surface, we shall deal with the total downward wave train following the first arrival which is associated with the near-surface and which exercises a considerable influence on the wave field as a whole.

When studying the direct wave, it is first of all necessary to guarantee a reliable coupling between the geophone and the borehole wall for which the natural frequencies of the geophone-wall system lie far outside the operating frequency range of the equipment.Experience shows that in the study of pulse spectra an upper limit of 150 Hz in recording bandwidth may prove inadequate in certain cases. This, in particular, is true of observations in high-velocity crystalline rock, which has a comparatively low absorption rate, and which is frequently encountered in seismic exploration for minerals. Here, the pulse may contain high-frequency components even at depths of 2 to 3 km. In such cases, it is advisable to use a band width from 5 to 500 Hz. Of special importance is the dynamic range of the equipment, which should allow the recording of the direct wave free from any non-linear distortion both near the source and at great distances from it. The stability of the instrumentation should be subject to rigorous control.

Controlling the Direct Wave

The strong influence that the shape of the direct wave exerts on the total wave field makes its control, and hence the methods to effect it, especially important. The choice of the excitation conditions for the direct wave and monitoring its reproducibility can be executed only by means of down-hole observations at depths where the direct wave has already become stable. Accordingly, to position control devices correctly, one has to assess the depth at which the direct wave has stabilized.

Studies of the stabilization process of the direct wave [89] enable quantitative relationships to be established between the parameters that determine the recording conditions and the structure of the near-surface. In particular, it has been demonstrated that the ratio (L_1/L_2), of the geometric spreading of the ghost reflection from the bottom of the low-velocity zone and of the direct wave from the source is solely a function of the charge depth below the low-velocity zone. This ratio for the ghost reflection from the surface is not only a function of the position of the source, but also of the near-surface parameters. At each detector position, it depends on the zone thickness, the depth of the source below the low-velocity zone, and the velocity change at the base of the zone. If the value $L_1/L_2 = 1.5$ is considered acceptable, then for a single-layered low-velocity zone

$$\Delta h_{\min} = 4\left(Z + \frac{h_1 v_1}{v_Z}\right)$$

where Z is the depth of the source below the base of the low-velocity zone, Δh is the depth of the detector relative to the source, h_1 and v_1 are the thickness and velocity of the low-velocity zone, respectively, v_Z is the velocity at the shot-point. Since in practice $Z \gg (h_1 v_1)/v_Z$, one may put approximately $\Delta h_{\min} = Z$, i.e., the smaller the low-velocity zone and the greater the velocity change across it, the smaller will be $\Delta h_{\min}$ for every fixed value of the source depth.

The control of the direct wave is of special importance in the study of absorption parameters (see Chapter 11).

In industrial practice, the choice of the shape of the direct wave is naturally determined by the objectives of the study, and by the seismic and geologic characteristics of the sub-surface, so that the depth at which the control devices are located should comply with these conditions. As seismic exploration develops based on the extensive use of the wave dynamics, the direct wave will be regarded as one of the factors initiating development of the wave process. This is already being done, specifically, in the study of absorption (see Chapter 11).

Specific features in the correlation of the direct wave

On an in-line vertical profile, the direct wave is usually recorded as a first arrival along the entire profile. Only in comparatively rare cases, that will not be dealt with here, refracted waves can be recorded as first arrivals in individual sections of the vertical profile [3]. The correlation of the first arrivals on the seismogram is the most reliable. This is because the background of ambient noise, not associated with the explosion, is very low in the borehole, and one can reliably determine from the seismogram the instant of arrival of the wave front, i.e., the instant the state of rest of the earth particles changes to that of motion. Many papers deal with the study of the arrival time of direct waves in VSP, and much experience has been gained. We note that in VSP the geophones are clamped and unclamped to the borehole wall from the surface, thus enabling clearer records of the first arrivals and hence more reliable information about the velocities to be obtained.

Of special interest in VSP is the study of not only the arrival times, but also of the shape of the direct wave, which initiates the whole wave process and carries valuable information about the Earth. The correlation of the direct wave depends greatly on the choice and constancy of the source of excitation condition. In the case of constant source excitation conditions, the shape of the direct wave usually changes smoothly, in spite of its complexity resulting from the superposition of other waves. No difficulties are normally experienced in executing phase correlation of such a wave. In cases where the direct wave has a simple shape, the traveltime curves of different phases of the signal are practically parallel to that of the first arrivals, and only at high values of l does pulse broadening sometimes become noticeable.

On *off-line* profiles, depending on the velocity distribution and the magnitude of l, the direct wave may be recorded as a later arrival over the entire vertical profile, or in individual sections, as well as possess a negative apparent velocity. The direction of arrival of the direct wave varies as l is increased. In individual sections, it approaches the horizontal. The apparent velocity of the wave increases, but the relative intensity of the vertical component drops, whereas for waves of other types (reflected, refracted) it may rise. This makes it difficult to discriminate the direct wave on the Z-component seismogram. Sometimes, the result may be that the first arrivals of the direct wave become barely detectable on the record with a much stronger refracted wave standing out against their background (Figure 27).

The conditions for correlation of the direct wave as later arrivals, depend on numerous factors but primarily on the intensities of the refracted and the transcritically reflected waves against whose background it is recorded. To a large extent the former is determined by the law of velocity variation in the underlying layer.

In the presence of an *appreciable velocity gradient* in the layer beneath the interface, the intensity of the refracted wave is high, and this makes correlation of the direct wave diffult.

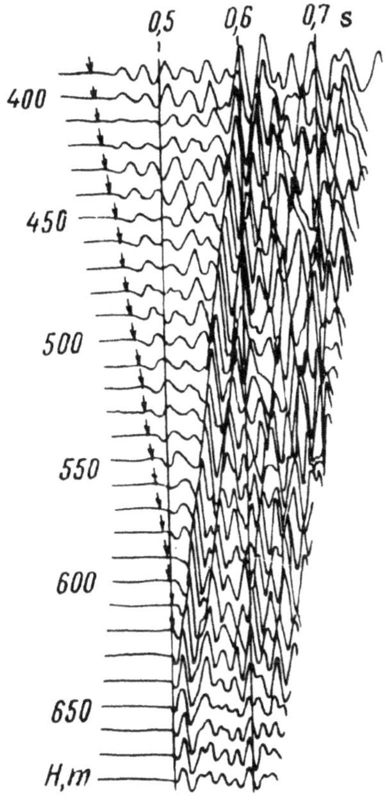

Fig. 27. Effect of the direct wave's direction of arrival on the intensity of its vertical component (shotpoint 1500 m, borehole Kletsko–Pochtovaya).

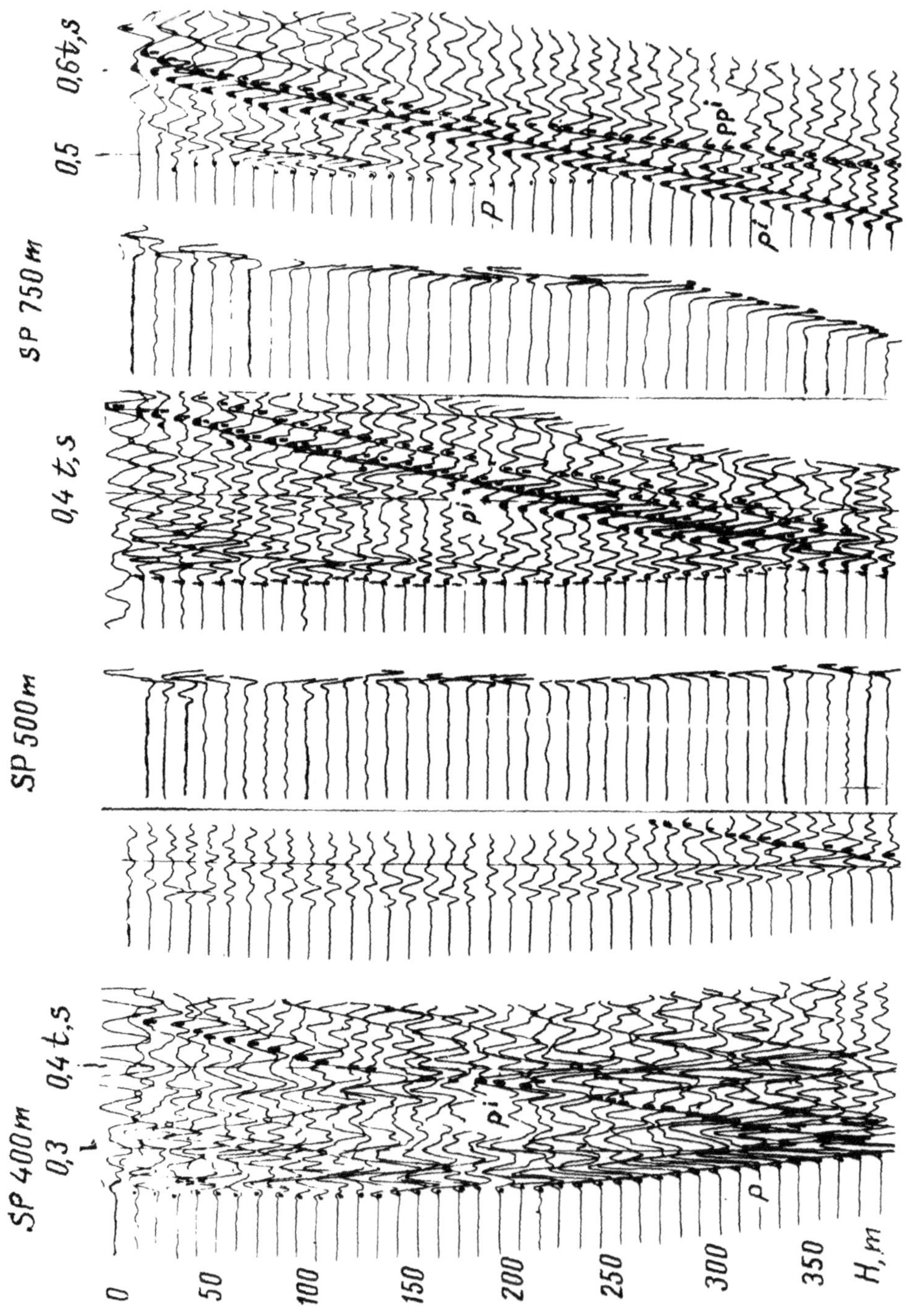
SP 400m
0,3
0,4 t,s
SP 500m
0,4 t,s
SP 750m
0,5
0,6 t,s
P
Pi
PPi
0
50
100
150
200
250
300
350
H,m

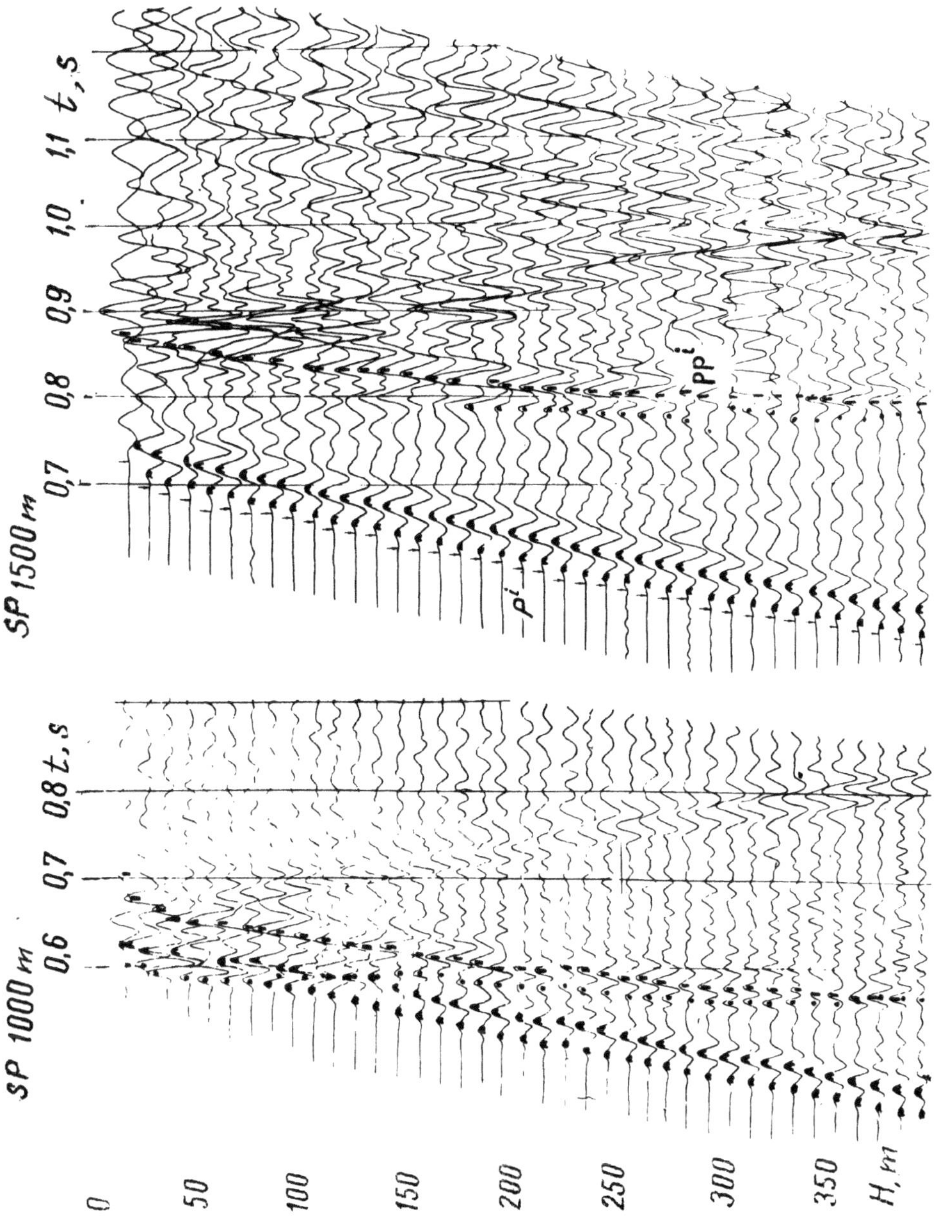

Fig. 28. Seismograms obtained from various shotpoints (borehole N90–a Kudinovskaya).

Consider the conditions for correlation of the direct wave along a vertical profile shot from five source points (Figure 28), for l = 400 m. On the seismogram recorded with high-sensitivities (left) the arrivals polarized upwards corresponding to a wave coming from above, having increasing amplitudes with depth are visible, this being the result of the direction of arrival of the wave drawing closer to the vertical. At low amplification (right), the recorded shape of the direct wave and its distorion due to interference with the reflected wave between 340 and 380 m are visible. For l = 500 m, the direct wave is replaced as a first arrival at the depth of 330 m and below by a wave refracted at the top of the Paleozoic. This is evident from the change in the polarity of arrivals and in the sign of the apparent velocities, especially on the high amplification channels (left). In the section of the profile contiguous with the interface, the direct wave is in the zone of interference with the refracted and the transcritically reflected waves, and cannot be discriminated. With a further increase in l, the refracted wave is recorded as a first arrival over a greater dept interval outside the zone of interference. In such conditions, the direct wave can sometimes be discriminated in subsequent arrivals between the refracted and the transcritically reflected wave. Thus, for l = 750 m, it is possible to track the direct wave after it interferes with the transcritically reflected wave.

From the remote shot-points (e.g., l = 1500 m), the refracted wave is recorded along the entire profile as a first arrival. However, the conditions for correlation of the direct wave in the subsequent portion of the Z-component record deteriorate because of its

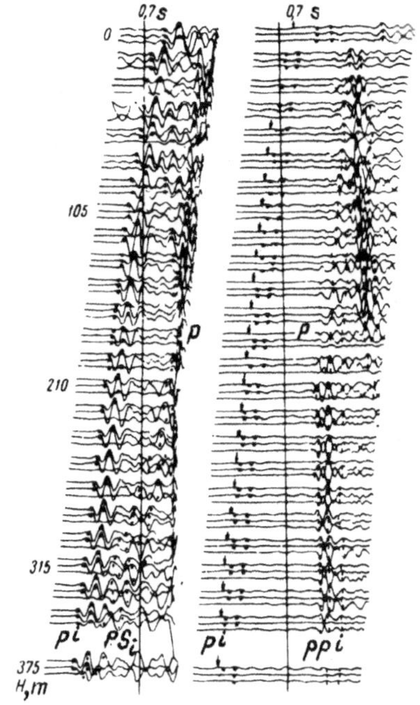

Fig. 29. Seismogram of a three-component symmetrical cluster (shotpoint 1500 m, borehole N90–a Kudinovskaya).

interference with the transcritically reflected wave and because of an adverse direction of arrival. The true amplitude ratio of the waves can only be calculated with the aid of three-component geophone records (Figure 29) – the intensities of the direct and the transcritically reflected wave are so much higher than that of the first refracted wave that one level of amplification is not enough to make joint tracking possible.

In the *absence of a discernible velocity gradient* in the *underlying layer*, or if its thickness is very small in comparison with the wave length, the intensity of the head wave drops drastically. In this case the direct wave can be reliably tracked in the subsequent arrivals down to the interface.

2. The Influence of Excitation Conditions on the Direct Wave

The direct wave pulse-form is determined in the vicinity of the source, and depends on the type of source itself (weight, shape, distribution of charge), on the structure of the earth in the region of the source (homogeneity, isotropy, lithology etc.) and on the point of observation. Numerous papers [90, 91, 92], both of a theoretical and experimental nature, deal with studies of the pulse shape as a function of the source. The result has been a conventional subdivision of the area in the vicinity of the source into three zones: the zone of the shock waves, the zone of plastic deformation, and the zone of elastic deformation. The separations at which elastic waves begin to appear, i.e., the limits of the zone of elastic deformation, are different for different media. A qualitative appraisal of this distance for clay-type earth materials can be made with the aid of the empirical formula $r = 0.13 \sqrt[3]{Q}$, where r is the radius of the sphere of deformation in metres, and Q is the charge weight in grams. As the charge size varies from several grams to several tens of kilograms, the radius of the sphere of inelastic deformation varies from fractions of a metre to the first units of metres. The zone of major interest for seismic studies is the zone of elastic deformation. In this zone the pulse shape changes with distance from the source, becoming stable only at a certain distance from it. According to experimental data, in a homogeneous medium this distance is approximately 0.75 to 1.0 wave length. The shape of the wave at the zone's boundary has been studied very little, but there has been considerably more investigation at distances of several wave lengths from the source. The results of analysis of the amplitudes and spectra of the direct wave, as functions of various explosion parameters obtained in homogeneous media cannot always be used in seismic exploration practice, since the latter employs explosions fired in the near-surface conspicuous for its great heterogeneity and complexity of structure. The results obtained in seismic exploration do not always agree with the theory.

A large amount of VSP work shows that the shape of the direct wave changes comparatively little after stabilising at a certain depth, the changes taking place mainly in sections adjoining the interfaces and in the result of superposition of subsequent waves. Here the shape of the combined wave is distorted. The simple short pulse expands, and new extrema appear that form separate waves as the distance from the interface increases. Subsequently the direct wave pulse recovers its simple shape.

THE EFFECTS OF CHARGE SIZE

VSP observations have established that the pulse shape varies regularly with the charge size (the period and duration of the pulse increased with increasing charge weight), which

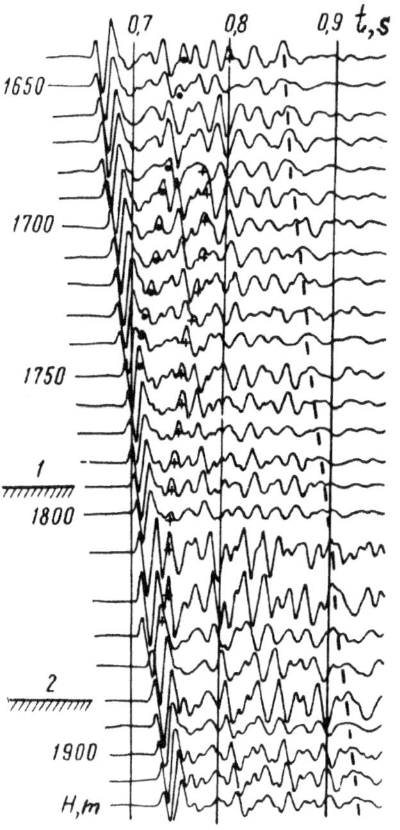

Fig. 30. Distortion of the shape of the direct wave in the process of formation of reflected waves (shotpoint 100 m, borehole N162, North Stavropol).

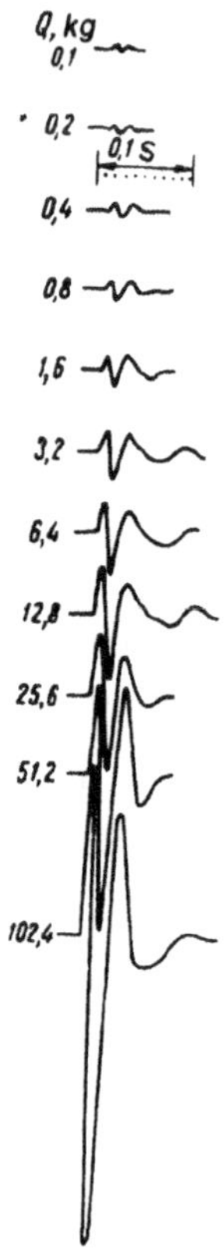

Fig. 31. Variation of the direct wave pulse following the variation of the charge (H = 1000 m, h = 20–40 m, band-pass 5–130 Hz, borehole N17 Kuschevskaya, courtesy E. Ya. Kutzenko).

was generally known before. VSP enables the relationship to be studied quantitatively and most rigorously. However, from the point of view of the character of the seismogram the dependence of the intensity of the first direct pulse on charge size is less interesting than the shape of the first pulse and the whole initial portion of the seismogram. Consider two examples (Figure 31, 32). In the former case, the absence of any major interfaces in the upper portion of the section permits a short simple pulse of the direct wave to be transmitted (Figure 31). When the charge size changes by three orders of magnitude, the shape of the first arrival changes comparatively little, the amplitude increases by about two orders of magnitude, and the period by a factor of two. In the second example, the heterogeneities in the upper portion of the section produce a somewhat more complex train of downward waves in the early portion of the seismogram (Figure 3(a)), but a hundred-fold increase in the charge size changes the pulse shape very little.

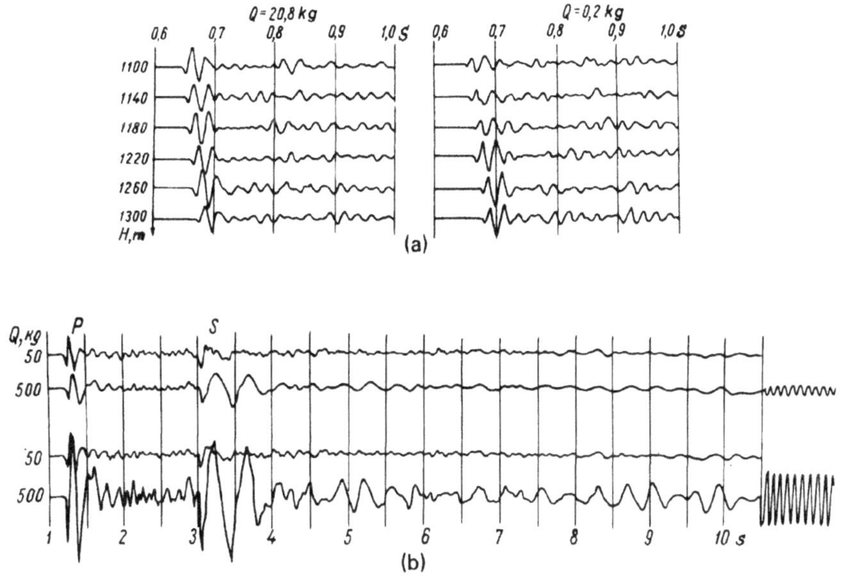

Fig. 32. Changes in the initial portion of the record following charge reduction: (a) borehole N1 Oktyabrskaya, Byelorussian SSR, courtesy S. V. Klushin; (b) variation of relative intensity of *P*- and *S*-waves following charge increase. *Z*-component record $H = 2000$ m, shotpoint 2800 m (borehole N10, Alma-Ata), $f_0 = 1$ Hz. Top-records of *P*-waves with normalised amplitudes, bottom-constant-gain records.

It cannot be assumed that source excitation conditions remain unaffected by a change in charge size. For a constant depth of charge, the conditions may change on account of changes in the shape of the charge, in the manner it has been buried in the hole, upheavals following the explosion, and changes in the frequency spectrum of the pulse at the source. However, a change in the charge has no decisive effect on the shape of the direct wave and the downward wave train when observed at great depths. At the same time, a change in charge size can considerably affect the character of seismograms observed on the surface. As the charge is increased, the amount of energy remaining in the upper portion of the geologic section increases out of all proportion, resulting in an increase in the number and the intensity of unwanted waves of various types that complicate the wave field, and distort the correlation of deep waves. This explains to a large extent the effectiveness of using single small charges with synchronous vertical stacking.

THE EFFECTS OF HETEROGENEITIES OF THE MEDIUM IN THE VICINITY OF THE EXPLOSION

In seismic exploration explosions are carried out in the most heterogeneous portion of the near-surface. Such heterogeneities as a rule transform the comparatively simple pulse formed by the source into a more or less complex train of downward waves. The mechanism for such transformation involves the interference of the primary pulse with various oscillations generated by the shallow heterogeneities.

It should be particularly emphasized that these heterogeneities can be the result of not just lithology but differences in the water content and the corresponding differences in acoustic rigidities. The presence of water-saturated layers results in the formation of powerful reflecting horizons from which high-intensity multiples are formed.

In the case of *explosions at constant depths*, the direct wave in the vicinity of the source has a comparatively simple shape; however, with increasing distance from the source additional extrema appear, and a redistribution of energy takes place among the phases. By degrees, one or even several pulses may arrive in a complex wave train. After that the shape of the downward wave changes little and ultimately ceases to change, so that it may be used in the studies of the dynamics of the wave field as a whole.

Theoretical and experimental studies carried out for certain typical models of the upper portion of the geologic section throw light on the nature of the variation, and the rapidity of the shape stabilization of the direct wave with depth in the case of normal incidence [89]. They demonstrate that the increased complexity of the direct wave is the result of the variation in the conditions of superposition of ghosts, which is determined primarily by the relative difference in the amount of geometric spreading of the direct pulse and the ghost. As the distance from the source is increased, the geometric spreading of these waves becomes equal, and as a result the shape of the downward wave experiences no more change with any additional increase in depth.

Changing the depth of the explosion, not only alters the properties of the medium surrounding the charge, but also changes the position of the charge with respect to the heterogeneities, and this in turn produces changes in the shape of the downward wave. This can be observed not only in the upper portion of the section, but at all depths. Theoretical studies of the influence of the upper portion of the section on the spectral composition of the direct wave, have demonstrated that an explosion beneath the low-velocity zone represented by a thin layer, attributes an interference characteristic to the direct wave. By bringing the explosion closer to the base of the low-velocity zone this effect is reduced, but cannot always be completely eliminated, since the direct wave propagating downwards is affected both by the wave reflected from the upper surface of the layer and by the waves formed inside it. Accordingly, any variation in the structure of the low-velocity zone may result in unstable source excitation conditions. In the case of comparatively thick low-velocity zones, the wave form may be a complex one. It usually becomes simpler as the low-velocity zone thins, provided the distance from the shot-point to the base of the layer remains short. The spectral characteristics of the upper portion of the section may strongly influence the spectral composition of the direct wave, which may determine the effectiveness of the studies.

Let us illustrate the effect of the excitation conditions and, primarily, the depth of the explosion on the shape of the direct wave in the case of a comparatively simple single-layer low-velocity zone. Figure 33(a) depicts the record of the direct wave obtained in a borehole at a depth of 1000 m with equipment having a band-width of 1 to 130 Hz. The amplitude of the record with the explosion fired at a depth of $h = 5$ m is very low. For charge depths within the low-velocity zone (10, 15 m), the record displays three extrema with a slight distortion in the tail portion of the record. When the charge is close to the bottom of the low-velocity zone ($h = 20$ m), a direct wave with a comparatively simple shape is obtained. As the charge depth gets below the low-velocity zone, the shape of the direct wave becomes strikingly more complex, the number

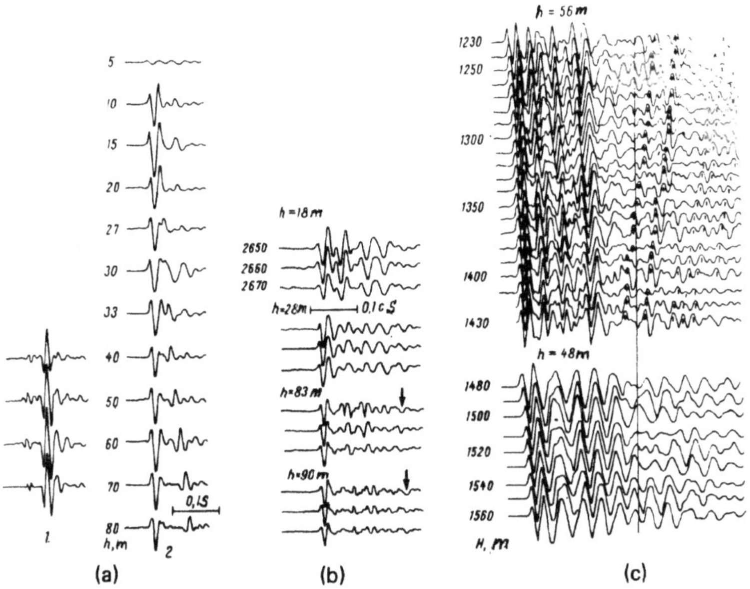

Fig. 33. Variation of the shape of the direct wave with h: (a) in conditions of a low-velocity zone with a simple structure for $H = 0$ (1) and $H = 1000$ m (2) ($l = 400$ m, $Q = 0.4$ kg, borehole 42 Staro-Minskaya); (b) in conditions of a multi-layered low-velocity zone ($l = 100$ m, $Q = 0.8$ kg, borehole N162 North Stavropol'); (c) in case of explosions under and inside the low-velocity zone ($l = 100$ m, $Q = 0.8$ kg, borehole N162 North Stavropol').

of phases grow, the record assumes an interference pattern, and the ghost separates out. The direct wave has the most complex form at $h = 33$ m, but already at $h = 40$ m the process of ghost formation has been alsmost completed, and with the further increase in depth it lags progressively behind the direct wave, which becomes a short simple wave varying little with depth. At $h = 80$ m, the shape of the ghost is almost identical with that of the direct wave, but has a phase shift of 180°, this being the result of reflection from a layer of lower acoustic rigidity. The amplitude of the ghost amounts to 0.6 to 0.7 times the amplitude of the direct wave.

In the case of a low-velocity zone with a more complex structure, the number of ghosts may be substantially higher, and consequently this interference may be more complex. For instance, in Figure 33b the records for charge depths of 18 and 28 m have distinct interference patterns and display great complexity and long duration of signal. Only for charge depths of 83 m and 90 m is the direct wave recorded in the form of a simple short pulse, although followed by a series of ghosts. The last weak low-frequency oscillation (marked by an arrow) is a reflection from the surface.

Sometimes a drastic change in the shape of the downward waves may take place with a comparatively small change in the explosion depth. This happens most frequently when the shots are fired close to sharp interfaces. Explosions in the immediate vicinity of an interface may generate downward waves of different shapes on either side of it. It will be seen from the individual sections of a vertical profile (Figure 33c), recorded with charges fired at different depths that there is an appreciable rise in the frequency and resolution of the record, when the charge is below the low-velocity layer (h = 56 m) compared with the charge in the low velocity layer (h = 48 m).

The spectrum of the downward wave, when there is interference between the direct signal and a ghost, can be regarded as the result of the passage of the pulse through a filter corresponding to a thin layer assuming the ghost only changes amplitude and polarity on reflection. In such circumstances consistent source excitation conditions must be maintained for the shape of the reflected waves to remain unchanged. If the

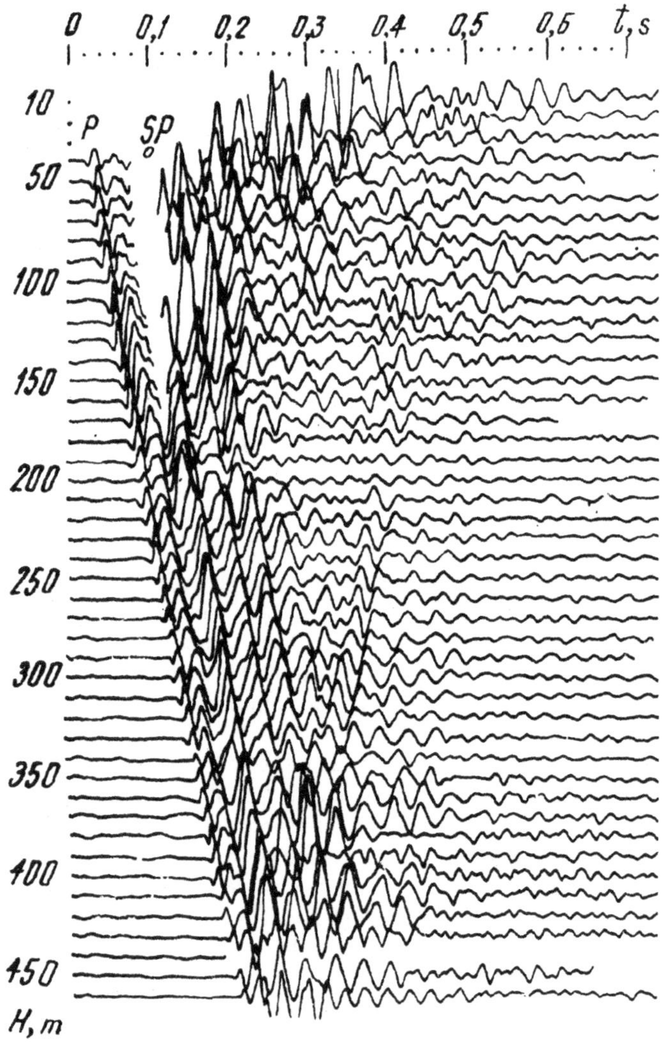

Fig. 34. Example of registration in the initial portion of a VSP seismogram of a converted reflected S_0P-wave [93].

reflection coefficient from the base of the low-velocity zone is presumed constant along the profile, consistent duplication of the shape of the downward wave can be achieved by ensuring that the interference of the direct wave with the ghost is the same. To do this, it is necessary to maintain a constant time lag between them, and to this end one should study the elevation difference of the reflecting interfaces above the explosion in detail and avoid changing the position of the charge with respect to them.

For different depths of the explosion, the shape of the direct wave can also change as the rock properties (lithology) of the medium surrounding the charge which mainly determine the pulse spectrum, alter. Explosions in hard rock create higher frequency spectra than explosions in unconsolidated rock. Shots fired in high velocity materials are accompanied by an increase in the dominant frequency of the pulse and by a corresponding decrease in its duration. The seismograms depicted in Figure 33c serve to illustrate the dependence of the spectral content of the record on the rock properties in the vicinity of the explosion.

In addition to longitudinal waves formed at the heterogeneities of the near-surface, converted P_0S- and S_0P-waves are also formed. Such waves are recorded with comparatively short time lags with respect to the initial P- and S-waves and their spectral content is similar to that of the initial waves: they form parallel alignments on the seismograms [93].

The S_0P-wave is recorded after the longitudinal wave. It may interfere with the latter forming a multiphase wavelet whose initial high-frequency phases correspond to the longitudinal wave, and whose subsequent lower-frequency phases correspond to the converted wave (Figure 34).

In the same way, the P_0S-wave is recorded with the velocity of the S-waves, but arrives somewhat in advance of the direct S-wave. It is distinguished by a higher-frequency spectrum, and may interfere with the direct S-wave forming a complex wave train with a high-frequency front end. In such conditions, the shape and the spectral composition of the direct waves greatly depend on the structure of the upper portion of the geologic section, on the position of the source, and on its spectral characteristics. Since the heterogeneities of the shallow part of the section exercise the greatest influence on the direct wave and the entire train of downward waves in the real earth, which as a rule is inhomogeneous, it is difficult to study the pulse at its origin, and we are compelled to regard the explosion and the earth together as a complex source. The existing strong dependence of the shape of the direct wave on excitation conditions enables it to be used in down-hole observations for rigorous control of the constancy of source conditions. Naturally, the change in the shape of the direct wave cannot fail to affect the shape of the reflected waves.

3. Studying the Shape of the Direct Wave

A seismic pulse travelling in a homogeneous and not completely elastic medium changes its intensity and shape. Its diminishes chiefly because of geometric spreading associated with the broadening of the wave front. However, such a reduction in intensity does not produce a change in the pulse's shape. The variations in the shape are associated with changes in both the amplitude and phase spectra with the distance. The variation of the amplitude is determined by the spectrum of the coefficient of attenuation. The variation of phase is the result of dispersion.

The real earth is not only incompletely elastic, but is also as a rule thinly layered and hence possesses filtering properties. A pulse travelling through a layered medium changes its shape, such changes taking place mainly in sections of the profile adjoining the interfaces (Figure 30). Outside any region of interference with subsequent waves, the direct wave again assumes a simple shape. The formation of direct waves in thinly layered media for the case of normal incidence has been studied theoretically, and it has been established that the direct wave loses its energy through reflection at the interfaces. However, the decisive effect is that exercised by other downward waves, which 'feed' it in the form of simple and multiple ghosts, returning to it a fraction of the scattered energy and thus reducing its attenuation. Accordingly, the theoretical treatement of direct waves in thinly layered media, even layers between 0.2 and 0.3 λ thick, should be considered.

In the example cited (Figure 30), at a charge depth of 90 m, it was possible to separate the ghosts from the direct wave, and the latter propagated deep into the Earth in the form of a simple short pulse. The shape of the pulse is well conserved. Only in individual sections in the vicinity of the interfaces at 1730 m and 1850 m, is the shape of the direct waves distorted by interference with the reflected waves. An interference pattern becomes noticeable only at some distance (40 to 60 m) from the interface. Having passed through the interface, the pulse recovers its shape, and because of that, even at great depths, it is often possible to obtain the direct wave in the shape of a simple short pulse.

Sometimes, despite a comparatively simple wave shape close to the source (as measured with control instruments), the records obtained at great depths are much more complex, this being possibly associated with the specific properties of the lithology. For example, in observations in a carbonate formation containing numerous interbedded terrigeneous deposits, we have not succeded in obtaining a direct wave of simple shape, or in separating it from the ghosts, despite the changes in charge depth. In all cases the direct wave is followed by a whole series of waves that interfere with it (Figure 33). The recorded shape of the series of waves as a whole remains practically unchanged over a sufficiently wide range of depths.

Hence, the chief factors influencing the shape of a propagating pulse are: absorption and dispersion due to non-ideal elasticity, reflection and refraction phenomena at interfaces and associated wave generation and the scattering of energy at sundry small heterogeneities.

A large volume of experimental data on seismic pulses, both down-hole and surface, has been accumulated [6, 7, 94, 95, 96]. Pulses recorded at distances of several wave lengths from the source have been obtained extensively by N. Ricker [92]. He produced short pulses of a comparatively simple shape and recorded them at various amplification levels over a wide dynamic range (about four orders of magnitude) demonstrating that a pulse does not display a sharp first break. The motion begins by degrees and because of that the apparent arrival time on the seismogram depends on the sensitivity of the instrumentation. The experimental data agree well with the theoretical.

In VSP work, simple pulses similar in shape to Ricker's are also sometimes observed. However, in the majority of cases the pulses are rather long and complex. Despite the large amount of work performed, data on the first arrivals so far obtained are inadequate to establish general rules. The upper portion of the geologic section is as a rule so inhomogeneous, and the nature even of its adjacent sections so greatly different, that it is unrealistic in practice to hope to analyze the variety of all possible shapes of the direct wave, and probably there is no value in doing so. It would be much more interesting to compare the shape of the direct wave obtained in different areas,

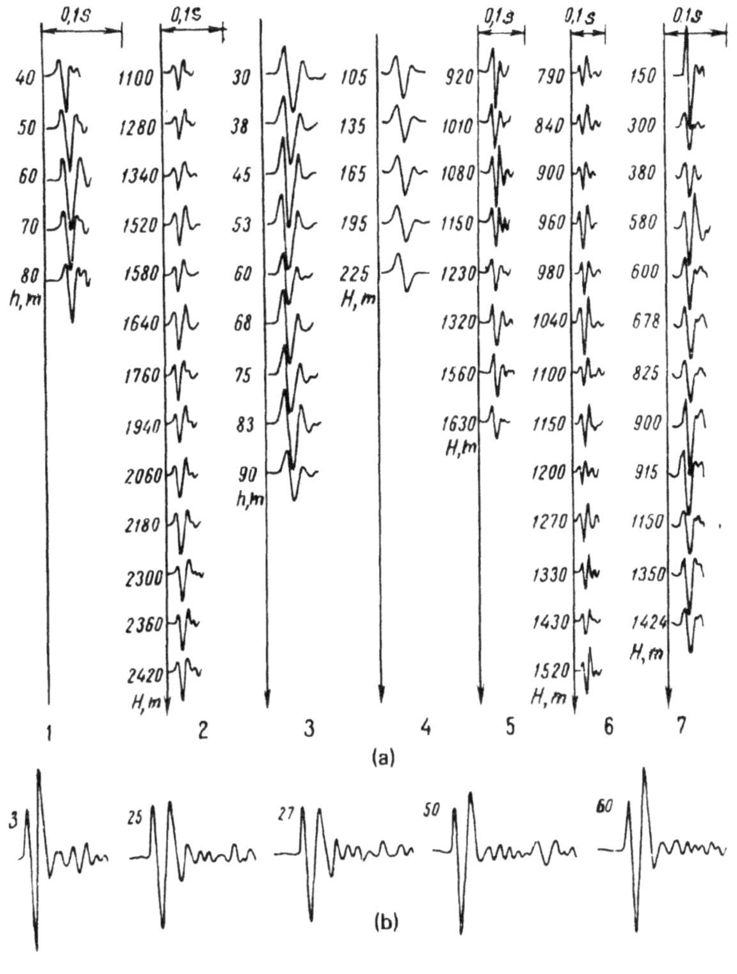

Fig. 35. Direct wave wavelet forms: (a) a comparison of direct wave wavelet forms for different recording depths (H) and charge depths (h). Wavelets 1 and 3 have been obtained at a constant recording depth H and at different charge depths. The remaining wavelets have been obtained at a constant charge depth h and at different recording depths H. 1 – H = 1000 m, borehole 42 Staro–Minskaya; 2 – h = 90 m, borehole 162 N. Stavropol'; 3 – H = 247 m Ricker's data [92]; 4 – McDonal's data [96]; 5, 6, 7 – data of Levin and Lynn [7]; (b) reproducibility of wavelet form in repeated shots. Shot ordifnals are shown on seismograms.

particularly when it is most simple and short. VSP experience shows that a short stable pulse, consisting of two or three extrema, can be obtained not only in different boreholes, but also in greatly differing areas, and even for charges shot in quite different lithologic conditions. Figure 35 depicts the simplest seismic pulse obtained with equipment free of distortion in the 15 to 130 Hz band in the Krasnodar and Stavropol provinces. Comparison of these pulse shapes with those obtained in other regions [7, 92, 96] shows them to be very similar. The only difference lies in the ratio of the amplitudes of the first and the third extrema. Special experiments have been carried out to evaluate this difference, in which a series of hundreds of repeated explosions of the same charge size at the same shot point were recorded. The observations demonstrate that the amplitude of the first and third extrema is not very stable, i.e, it can vary within *the same series* of recordings (Figure 35b). Analysis of the theoretical and experimental pulses shows that the extrema are not symmetrical, and that the left and the right slopes of the pulse are different. For the Berlage pulse, $F(t) = a_0 t e^{-\alpha t} \sin \omega t$, the abscissa of its extremal value can be found from the equation [97]:

$$\left(\frac{\alpha}{\omega} - \tan \gamma\right)\left(\frac{\pi}{2} - \gamma\right) = 1,$$

where γ is the displacement in degrees of the abscissa of the extremal value from the middle point. γ increases with the increase of the number in the series. If the area of the extremum is divided in two by the ordinate of the middle point, the ratio of the areas S_1/S_2 will in all cases be seen to increase as α/ω increases.

The pulses recorded in VSP work have a wide amplitude spectrum. When choosing analytical expressions to describe the pulses observed, we paid special attention to the sharpness of the pulse arrival, this determining the number of the displacement derivative that experiences discontinuity in the course of the transition from the state of rest to the state of motion. With increasing distance from the source, the arrivals become progressively less sharp. Whereas in the immediate vicinity of the source either the displacement itself, or its first derivative (the velocity) experiences discontinuity; at great distances the initial velocity of the displacement is zero, and discontinuity is

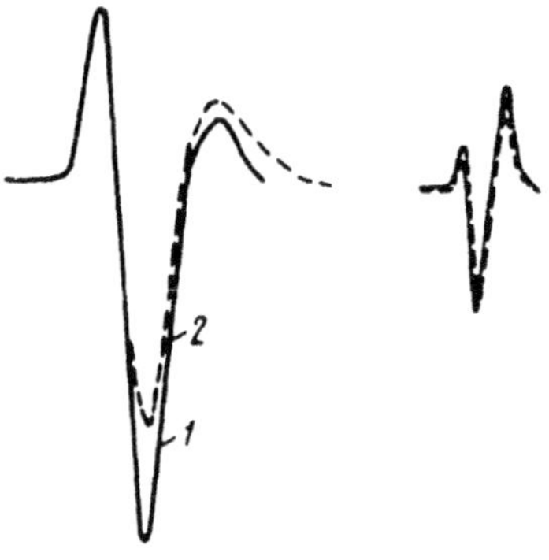

Fig. 36. Comparison between the observed (1) and the computed (2) pulses.

experienced by the acceleration or even higher displacement derivatives.The pulse obtained in borehole N 162 can well be approximated with Berlage's expression for $t \geqslant 0$ [98]. Typical for such a pulse is the discontinuity of the nth order, and the fact that it is not the first extremum, but the following, which has the maximum value. Comparison between observed pulses in various media and the theoretical shows them to be similar. For example, both pulses perfectly coincide at three extrema and at three zero crossing points (Figure 36). The shape of the pulse recorded in borehole N 42 (Figure 36, right) is most similar to that of a Ricker' pulse (Figure 36, left). All the derivatives of the theoretical Ricker' pulse are continous, and its amplitude rapidly decreases with time (because of the exponential member). Accordingly, its duration is limited in practice. A pulse of a similar shape has been considered in detal by N. N. Puzyrev [54].

When the shape of the direct wave is consistent and unchanging, this indicates that the charge depth remains unchanged also.

The structure of the direct wave

Recent developments in production seismics and studies of physical rock properties have brought about a sharp increase in the volume of work carried out for the purpose of analysing the direct wave and especially its dynamics. While development of PM–VSP enabled the polarization of oscillations to be used to gain additional independent information about the earth, to study polarization one had to effect substantial improvements in the measurement techniques and in the technology of data processing, and this in turn enabled new data on the structure of the direct wave to be obtained. It had been demonstrated already in the course of the earliest PM development work [2] that the polarization of the direct wave is as a rule not strictly linear. Polar correlation and controlled directional reception in terms of the displacement directions (CDR–I), have demonstrated that the main cause of a non-linear wave polarization is its complex structure, which is not always evident from the records of the Z-component. Analysis of a large volume of PM-VSP data obtained in recent years in various seismic conditions in both mining and oil-producing regions has enabled the structure of the direct wave, which as a rule is formed as the result of superposition of several pulses, to be studied in detail. By way of an illustration, consider data obtained in several regions with different seismic conditions.

A large volume of VSP work has been carried out in several provinces of Central Kazakhstan by KazVIRG in collaboration with IPE.

In the Nura-Taldy region the geologic section is represented by a highly metamorphosed and high-velocity formation of sandy slate of Silurian age, 400 to 600 m thick, lying on top of Permian intrusive granites. The shape of the direct waves on most seismograms recorded with a three-component symmetrical geophone cluster at three amplification levels (Figure 37a) is relatively simple. However, the shape of the pulse is seen on polar seismograms (Figure 37b) to depend greatly on the orientation of the oscillations in space. Since the direct wave is recorded against a quiet background, its tracking component coincides with the full vector (P-component) [2]. The Z-component record depends on the wave's direction of arrival. In the case of a near-vertical arrival (from nearby shot points), the Z-component record practically coincides with, or is very similar to the P-component record (seismograms 1–4). In the case

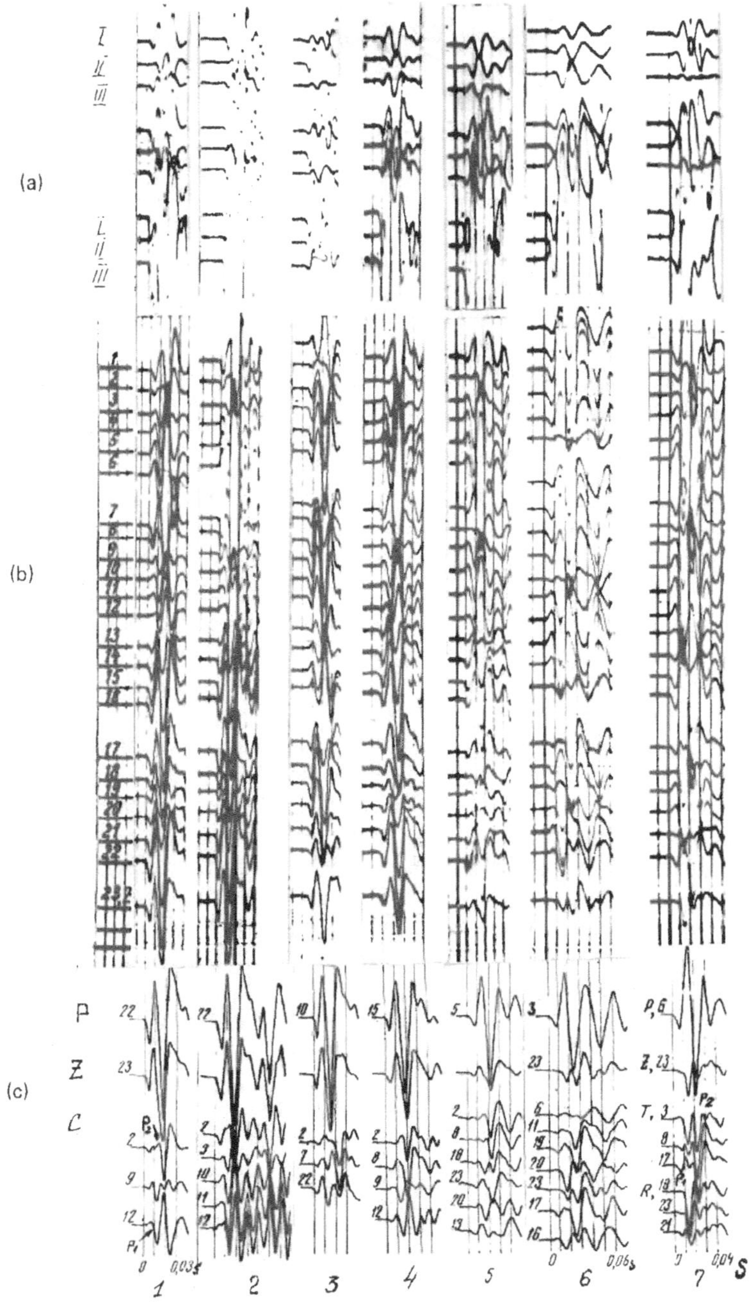

Fig. 37a-c.

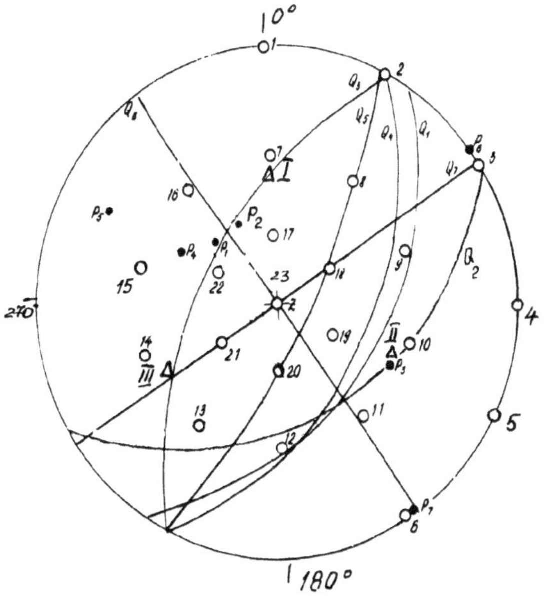

Fig. 37d.

Fig. 37a-d. PM–VSP seismograms of various components for seven direct wave wavelets (orientation of oscillation components in space shown in Figure 39d); (a) three-component seismograms obtained with a symmetrical three-component cluster I, II, III are shown on three gain levels; (b) overall polar seismograms (components 1–23); (c) planar CDR–I polar seismograms. In addition to the components located in the *P*-wave zero displacement plane, the figure also depicts seismograms of components coinciding in direction with the full vector (*P*) and the vertical (*Z*) component.

of a near-horizontal arrival (from remote shot points), the first wave is recorded weakly on the *Z*-component record or is not recorded at all (seismograms (5–7)). To facilitate comparison, the records of the *P*- and the *Z*-components on CDR–1 seismograms (Figure 37c) are shown side-by-side. The shape of the first arrival is fairly simple, and the pulse is short. The shape of the pulse recorded on different traces will be seen to be different, this being one of the main characteristics of a complex oscillation. Other pulses can be reliably discriminated on the components, which are orthogonal to the displacement direction, where the amplitudes of the first extremum vanish. Such pulses are clearly visible on the CDR–I seismograms, which represent a set of components located in the zero-displacement plane of the first arrival (the orientation of the full *P*-vector and of the CDR–I components is shown on the stereogram (Figure 37d)).

On seismogram (Figure 37c) the direction of the *P*-component is close to that of the *Z*-component and coincides with component 22 of the polar seismogram. The first extremum of the record of component 12 is practically suppressed, and the pulse P_1, whose arrival displays a time shift of 0.004 s with respect to the first arrival of the *P*-wave is recorded. The pulse P_2 with a sharp front displaced by 0.011 s with respect to P_1 is clearly visible on component 2.

A similar picture is observed on the other seismograms. Thus, the displacement

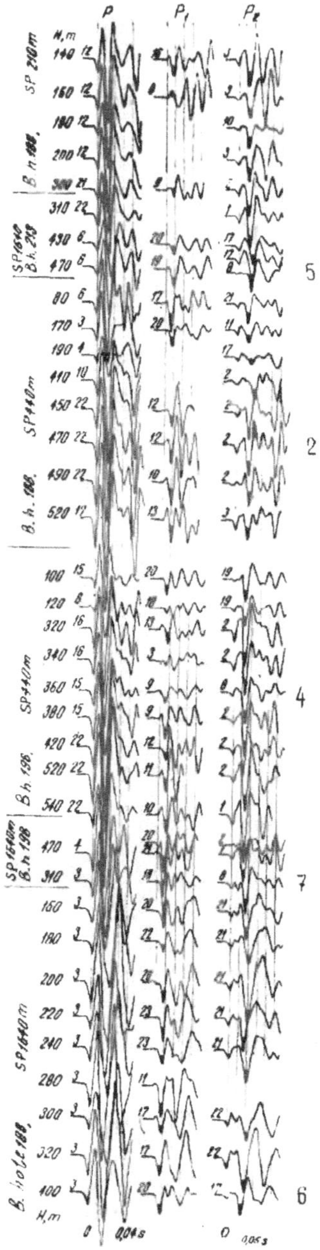

(a) (b)

Fig. 38. PM–VSP positional seismograms of tracking components (component ordinals shown on traces, and their orientation on stereo net (Figure 39d)) obtained in several boreholes in the same area from various shotpoints: b/h 188, SP 210 m; b/h 213, SP 164 m; b/h 188, SP 440 m; b/h 196, SP 440 m; b/h 198, SP 164 m; b/h 188, SP 164 m: (a) p wavelet; (b) p_1 wavelet tracking components; (c) p_2 wavelet tracking components.

direction on seismogram, (Figure 37c) is close to the horizontal and coincides with component 11 of the polar seismogram. The pulse P_1 with a sharp front is recorded with a time lag of about 0.010 s with respect to the first P-pulse on components 18 and 23 on which the first extremum of the P-wave is not recorded. Next, the pulse P_2 stands out on components 18 and 17 on which the pulse P_1 is considerably suppressed. Both pulses are approximately of equal intensity, which is just twice less than that of the P-wave. Oscillations of sufficient intensity are present also on the other CDR–I seismograms.

The oscillations identified on point seismograms can be tracked along the line of observations. This can be seen from the composite seismogram (for the whole area) (Figure 38) containing the records of the P-component (left) the tracking components of the P_1 oscillations (middle) and the P_2 oscillations (right) for different boreholes and different shot-points.

Consider observations made on fixed-depth profiles. A spatial system has been worked out in the Zhairem field in accordance with which observations were carried out at three geophone levels: 240 m, 460 m, 760 m, and the shotpoints were located on the surface along six profiles, which intersect at the top of the borehole. The length of the principal profiles was 1000 m, and the distance between adjacent shot-points was 50 m (see Figure 39d). Oscillations in the zero-displacement plane for the P-wave are identifiable on CDR–I seismograms (Figure 39b).

The wave recorded first on component III (H = 240 m, Figure 39c) along the entire profile is the P-wave. The first extremum on the component-III record from nearby (100 to 400 m) shot-points still corresponds to the P-wave, but already from shot-points 450 to 750 m the first to be recorded is the P_1 pulse that has a time shift with respect to the first extremum of the P-wave. This is clearly illustrated by the CDR–I seismogram obtained from shot-points 750 m (in the lower left corner, Figure 39c). The displacement direction of the P-wave will be seen from the seismogram to be horizontal (component 6), and the zero-displacement plane of the P-wave to be vertical. The records of components 3, 18, Z, 21, III, which lie in this plane, show clearly that the P-pulse is not recorded, but that the pulse that arrives is the P_1 having the time shift with respect to the P-pulse. The P_1 pulse can be reliably tracked in the interval of the profile from shot-point 450 m to shot-point 750 m.

A similar picture is also observed at the depth 460 m (Figure 39c). The first longitudinal wave can be reliably tracked on the Z-seismogram in the interval from shot-point 200 m to shot-point 700 m. On the component-III seismogram, the P-wave is recorded only from nearby shot-points (200 to 450 m). From shot-points 500 to 700 m when the first extremum of the P-wave cannot be recorded on component-III, the first pulse seen on the seismograms is the P_1. The displacement direction of the P-wave will be seen from the CDR–I seismograms from shot-points 780 m to make an angle of 30° with the vertical, and to coincide in direction with component 18. The components lying in the zero displacement plane are 6, 13, III, and 14. The first arrival of the P-wave is not recorded on those components, and the pulse P_1 appears with a time shift.

A similar picture can be observed at the depth of 760 m. A high-intensity P_1-pulse is detectable and reliably tracked on the seismograms of the component-III (from shot-points 400 m to 700 m) on which the P-wave pulse does not record at all, or is appreciably attenuated.

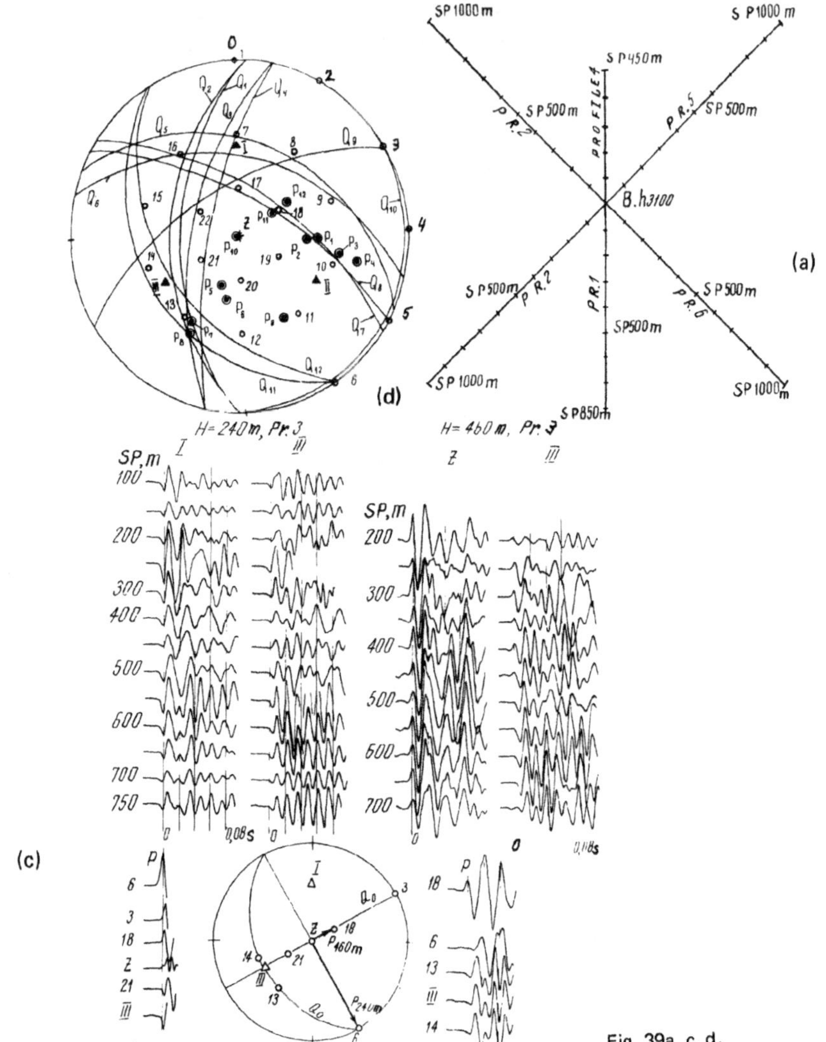

Fig. 39a, c, d.

Fig. 39a-d. Studying the direct wave wavelet form in level observations: (a) observation system. Solid lines are surface profiles of shotpoints. Level PM–VSP observations were carried out at three levels (H = 760, 460, and 240 m) in borehole 3100 located at the cross-point of shot profiles; (b) planar CDR–I seismograms of 12 direct wave wavelets obtained at H = 760 m in case of shots fired from different profiles at different offsets. In addition to seismograms located in the vicinity of the P wave zero-displacement plane ($Q_1 - Q_2$ on the stereo net), the figure also depicts P- ($P_1 - P_2$ on the sereogram) and Z-components; (c) level seismograms of I- and III-components of the original coordinate system obtained at a depth H = 240 m from shotpoints (100–750 m) arranged along profile 3 (left, top). Bottom – CDR–I seismogram of a direct wave from SP 750 m, H = 240 m. Right, top – level seismograms of components Z and III obtained at H = 460 m from SP (200–700 m) arranged along profile 3. Bottom – CDR–I seismogram (orientation of components shown on stereogram on the left) for SP 700 m; (d) orientation of CDR–I seismogram components and of zero-displacement planes $Q_1 - Q_{12}$ shown in Figure 39b.

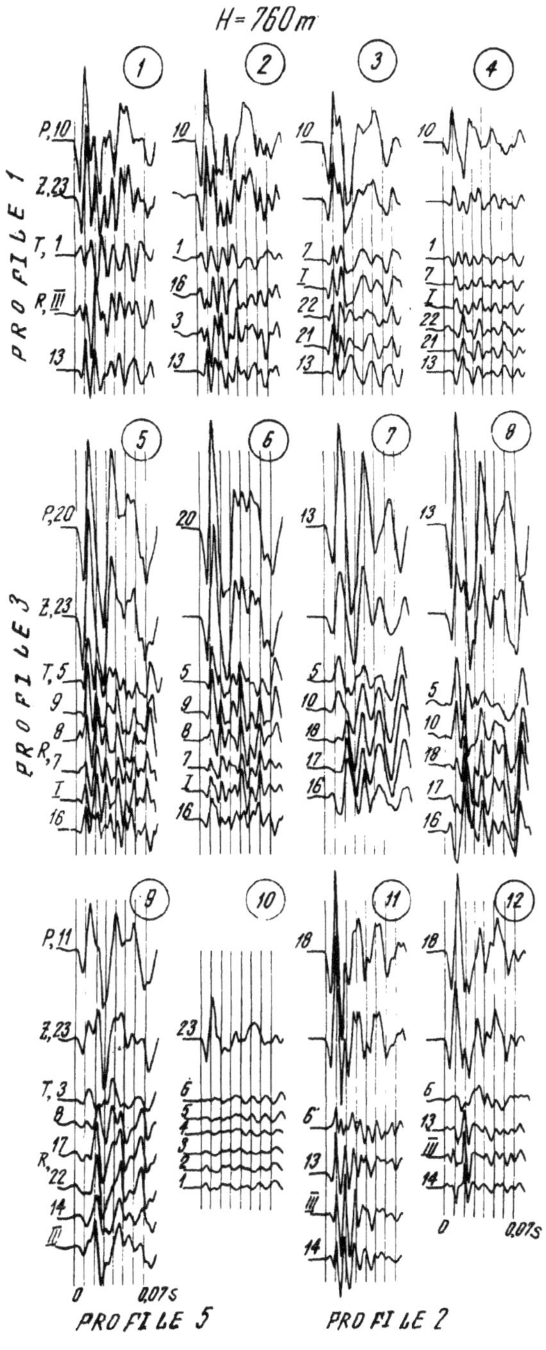

Fig. 39b.

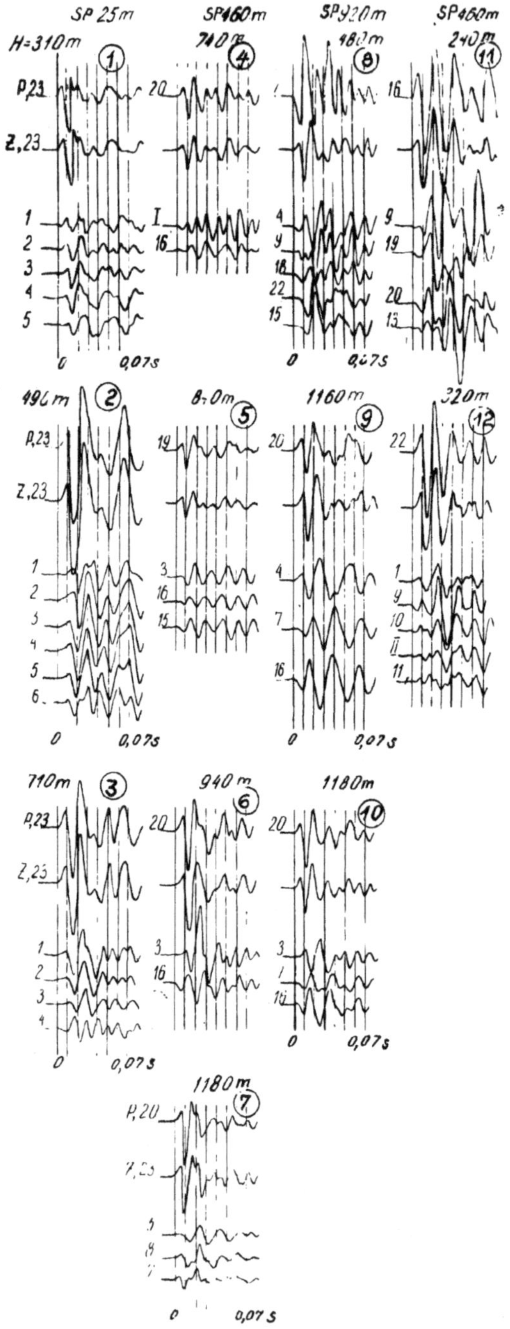

Fig. 40a.

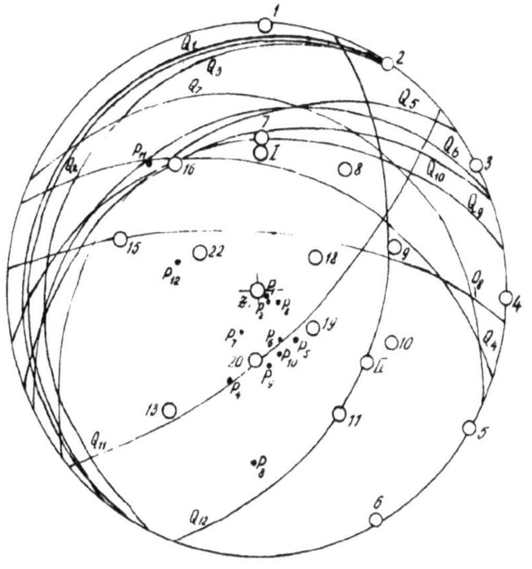

Fig. 40b.

Fig. 40a-b. Observations in oil-rich regions: (a) planar CDR–I seismograms obtained at different depths H and from different shotpoints (l); (b) orientation in space of seismogram components and of zero-displacement planes.

A similar seismogram (Figure 40) was recorded in the Tenniz depression containing characteristic sedimentary-type deposits. High-intensity oscillations P_1, P_2 are identifiable everywhere on the seismograms, even on those with a P-wave of a very simple shape (H = 710 m, shot-point 1, l = 25 m; H = 940 m, shot-point 3, l = 460 m; H = 1180 m, shot-point 4, l = 920 m; H = 320 m, shot-point 5, l = 460 m.

Hence, the pulse of the direct wave observed in the borehole is, as a rule, a complex one and consists of several pulses each having a different orientation in space. The pulses can be reliably identified in the zero-displacement plane of the preceding pulses, and often have a sharp front. Since the pulses interfere with one another, it is difficult to determine the displacement direction, but pulses can be said to be present polarized like both transverse and longitudinal oscillations with respect to the P-wave. The intensity of the P_1 and P_2 pulses is only 2 to 3 times less than that of the P-pulse. Their recorded shape in different conditions is not the same. In addition to pulses in the seismic exploration frequency band, very high-frequency pulses (up to 150 Hz and higher) are also often observed. The pulses can be quite reliably tracked on seismograms recorded on fixed depth and vertical profiles, and may be different on records of initial components which are differently oriented with respect to the direction from the shot-point. Pulses are also observed whose displacement directions are practically orthogonal to the displacement of the P-pulse, and whose shape closely resembles that of the P-pulse, but with a time-lag. In many cases, the typical symmetrical pulse shape (seismograms 6, 7, 9) is explained by an in-phase superposition of pulses.

It should be stated that the complexity of the direct wave has been observed before. The complex shape of the direct wave has been shown in [92] to be the result of the interference of three closely spaced pulses of practically idential and typical shape, each of which has a smaller amplitude than the preceding one. Such a series of pulses is similar to one produced by pulsations of a gas bubble in water. The pulses are presumed to be the result of vibrations of the shot-hole walls. On account of the small δt's separating them, it is not possible to resolve them in time. Unfortunately, the available conclusions are founded only on Z-component records, and are unable to reflect the complexity of the formation process of the direct wave.

Studies of the pulse structure and shape assume great practical importance at the present time in addition to their theoretical importance. However, despite the great volumes of PM–VSP work performed, the available data are inadequate for an unambiguous determination of the nature of the individual pulses, and the formation processes of both the pulses and the direct wave. In this context, one will have to study the conditions of excitation and reception, including measurement techniques (the reactions of clamping devices, column vibrations etc) and the structure of the Earth both in the vicinity of detector points and along the wave path. Special PM–VSP experiments should be performed to clarify these problems.

The requirements that the direct wave will have to satisfy are determined by the goals set by the studies. For the study of the dynamics of the direct wave, for the solution of many other problems in the physics of the propagation of seismic waves in the real earth, the direct wave should be produced in the form of a short, simple and high-frequency pulse with a wide spectrum. This may prove to be of such decisive importance as to justify the execution of special experiments in each specific case. However, in other studies, it is more expedient and less costly to obtain a downward wave of a somewhat more complex shape, but without ghosts (Figure 30, h = 15 m, 20 m, 27 m). In such cases, the shape of each individual wave on the seismogram will be more complex than could be obtained with a simple direct wave pulse, but the structure of the seismogram as a whole, because of the substantially smaller number of waves recorded on it, will be appreciably simpler.

CHAPTER 6

THE VSP METHOD FOR LONGITUDINAL REFLECTED WAVES

From its very inception, the seismic method of longitudinal reflected waves has been and still remains the principle one in the complex of geophysical methods of oil and gas exploration. At present, the method of longitudinal reflected waves is the main one for locating structures for deep drilling. However, the effectiveness of the method is substantially reduced with the increase in exploration depth and with use in increasingly complex areas. It has been established as a result of VSP carried out over greatly varying geologic structures that in practically all areas and at all accessible depth, there are horizons in sedimentary beds with which reflected waves of sufficient intensity are associated. However, it is impossible to identify and track them on the surface, because of the high intensity of various unwanted waves of which the most important are the multiples. Practically all efforts at perfecting the reflection method during the last 40 years have been aimed at a more effective suppression of multiples, such as the common-depth-point method.

Despite a colossal volume of experimental and theoretical work on multiples performed in the last 20 years, they increasingly limit the potentialities of the method of longitudinal reflected waves. This is the outcome of a continuous increase in the required depth of exploration such that the intensity of primary reflected waves is reduced to a level comparable to, or in many areas even below, that of the multiples associated with reference interfaces in the upper portion of the geologic section.

There are three separate and fundamental aspects to the problem of the multiples: the study of their kinematic and dynamic characteristics, their discrimination on the seismograms and their suppression. The effectiveness of all the methods of multiples suppression depends to a great extent on the knowledge of their parameters, their nature and the mechanisms of their origin. The difficulties involved in the study of multiply reflected waves in surface observations are occasioned by the fact that the kinematic and dynamic characteristics of singly and multiply reflected waves observed along a horizontal profile are, in practice, quite similar, including the region of their recorded times.

As a rule, it is possible to establish the presense of multiples on the seismograms, but it is not always possible to determine unambiguously their nature, the reflecting interfaces with which they are associated or their relative intensity.

1. Longitudinal Primary and Multiple Reflected Waves in VSP

VSP enables the process of orientation and propagation of primary and multiple waves to be observed and studied, and the effects of various geological elements on the formation of the wave field to be evaluated.

The conditions for the correlation of reflected waves on a vertical profile are, as a rule, more complicated than in surface observations. In VSP the relative intensity of singly reflected waves, as compared with that of the multiples, may be less than in surface observations, because down-hole one can observe multiples that have experienced fewer reflections than waves observed at the surface.

The traveltime curves of the direct and the reflected waves obtained in VSP have some common points. The mutual interference of those waves can be classified as belonging to the convergence type. The distance, starting from which the reflected wave can be tracked, and consequently the accuracy of stratigraphic correlation of seismic horizons, depends on the dimensions of the zone of interference. The dimensions of the zone of interference are determined both by the duration of the incident wave and by the location of the shot-point.

The dimensions of the zone of interference are smallest on in-line vertical profiles.

The amplitude of the useful signal drops with recording time, whereas the background noise rises, and this, as a rule, worsens the long-time wave correlation.

The amplitudes of longitudinal waves on the records of vertical components observed along off-line profiles decrease with offset distance. At the same time the amplitudes of converted and transverse waves rise. Such waves do not always produce continuous line-ups, however their intensity may be high enough to distort the correlation of the longitudinal waves.

The correlation of deep reflected waves in the initial portion of a VSP seismogram is usually distorted by downward waves associated with the upper portion of the section, and which occupies a substantial interval of the record after the direct wave. In fact, in observations at all accessible depths, we have been unable to get rid of the effects of the upper portion of the geologic section. High values of reflection coefficients typical of the shallow interfaces and especially of the bottom of the low-velocity zone, are the cause of the high intensity of downward waves that may greatly exceed that of singly reflected waves. It should, however, be pointed out that in many cases the waves and sometimes even the phases can be identified reliably in sections preceding and following the zone of interference thanks to the comparatively small dimensions of the zone.

The correlation of waves in the case of interference between primary and multiple upward reflected waves presents much greater difficulties. When a wave reflected from a deeper-lying interface and propagating towards the surface enters the region of upward travelling multiples, a parallel superposition of the interfering wave forms takes place. In the case of horizontally layered media, the apparent velocities of such waves will be the same for an in-line profile. In the case of low-dip interfaces or off-line profiles, the velocities will be almost equal. The discrimination of interfering waves in conditions of parallel superposition is fraught with great difficulties. At the same time it is frequently highly important in practice, particularly for the determination of the nature of the waves and for stratigraphic correlation of multiples, to establish such interference. The dynamic features of the wave (sudden changes in the intensity and in the periods of oscillation) should prove of substantial value for the detection of such interference.

The results obtained in the study of reflected waves can be summarized as follows:

The primary and multiple reflected waves are the major wave types recorded in the later portion of the seismogram on vertical and horizontal profiles. The dominant waves are associated with quite definite elements of the geologic section (in most cases thin layers and stacks of layers) and can be stratigraphically correlated. Combined reflected waves have a much lower amplitude and are observed in media with a weak velocity differentiation.

The shapes of reflected waves in thinly-layered media may be quite different from that of the incident wave and are greatly dependent not only on the frequency spectrum of the incident wave, but also on the structure of the reflection series [46]. In such conditions the duration of the reflected wave increases sharply. Even a slight change in the shape of the incident pulse may cause a substantial change in the shape of the reflected wave. Waves reflected from different layers and series of reflectors may differ, with the ghosts differing from the singly reflected wave. Because of the dependence of the magnitude of the reflection coefficient on the angle of incidence, the shape of the reflected wave may change with the distance (with increasing angle of incidence), even if the pulse of the incident wave and the structure of the heterogeneous series of reflectors remain constant. In the course of a detailed study of the process of formation of reflected waves it is not always possible to approximate the heterogeneous series of reflectors by a homogeneous layer, and one has frequently to take into account the internal structure of the reflection series.

In the formation of the multiples wave field, interfaces located in the upper portion of the geologic section, and particularly the low-velocity zone, play a major part. Even one or two such interfaces may be enough to determine broadly the wave field both on the surface and at interior points of the medium. The chief obstacle for the discrimination of useful singly reflected waves is the presence of partial multiples that have experienced intermediate reflection at interfaces in the near-surface.

In individual areas, multiples can originate at great depths. Basement may also serve as a reflecting interface, even if it is located at a comparatively great depth (e.g. 7 to 9 km). Multiples may be observed at internal points of the medium, even when no regular waves are recorded on the surface seismograms. The absence of regular waves on surface seismograms is not, as a rule, proof of the absence of reflecting horizons in the geologic section.

The reflecting properties of the low-velocity zone exercise a profound effect on the intensity of the multiples. The bottom of the low-velocity zone is the most sharply defined interface in the section and can turn a substantial part of the upward travelling energy back into the depth of the medium. The dominant multiples experience intermediate reflection at the bottom of the low-velocity zone. In conditions when the velocity in the near-surface varies monotonously, and the low-velocity zone is not clearly defined, the intensity of the multiples is usually comparatively low, even if the section contains powerful reference interfaces. Reflections from the surface as compared with those from the low-velocity zone display lower intensity and a higher content of low-frequency components.

The overwhelming part of the energy recorded on surface seismograms is associated with waves that originate at the interfaces in the upper portion of the gelogic section. Even in conditions of weak velocity differentiation and in the absence of sharp interfaces a large number of unwanted waves of different origin may be observed on the surface seismograms. Among them, the longitudinal and transverse bent waves and their multiples are of major importance. These waves owe their presence to the existence of a vertical velocity gradient in the medium as well as to source-excited transverse waves whose intensity in individual cases exceeds, or is comparable to, that of the longitudinal waves.

Surface waves decay rapidly with depth, and it is practically impossible to detect them on VSP seismograms at a depth of around one wave length. One should take account of the transformation of a part of the energy of body waves into the energy of the secondary surface waves, and of the energy of the surface waves into the energy of body waves [29, 100]. Such transformations have been studied inadequately. There is a need for future special research aimed at studying wave fields associated with the upper portion of the geologic section and of their effect on surface seismograms.

Studying the parameters of the primary and multiple reflected waves, one is able in each particular case to evaluate objectively the potentialities and the effectiveness of various methods of suppressing multiples and to choose optimum observation systems.

2. The Effect of the Near-Surface and the Low-Velocity Zone on the Field of Primary and Multiple Reflected Waves

Large-scale VSP observations carried out in various geologic conditions have demonstrated that the upper portion of the section, which, as a rule, contains sharp interfaces, particularly the low-velocity zone and the surface, plays a major part in the formation of the wave field. All the parameters of the low-velocity zone that are important for longitudinal waves differ sharply from those of underlying rock; the difference in the parameters of the low-velocity zone and of the underlying zone that are important for transverse waves being slight. This is illustrated by Table IV which shows parameters of the low-velocity zone and the underlying rock in the Leningrad area in the Krasnodar province [101].

The low-velocity zone will be seen from the table to be a strongly absorptive and dispersive medium for *P*- and *S*-waves. The attenuation factor for the *P*-waves in the low-velocity zone is two orders of magnitude higher than in the underlying rock, while the corresponding values for transverse waves are identical. The parameters shown are typical for many platform regions. The low-velocity zone serves as a low pass filter for longitudinal waves and has a transmission maximum in the 30–40 Hz frequency band.

Effect of the Low-Velocity Zone

Data obtained in VSP show that the effect of the near-surface on seismic records is as a rule, underestimated: whereas the absorption properties of the low-velocity zone have a small effect on the frequency spectrum of waves associated with great depths or recorded at long distances, the reflecting properties of the near-surface and the low-velocity zone can decisively affect the number of waves and the entire structure of the seismogram. The low-velocity zone is, as a rule, the most sharply defined interface in the section (the discontinuity in the velocities may amount to 3 or 4 times), and most intense reflected waves are associated with it.

The reflecting properites of the low-velocity zone manifest themselves first of all in the formation of ghosts, which may greatly complicate the recording of the primary incident wave (see Chapters 3 and 5). However, the major effects of the reflecting properties of the low-velocity zone and of the surface are on the formation of full or partial multiples. The reflecting properties of the low-velocity zone determine the part of the upward travelling energy that will be reflected back into the medium, and this to a substantial extent determines the intensity of the multiples. If the base of the low-velocity zone constitutes a sharp interface, the intensity of the multiples may be

TABLE IV

Parameter	Low-velocity zone	Base of Low-velocity zone	Underlying rock
V_P	200–1200 m s^{-1}	–	1700–1800 m s^{-1}
V_S	100–200 m s^{-1}	–	200–500 m s^{-1}
Increase in V_P	–	from 800–1200 to 1700 m s^{-1}	–
Increase in V_S	–	none	–
V_P gradient	0.15(0.20)–0.025(0.033) $\frac{1}{M}$	–	0.00011–0.00012 $\frac{1}{M}$
V_S gradient	0.1–0.2 $\frac{1}{M}$	–	0.03–0.012 $\frac{1}{M}$
P dispersion (f = 30–140 Hz)	+9 to 11%	–	+1.8%
S dispersion (f = 10–60 Hz)	+8%	–	+8%
$\bar{\alpha}_P$ (f = 50 Hz)	0.01 to 0.1 m^{-1}	–	0.001 m^{-1}
$\bar{\alpha}_S$ (f = 50 Hz)	0.2 m^{-1}	–	0.2 m^{-1}

comparable to that of the reflected waves even in the absence of reflecting marker interfaces in the geologic section. At the same time, if the low-velocity zone is not sharply defined, the relative intensity of the multiples will be lower even in the presence in the geologic section of reflecting marker interfaces, and singly reflected waves will prevail on the seismograms. It should be pointed out that the effect of the low-velocity zone on the multiples field depends greatly on the frequency spectrum of the waves. For waves of long period that 'do not see' the low-velocity zone, the part of the reflecting interface is played by the surface. This is particularly true in earthquake seismology.

Here are several examples to illustrate the effects of the low-velocity zone and the surface. When the low-velocity zone and the surface do not act as strong reflecting interfaces, the waves recorded on the vertical profile consist mainly of upward reflected waves whose continuity is determined largely by the surface conditions including terrain (Figure 41). However, such cases are comparatively rare. In this respect, of special interest are marine observations where, although there is no low-velocity zone in the traditional sense of the word, the surface of water itself acts as an ideal reflecting interface. Here practically the entire energy is reflected downwards, and waves recorded on VSP seismograms consist mainly of downward waves against which it is difficult to detect upward waves (Figure 42a): the upper part of Figure 42b depicts for the sake of comparison a seismogram obtained in a similar geologic section, but on land. In the latter case the downward waves are much less numerous, and it is possible to identify upward reflected waves.

On the seismogram of Figure 42c the base of the low-velocity zone in conjunction with marker interfaces at 1400 m and 1500 m has been responsible for a large number of full and partial multiples occupying the interval up to 2.25 s. We have seen a similar picture in the Staro–Minskaya area.

To illustrate reflection from the surface, consider the seismograms of components in a local coordinate system obtained in the upper portion of the geologic section (see

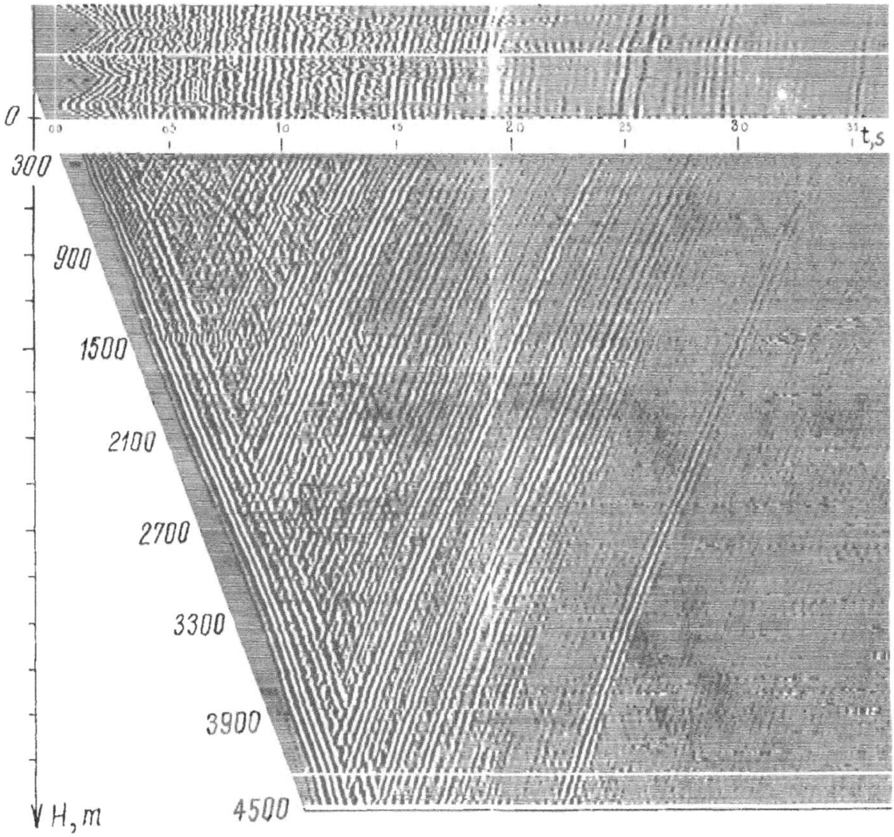

Fig. 41. VSP seismograms (Z-component) when the low-velocity zone does not constitute a sharp reflecting interface (Potvar inflection, W. Pakistan) [34].

Figure 54) along an off-line profile (l = 500 m). All longitudinal and converted waves incident on the surface are reflected downwards.

The intensity of the multiples decreases with an increase in the thickness of the low velocity zone, this being due to the decrease in the magnitude of the velocity increase at the base of the low-velocity zone. This correlation will be clearly visible, if one compares the charts of the low-velocity zone thickness and the level of multiple activity drawn up from VSP data for the north-western Ciscaucasian region (Figure 43).

Qualitative relationships between the low-velocity zone parameters and those of multiples associated with the low-velocity zone are determined by the spectral characteristics of the low-velocity zone. Accordingly, studies of these parameters are important for the understanding of the mechanism of formation and the characteristics of the multiples. Observations in shallow boreholes make it possible to study downward waves and from them to determine the intensity of the multiples.

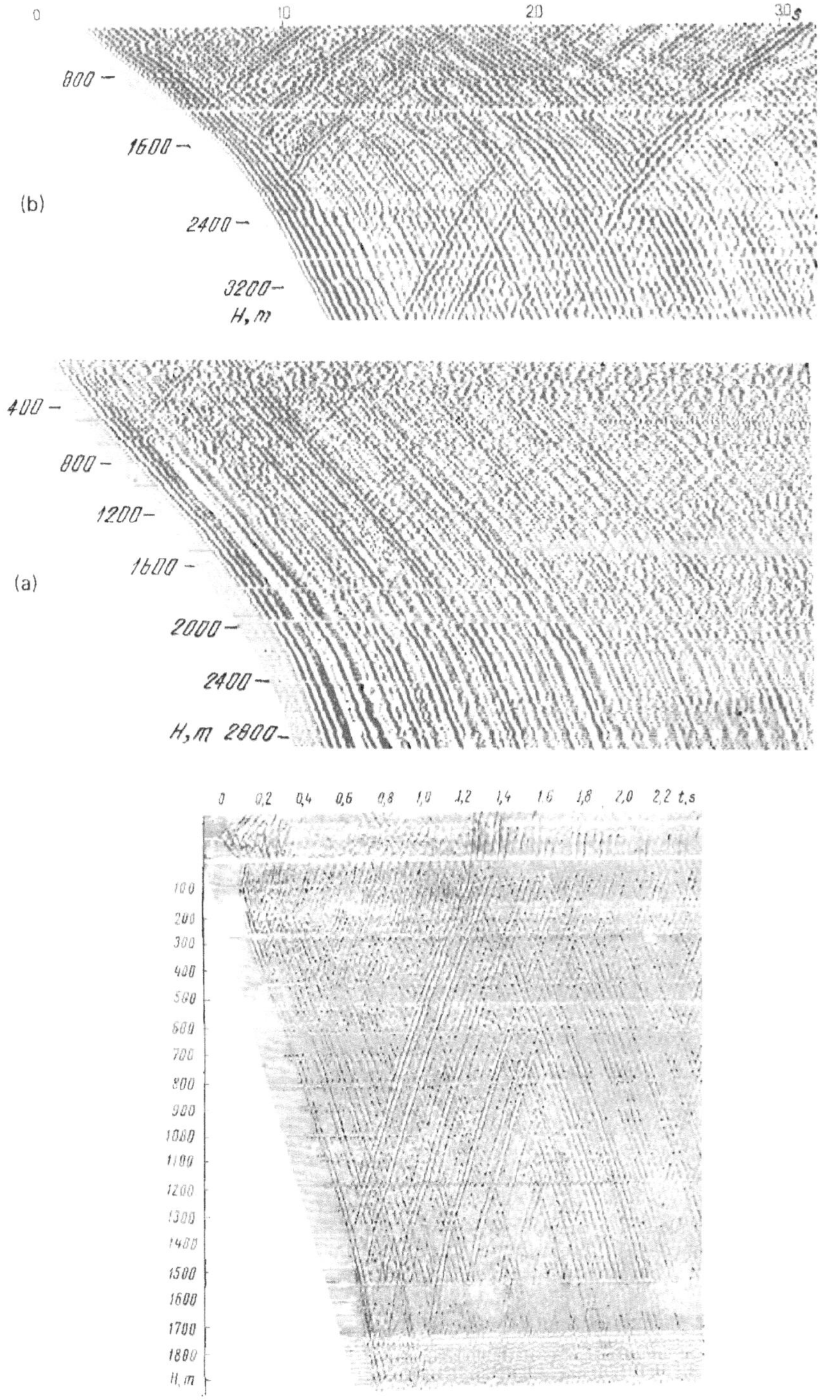

Fig. 42. VSP seismograms when the low-velocity zone constitutes a sharp interface (A. N. Amirov's data) (a) Black Sea; (b, c) Crimean Steppes.

(a)

(b)

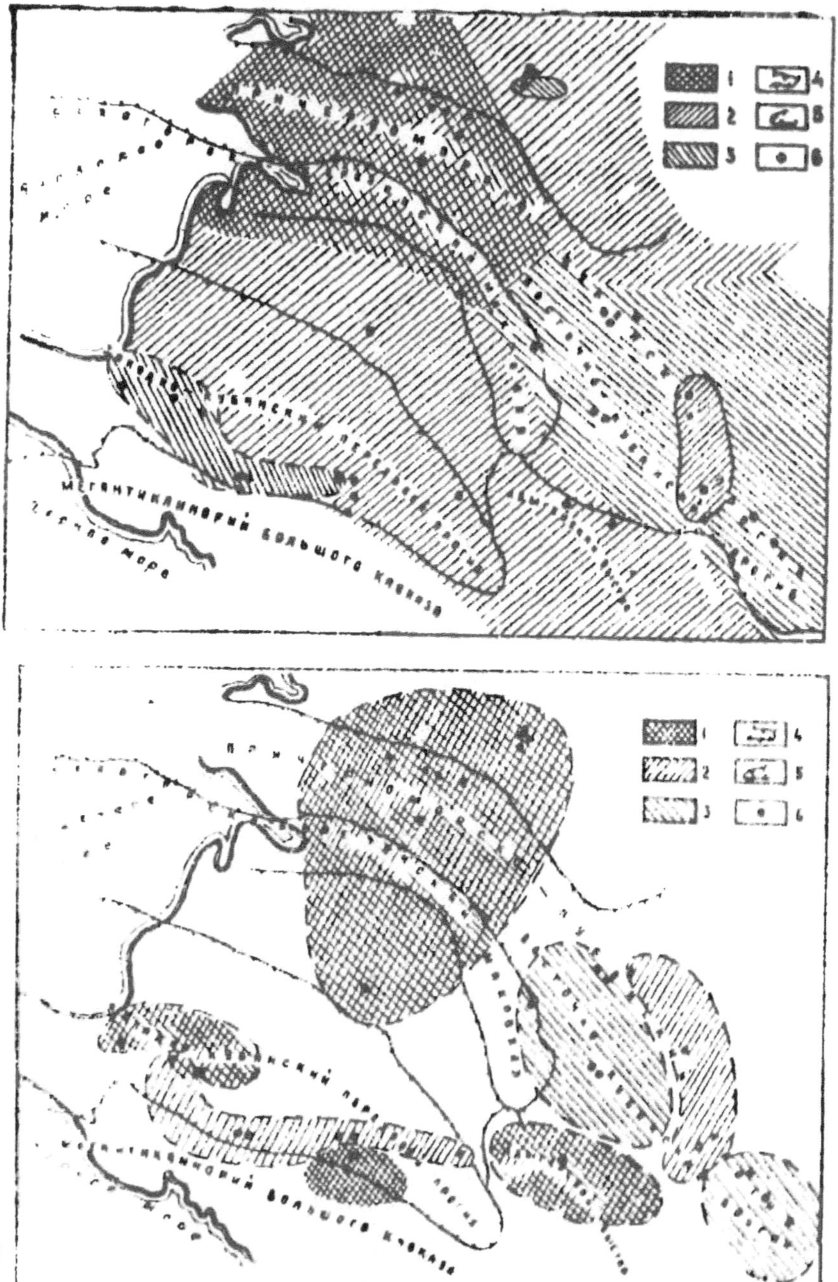

Fig. 43. Intensity of multiples (a) as a function of the thickness of the low-velocity zone (b) in north-western Ciscaucasian region [102]. (a) 1 – regions with a high level of multiples, 2 – with a medium level, 3 – with a low level, 4 – boundaries of structures of the 1st order, 5 – platform folding, 6 – deep boreholes; (b) 1 – thickness of the low-velocity zone does not exceed 15 m, 2 – from 15 m to 30 m; 3 – not more than 30 m; 4, 5, 6 – see under (a).

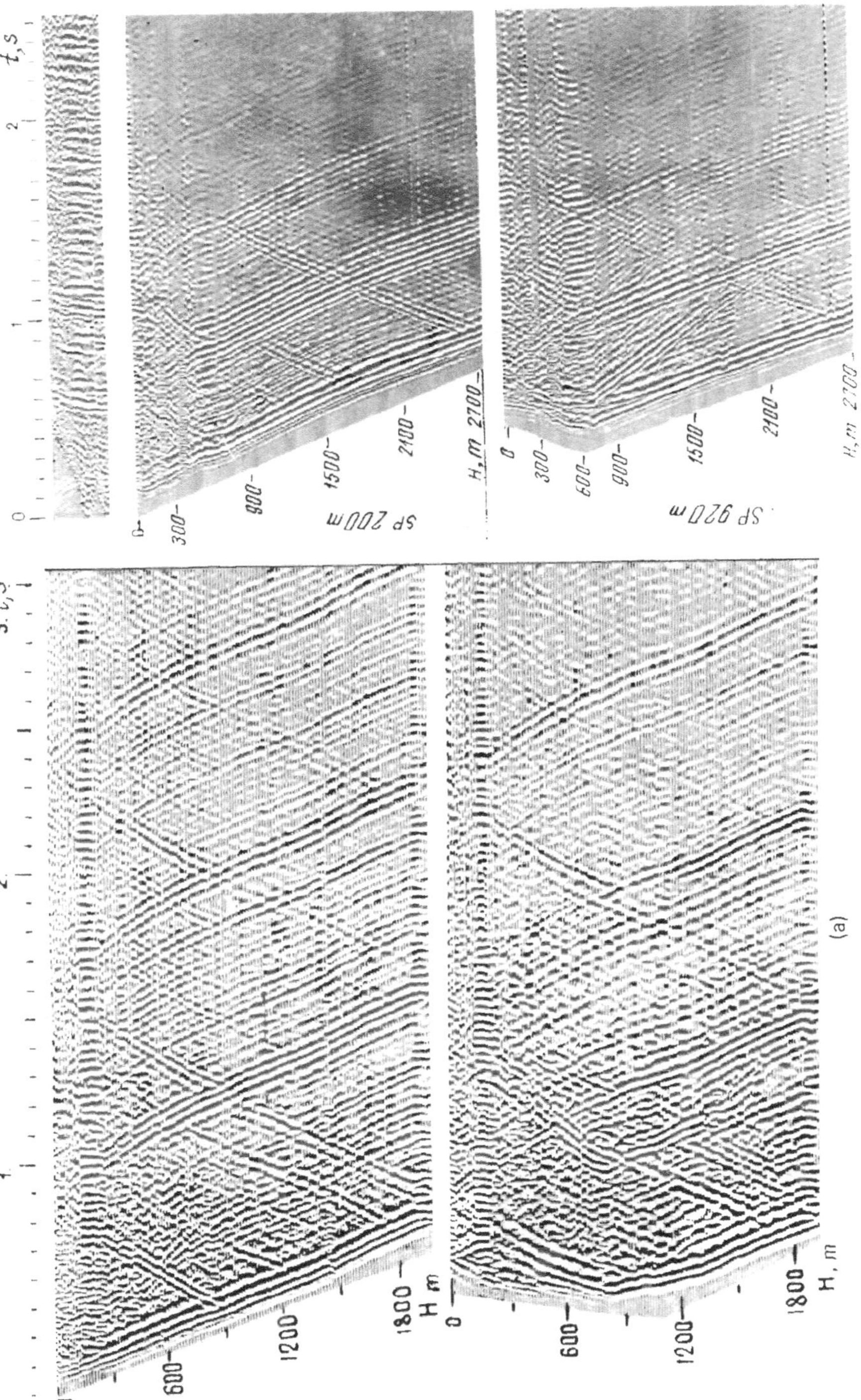

Fig. 44. VSP seismograms from two shotpoints: (a) Warsaw syncle, Lambe–Welkie, borehole N1; (b) Leba, borehole N3 [34].

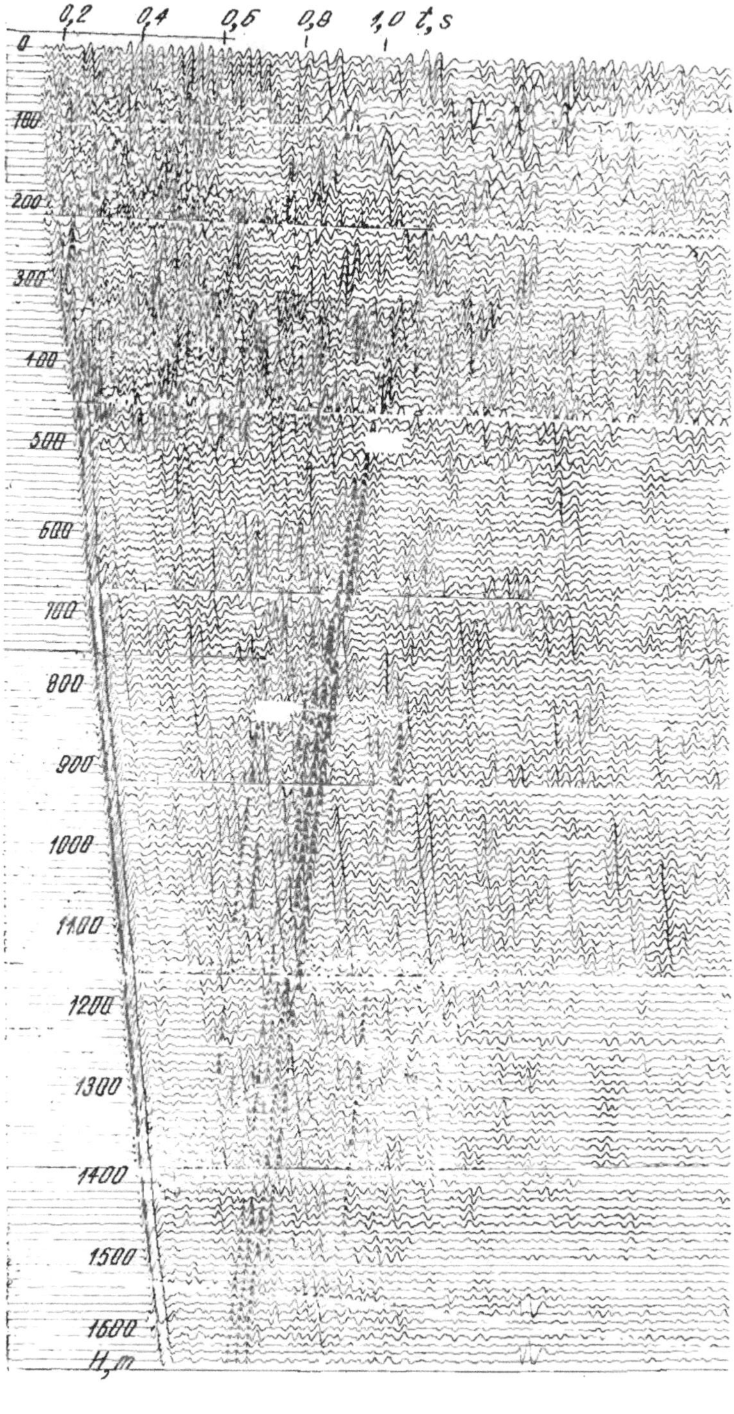

Fig. 45. VSP seismogram (Oktyabrskaya, borehole N1, courtesy S. V. Klushin).

EFFECT OF THE UPPER PORTION OF THE GEOLOGIC SECTION

In addition to the low-velocity zone, an important part in determining the structure of the wave field is played by the entire near-surface, which usually contains sharply defined interfaces. The wave field is greatly dependent on the relative disposition of such interfaces [34]. The presence in the upper portion of the section of several sharp interfaces located at greatly different depths results in the formation of a complex wave field containing numerous types of multiples recorded both in VSP and surface observations. In such conditions deep waves either cannot be tracked at all, or their tracking involves considerable difficulties (Figure 44a).

Where only one or two sharp interfaces separated by a short distance are present in the near-surface such as is usually encountered on platforms when sedimentary deposits corresponding to certain geologic periods are missing, a different situation exists. In this case, deep reflections can be tracked only up to the upper reflecting interfaces where their correlation is distorted by interference (Figure 44b). Wave fields of this type are quite common, when the depth of the upper reflecting layers is, as a rule, between 200 and 600 m. At such interfaces the correlation of waves is substantially improved. Similar fields can also be obtained in salt beds traversed by greatly dislocated deposits that obstruct the correlation of deep waves. In such conditions deep waves can be tracked only by VSP (Figure 45).

3. Primary and Multiple Reflected Waves in Certain Types of Geologic Sections

We shall illustrate the potentialities of VSP in the study of primary and multiple waves by citing examples of observations in two types of sections: a low-velocity one containing individual thin layers of higher velocities, and the other consisting of a high-velocity carbonate section interspersed with layers of low-velocity terrigeneous deposits.

SECTIONS CONTAINING INDIVIDUAL THIN HIGH-VELOCITY LAYERS

We shall consider primary and multiple waves in media with thin high-velocity layers, making use of the example from the Staro-Minskaya area where the geologic section contains several thin layers. The first sub-section (H_v = 270 m) consisting of sandstones with $v_P = 2500 \text{ m s}^{-1}$ and 11 m thick is located in a continuously-changing sandy clay bed with velocity $v_{av} = 1800 \text{ m s}^{-1}$ (Figure 46). The second sub-section – a stack of thin limestone interlayered with terrigenous deposits of Paleocene age – is located in sandy clay rock at a depth of 1334–1380 m (interface 3 in Figure 46). The principal information about the layers is contained in Table V.

The presence in the upper portion of the cross section of two thin layers with a large velocity difference, extending over the entire area, exercises a decisive effect on the wave field observed both on surface and down-hole seismograms. Observations at great depths outside the recording time interval of the multiples have enabled reflected waves associated with horizons located at great depths, and of considerable geologic interest, to be identified.

The determination of the nature of waves on surface seismograms and the discrimination of singly reflected waves associated with great depths has been performed on the basis of VSP data. The PP^1 wave reflected from the first thin layer (H_v = 270 m)

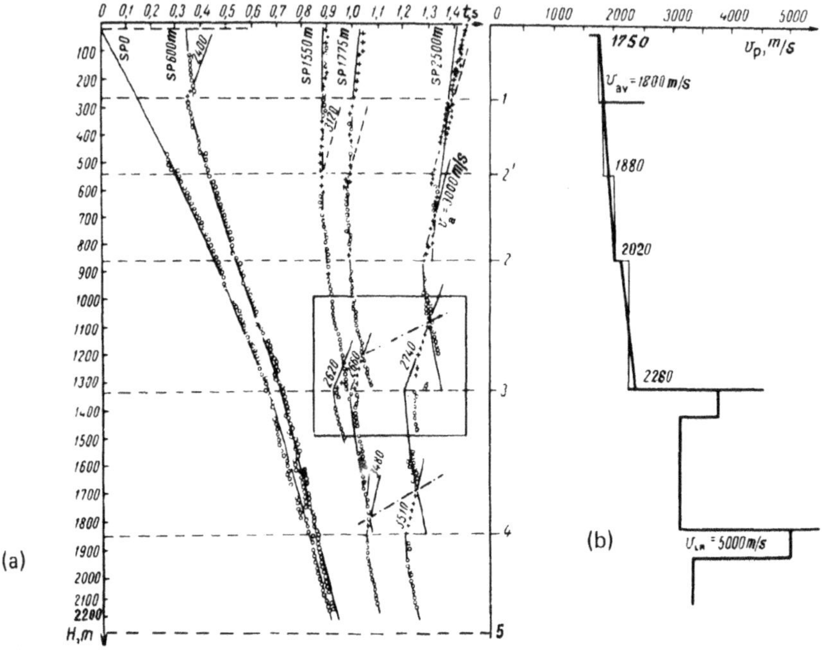

Fig. 46. Traveltime curves of the first (direct, head and transmitted) waves (a) and the velocity model (b) constructed from VSP data (borehole N42, Staro–Minskaya). The traveltime curves inside the rectangle are shown in Figure 65 on a larger scale.

TABLE V

No. of layer	H_v (m)	d (m)	V_{P2} (m s^{-1})	V_{P1} (m s^{-1})	V_{P3} (m s^{-1})
1	270	6–11	2200–2600	1800	1800
2	1330	25	4890	2500	3000

No. of layer	$V_{P_1\,av}$ (m s^{-1})	$\frac{V_{P_1\,av}}{V_{S_1\,av}}$	$\frac{V_{P_1\,av}}{V_{P_2\,av}}$	$\frac{d}{\lambda_2}$	$\frac{H_v}{\lambda_1}$
1	1600	3.5–4	0.7–0.8	0.12–0.32	8
2	2000	2–7	0.49	0.20–0.24	2–6

Note: Subscript 2 refers to the average parameters of a series of layers, subscripts 1 and 3 – to average individual layer parameters (layers about a wavelength thick), the former directly above and the latter directly below the series.

(Figure 46) on the record of the vertical component has an intensity no less than that of the direct wave and a comparatively high-frequency spectrum. The PP^3 associated with the Paleocene top ($H_v = 1434$ m) is also a reflected wave of very high intensity. It is reliably recorded on seismograms from all shot-points in the range 100–2500 m and is identified with the wave on the horizontal profile corresponding to the time $t = 1.3$ s. At great distances (SP 2500 m) the PP^3 wave is recorded as a transcritically reflected wave. Waves reflected from other interfaces exhibit comparatively low intensities and can be tracked on the vertical profile only in individual sections close to the interfaces. Away from the interfaces the correlation of these waves disappears.

The following conclusions can be drawn from the graphs of the amplitudes of the direct, the primary and multiple waves along the vertical profile (Figure 47).

(1) The amplitudes of all waves vary smoothly. The amplitude of the direct wave decreases by about 30% within the depth range from 800 to 1200 m. The decrease in the wave amplitudes enables the effective absorption to be determined.

(2) Reflection causes an abrupt change in the wave amplitude. It follows from the graph that the jumps in amplitudes are greater in case of reflection from the lower interface than from the low-velocity zone. A comparison between the amplitudes and the spectra of the incident and the reflected waves enables the reflection coefficient to be determined and thus the intensity and the shape of the reflected wave to be assessed (see Figure 117).

(3) The intensity of multiples decreases drastically with an increase in multiplicity. However, every multiple prevails in a narrow time interval where it is recorded. This means that the intensity of the multiples exceeds that of singly reflected waves, and for this reason by increasing the recording time it will be possible to improve the effectiveness of discrimination of singly reflected waves from deeper interfaces.

Spectral characteristics of the multiples on the vertical profile are of special interest (Figure 48). The waves with the highest frequencies are those that arrive from the nearest shot-point. An increase in the distance from the source and in the depth of the point of observation brings about a reduction in the recorded frequency, although only a slight one.

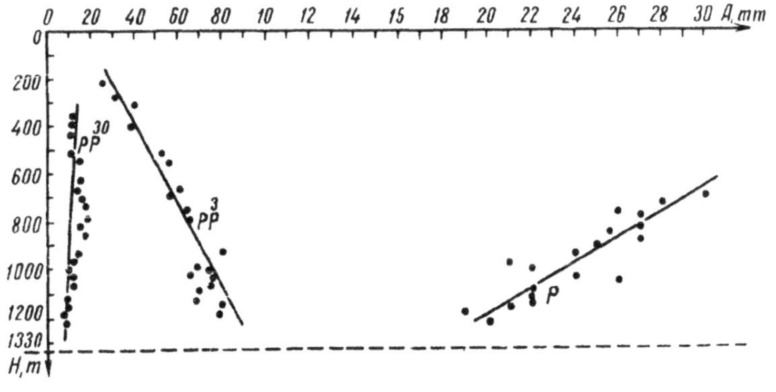

Fig. 47. Plots of amplitude vs depth variations of the Z-component oscillations for the direct P-, the reflected PP^3- and the multiple PP^{30}-wave (shotpoint 100 m).

Comparing dominant frequencies of waves of various multiplicity we may draw the following conclusions:

(1) For all shot-points, waves of different multiplicity have different dominant frequencies (Figure 48). The direct wave has the highest frequency, its visible frequency being 70 Hz. The dominant frequency decreases with increasing multiplicity. Thus, for the PP^3 wave the average value of the dominant frequency is about 64 Hz, for the PPP^{30} about 50 Hz, and for the $PPPP^{303}$ about 47 Hz. Reflection from an interface is accompanied by an abrupt change in the dominant frequency of a wave. The visible frequency varies smoothly during the course of the wave's propagation between interfaces in the observation range of 750 m.

(2) The reflection of waves of different multiplicities from the same interface is accompanied by different changes in their spectral composition.

(3) The change in the frequency of the multiples with the distance l is small. Thus, in the range from 300 to 1200 m the decrease for a twice reflected wave does not exceed 3–6 Hz. The same holds for waves of higher multiplicities.

In addition to the principal multiples, the seismograms contain a great number of

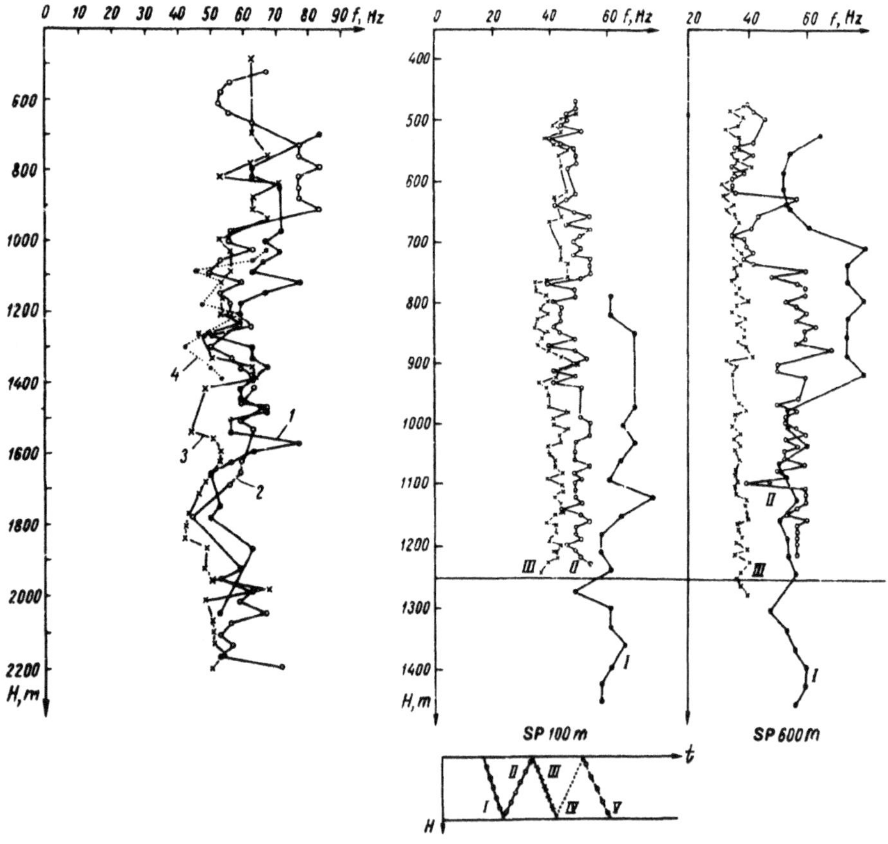

Fig. 48a.

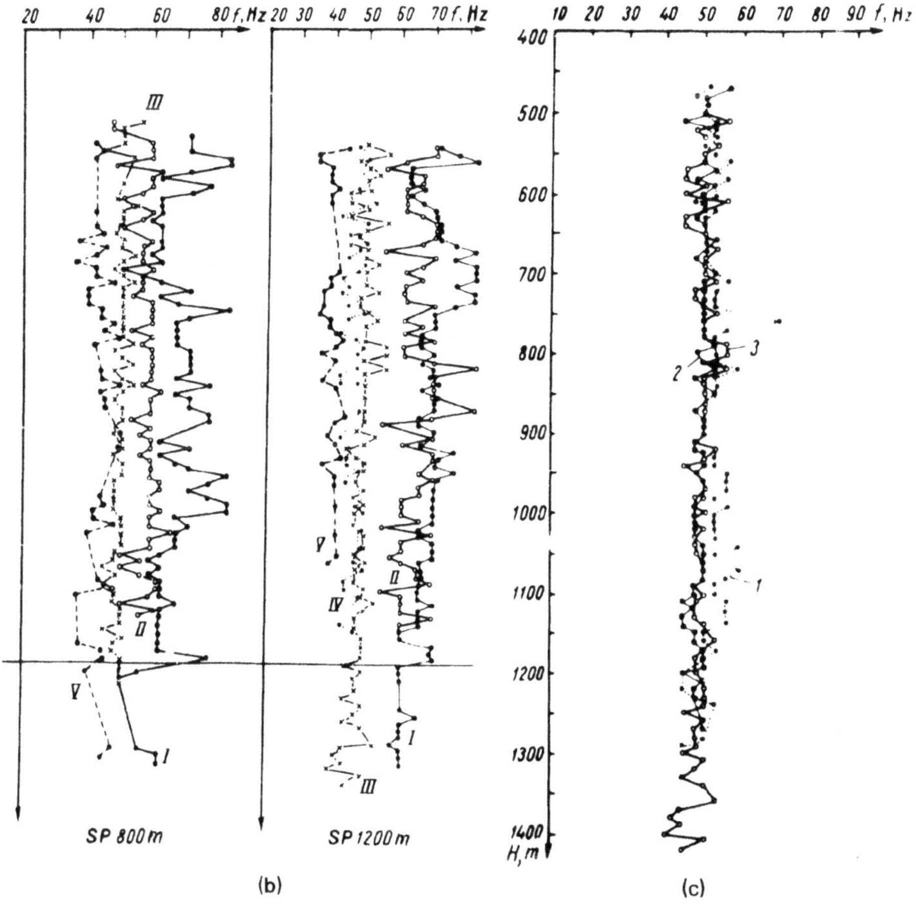

Fig. 48a-c. Plots of dominant frequencies (borehole N20 Staro–Minskaya): (a) for the first wave: shotpoint 100 m (1), 600 m (2), 1200 m (3), 2500 m (4); (b) for the primary wave and multiple waves of different multiplicities; (c) for doubly-reflected wave shotpoint 300 m (1); shotpoint 800 m (2); shotpoint 1200 m (3).

combined multiples of various types produced as the result of interference. At times greater than 4 s nothing but the multiples from two thin layers can be identified on the seismograms. Hence, the presence in the cross-section of two thin layers of higher velocity plays a decisive part in the formation of the wave field, so that it is impossible to identify on the horizontal profile any wave associated with deep interfaces.

Waves reflected from deep interfaces

Observations performed at great depths enabled waves reflected from deep interfaces to be detected. The most stable waves are associated with two interfaces located at depths of 180 m (PP^4) and 2310 m (PP^5). The wave PP^4 cannot in practice be tracked along the in-line profile. Above the Paleocene layer ($H_v < 1334$ m) the wave PP^4 is contained in the region where it interferes with multiple reflected waves along the entire profile. Below the Paleocene layer the wave correlation is repeatedly distorted by converted and multiple downward waves. The wave can be identified only in the immediate vicinity of an interface. The relative intensity of the wave increases with an increase in l. The wave is dominant on records of the Z-component and can be tracked along a substantial section of the vertical profile not only below the Paleocene layer.

The intensity of the PP^5 wave associated with the top of the Jurassic–Triassic is higher. It can be recorded close to an interface from the nearest shot-points (SP 100 m and 600 m), but above the interface its correlation is disrupted. From SP 2500 m, the PP^5 wave can be identified only in very short sections of the profile, where converted downward waves are not recorded. It is impossible to track the PP^5 wave and the PP^4 wave along the vertical profile in the region of the multiples. Owing to the great geologic interest in the PP^5 wave, special experiments have been carried out with the aim of assessing the possibility of discriminating and tracking it along the vertical profile in the region of the multiples. The study of the spectra of primary and multiple reflected waves leads to the conclusion about the expediency of employing frequency discrimination to suppress the multiples. The presence of peak frequencies of 45–55 Hz (see Figure 117) in the spectrum of reflection coefficients from the thin layer (interface 3 in Figure 46) with which the principal multiples are associated, encouraged an attempt to discriminate waves associated with the top of the Jurassic–Triassic in a much lower frequency band.

On the low-frequency seismogram (Figure 49), the P^5 wave reflected from the top of tje Jurassic–Triassic ($H_v = 2310$ m) can be reliably tracked in the section of the profile that contains the multiples. Its visible frequency on the record is 12 to 14 Hz. The multiples are much less numerous on low-frequency records. Accordingly, low-frequency waves can be more reliably tracked along the observed section.

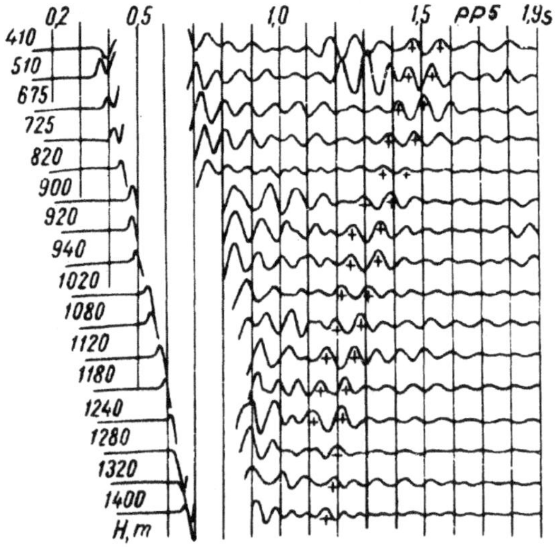

Fig. 49. A reflected PP^5-wave recorded by a low-frequency instrument.

It should be pointed out, however, that it is not always possible to attenuate multiples and to improve the correlation of the reflected waves by changing to lower frequencies. Large volumes of VSP data obtained in the Moscow syneclees which features a high-velocity thin-layered section, with strong velocity differentiation in the upper portion and weak differentiation at depth, have demonstrated that the change to a high-frequency band improves the discrimination of reflected waves associated with pre-Devonian deposits at great depths. The structure of the wave field, the relative intensity of individual waves and the recorded time intervals depend essentially on the recording frequency band. This does not apply to the top of the crystalline basement from which reflections can best be tracked in the lower frequency band.

HIGH-VELOCITY CARBONATE SECTIONS

In a carbonate section, the multiples may be widespread when regular waves cannot be identified on surface seismograms at all. The multiples can be associated not only with interfaces in the upper portion of the section, but also with those located at great depths. Consider a wave pattern typical of the Russian platform. It is a well-known fact that, because of extremely complex seismologic conditions obtaining on the right bank of the Volga, in the Saratov and the Volgograd regions, the effectiveness of seismic studies of the terrigeneous Devonian deposits that were of most geologic interest, remained inadequate for a long time, despite the great volume of work performed. This is explained primarily by the small number of regular waves recorded on surface seismo-

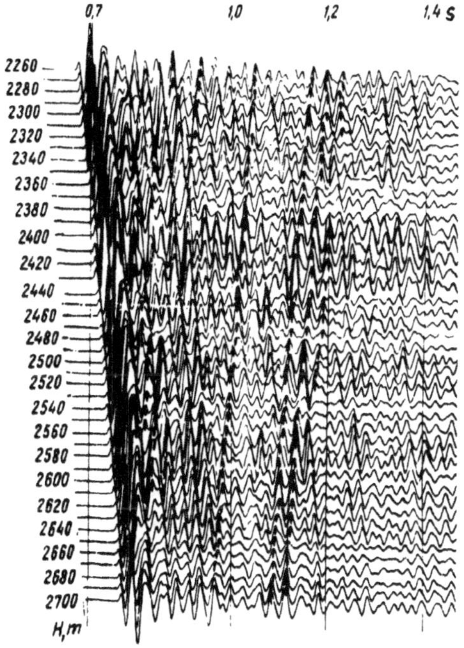

Fig. 50. The initial portion of a VSP seismogram (shotpoint 70 m, h = 125 m, borehole N90 Kudinovskaya).

grams if one excludes the stable dominant wave reflected from the top of the Paleozoic, where the terrigeneous bed is superceded by carbonate deposits.

In many cases, this situation has been attributed to a thinly layered structure, to the variability of the thickness of the thin layers contained within the Devonian deposits, to the roughness of the interfaces for seismic reflection frequencies or to the displacement of the dominant seismic frequencies to extremely high values that are than scattered and absorbed in the process of propagation to the surface. Such conclusions that deny any chance for any future reflection-method work are obviously dangerous. On account of the great importance of the problem for the national economy, we deemed it interesting and valuable to study the causes of the lack of reflected waves at the surface by making use of VSP techniques.

Observations carried out almost on top of the terrigeneous Devonian bed (at depths down to 2700 m) have demonstrated, despite the complexity of the wave field, that waves reflected from several interfaces contained in the terrigeneous bed (Figure 50) can be identified down-hole, although the recording conditions for such waves on the vertical profile are unfavourable, because of interference from numerous downward waves. Such waves have been tracked along horizontal profiles where the wave field is somewhat simpler than on the surface.

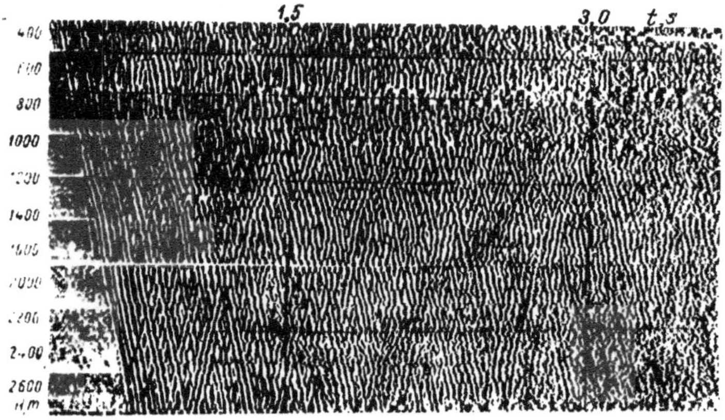

Fig. 51. A VSP seismogram depicting primary and multiple waves from deep interfaces in a sedimentary bed (shotpoint 70 m, borehole N90 Kudinovskaya).

Waves associated with deep interfaces consist not only of primary, but also numerous multiply reflected waves (Figure 51). In fact, all downward waves (and they are present on the seismograms in great numbers) are reflected by deep interfaces and return to the surface. On account of the abundance of waves, they are all located in an area of mutual interference, and it is rather difficult to decipher the resulting wave pattern. The only point that can be made is that all singly reflected waves associated with sedimentary beds are distinguished by higher frequencies. The first intense low-frequency wave is apparently the wave reflected from the crystalline basement located at a depth of 3400 m. Most multiple upward waves recorded after it display a comparatively low-frequency spectrum.

4. Wave Fields Associated with the Upper Portion of the Geologic Section

The upper portion of the section is the most heterogeneous part, and many waves other than multiples that have a high intensity and decisively influence the effectiveness of the seismic reflection method originate in it. Accordingly, the study of waves originating in the upper portion of the section is not only of scientific, but also great practical interest. Consider wave fields associated with the near-surface for two separate conditions, one with sharp interfaces and one with a weak velocity differentiation. Such structures are typical of many regions.

UPPER PORTION OF THE SECTION WITH SHARP INTERFACES

In many regions, the upper portion of the section contains one or more sharp interfaces, e.g., for the Russian platform, such an interface is the top of the Paleozoic at which the transition from terrigeneous to carbonate deposits takes place. The interface is stable over a tremendous area, however, its depth does not remain constant varying by several hundred meters. A combination system has been shot consisting of both horizontal and vertical profiles. The latter has been shot from eight points in the range 0 to 2000 m. Observations in a specially drilled hole that penetrated the top of the Paleozoic, at the depth 374 m, have been carried out with the aid of a three-geophone (Z-component) probe and a three-component symmetrical geophone cluster. The horizontal profile was made with the aid of three-component XYZ geophones. The observations demonstrated that the principal elements of the geologic section consist of the base of the low-velocity zone ($H_v = 16$ m), where an abrupt change in the velocity from 560 to 1800 m s^{-1} takes place, and the top of the Paleozoic ($H_v = 374$ m) with a velocity change from 1880 to 4000 m s^{-1}. A sandy clay bed contained within those boundaries is weakly differentiated with respect to both longitudinal and transverse waves. Two thin interbedded sandstones with comparatively small velocity variations also manifest themselves in the section.

A large number of various wave types were recorded on the VSP seismograms. Let us dwell on the details of the early part of the seismogram, before total mutual interference of the waves takes place. The first wave to be recorded on the VSP from the nearest shot-points (SP 40 m and 200 m) is the direct downward wave, followed by a sequence of multiply reflected waves associated with the near-surface (Figure 52). The total downgoing wave field includes a large group of waves which have been reflected from the top of the Paleozoic and again, on reaching the base of the low-velocity zone, have been reflected downwards forming a continuous wave train. With increasing offset distance, the wave pattern in the initial portion of the record becomes more intricate. The direct wave from SP 400 m in the section of the profile contiguous with the top of the Paleozoic (H = 360–380 m) is superceded as the first arrival by the refracted wave. With increasing offset distance (SP 500 m–750 m) the refracted wave is recorded as the first arrival over an increased depth range (H = 230–380 m). The refracted, the direct and the transcritically reflected waves all interfere in the initial portion of the record. The intensity of the transcritically reflected waves is sufficiently high to make it almost impossible to detect the direct wave against their background.

The first wave to be recorded from SP 1000 m and 1500 m along the entire profile is the wave refracted in the Paleozoic. The direct wave cannot be identified. The

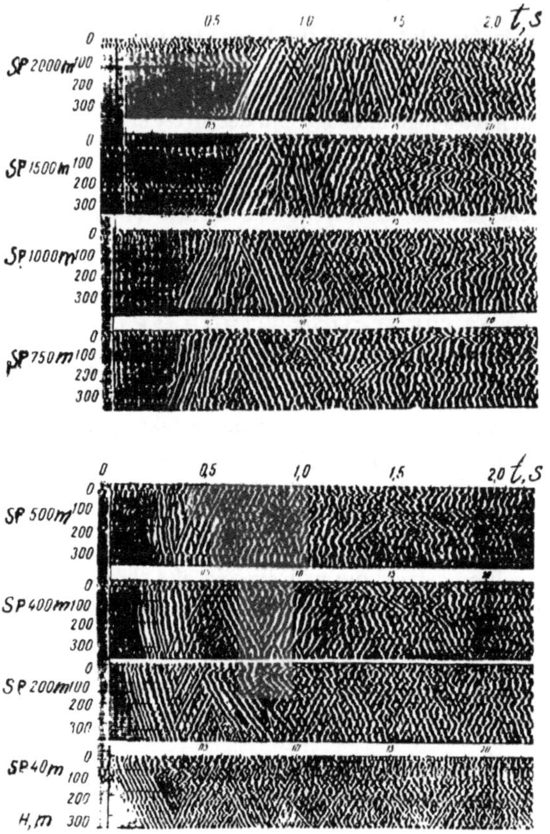

Fig. 52. VSP seismograms (borehole N90–a Kudinovskaya).

transcritically reflected wave has high intensity and generates multiples. A further increase in offset (SP 2000 m) extends the interval in the early part of the seismogram occupied by refracted waves, with the transcritically reflected and associated waves recorded at longer times.

The subsequent portion of the record shows a complex wave field resulting from the superposition of waves of a great number of different types. In addition to longitudinal waves, converted and transverse waves can also be identified on the seismograms of the Z-component.

Hence, the presence in the upper portion of the section of a sharp interface results in the formation at short offset distances of a set of reflected and multiple waves associated with the direct wave. At long distances the intense secondary waves in the initial part of the record are associated chiefly with the wave refracted in the top of the Paleozoic. On the other hand, the direct wave propagating at a low velocity forms its own set of secondary and multiple waves in the later part of the record. All such waves interfering with one another form a complex train of downward waves that are multiply

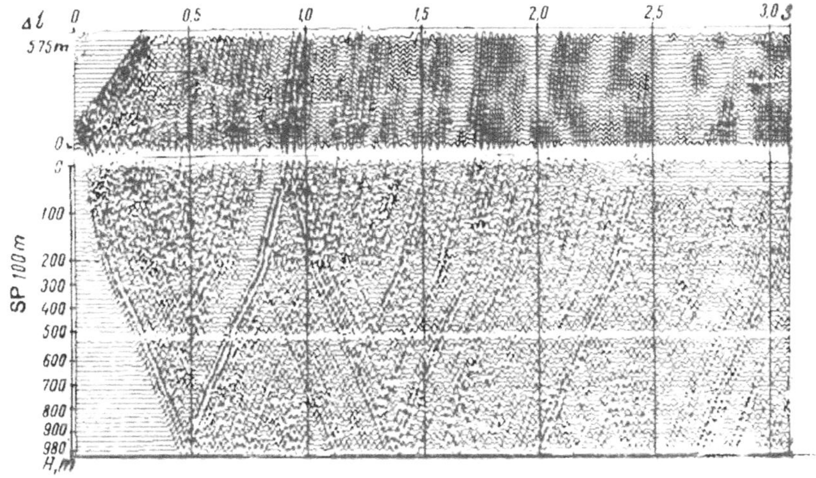

Fig. 53. Surface seismograms (0–575 m) and VSP seismograms recorded in a locality with a deep-lying Paleozoic top (according to [103]).

reflected at the interface, and greatly complicate the discrimination of primary reflections associated with great depths.

It should be noted that in areas where the top of the Paleozoic is comparatively deep, it also excercises a substantial influence on the wave field. This influence is not however, the same as in the case of a shallow interface. Where the top of the Paleozoic is located at a depth of about 1000 m, the wavefield includes singly reflected waves and full and partial multiples associated with the principal and the intermediate interfaces (Figure 53). These waves dominate the surface seismograms occupying almost the entire recorded time interval.

The polarization method opens up great possibilities for the study of wave fields typical of the upper portion of the geologic section. The seismograms of Figure 54 show a complex wave field formed by the superposition of a great number of waves of various types. The first longitudinal wave can be reliably tracked on the *R* and *X* components. The longitudinal reflected waves in the initial and the subsequent portions of the record can be reliably tracked on the *R*, *Z*, 17 components. The converted reflected waves occupy the whole of the initial portion of the record of the *R*, 12, *X* components. The recording of the direct transverse waves is at its optimum on the *R*, *Z*,17 components. The reflected transverse waves in the subsequent portion of the record are recorded with the aid of the *R*, *Y* components.

Thus a complex field of unwanted waves is built up in the upper portion of the section which limits the effectiveness of seismic studies in the area.

THE UPPER PORTION OF THE GEOLOGIC SECTION WITH WEAK VELOCITY DIFFERENTIATION

The upper portion of the geologic section exerts a strong influence on the wave field even in the absence of sharp interfaces, e.g., when there is a continuously-changing medium

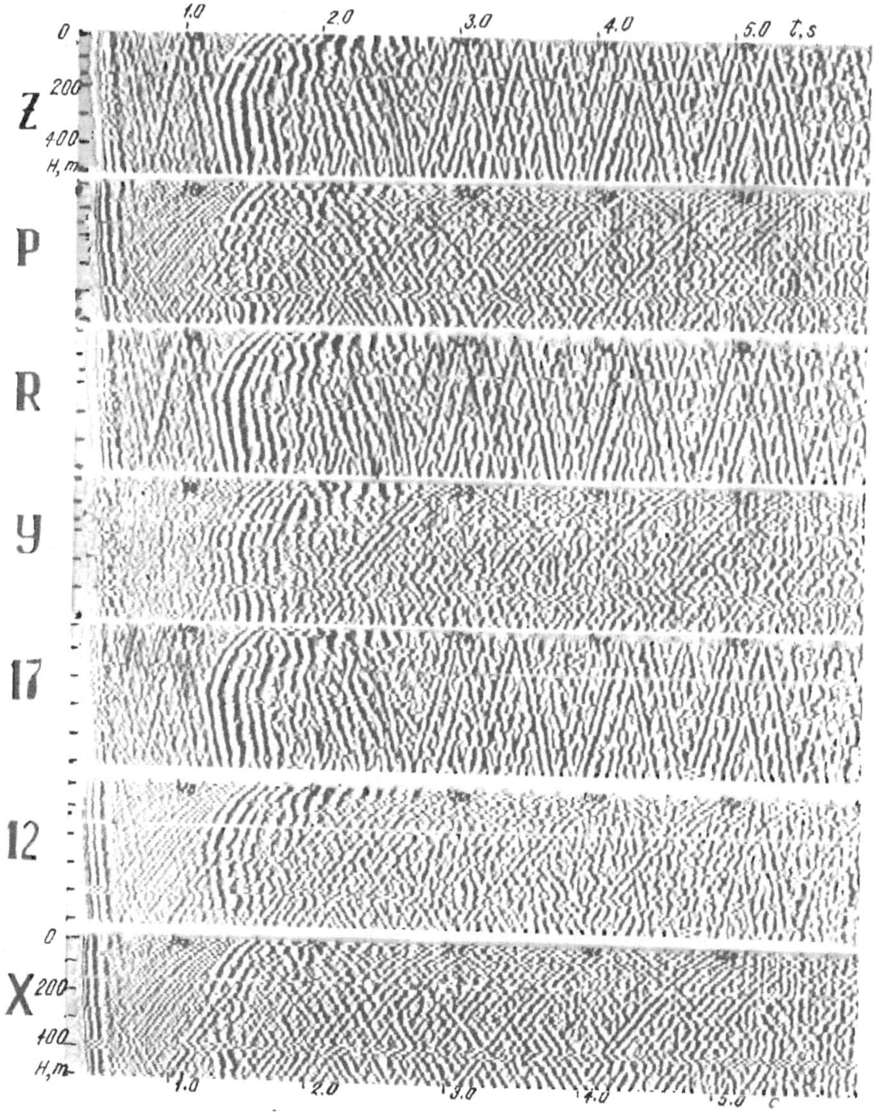

Fig. 54. PM–VSP seismograms illustrating a wave field in the upper portion of the geologic section, $l = 560$ m.

or a medium with weak velocity differentiation. Let us illustrate the point making use of observation data obtained in Central Asia, where a monotonous increase in velocity with depth and the absence of sharp interfaces are typical of extensive folding of terrigeneous deposits. Still, a complex wave field produced by curved longitudinal and transverse waves and their multiples is recorded on the seismograms making discrimination of reflected waves impossible.

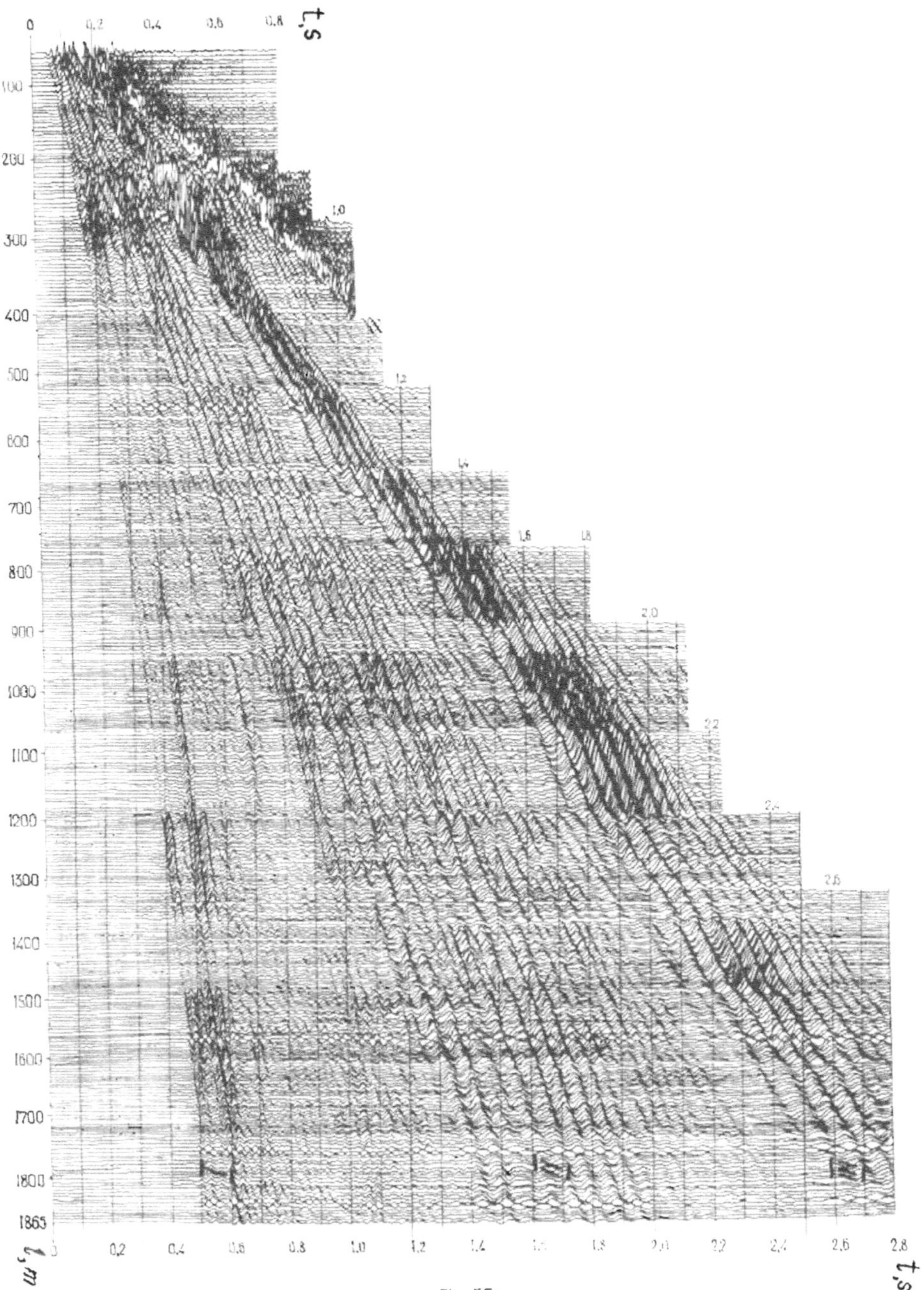

Fig. 55a.

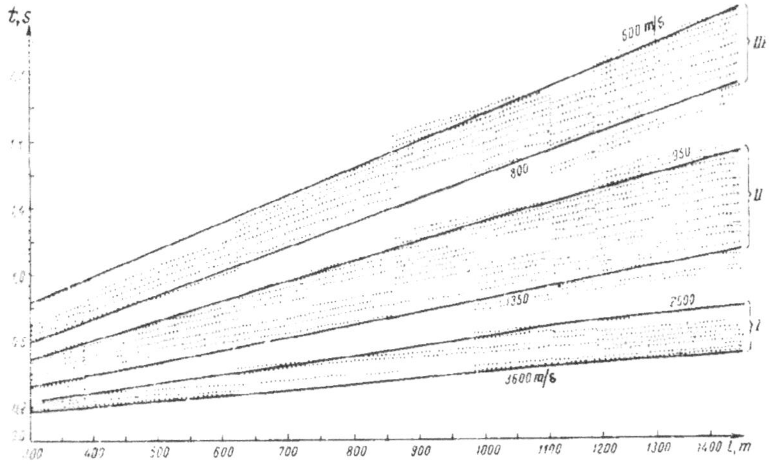

Fig. 55b.

Fig. 55a-b. Seismogram recorded along a horizontal profile (a) and traveltime curves of three dominant groups (I, II, III) of unwanted waves (b) (Kaptut, Uzbek SSR).

VSP data have been obtained in areas located on the Naryn monocline (in the centre and on the southern slope) as well as on the western fringe of the Fergana valley [104]. In the majority of regions the upper portion of the section is represented by loessy loam changing to conglomerates with individual lenses, and thin interlayers of clays, flint-stones, sands and siltstones.

The longitudinal wave velocities in various regions vary in the low-velocity zone from 400 to 700 m s^{-1}, and in the bed underlying the low-velocity zone – from 1200 to 2000 m s^{-1}. The upper portion of the section is not differentiated with respect to the velocity of transverse waves; the wave velocities increase smoothly with depth from 800–900 m s^{-1} at a depth of 50 m to 1000–1100 m s^{-1} at 200 m. The velocity ratio of the longitudinal and transverse waves varies in the continuously changing bed from 2.2 to 2.5 in the upper part to 2.0 at a depth of 400 to 500 m. This ratio changes little with further increase in depth. The value of the coefficient of velocity increase for longitudinal waves, β, for various regions is equal to about 0.0010 m^{-1}. For transverse waves $\beta = 0.0015$ m^{-1}.

Observations were carried out in a specially drilled parametric hole. Detailed observation systems in the vertical plane (see Chapter 2) were devised, enabling zone contours for each principal group of waves to be drawn, and their time fields to be constructed. VSP observations were carried out with the aid of a symmetrical three-component geophone cluster with a natural frequency of 15 to 17 Hz. The principal feature of the wave fields in all the explored regions was that they all belong to the same type. A characteristic feature of surface seismograms is the presence of a large number of clearly defined waves that can be subdivided into three groups (Figure 55) according to their times of arrival, apparent velocities, visible recorded frequencies and to other parameters.

Waves in the first group were recorded by the vertical and the horizontal geophones

TABLE VI

Region	l (m)	V_a (m s^{-1})
Akbel'	50–3000	2400–3600
Kaptut	50–2000	2200–4000
Kapchagai	50–2000	2800–3200
Dzhel'tmess	50–2000	2200–3100

and constitute the first arrivals and the early part of the record. The number of waves making up the group is not the same in different regions. The traveltime curve of the first arrivals is smooth. Changes in the types of waves in the first arrivals along the profile can be detected only in the immediate proximity of SP 50 to 100 m. The apparent velocities of the first arrival increase with offset distances (Table VI). For subsequent waves the apparent velocities diminish. Thus, in the Akbel region at the distance l = 400 to 500 m, the apparent velocity measured from the initial phases (at the time t = 0.3 s) is equal to 2400 m s^{-1}, dropping to 1600 m s^{-1} for the last visible phases at the time

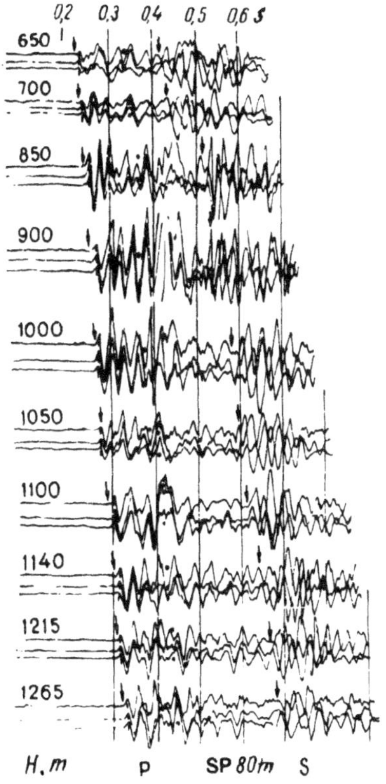

Fig. 56. Seismograms of a three-component cluster along a vertical profile (Airitan, Tadjik SSR).

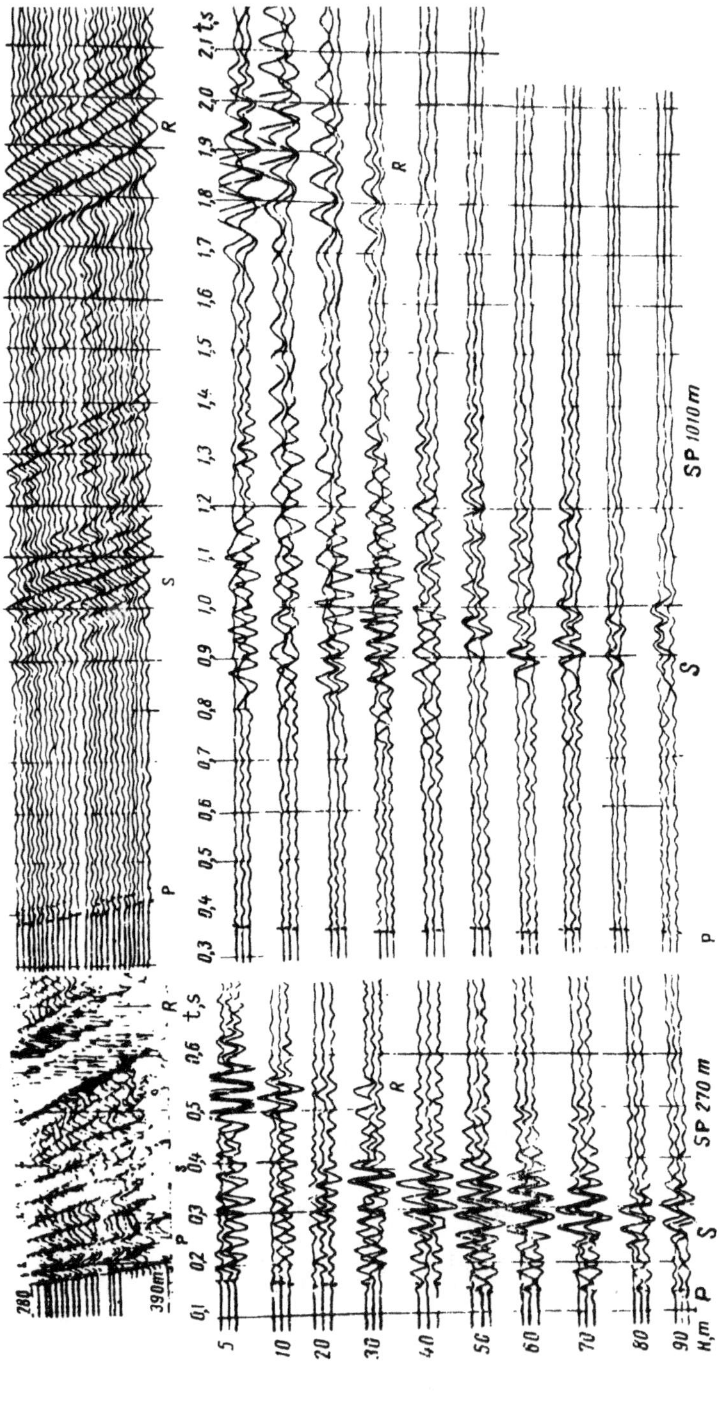

Fig. 57. Seismograms of the Z-component along a horizontal profile and of components of a three-component symmetrical cluster along a vertical profile (on the left SP 270 m, on the right SP 1010 m).

$t = 0.6$ s. For $l = 2000$ m, the apparent velocity drops from 3100 m s^{-1} at $t = 0.9$ s to 2500 m s^{-1} at $t = 1.5$ s.

The intensity of the waves in the first group is the lowest on the Z-component seismogram, and changes little with an increase in offset (Figure 56, 57). The wave frequencies close to the shot-point vary in different regions within the limits 50 to 80 Hz, gradually decreasing with increasing offset from the shot-point, to approximately 30 Hz at a distance of 3000 m. The shape of the record along the vertical profile is stable. Visible frequencies of the first arrival for various regions decrease with depth from 70 Hz at $H = 150$ m to 50 Hz at $H = 1265$ m. At greater offsets, the record of the first group of waves becomes more complex with the appearance of additional phases (Figure 58). Linear polarization of the first arrival is usually retained in the entire range of distances and depths.

Analysis has shown the first wave group to consist of longitudinal curved waves and their multiples propagating in a continuously changing layered medium. Near the shot-point, such waves interfere with the reflected waves and their multiples associated with the bottom of the low-velocity zone and the upper portion of the section. The effect of the upper interfaces diminishes with increasing distance from the shot-point.

Apart from the longitudinal waves, the first group also includes converted SP-type waves originating in the upper portion of the section. Such waves are distinguished by greater intensity than the first arrival and by a lower recorded frequency.

Waves of the second group occupy a large portion of the seismogram and for this reason present the maximum interest. Wave kinematics on the horizontal and the vertical profiles agrees well with theoretical calculated results for curved-path transverse waves and their multiples. The values of v_a on the horizontal profile increase smoothly with an increase in offset, e.g., in the Akbel-region they grow from 1000 to 1800 m s^{-1} in the range from 100 to 3000 m (Figure 55). With increasing distance from the shot-point, the number of phases increases, the group broadens, and the waves can be separated (Figures 56, 57). The apparent velocities diminish gradually to 1100 to 1200 m s^{-1} as subsequent phases are considered. The waves are linearly polarized. The displacement directions are orthogonal to those of longitudinal waves. The frequencies of the waves in the second group are lower than those of the first, and vary with the distance from the shot-points in all areas, from 30 to 35 Hz for $l = 100$ to 300 m to 20 Hz for $l = 2$ to 3 km. The change in the wave frequencies with depth is very slight. There is no visible dispersion.

Close to the shot-points, the wave intensities of the second group on Z-component seismograms exceed those of the first group by about one order of magnitude. They diminish with increasing distance from the shot-point, and at a distance of 2 to 3 km the intensities of the waves of groups I and II equalize (Figure 55). At small depths, the intensity of group II waves is 3 to 4 times that of group I. With increasing depth this ratio diminishes, and at a depth of 1200 m the intensities of waves belonging to both groups become practically equal. The shape of the record remains stable along the vertical profile. In contrast to the waves of group II, the recorded frequency of group I waves has been found to depend on the charge size. Thus, a change in the charge from 0.5 to 16 kg resulted in the recorded frequency decreasing from 35 to 25 Hz for $l = 300$ to 400 m. The intensity ratio for the waves of groups II and I varied correspondingly from 2 to 3 for a 0.5 kg charge, to 10 to 12 for a 16 kg charge. In downhole observations,

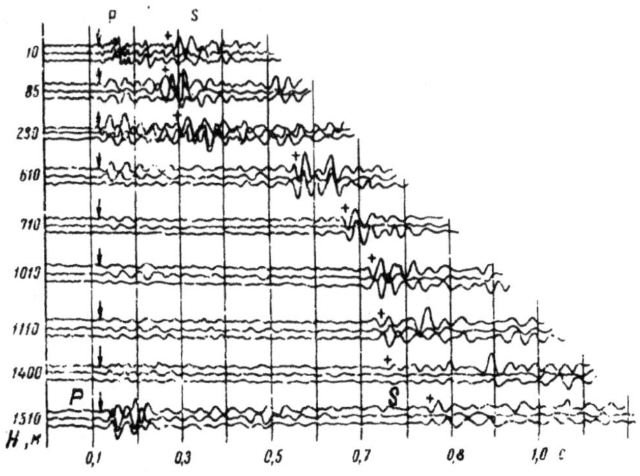

Fig. 58. Seismograms of components of a three-component symmetrical cluster along a level profile (H = 200 m) (Akbel', Tadjik SSR).

group II waves were recorded everywhere at all accessible depths. The vertical travel-time curves of group II waves in all regions belong to the same type and correspond to a continuously changing medium.

The group II waves constitute the principal unwanted waves obstructing the recording of longitudinal waves associated with deep horizons. For a long time the origin of group II waves remained obscure. Analysis of VSP data has shown group II to consist of direct transverse waves and their multiples. Their presence is mainly determined by the formation of intense transverse waves in the course of an explosion and by the increase in velocity in the upper portion of the section.

Waves in the third group depend greatly on the low-velocity zone. In some areas they dominate the seismograms and in others their intensity is so low that they cannot be identified at all. On surface seismograms, group III waves are fan-like diverging oscillations with great dispersion (Figure 55). Their apparent velocities vary from 1000 to 2000 m s^{-1} in the initial portion to 200 to 300 m s^{-1} in the tail end of the wave train, increasing slightly with the distance. The dominant recorded frequencies in the initial portion are 10 Hz. The waves are polarized in the vertical plane.

Typical of group III waves is a drastic drop in intensity with depth, so that at a depth of about 40 m they can practically no longer be identified (Figures 56, 57). The attenuation is more rapid in the case of the high-frequency components. The spectral band width decreases with the depth, shifting to the lower frequency end. Long waves penetrate to greater depths. The thickness of the layer in which the group III waves can be recorded in practice amounts to about three quarters of the wave length.

Group III waves are surface waves of the Rayleigh type. It should be noted that the field of low-velocity and low-frequency unwanted waves may be formed by various harmonics of Rayleigh surface waves, with phase velocities varying from that of Rayleigh surface waves on the surface to the maximum velocity of transverse waves in the Earth. The intensity of such waves on surface seismograms is determined by the frequency, the order of the harmonic, the depth and the type of source. The highest harmonics are the most intense, and, in contrast to the fundamental frequency, their intensity decays slowly with the depth of the charge.

In addition to the three main groups, the seismograms also contain less intense regular unwanted waves, which can be detected only after the dominant waves have been eliminated, as well as any random background noise. The origin of the latter is not discussed herein. A similar wave pattern is typical in Antarctica. Study of the properties of the wave field produced information necessary for the computation of geophone grouping parameters. Multi-elements groups enable the level of unwanted waves to be substantially reduced, and seismic exploration to be carried out in regions formerly unsuitable for seismic observations. However, the longitudinal reflected wave method remains inappropriate in many areas. To improve the effectiveness of seismic exploration in such conditions, it is advisable to use the polarization method, simultaneously employing both longitudinal and transverse waves. In 1966, in two other areas in the Fergana valley, special experiments were carried out with the aim of detecting reflected *SS* waves on the surface. Despite the comparatively limited extent of the experiments, horizontal surface geophones have been able to record reflected SS waves at times corresponding to the arrival of waves reflected from deep horizons. The polarization method may be expected in such conditions to substantially simplify the observation procedure and make possible the exploration of many areas that are still unsuitable for seismic exploration.

5. Experience Gained in the Study of the Wave Field in Mining Areas

VSP observations have gained special importance in seismic exploration for ores. High-velocity and weakly dislocated media, high-dip and intricately shaped interfaces are responsible for extremely complex wave fields typical of the mining areas. In such conditions, the analysis of the wave field from surface observation data is fraught with great difficulties, and in many cases, cannot be performed unambiguously. This is primarily true of the initial part of the seismogram, and for this reason, seismic exploration for ores provides scanty information about the upper portion of the geologic section (down to 1000 to 1500 m) that is of most geologic interest. The above-mentioned factors explain the existing slow progress in seismic exploration for ores.

The existence of a large number of boreholes and comparatively shallow depths, usually characteristic in mining exploration, facilitate the use of VSP. Technical difficulties result mainly from the small borehole diameters. A large volume of VSP (more than 10000 m) carried out in holes of 59 mm diameter drilled with diamond drills in Khibin mountains [105], enabled the wave fields to be studied in detail, and the reflected waves (longitudinal, transverse and converted) associated with various petrographic differences of alkaline deposits in the rock mass to be identified. Simultaneously the presence of stable reflecting horizons has been demonstrated.

The PM–VSP method created new opportunities for the development of seismic exploration for ores, as shown by the large amount of PM–VSP observations carried out for mining purposes in Central Kazakhstan. This work marked the beginning of the development of the polarization method [2]. On the basis of experience gained in the course of the development and employment of PM–VSP, it is possible to formulate certain special features of waves observed in ore-rich areas. In ore-rich areas one is able to record a large number of waves associated with both lithological interfaces and with tectonic faults. It is not always possible to determine uniquely the nature of the reflecting interfaces without detailed PM–VSP studies. Reflected waves of various types associated with the same interfaces, both sub-horizontal and sub-vertical, can be identified and tracked on PM–VSP seismograms. To determine the origin of waves associated with the vertical interfaces, one must have data on wave polarization in addition to kinematic data. In many cases, PM–VSP data may prove more valuable for mapping sub-vertical interfaces than data obtained from surface observations. By combining VSP with surface observations, it is possible to ensure the effectiveness and reliability of the geologic results of seismic exploration for ores.

The orientation of the tracking components of various wave types corresponds mainly to the direction of approach of the waves, enabling, in many cases, the sidewaves that greatly distort the section when interpreted in the vertical plane to be identified and ignored. The appropriate sections usually take the form of a family of intersecting interfaces, which may lead to gross errors in the geologic interpretation.

A large volume of PM–VSP work has been carried out in the iron-mangese and barium-polymetallic mining area of Zhairem located in the Taimin syncline. The region features an exceptionally complex wave field containing numerous waves of different types, with different apparent velocities. Let us illustrate the point by analyzing PM–VSP data obtained on the vertical (PM–VSP) and the surface (PM) reflection profiles shot from the same point. It follows from the fixed component seismograms (Figure 59) that different waves, whose tracking component regions are shown on the stereographic net (Figure 59b, top) are recorded with the aid of different components. The stereogram shows the regions of the wave tracking components to occupy practically the entire space. The most complex wave field has been recorded on the Z-component. Analyzing the records of other components, one is able to separate various waves, e.g. the reflected longitudinal *PP* waves with a larger (v_a +) and a smaller (v_a –) apparent velocity; PP_{vc} waves reflected from the vertical contact; converted reflected and transmitted *PS* waves. The tracking regions of the corresponding waves are found to be well correlated along the vertical and the horizontal profiles shot from the same point, if the top and the bottom diagrams in Figure 59b, are superimposed on each other in the direction of the vector P (bold arrow), i.e. in the direction of propagation of the refracted wave associated with the bottom of the weathering layer ($H_{in} \sim 100$ m). The difference between the directions of the vector P in the top and the bottom diagrams is due to different orientations of the three-component clusters used in downhole and surface observations.

The tracking regions of the first longitudinal (P), and the longitudinal *PP* waves (bold solid line), generally extend in the direction of the plane of the rays. *PP* waves with high apparent velocities on the vertical profile (v_a +), the PP_{vc} waves and the *PS* waves all display great azimuthal deviations. Accordingly, the latter are well correlated on the Y-component (Figure 59a). Analysis of the wave field has demonstrated that the orientation of the regions of tracking components for longitudinal reflected waves associated with sub-horizontal interfaces depends substantially on the presence of unwanted waves. The PP_{vc} and the *PS* waves of highest intensity can be well tracked on those components that coincide in direction with that of their propagation.

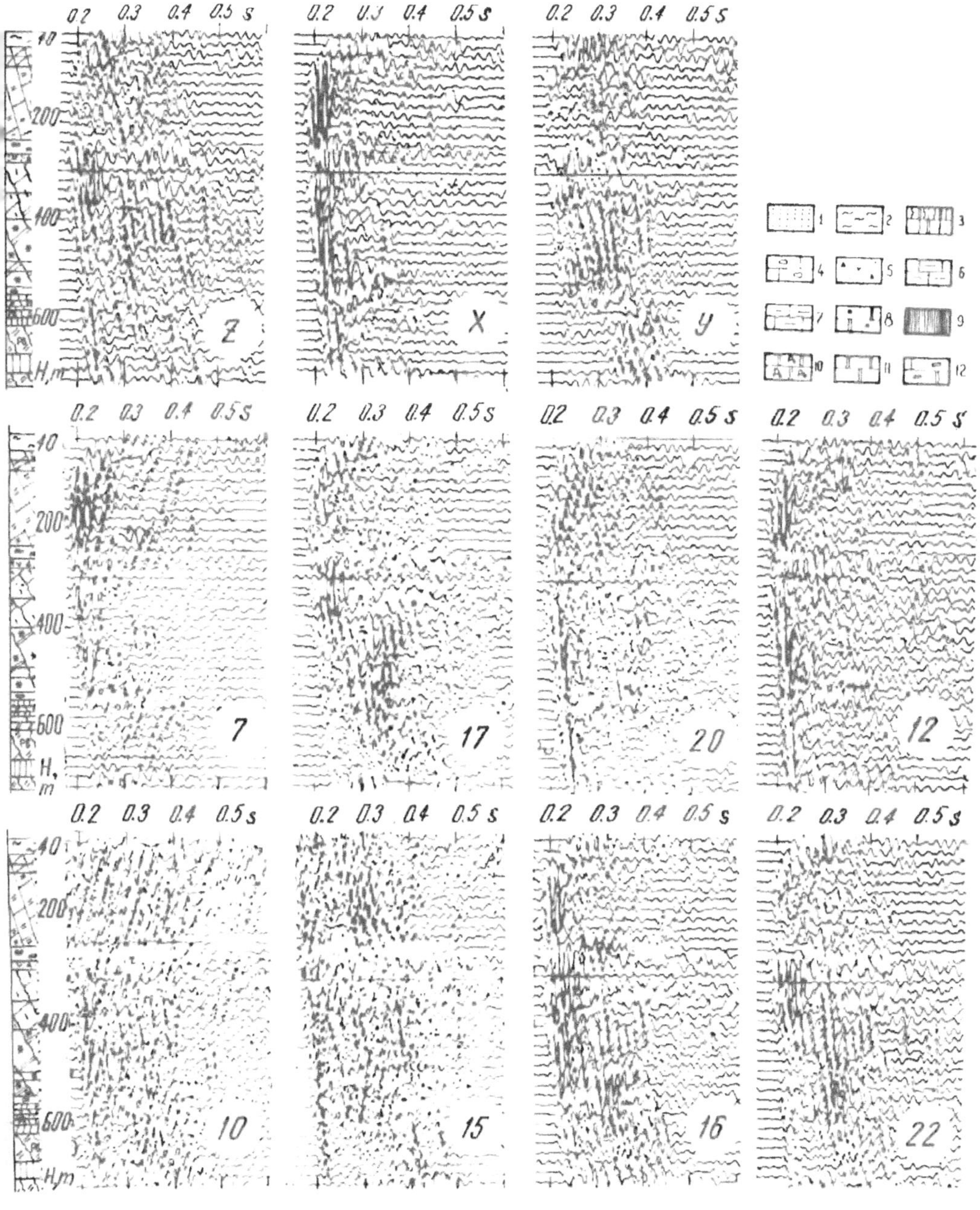

Fig. 59a.

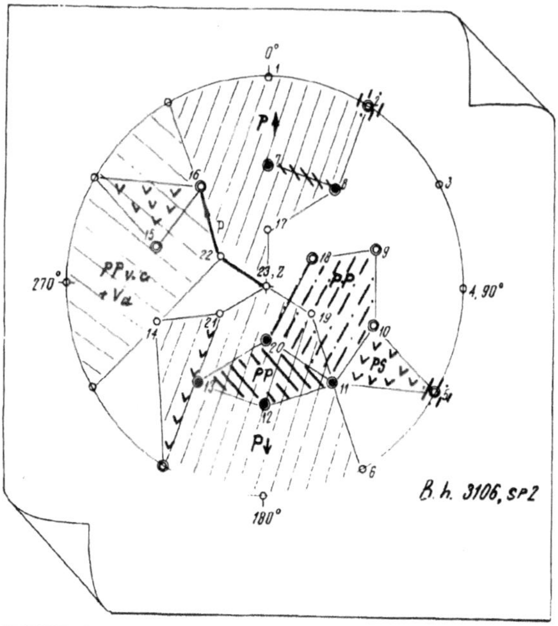

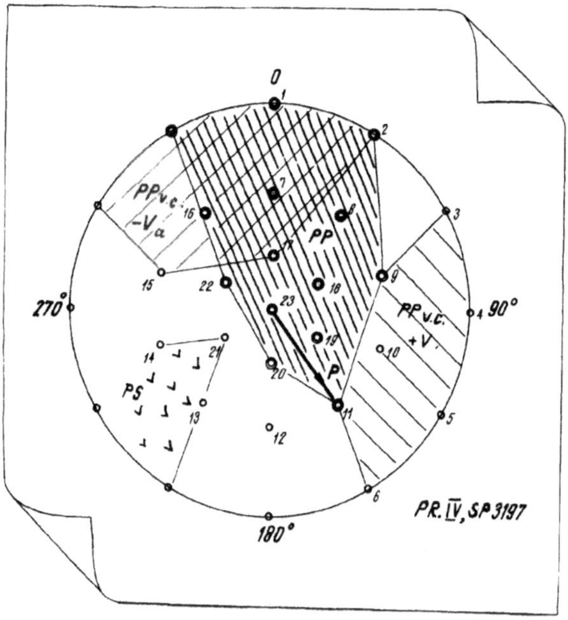

Fig. 59b.

Fig. 59a-b. PM–VSP seismograms recorded in the Zhairem ore fields (bortehole N3106, $l = 660$ m) (courtesy R. N. Khairutdinov): (a) set of fixed components; (b) stereograms of tracking components of waves of various types on a vertical (top) and a horizontal (bottom) profiles: 1 – sands, argillaceous sands, 2 – Aral suite clays; 3 – quartz limestones of the upper horizon of upper Tournaisian substage, 4 – horizon of nodular limestones; 5 – subnodular horizon of tufaceous limestones and tufaceous argillites; 6 – horizon of 'coal' limestones; 7 – quartz-carbonate rock; 8 – red quartz rock of the upper horizon of the upper Famennian; 9 – iron ore horizon; 10 – lead-zinc horizon; 11 – clay-quartz-carbonate rock of the productive bed of the upper Famennian; 12 – quartz limestones with concretions – lower horizon of the upper Famennian.

6. Relative Intensity of Multiples Obtained from VSP Data

The potential of the seismic reflection method, as has been demonstrated in VSP studies, is often determined by the effectiveness of the methods used to suppress multiples. In order to compute and design optimum observation systems, one should know the parameters of the multiples, particularly their relative intensities, i.e., the signal-to noise ratios. A qualitative assessment of the intensity of the multiples can be derived on the basis of the downward wavefield observed on VSP seismograms. However, in order to compute common depth-point-systems, quantitative estimates must be available. It is worth noting that the common-depth-point method enables the multiples to be evaluated during the processing stage.

VSP primarily offers new possibilities for the study of multiples, including the means to extract data independently and reliably on their relative intensity on the surface. It took only a few years to develop about ten methods of evaluating the intensity of the multiples, each distinguished by its own particular limitations and simplifying assumptions concerning the earth and the nature of the multiples, and by the initial data, measured parameters and methods of processing.

A detailed analysis of these methods has been carried out under the supervision of G. I. Petrashen', T. I. Vavilova, and N. I. Berdennikova. We shall consider here only the principal features of the more general methods. The idea of employing VSP data to determine the relative intensities of multiples was first put into practice by I. M. Muzyka, who suggested a method for computing the intensity of partial multiples applicable to thick-layered media containing sharp interfaces [102]. In principle, a rigorous solution of the problem of determining the relative intensities presents great difficulties. However, by introducing simplifying assumptions, one is able to solve the problem approximately, with an accuracy that is adequate for all practical purposes.

These methods can be conveniently subdivided into two groups. The first applies to thick-layered media, in which case it is possible, in principle, to obtain an idea of the earth model, the principal multiple generating interfaces and the trajectories of the multiples [106]. Actually, the real earth and, particularly, sedimentary beds are made up of thin layers, and waves originating and propagating in them turn out to be combined waves. For such media, the second group of methods were developed, which do not require a knowledge of the Earth model and the ray-path pattern [107, 108]. However, all these methods require down-hole measurements of the parameters of various waves and their subsequent conversion to surface parameters, in order to determine the relative intensities of the multiples on the surface with respect to the simultaneously recorded primary reflections. The general schematics of the second group of methods was devised by G. N. Gogonenkov [109] for a thin-layered medium with almost horizontal bedding. Let us consider this method in more detail. The method makes use of VSP and surface data for normal incidence. The complex interference pattern formed as the result of superposition of the primary reflections and the combined wave field of the multiples arriving simultaneously at a given point are compared with the useful reflected wavefield. The parameters of the useful reflected wave close to the inteface, where it is least distorted, are converted to surface parameters. To this end, it is assumed that there is a reference reflection with a dominant amplitude both on the surface and the borehole data, and that the propagating waves do not change their form. This obviates the need for information about the ray-path pattern of the multiples.

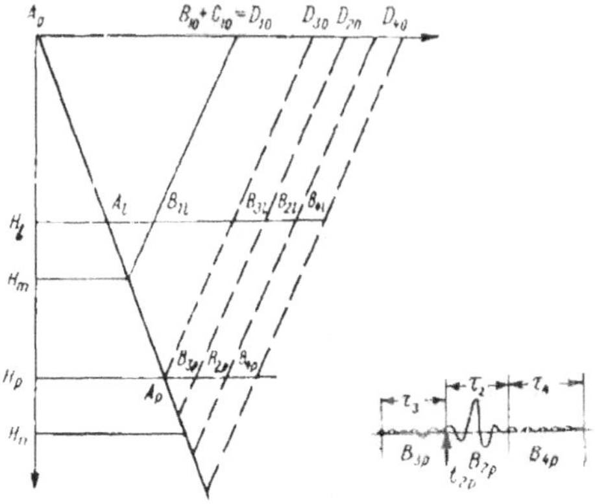

(a)

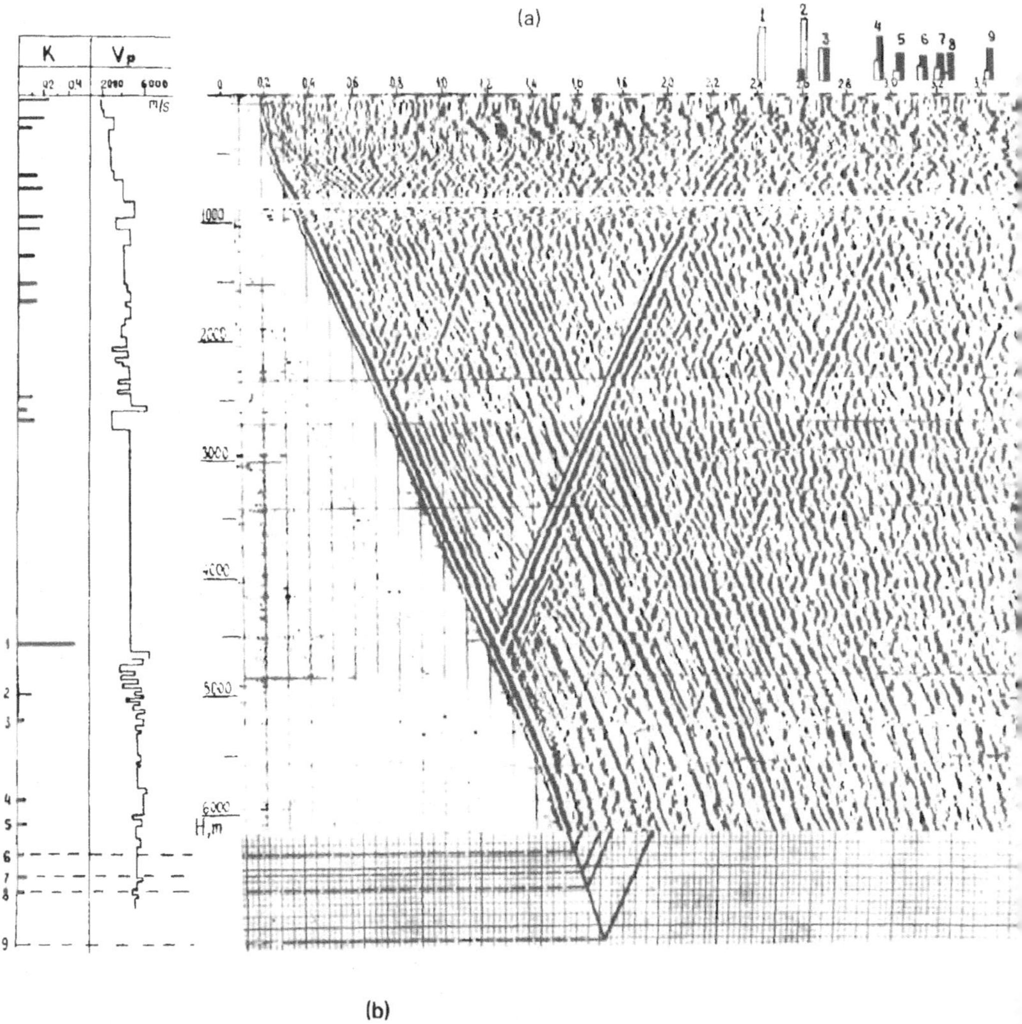

(b)

Introducing the notation A, B, C for the amplitudes of the direct, the primary reflection and the multiple wave, respectively, the following parameters can be determined from a VSP seismogram:

(a) The amplitude of the direct wave (A_l and A_p) at two depths (H_l and H_p) (Figure 60).

(b) The amplitude B_{1l} of the reference reflection at the depth H_l.

(c) The amplitude B_{1p} of the useful reflected wave at the depth H_p is measured in the entire recorded time interval τ_2 (the first suffix indicates the number of the interval of registration, and the second the observation depth).

Before the amplitudes of the reflected waves are read off, the downward waves are suppresed by means of directional summation or subtraction.

The following are measured on surface seismograms:

(a) The amplitude B_{10} of the reference reflected wave on the surface.

(b) The complex interference wave field $D_{20} = B_{20} + C_{20}$ in the interval of the 'useful' wave τ_2 and the wave fields $D_{30} = B_{30} + C_{30}$ and $D_{40} = B_{40} + C_{40}$ in the intervals τ_3 and τ_4, respectively.

Consequently, the problem reduces to the determination of the B_{20}/C_{20} ratio, of B_{1p}, B_{2p}, B_{3p}, and B_{4p} waves from VSP seismograms, and D_{10}, D_{20}, D_{30}, and D_{40} from surface seismograms. The wave B_{2p} is represented by the average recorded intensity as determined by the standard deviation

$$B_{2p} = \frac{1}{n}\sqrt{\sum_{t_2}^{t_2+n\Delta t} (B_i - B_{av})^2},$$

where t is the sampling rate, ($n\Delta t = \tau_2$), B_i is the amplitude recorded with a sampling rate Δt, and B_{av} is the average amplitude on the seismogram in the interval τ_2. For the conversion of B_{2p} and B_{20} one makes use of the surface (B_{10}) and the down-hole (B_{1l}) amplitude values of the reflected reference wave:

$$B_{10} = \frac{A_{00}v_0}{2\sum_0^m H_i v_i} \exp\left(-2\sum_0^m H_i\alpha_i\right)\prod_0^m (1 - K_i^2)K_m G_0, \tag{29}$$

where A_{00} is the intensity parameter of the source obtained from the surface seismogram; G_0 is the conversion coefficient; α_i is the absorption coefficient in the ith layer; K_i is the coefficient of reflection from the interface between the ith and the $(i+1)$th layers.

$$B_{1l} = \frac{A_{0l}v_0}{\sum_0^m H_i v_i + \sum_l^m H_i v_i} \exp\left[-\left(\sum_0^m H_i\alpha_i + \sum_l^m H_i\alpha_i\right)\right] \prod_0^m (1 + K_i)\prod_0^m (1 - K_i) K_m, \tag{29'}$$

← Fig. 60. Determination of the relative intensities of the multiples [109]: (a) ray-path schematics and an example of the selection of recording intervals of upward B_{2p}–B_{4p} waves; (b) computation results. Values of reflection coefficient (K) and $V_p(H)$ curves are shown to the left of VSP seismograms. Columns on top indicate the intensities of the singly-reflected wave (light) and of the multiples (dark).

where A_{0l} is the source parameter at the depth H_l. The ratio of the amplitudes of the reference wave on the surface and at the depth H_l will be:

$$\frac{B_{10}}{B_{1l}} = \frac{A_{00}}{A_{0p}} \frac{\sum_0^m H_i v_i + \sum_p^n H_i v_i}{2\sum_0^n H_i v_i} \exp\left(-\sum_0^p H_i \alpha_i\right) \prod_0^l (1 - K_i) G_0, \tag{30}$$

and the ratio of amplitudes of the useful signal

$$\frac{B_{20}}{B_{2p}} = \frac{A_{00}}{A_{0p}} \frac{\sum_0^n H_i v_i + \sum_p^n H_i v_i}{2\sum_0^n H_i v_i} \exp\left(-\sum_0^p H_i \alpha_i\right) \prod_0^p (1 - K_i) G_0. \tag{31}$$

The amplitude of the useful surface signal is obtained by dividing Equation (30) by Equation (31):

$$B_{20} = B_{2p} \frac{B_{10}}{B_{1l}} \frac{A_{0l}\left(\sum_0^n H_i v_i + \sum_p^n H_i v_i\right) \sum_0^m H_i v_i}{A_{0p}\left(\sum_0^m H_i v_i + \sum_l^m H_i v_i\right) \sum_0^n H_i v_i} \exp\left(-\sum_l^p H_i \alpha_i\right) \prod_l^p (1 - K_i) \tag{32}$$

The last two multipliers in Equation (32) are unknown. To find them, one may make use of the amplitude ratio of any reliably tracked wave at points H_l and H_p. In the case of a direct wave

$$A_l = \frac{A_{0l} v_0}{\sum_0^l H_i v_i} \exp\left(-\sum_0^l H_i \alpha_i\right) \prod_0^l (1 + K_i);$$

$$A_p = \frac{A_{0p} v_0}{\sum_0^l H_i v_i} \exp\left(-\sum_0^p H_i \alpha_i\right) \prod_0^p (1 + K_i).$$

Should H_l and H_p be chosen so as to make $v_l = v_p$. one would obtain:

$$B_{20} = B_{2p} = \frac{B_{10}}{B_{1l}} \left(\frac{A_{0l}}{A_{0p}}\right)^2 \frac{A_p\left(\sum_0^n H_i v_i + \sum_p^n H_i v_i\right) \sum_0^m H_i v_i \sum_0^p H_i v_i}{A_l\left(\sum_0^m H_i v_i + \sum_l^m H_i v_i\right) \sum_0^n H_i v_i \sum_0^l H_i v_i}.$$

It is important to guarantee identical conditions of source excitation and instument setting at $H = H_p$ and $H = H_l$, i.e., $A_{0l} = A_{0p}$. Then expressing B_{20} in terms of the standard deviation, we obtain

$$B_{20} = B_{2p} = \frac{B_{10}}{B_{1l}} \frac{A_p}{A_l} \frac{\left(\sum_0^n H_i v_i + \sum_p^n H_i v_i\right) \sum_0^m H_i v_i \sum_0^p H_i v_i}{\left(\sum_0^m H_i v_i + \sum_l^m H_i v_i\right) \sum_0^n H_i v_i \sum_0^l H_i v_i} \frac{p_p v_p}{p_{pl+1} v_{pl+1}}$$

The right-hand side of the last expression contains no unknowns and may be used to find B_{20}, the standard deviation of the single reflection from deep horizons on the surface, and subsequently the quantity C_{20}.

The above consideration is not meant to be an exhaustive one. It simply illustrates the computation procedure. The method described automatically takes into account the entire wave field of the multiples recorded in the time interval of the useful wave. This refers not only to individual maximum recorded amplitudes, but to the amplitudes of all phases. Gogonenkov's method [109] is quite effective in media in which the waves propagate without the pulse being distorted. In cases when the visible recorded frequency is not conserved such computation should be performed for individual spectral components and the amplitude spectra of all waves should be computed beforehand.

A great volume of computation has been performed in accordance with the procedure described above, and has enabled the intensities of the multiples to be determined, and the optimum observation system in platform conditions to be selected. Figure 60 depicts VSP seismograms schematically representing the intensities of simultaneously arriving singly reflected and multiple waves. The wave corresponding to the sharp interface (1) has been chosen as the reference wave.

The seismogram (Figure 60) shows the intensity of the reflected wave 2 associated with the interface at the depth 5000 m to be 5 times that of the unwanted waves, whereas the wave 3 following it is comparable in intensity with the unwanted waves. The intensity of all subsequent waves associated with interfaces in the depth interval 5800 to 7000 m is appreciably below that of the multiples, this ratio diminishing with time from 0.5 for wave 4 to 0.25 for wave 9.

G. N. Gogonenkov's approach described above [109] was generalized to include the case of the useful signal whose shape changes in the process of its propagation along the ray path. This change is due to the superposition on the direct wave of short-term multiples formed in the result of reflection and refraction in thin layers.

To take account of the changes in the shape of the record, instead of measuring the amplitudes of the waves (the reference wave B_1 on the surface and in the borehole, and the useful wave in the borehole) we must measure the spectra of the signals. Comparing the spectra of corresponding waves, one can determine different filtering characteristics of the rock in particular parts of the section, and this enables the shape of the signal at various points on the surface to be determined.

There are methods of appraising the energy of the useful signal and noise, without prior knowledge of the shape of the useful signal [107]. The computation of energies has certain advantages, because it (a) improves the accuracy of the results due to the extended range of measured energies, (b) there is no longer any need for the conservation of the shape or the frequency of the oscillations, and (c) the energy transmission coefficient (in contrast to the amplitude coefficient) is the same both for downward and upward waves. When determining the wave's energy from a seismogram, one can compute the energies for individual half-periods of oscillations approximated by half-periods of sine waves. In this case, the energy of every half-period will be determined by the formula

$$E_i = \int_0^{T_0/2} \left(a \sin \frac{2\pi}{T} t\right)^2 \mathrm{d}t = \frac{a^2 T_0}{4},$$

where a is the amplitude, T_0 is the period of the sine wave, and the energy of the total pulse will be equal to the sum of all E_i

$$E = \sum_{i=1}^{n} \frac{a_i^2 T_{0i}}{4},$$

where n is the number of half-periods.

To illustrate the method, we present in Table VII the results of computation of the relative intensities of multiples in a borehole in Ochrsk in the Perm' province [107]. The 4441 m deep borehole penetrated the top of terrigeneous deposits of Konkian age K_n, Eifelian carbonates E_{if}, the top of dolomites R_{k0}, and a stack of sandstones within the dolomites R_{k1}.

It follows from the table that with increasing depth the intensity of the multiples, compared to that of the singly reflected waves recorded at similar times, increases, and this calls for substantial sophistication in the observation systems.

To sum up, in media where the waves do not change their shape, amplitude measurements may be used; in the case of changing shape, one may study the wave form either with the aid of spectral representations or, when the form itself is irrelevant, with the aid of energy estimates.

TABLE VII

Reflecting interface	K_n	E_{if}	R_{k0}	R_{k1}	R_{k2}
Relative intensity of multiples	0.90	0.85	2.61	2.98	3.00

7. Low-Velocity Zone Studies and Sub-zonal Observations

With the perfection of methods of experimental studies of wave fields in the real earth and with improved facilities for their dynamic interpretation, the effect of the upper portion of the section on the wave field is seen to be increasingly clear. Whereas earlier this effect was chiefly associated with waves propagating in the upper portion of the section along the surface, VSP has subsequently demonstrated that the low-velocity zone exercises a strong influence on waves propagating into the depth of the Earth, and on those arriving at the surface from great depths.

The low-velocity zone acts as a low-pass filter whose special characteristics are determined by the low-velocity zone structure and properties. In individual cases, the filtering action of the low-velocity zone is so effective that it may be comparable to that of the section as a whole. To improve the resolution of the reflection method in surface observations, compensation is needed for the filtering action of the low-velocity zone. However, to this end it will be necessary to study, in addition to the velocities and thicknesses, also the reflecting and absorbing properties of the low-velocity zone and of the upper portion of the section, and in particular, their spectral characteristics.

The use of the polarization version of up-hole shooting enables information to be obtained not only about the velocities of longitudinal waves, but also of transverse waves. Much greater difficulties are involved in the determination of the absorption coefficient. This is because, when use is made of direct waves, the absorption coefficients obtained from observations in shallow boreholes are valid for waves of substantially higher frequencies than those of waves penetrating to great depths. This is of no importance when velocities are being determined, since the dispersion of longitudinal waves in the seismic frequency band is insignificant. However, the absorption coefficient depends greatly on the frequency, and this dependence has as yet not been adequately studied. Accordingly, α has to be determined directly in the operational frequency band. Moreover, the effective absorption coefficient $\bar{\alpha}$ that characterizes the attenuation of the amplitudes in a heterogeneous anisotropic medium depends on the direction of propagation (with account taken of the directions of passage of waves across the interfaces), so that it has to be determined for each wave type.

The most authentic data about the low-velocity zone can be obtained from studies of the ghosts of the direct wave. The ghosts of the direct wave cannot be recorded in surface observations, and ghosts of the reflected waves are recorded in the later portion of the record against a background of unwanted waves. In VSP, the ghosts immediately follow the direct wave, thus greatly improving the accuracy of the determination of the reflection coefficient. However, the study of the reflecting properties of the upper portion of the section using a fixed shot-point and a vertical profile involves great difficulties resulting from such factors as small recording intervals of the reflected wave, shallow depths, and the consequent need to perform difficult observations along an in-line profile using charges suitable for the reflection method, as well as high-level surface waves.

Several possibilities exist for the study of absorption and reflection in the upper portion of the section, which are associated with the method of variation of charge depths [17]. Observations at great depths provide the following advantages: no signal overscaling of channels when comparatively powerful charges are used; shot-points can be displaced from the borehole without introducing substantial changes in the direction of arrival of the waves; waves can be tracked across long sections of the profile. Such factors in VSP make it expedient to employ the variations of the amplitude of the direct wave and its ghost produced by the variation in charge depth. Such waves can be reliably identified and tracked by VSP. The parameters $\bar{\alpha}$ and K (reflection coefficient) in the upper portion of the section were determined for a fixed point of reception in the depth interval 800 to 1500 m. Charges similar to those used in reflection work were exploded at variable depths. The direct wave and its ghost are clearly visible on the seismogram (Figures 30a, d). Their phases are identical if phase reversal is taken into account. The average value $\bar{\alpha}$ can be obtained from the slope of the graph of the relative

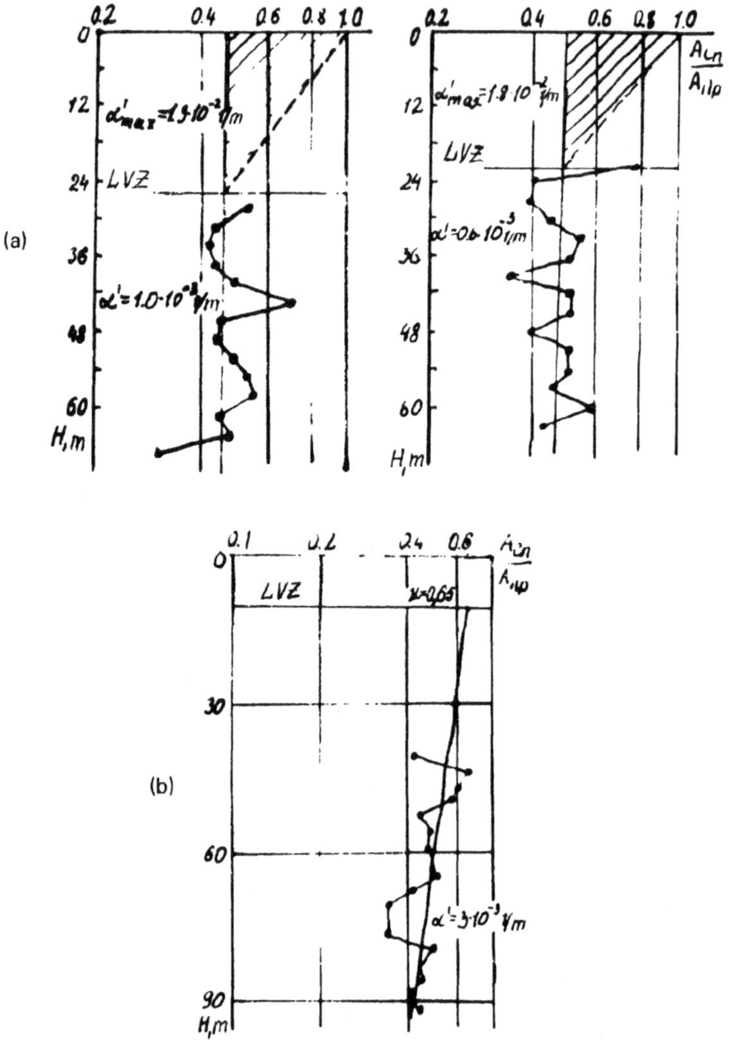

Fig. 61. An example of determination of reflection coefficient K from the base of the low-velocity zone from data obtained with the variable charge depth method in Pamuk (a) and Timbulak (b) areas [17].

amplitudes of the ghost and the direct wave versus charge depth constructed in semi-logarithmic coordinates, whith the segment cut by the graph on the abcissa yielding the average values of K. Figure 61 shows expamples of the determination of $\bar{\alpha}$ and K in certain sections, borrowed from [17]. For the Pamuk area, K from the surface is equal to 0.50 ± 0.03; in another area $K = 0.70$; the maximum value of $\bar{\alpha}$ in the low-velocity zone in the Pamuk area is $\bar{\alpha}_{max} = 1.6 \pm 0.3 \times 10^{-2}\ m^{-1}$; for Urtabulak $\bar{\alpha} = 2.0 \pm 1.0 \times 10^{-2}\ m^{-1}$. For reflected waves in surface bedrock in the Pamuk area, $\bar{\alpha}_{max} = 2.5 \pm 1.5 \times 10^{-3}\ m^{-1}$. The absorption coefficient for the direct wave measured with the aid of uphole shooting is greater by about an order of magnitude than that of the reflected waves. Such sharp differences are due to higher frequencies (90 Hz) of the direct wave, and to energy losses at internal interfaces in the thin-layered bed, which are smaller in the case of the reflected wave having a frequency of 50–60 Hz.

It should be pointed out that absorption in the upper portion of the section can exercise a strong influence on the relative intensity of the waves associated with deep interfaces. Thus, for the area of N. Kamashi being considered here, a 1.5-fold change in the absorption coefficient in the upper portion of the section (averaged over the layers from 1.0 to $1.5 \times 10^{-3}\,\mathrm{m}^{-1}$) results in a 3.5-fold reduction in the ratio of surface intensities of the doubly and the singly reflected waves associated with Bukhara sandstones [17]. For the Pamuk area an approximately two-fold change in $\bar{\alpha}$ (from $\bar{\alpha} = 3.0$ to $1.4 \times 10^{-3}\,\mathrm{m}^{-1}$) results in a five-fold change in the amplitude ratio. Hence, the level of intensity of the multiples associated with the upper portion of the section greatly depends on its absorption properties. The procedure for studying $\bar{\alpha}$ and K should be planned on the basis of data from a detailed VSP profile.

Detailed studies of the reflecting properties of the low-velocity zone in various seismologic conditions, and the establishment of a correlation between the low-velocity zone structure and its reflecting properties (with account taken of surface effects) are of essential importance in the study of the formation of the multiples.

SUB-ZONAL OBSERVATIONS

Because of the strong influence of the low-velocity zone, there is increasing need for sub-zonal observations. This applies primarily to the high level of noise associated with the low-velocity zone, as well as its high absorption. The reasons why it is expedient to employ sub-zonal observations are as follows:

(a) By eliminating the influence of a highly absorbing low-velocity zone, it is possible to raise the frequency spectrum of the waves and thereby substantially improve resolution.

(b) The effectiveness of frequency discrimination can be improved. The spectra of all waves observed on the surface are band limited and are little influenced by, or even practically independent of, the spectra of the source. A discrimination of waves according to their frequencies takes place below the low-velocity zone, with higher frequencies prevailing in the spectra of deep waves. Great absorption in the low-velocity zone accounts for the lower frequency spectrum of the multiples compared with that of singly reflected waves. Moreover, surface waves below the low-velocity zone have a lower frequency spectrum than those on the surface.

(c) The signal-to-noise ratio is higher, because of the reduced level of unwanted waves excited by the explosion and propagating in the upper portion of the section. The latter include surface and channel waves of various types, whose intensity rapidly drops with the depth.

Figure 62 depicts the spectral characteristics of waves observed on the surface and below the low-velocity zone computed from acoustic well survey data [110].

The fall-off at high frequencies is the effect of inhomogeneities and absorption. Whereas the peak frequency of the waves at the base of the low-velocity zone was about 50 Hz, its value on the surface was only 30 Hz.

A great volume of VSP work has demonstrated the potential and the advantages of

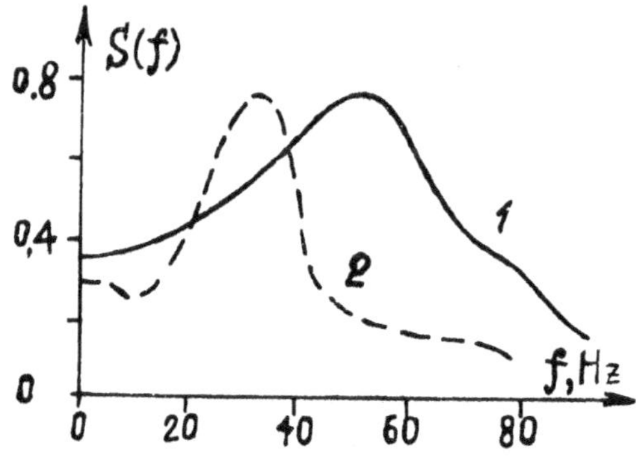

Fig. 62. Wave spectra at the base of the low-velocity zone (1) and on the surface (2) [110].

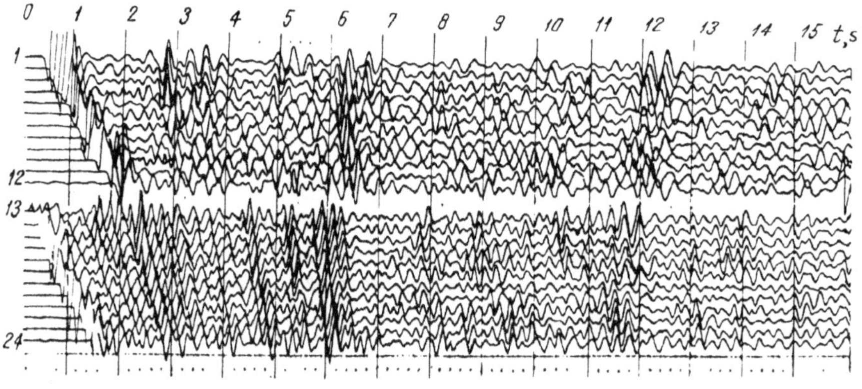

Fig. 63. Seismograms recorded on the surface (traces 1–12) and below the low-velocity zone (traces 13–24) for a peat bed of variable thickness (Pavlovo moor, h = 15 m, charge 5.2 kg [112]).

the VSP reflection method (see Chapter 5) to solve structural problems. However, such studies require observations at great depths, and this limits the studies chiefly to sections in the immediate vicinity of an interface. In this respect the potential for observations below the low-velocity zone is substantially greater [111].

Sub-zonal observations in a form of the surface seismic method using buried geophones have been proposed for the study of extensive marshland in the Western Siberian depression, where the low-velocity zone consists of peat [112]. Seismograms obtained with buried geophones exhibit substantially higher frequencies and better record resolution, improved wave correlation (Figure 63) and greater accuracy of wave correlation at conjugate points. It should, however, be noted that the effectiveness of sub-zonal observations may vary in different conditions, being determined chiefly by the relationship between the filtering properties of the low-velocity zone and the rest of the section. The effectiveness is higher the less pronounced are the different filtering properties.

The principal goal in the initial stages of sub-zonal observations is to perform the studies in greater detail and to utilize the dynamic features of the record. In recent years, owing to development of PM, observations below the low-velocity zone have gained special importance, since they enabled the effectiveness of wave discrimination according to polarization to be improved, and thus the potential of the polarization method to be fully realised [2]. Sub-zonal observations with the aid of PM enable the effectiveness of the studies to be improved, and some of the problems to be solved with the aid of the VSP reflection method in comparatively deep boreholes.

CHAPTER 7

THE VSP METHOD FOR REFRACTED WAVES

The waves recorded in the early portion of the seismogram at offsets great as compared with the depth of the interface consist mainly of refracted (head*) and transcritically reflected waves. The study of the origin and the relative intensities of such waves is of great interest for both theory and practice, since this factor is of decisive importance for the exploration potential of one of the principal methods – that of the correlational method of refracted waves. The difficulties associated with such studies are due mainly to the ambiguity of interpretation of the wave's origin arising from the similarity between the kinematic features of the head, the curved-path and the reflected waves recorded on the surface at great distances from the energy source, so that they cannot be used to identify the wave type reliably. Accordingly, it was presumed that the end could well be served by VSP. Down-hole observations at large offset distances from the shot-point have the following advantages over surface observations [14]:

(1) The study of the wave field in the medium underlying the refracting interface provides the most authentic data about the mechanism of energy return and, hence, about the origin of the wave recorded above the refracting interface.

(2) The recording of waves as first arrivals in the immediate proximity of the refracting interface and near the point of emergence enables the reliability of stratigraphic correlation to be improved.

(3) The employment of such a sensitive parameter as the trajectory of particle motion of the medium for determining the wave's origin.

(4) The rapid decrease in the seismic background noise with depth enables a much higher useful instrumentation sensitivity to be utilized, so that waves of very low intensity can be studied.

(5) A chance to compare the intensity of various waves observed in the immediate proximity of the interface, which is important, because the intensity ratio may change with the distance from the interface.

The correlation of head and curved path waves in VSP is characterized by the fact that, because they have similar or equal apparent velocities, such waves form extensive zones of interference. However, they differ sharply in intensities, and this occasions a substantial difference in their conditions for correlation. Curved-path waves, which dominate the early portion of the record can be reliably tracked in both the first and later arrivals; head waves can be identified and tracked in the first arrivals.

Waves reflected at large angles from the boundaries of a layer or from whithin it, when observed at large offset distances from the source can approach the head wave quite closely and interfere with it, exercising a strong effect on its parameters. In this respect, it is of major importance that VSP be supplemented by observation systems and theoretical computations of the wave fields devised on the basis of detailed studies of

* The term head wave as distinct form a curved-path wave is used to describe a wave whose energy returns to the surface as the result of diffraction.

the velocity distribution. A specific feature of VSP, as used in the refraction method is the need to employ the polarization method to study the total vector of oscillations of each wave. The recording and study of just the vertical components of the record, in the case of waves approaching from directions close to the horizontal, can substantially distort the true picture of the relative intensities of different wave types, and lead to erroneous conclusions as to their origin.

A large volume of VSP work in a variety of media and structures has been carried out to study waves recorded in the early portion of the record. The first VSP studies that we carried out in the Ukraine in 1959, and in the Kuibyshev province in 1960 [15], demonstrated that head waves associated with thin layers are so weak that it is practically impossible to detect them on the surface seismograms. These results were subsequently corroborated by studies carried out in the Krasnodar province. Waves associated with individual thick high-velocity layers were studied in the Stavropol' province. Wave fields typical of carbonate sections were studied in the Volgograd province. In recent years, theoretical and experimental studies of waves associated with the crystalline basement [113] and with thick layers were carried out on the basis of VSP. The effect of individual elements of the geologic section on the wave field observed at large offset distances from the shot-point has also been studied [21], and principles have been formulated for the construction of models corresponding to the observed wave patterns. The following are the principal results:

(1) The intensity of transcritically reflected waves is always higher than that of the head waves originating at the same interface, the difference in the intensities being as much as two orders of magnitude or more. For this reason, head waves on surface seismograms cannot be identified in either the later or even the early arrivals [13], in the presence of transcritically reflected waves. Thin high-velocity layers play a decisive part in the formation of the wave field at comparatively short distances from the shot-point. With increasing separation, the effect of thin layers diminishes, and curved-path waves in the bed underlying the layer become dominant. The relative intensity of the head wave increases as the layer thickness increases, but it still remains appreciably less than that of the transcritical wave.

(2) It has been demonstrated, by means of direct observations in the underlying layer, that in the earth there is generally an increase in velocity with depth, and that it is not a head wave that is recorded first on the seismograms but a curved-path wave. Refraction and transcritical reflection play a decisive part in determining the intensities of the waves, and act as the chief mechanism of transporting energy back to the surface.

(3) In observations at ranges covered by the refraction method, as well as at ranges covered by the reflection method, the upper portion of the section can excercise a substantial effect on the wave field observed on surface seismograms. Observations carried out at great depths both below the source and at appreciable distances from it, enabled the reflected-refracted and the refracted-reflected waves to be separated. This supports the theory that waves recorded in the early portion of the seismogram with velocities similar or equal to those of first arrivals may be reflected-refracted waves associated with the upper portion of the section.

It should be remembered, that unfortunately, only Z-component seismograms have generally been recorded in the VSP refraction studies, and this greatly limits the scope of wave analysis.

1. Waves Associated with Thin High-Velocity Layers

Problems involving thin high-velocity layers are of great interest, because, on the one hand, they determine the exploration potential of the refraction method, and, on the other, erroneous ideas have for a long time existed on the subject. Instances of the recording of head waves originating from layers whose thickness was tenths or even hundredths of a wave length have been described in literature [85]. Solid-fluid models contributed to the conclusion about the existence of head waves from layers less than 0.1 wave length thick. However, subsequent research on solid models and theoretical studies have demonstrated that head waves from thin layers are not possible. The great difficulty of obtaining quantitative data from theoretical computations adds special importance to experimental studies of waves associated with thin high-velocity layers in the real earth. VSP was employed in several areas to study waves from thin layers. Consider data obtained in the Staro-Minskaya area in the Krasnodar province.

Intensities of Head and Transcritically Reflected Waves

Compare the intensities of the head and the transcritically reflected waves associated with two thin layers located at depths $H_v = 270$ m and 1334 m (Figure 76). Reflections from these layers dominate the surface seismograms both near the shot-point, and at offset distances which greatly exceed the depth of the interface. The primary and multiple reflections from those two thin layers almost completely determine the entire surface seismogram. However, head waves from thin layers could not be detected on the surface seismograms, despite the great volume of work specially undertaken for this purpose [13].

In down-hole observations, the high useful instrument sensitivity, and the opportunity to carry out observations in the immediate vicinity of the layers, enabled the head waves associated with them to be recorded, and the intensities of the head and transcritically reflected waves for each layer to be compared.

First Layer ($H_v = 270$ m)

The Z-component seismograms in the section of the profile intersecting the first layer, starting from the depth 220 m down to the interface at 270 m record on the maximum sensitivity channels, a weak 'down' arrival of the head wave PPP (Figure 64) with a negative apparent velocity, before a clearly defined 'up' arrival of the direct wave. The wave's intensity and the direction of the particle motion experience a sharp change as the interface is crossed. The first wave to be recorded on the record below the interface is the transmitted wave. On the vertical traveltime curve (Figure 46), there are two characteristic reversals of sign of the apparent velocities: the positive apparent velocity of the direct wave changing to the negative velocity of the head wave at $H_v = 220$ m, and the negative apparent velocity of the head wave changing to the positive velocity of the transmitted wave at $H_v = 270$ m. The head wave can be discriminated in sections where it is recorded first. The high-intensity direct wave is also recorded reliably in subsequent arrivals.

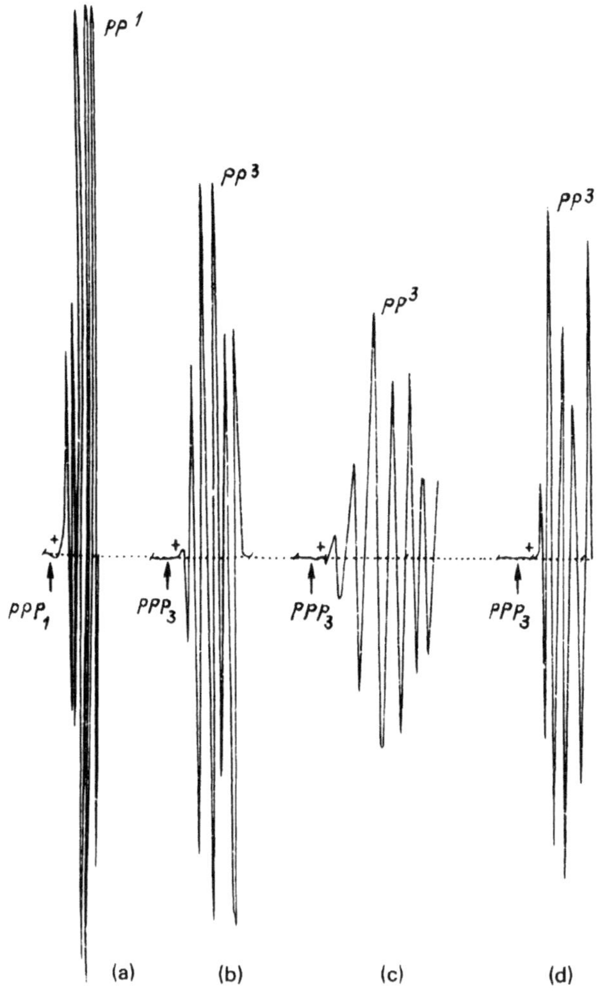

Fig. 64. Intensity ratios of head and transcritically-reflected waves (times of first arrivals are shown by arrows, head waves are indicated by crosses) (bortehole N42 Staro–Minskaya): (a) H = 260 m (shotpoint 600 m); (b) H = 1260 m (shotpoint 1775 m); (c) H = 1150 m (shotpoint 2500 m); (d) H = 1180 m (shotpoint 2500 m). The index ordinal corresponds to that of the interface, the top index to reflected waves, and the bottom one to the refracted waves (see Figure 24).

The transcritically reflected PP^1 wave on VSP seismograms in the immedite proximity of an interface cannot be separated in a pure form, because of interference with the direct wave. A qualitative estimate of the intensities of the PP^1 and PPP_1 waves demonstrates that the intensity of the transcritically reflected wave is over 200 times that of the head wave. The ratio of intensities of the PP^1 and the PPP_1 waves can be clearly seen on the seismogram (Figure 64) recorded with the aid of a wide dynamic range instrument and compiled from records obtained at different amplification levels. Naturally, it is practic-

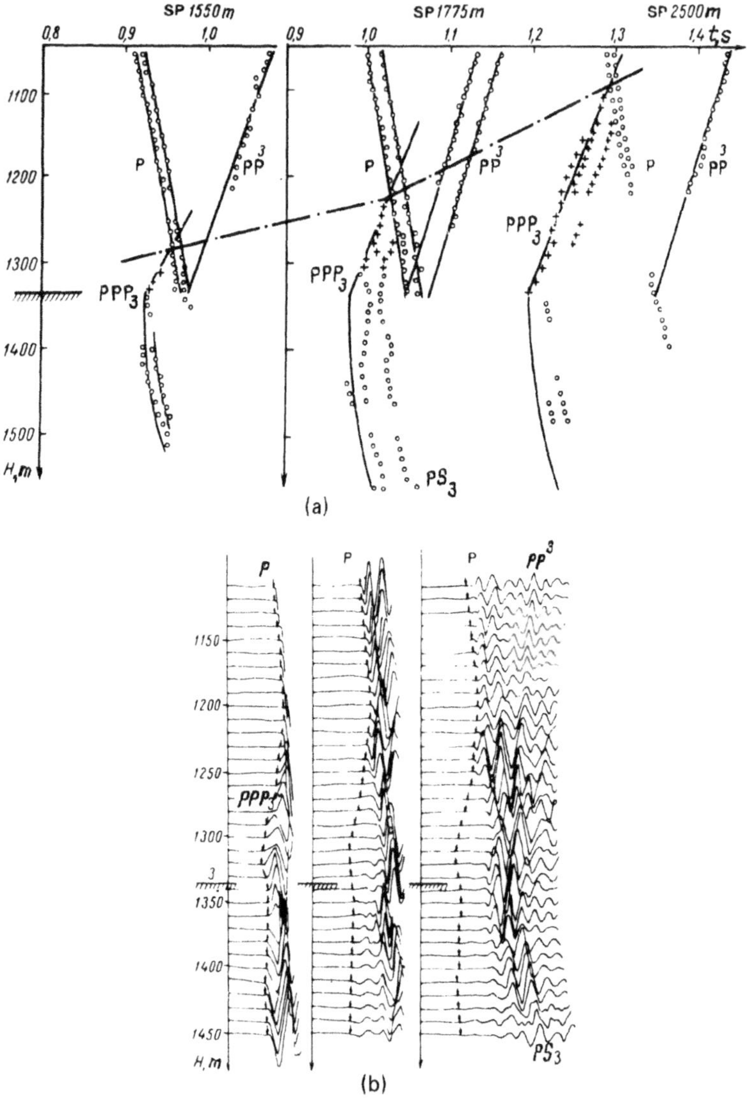

Fig. 65. Vertical traveltime curves (a) and seismograms from shotpoint 1775 m (b) of direct, head and transcritically-reflected waves in the region of interface 3 (borehole N42 Staro–Minskaya).

ally impossible to separate such a head wave on the surface, and the visible first arrivals on the surface seismograms are those of the transcritically wave.

SECOND LAYER (H_v = 1334 m)

The head wave PPP_3 is trackable only in the first arrivals in the form of weak oscillations on maximum gain channels.

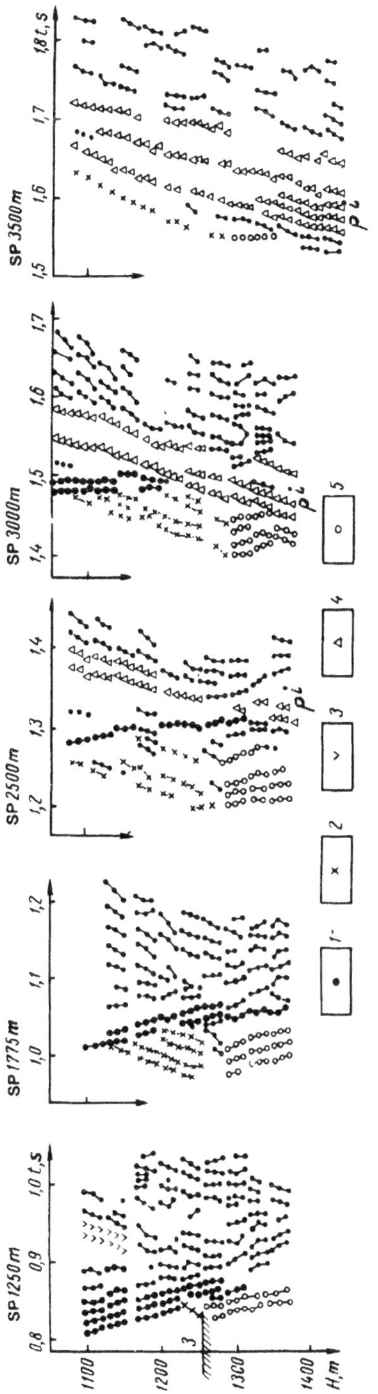

Fig. 66. Vertical traveltime curves of direct (1), head (2), transcritically-reflected (3), bent (4), and transmitted (5) waves (borehole N20 Staro–Minskaya).

The points where the direct wave changes to the head wave are clearly visible on the seismograms (Figure 65b) as the points where the sign of the arrival changes, and on summary traveltime curves (Figure 65a) on which they are connected by dashed lines. The region where the PPP_3 head wave is recorded as a first arrival increases with the increase in the separation from the shot-point, the velocity of propagation of the wave along the interface decreasing from $v_{in} = 4550$ m s^{-1} for *SP*1550 m to $v_{in} = 3900$ m s^{-1} for *SP* 2500 m. A possible explanation for this is either uncertainty in the recording of the instant of the wave's arrival, or the interference nature of the wave.

The transcritically reflected wave PP^3 in the vicinity of the interface is located in the zone of interference with the direct wave. It follows from records on which the transcritically reflected wave has already left the zone of interference (Figure 65) that its intensity is great, and that it is recorded on minimum sensitivity channels. It is difficult to make a quantitative estimate of the head wave's intensity or to discuss its shape, because it always remains in the zone where it interferes with the direct wave. The only point that can be made is that its first extremum has a lower frequency than that of the direct wave. As a qualitative estimate, the intensity of the PPP_3 head wave is about 150 to 200 times less than that of the transcritically reflected PP^3 wave (Figures 64, 65). Despite special experiments, such waves could not be detected in surface observations at all, even after they have been discovered in VSP.

THE EFFECT OF A THIN LAYER ON THE WAVE FIELD

It would be interesting to assess the variation in the effect on the wave field of a thin-layer with increasing offset distance from the shot-point. With this aim in view, we carried out observations of waves excited in shot-points at offset distances between 900 and 4000 m in a section of the vertical profile intersecting the thin layer of interest at a depth of 1250 m. The increase in the offset distance from the shot-point extends the observation interval in the bed underlying the thin layer. We recorded the vertical oscillation component. PM observations were carried out at great offsets, and total oscillations vectors were studied.

The initial portion of the record on vertical profiles is made up of the direct, the head and the transcritically reflected waves from the thin layer, as well as the curved-path wave in the bed underlying the thin layer (Figure 66). The head waves are discernible in the first arrivals from *SP* 1250 m, 1775 m, and 2500 m. The head wave from SP 1250, located near the origin, is recorded only in a 20 m interval from 1240 to 1260 m. The region over which the head waves are recorded grows with the distance from the shot-points.

At longer distances, the waves arriving from below gain in importance on vertical seismic profiles. The relative intensity of the *Z*-component drops owing to the direction of arrival tending to the horizontal. The dominant wave on the seismograms from *SP* 3000, 3500, and 4000 m is the P^i wave with a velocity $v_a = -3700$ m s^{-1}, which can be reliably tracked on the vertical profile. The relative intensity of the P^i wave increases with the offset*. Studies of the kinematics of the P^i wave, in particular, fixed geophone traveltime curves, reveal a regular increase in apparent velocity typical of curved-path waves.

* A detailed analysis of the nature of waves at great distances from the shot-point has been carried out by a member of Leningrad University staff G.G. Pogonyailo [20, 22]. This is a short summary of her results.

It has been demonstrated as the result of studies of the changes in the intensity ratios of various wave types with distance from the source, that the dominant wave at distances of 1250 m and 1775 m is the direct wave, whose intensity exceeds that of the head wave by about two orders of magnitude. The dominant wave at distances of 2500 m and 3000 m is the curved-path wave whose intensity is 5 to 10 times that of the direct wave, and 150 times that of the head wave. On the records from *SP* 3500 this latter ratio is equal to 10, and this fact casts doubt on the correctness of the interpretation of the first arrival recorded at this distance as being a head wave.

The use of computer correlation on the vertical profile [20] enabled the direct and the curved-path waves to be separated in the region of their interference, and their apparent velocities of -3400 to $-3700\ \mathrm{m\,s^{-1}}$ for the curved-path wave, and $+4800\ \mathrm{m\,s^{-1}}$ for the direct wave, to be determined. Computer correlation also enabled records of various waves in averaged form to be obtained. The direct and the curved path wave each have several extrema, and if one takes the 180° phase reversal of the curved-path wave into account, their records will be of identical shape.

Theoretical computations of wave fields,* utilizing formulae of the ray-path method, have been undertaken to ensure a more reliable identification of the P^i wave type from possible variants of transcritically reflected and head waves, associated with a deeper interface, or from curved-path waves. Wave amplitudes along horizontal fixed depth profiles have been computed. According to calculations, at distances $0 < l < 1500$ m the most intense wave in the later portion of the record is the reflected wave. Near the shot-point, it is 3 to 10 times weaker, depending on the model chosen, than the direct wave, but it decays less and becomes the dominant wave as the distance from the shot-point is increased. In the region of the origin, the amplitude of the transcritically reflected wave exceeds that of the direct wave. With increased offset (2500 m) the curved path wave becomes the dominant wave on theoretical seismograms. The observed curves of the waves being considered here are in fair agreement with the theoretical.

The curved-path wave P^i on surface seismograms in the Staro–Minskaya area, is recorded as the first arrivals in the range from 4000 to 6000 m, where it is superceded by the curved-path wave in the basement.

THE RELATIVE INTENSITIES OF HEAD AND TRANSCRITICALLY REFLECTED WAVES IN THE CASE OF LAYER THICKNESSES COMMENSURATE WITH THE WAVE LENGTH

The relative intensity of the head wave grows with the increase in the layer's thickness, however, it still remains much less than that of the transcritically reflected wave. Homogeneous layers of comparatively great thickness in the real earth are much more rare than thin ones. In the majority of cases, thick layers are not homogeneous, and, as demonstrated by detailed velocity measurements, generally consist of one or several assemblages of thin layers. In the profile of the N20 borehole in the Staro–Minskaya area, Figure 67, the 900 m deep interface can be assumed to be the interface between 'medium' thickness layers. The velocity contrast is about 0.9 ($v_{l_1} = 2040$, $v_{l_2} = 2310\ \mathrm{m\,s^{-1}}$), the ratio of the wavelength to the layer thickness is $\lambda/d = 5.5$. First arrivals of the waves and their sign reversals are clearly visible on the records obtained at maximum gain. In the depth range from 580 m to 670 m the seismograms show 'up'

* A set of VSP, sonic, and *P*- and *S*- wave logging data was employed to choose an earth model.

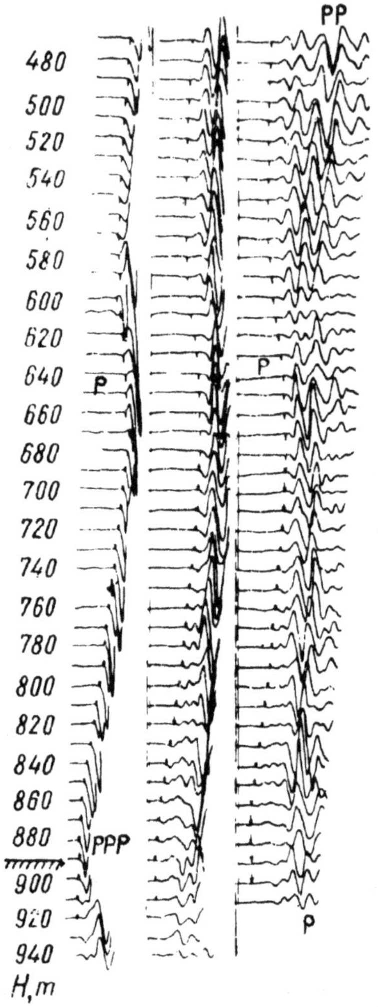

Fig. 67. Seismograms of head waves associated with the interface at the depth of 900 m at three amplification levels (shotpoint 2500 m, borehole N20 Staro–Minskaya).

breaks corresponding to the direct wave. From 680 m and deeper, distinct 'down' breaks for the head wave are visible. On medium gain records, the head wave is recorded as a weak low-frequency oscillation accompanied by an oscillation of higher frequency. Its apparent velocity is negative. Minimum gain seismograms show an intense interference pattern of the direct and the reflected wave following the head wave. As the distance from the interface increases, the reflected wave separates from the direct wave and starting at the depth of 520 m is recorded outside the region of interference. The head wave is weaker by about an order of magnitude than the transcritically reflected wave. It is impossible to resolve the head wave against the background of the transcritically reflected wave on the vertical profile.

A similar picture is observed in borehole 42. In both boreholes the relative intensity of the head wave associated with the 900 m deep interface substantially exceeds that of the head waves associated with the thin layers in the section, due mainly to greater layer thickness. However, the influence of refraction on the wave's intensity should not be excluded.

Thus, results of the down-hole observations considered above have demonstrated convincingly that head waves associated with thin layers are very weak and are practically not detectable on surface seismograms. Accordingly, such waves can only be studied with the aid of VSP employing wide dynamic range equipment. These VSP results are in good agreement with theoretical computations [114] that show that head waves even under the most favourable conditions are weaker than transcritically reflected waves, their intensity rapidly decreasing with a decrease in the layer's thickness.

Although head waves from a thin layer can be recorded in VSP, it is possible to track them only in the immediate vicinity of the interface. The relative intensity of the head and reflected waves as a function of the layer's thickness has been studied on solid models with a hard contact between the layers, and it has been demonstrated that the ratio of the amplitudes of the reflected and the head waves changes in accordance with the layer's thickness due to the decrease in the intensity of the head waves. The amplitude ratio increases with an increase in distance and diminishes with an increase in the layer's thickness. Recent erroneous estimates of the intensity of head waves associated with thin layers can largely be explained by applying the results obtained with solid-fluid models to the real earth.

2. Waves in Carbonate Sections in Platform Regions

Carbonate sections typically display numerous sharp interfaces between alternating high-velocity carbonate and low-velocity terrigeneous deposits. The study of the wave field recorded with the refraction method in such media is of considerable interest, one important reason being that such waves are typical of extensive areas of the Russian platform.

The combined observation system (Figure 14b) has been calculated taking account of the velocity function constructed along an in-line VSP profile. It has been designed for analyzing wave fields with a range of offset distances from 0 to 32 km.

It follows from the combined traveltime curves (Figure 68) that waves recorded on the surface in the early portion of the record in the offset range from 0 to 11 km, can be subdivided into three groups. The first group of waves PPP_1 is associated with the high-velocity layer located in the Middle Carboniferous (l = 125–270 m) and consists of a wave refracted at the top of the Paleozoic, as well as a transcritically reflected and a multiple reflected-refracted wave. The waves recorded in the subsequent portion of the record include single and multiple curved-path waves in the Paleozoic bed, as well as waves reflected from sharp interfaces represented by thin limestone layers located at depths from 390 to 520 m.

The second group of waves P^2, consists of single and multiple curved-path waves in the second limestone layer (in Lower Carboniferous, 640–850 m), as well as waves reflected from interfaces down to the top of the terrigeneous Devonian.

The third group P^3 consists of curved-path waves in the basement. Observations

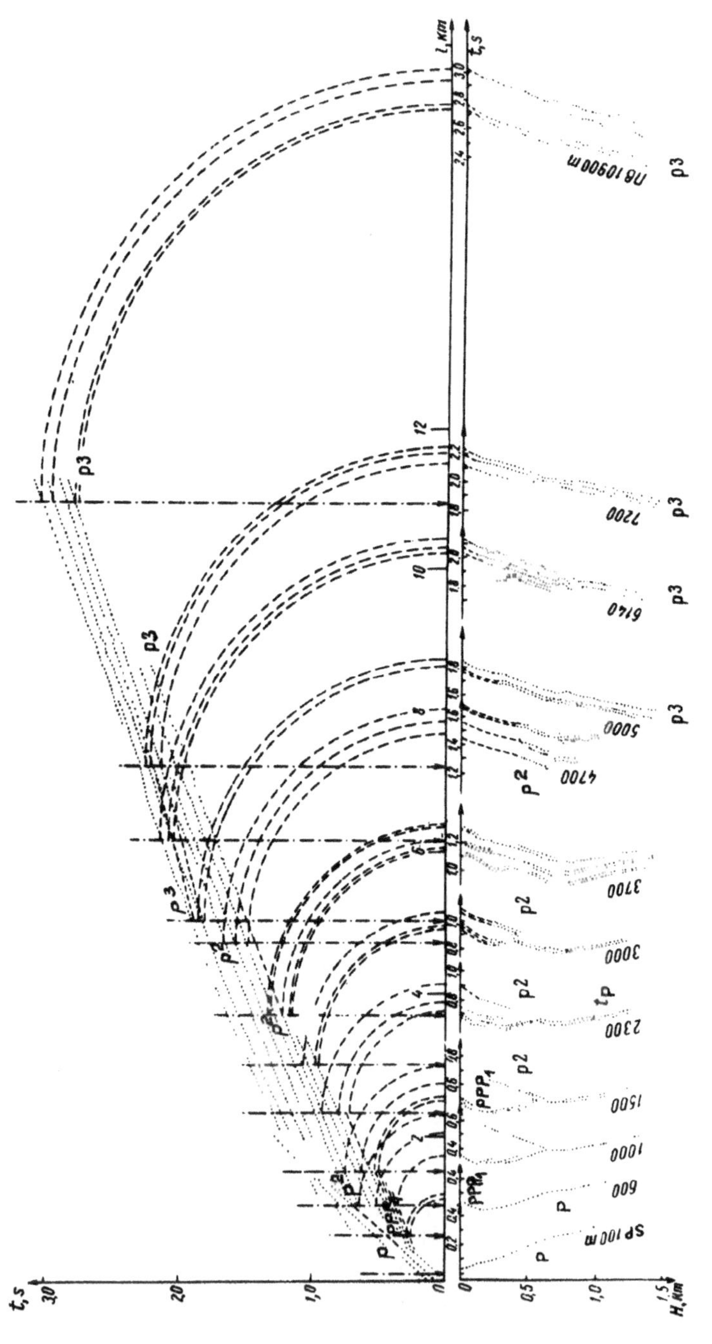

Fig. 68. Combined traveltime curves of waves in the initial portion of the seismogram (borehole N1 Kletsko–Pochtovaya).

along the vertical profile demonstrate that the P^3 group also consitutes a complex interference pattern, which includes the curved-path wave itself and multiple curved-path waves. The conversion from the waves of one group to another takes place as the result of attenuation of the waves in the first arrivals.

Three groups of refracted waves are registered on vertical profiles, just as on surface profiles. The first wave, with a negative apparent velocity, is recorded from SP 600 m, 1000 m, and 1500 m. The refracted wave associated with the second layer becomes trackable on the vertical profile only starting at *SP* 1500 m. It is recorded in the first arrivals over a gradually broadening zone with increasing separation, and from *SP* 4700 m it becomes the first recorded wave up to the surface. Multiple curved-path waves associated with the second layer have been recorded on the vertical profile also from *SP* 1500 m, with the first curved-path wave dominating at distances up to 3000 m. At greater separations the intensity of multiple curved-path waves becomes greater. The wave associated with the crystalline basement appears on the vertical profile shot from *SP* 5000 m. All waves approaching the surface are reflected from the base of the low-velocity zone and subsequently from the top of the Paleozoic, forming a complicated combined wave field.

The nature of waves that constitute the first arrivals on surface seismograms is best considered using the waves associated with the second layer H_v = 640 to 850 m in Figure 69 as an example. The two dash-dot lines on the overlapping traveltime curves of first events from *SP* 1020 m, 1500 m, and 2300 m depict the line of conversion from the direct wave to the refracted, and the line corresponding to polarity reversal of the apparent velocity of the refracted wave. The two lines serve as boundaries enclosing the region of negative apparent velocities, i.e., the region of direction of approach of the refracted wave from below (shaded area).

The conversion from the direct to the curved-path wave is clearly visible on the Z-component seismograms (Figure 69b). From *SP* 1020 m, this conversion takes place in the immediate proximity of the interface. From *SP* 1500 m, the curved-path wave is recorded as the first event in a 50 m long section of the profile above the interface. The seismogram from *SP* 2300 m has been recorded at a low gain, and on it the curved path wave is recorded and can be identified in the entire recorded interval. The intensity of the curved-path wave from *SP* 2300 m increases somewhat as the wave from the refracting layer enters the overlying layer on account both of the increase in the velocity and of the wave's direction of arrival approaching the vertical. The form of the record is preserved below the refracting interface, the only difference being a smooth decrease in the amplitudes, phase inversion and a similarly smooth increase in the amplitudes. All these features point to the conclusion that the wave below the 640 m interface is a curved-path wave in the Middle Carboniferous limestone bed.

As to the nature of waves associated with great depths, difficulties have arisen in the identification and stratigraphic correlation of such waves, because of the lack of observations carried out directly in the beds producing them and in the basement. To make possible a more reliable determination of the nature of waves at great distances from the shot-point, the wave field needs to be computed. However, certain specific features arose when the model of the medium was being selected [21]. Let us dwell on this point.

When constructing a model of the earth, one has first of all to select the principal

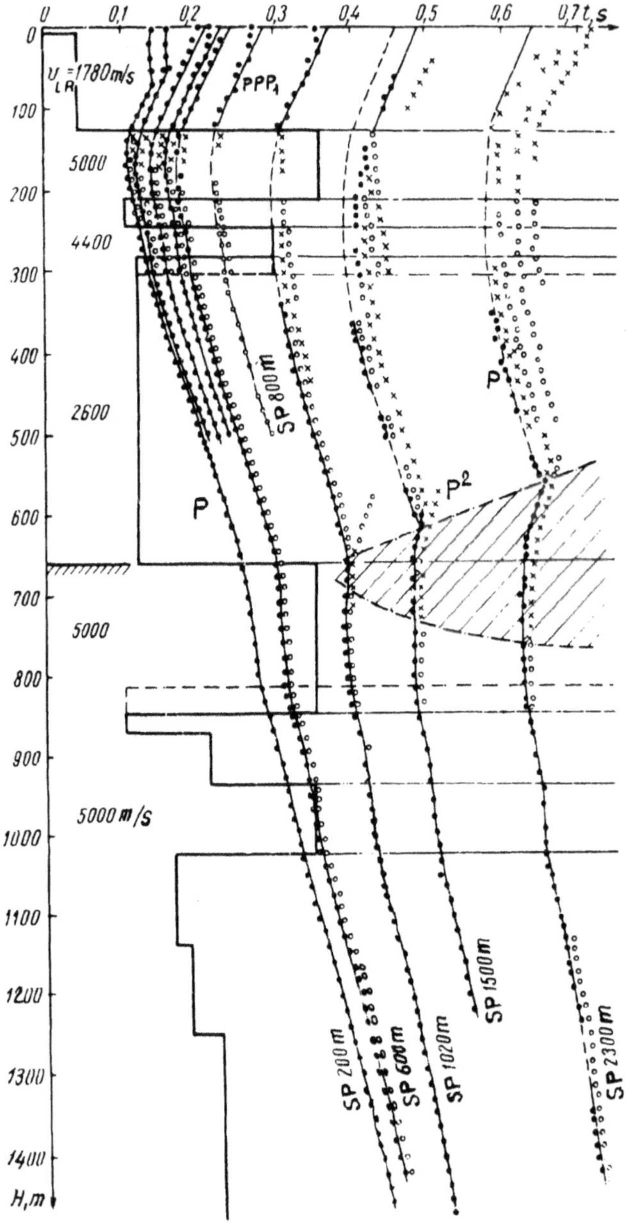

Fig. 69a.

Fig. 69a-b. Vertical traveltime curves (a) and seismograms (b) of the first waves (borehole N1 Kletsko–Pochtovaya). The velocity cross-section is schematically represented on the traveltime curves.

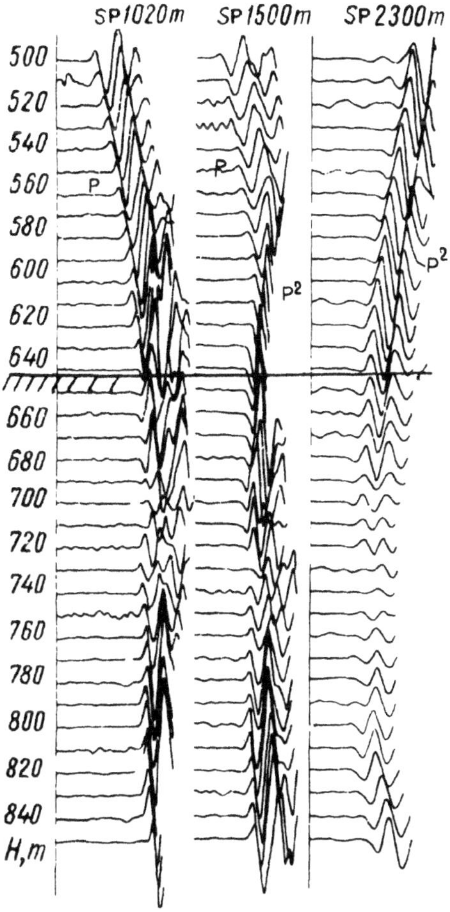

Fig. 69b.

elements of the geologic section that excercise a decisive influence on the wave field. The amount of detail into which the velocity distribution will have to be subdivided is determined by its contribution to the wave field. In addition to the detailed study of the earth's structure at great depths, one should take into account the structure of the overlying beds and, particularly, of thin high-velocity layers that may greatly influence the characteristics of deep transmitted waves observed at large offset distances and at near- critical and transcritical angles.

Making use of VSP, surface observations and resistivity logging data, as well as any other available information about the section, we were able to construct models of the earth's kinematic and dynamic wave field characteristics, which were in agreement with experimental data obtained in short offset observations, i.e., at normal or near-normal incidence. Such a model comprised 40 layers. However, as the offset distance was increased even to 1 or 1.5 km, the theoretical results no longer coincided with the observed: the computed arrival times and wave amplitudes were considerably smaller than the experimental ones (the amplitudes were smaller almost by an order of magnitude). To bring the results into accord, a new model was required. The new model comprised 29 layers, in which the velocity in the upper portion of the section had to be reduced, and a thin-layered limestone bed between 600 and 750 m had to be replaced by a continuously changing medium. This model agreed with observations in the range 0.5 to 3.0 km. Beginning with 3 to 4 km offsets, one has to introduce a new model with a lower velocity in deeper-lying thin layers. The new model was effective in the 4 to 8 km range. To account for the waves observed at the bottom of the profile and associated with the basement, a model with 9 layers corresponding to thick rock units had to be constructed.

Calculations of the wave fields at various offsets from the source and their comparison with the observed fields, made with the aim of interpreting deep waves associated with interfaces beyond the scope of observations, have demonstrated that it is impossible to describe the wave field in a thin-layered medium at different source offsets by means of a single model. As the separation from the source is increased, one has to introduce changes into the model in the direction of lower velocity for thin high-velocity layers, greater thickness and greater absorption coefficients and smoother velocity gradients, i.e., the parameters must be functions of the offset distance. Such modifications to the model enabled a generally satisfactory correspondence between the computed and the observed amplitudes for great separations to be obtained, if one ignores the more rapid decay of reflected waves. If the wave fields in media composed of high-velocity thin layers are calculated for large offsets with the aid of formulae of the zeroth approximation of the ray-path method, the class of models has to be changed with the offset.

It has been demonstrated by numerous calculations [21] that the main reason for the need to adapt the model for increased separation is the effect exercised on deep waves by groups of thin high-velocity layers. The arrival times of internal multiples originating in them practically coincide with the arrival time of the singly reflected waves. Reflected waves, too, when complicated by short-term multiples, increase in intensity with offset as compared with the singly reflected wave (100 to 1000 times for $l > 5$ km in the case being considered). They all reinforce in-phase forming a single combined wave. In the formation of the wave field at all distances, the major part is played by waves reflected from sharp interfaces and complicated by short-term multiples appearing at long distances inside high-velocity layers. The most intense short-term multiples originate in the layers with highest velocity, the magnitude of the velocity contrast affecting the intensity of waves only in the nearest zone. The passage of transmitted singly reflected waves through groups of thin high velocity layers is accompanied by an increase in their intensity by 3 or 4 orders of magnitude, owing to in-phase superposition of internal multiples. At large offsets the wave field associated with such groups of thin layers takes a comparatively simple form due to the superposition of waves with idential v_a and a stable wave shape.

Experiment has failed to uphold the possibility of using the zeroth approximation of the ray-path method to describe the wave fields in a thin-layered medium, using large offsets, when based on a model constructed for a short offsets. We whould also like to point out one feature of the wave fields observed in the presence of alternating carbonate and terrigeneous deposits. The latter consitute low-velocity layers, and act as waveguides in which the energy is concentrated in the form of numerous waves propagating in them. This is clearly visible on the seismogram obtained in the platform conditions in the Volgograd province (see Figure 90). VSP creates the possibilities for detailed studies of the wave propagation process in wave guides. This is especially important, because such observations may serve to construct models for studying wave guides located at great depths in the Earth's crust and in the mantle. Also, the low-velocity channel under the permafrost may serve as such a model.

3. On the Origin of Waves Associated with a Thick Layer

The real earth is as a rule grossly heterogeneous, and it is very difficult to find a homogeneous layer of adequate thickness. The crystalline basement conventionally regarded as a half-space could serve as a medium suitable for such studies. However, there are, unhappily, very few holes drilled deep enough to reach the crystalline basement. In the majority of cases, the boreholes only come close to the basement and only rarely penetrate its top. The appropriate medium for the study of waves associated with thick layers was found in the deposits of the Jurassic-Triassic complex in the Stavropol province. According to resistivity logging data, the geologic section in the depth range 1880 to 2700 m appeared to be relatively homogeneous. A considerable portion of the complex (over 900 m) was revealed by the North–Stavropol borehole N162. Lithologically, the deposits are represented by a variegated siltstone bed. A thin layer of high-velocity chalk with velocity $v_l = 5000 \text{ m s}^{-1}$ above the top of the bed at $H_v = 1790$ m, greatly complicated the wave pattern, and made it difficult to study the refracted wave associated with the basement in the vicinity of its point of emergence.

The refracted waves were recorded along off-line profiles with horizontal distances between the shot-point and borehole lying in the range 1200 m to 14 100 m (Figure 70).

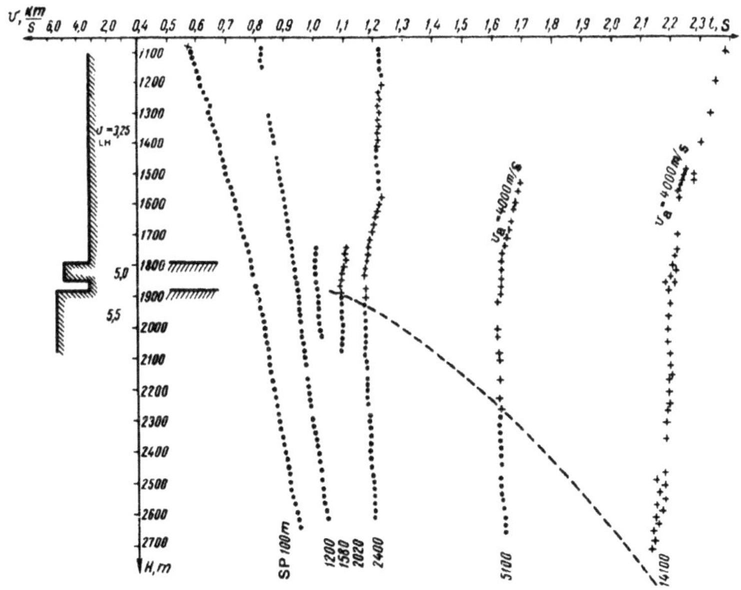

Fig. 70. Vertical traveltime curves in the section crossing the top of Jurassic–Triassic deposits.

Only one section of the vertical profile of interest to us was shot from *SP* 1580 and 2020 m for making detailed observations of the refracted wave in the vicinity of the point of emergence. The refracted waves (along the thin layer and along the top of the bed) from *SP* 1580 m are not recorded separately (Figure 70) and only just appear in the zone of interference with the direct wave. The high-gain summary seismogram from *SP* 1580 m (Figure 71), which depicts the first arrivals, shows that in the depth range 1750 to 1760 m the 'up' direction of arrival, corresponding to the direct wave (circles O in Figure 70) changes to an equally sharp 'down' event, corresponding to the refracted wave (crosses + in Figure 70). At the interface with the thin layer at $H_v = 1790$ m, the arrivals again change sign owing to the interchange from the refracted to the transmitted wave. A similar picture is observed near the top of the basement at $H_v = 1880$ m, where in the depth range 1860 to 1870 m the first arrivals break 'down', corresponding to the refracted wave. In crossing the interface the refracted wave is again replaced in the first arrivals by the transmitted wave, which is recorded as the first arrival down to the end of the profile with 'up' breaks.

Observed from other shot-points, these waves are not discrete, and only one wave, probably a composite, is recorded in the first arrivals. The intensity of the wave associated with a thin layer could be expected to decay rapidly with distance and not to exert an appreciable influence on the intensity of the primary wave at great offset distances.

From *SP* 2400 m, the conversion from the refracted to the direct wave detectable from the change of polarity of the arrivals takes place at the depth of between 1570 and 1580 m, whereas from *SP* 14 100 m the refracted wave is trackable as a first arrival along the entire section of the vertical profile above the interfaces.

Of major interest in the determination of the origin of the wave recorded as the first event, is the study of the polarity of motion in the basement bed. The following relationship is observed with an increase in the offset, the depth at which conversion takes place increases. From *SP* 2020 m the conversion is at a depth of 1890 m; from *SP* 2400 m – at 1920 m, and from *SP* 5100 – at 2230 m (Figure 70). It is difficult to determine the reversal of signs unambiguously from *SP* 14 100 m because of low amplitudes. The apparent velocities of the refracted wave above the top of basement are negative and equal to 4000 m s^{-1} from all shot-points. The apparent velocities of transmitted waves in the section of the profile from the top of basement to the point of sign reversal tend towards infinity. The locus of points of sign reversal is shown in Figure 70 by a dotted line. All the features described above are typical of a medium with a monotonously increasing velocity with depth.

The characteristics of the record enable certain conclusions to be drawn as to the nature of the wave first recorded on the vertical profile. According to its kinematics, the wave detected above the refracting interface can be interpreted both as a head and as a curved-path wave. The values of the velocity along the interface determined from the vertical traveltime curves of the first arrivals vary in the range 5100 to 5400 m s^{-1} and cannot surve as a criterion for establishing the nature of the waves. Such variations do not conform to any law, and represent the systematic error originating from the inaccuracies of velocity determination. Theoretical traveltime curves of the head and the curved-path waves have confirmed that their kinematics are practically identical, and as a result both traveltime curves quite satisfactorily approximate the observed data. It

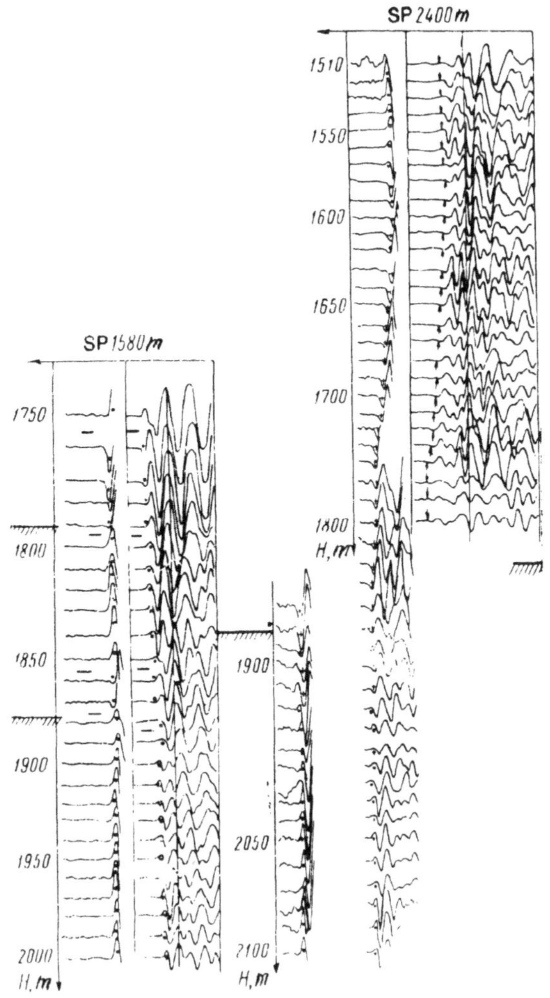

Fig. 71. Seismograms recorded in the section crossing the top of the Jurassic–Triassic bed.

was possible to use the dynamic features of the waves at longer distances because of the significant differences in the charge sizes (from 2 to 50 kg).

In this connection, it was interesting to consider the characteristics of waves recorded below the refracting interface and inside the basement. Here a wave with a negative apparent velocity is recorded. An increase in the distance from the interface is accompanied by a decrease in the amplitudes, a reversal in the polarity of motion, and an apparent velocity change from negative to positive on the Z-component record. This inversion takes place at the minima of the traveltime curve of the first arrivals whose depth increases with the distance from the shotpoint to the profile.

By studying the locus of polarity change of the event, we have been able to determine

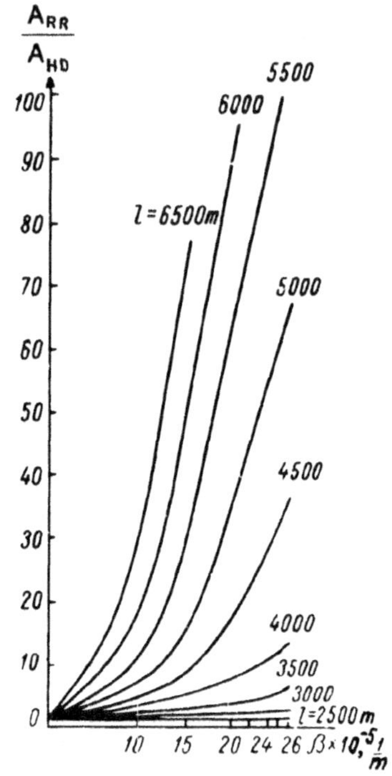

Fig. 72. Effect of curved-ray refraction on the intensity of refracted waves.

the coefficient $\beta = 0.00005\ \mathrm{m}^{-1}$, assuming a linear law of velocity increase. An important part in the determination of the wave's nature may be played by comparison of the shape of records obtained at different points of the profile. From *SP* 2400 m, records of waves that have been obtained above the refracting interface have been compared with those from immediately below the interface and at a relatively great distance from it, in the section of the profile where the wave has a positive apparent velocity. At two of the former points, the waves arrive from below, the records are in-phase, and their shape is identical. At a deeper point, the shape of the record is identical to that obtained above the interface, but the phase is inverted by 180°. From *SP* 5100 m, records obtained above the upper refracting interface, between both interfaces, immediately below the basement top and in sections where the wave is recorded with a positive velocity have all been compared. The shape of the first event is preserved at all points. For the last two points a phase inversion of 180° takes place.

The shape of first events above the refracting interface, below it and in the section where direct transmitted waves are recorded, suggests that the wave recorded above the interface is not a head wave, but a curved-path wave. Similar results have been obtained earlier in the studies of the nature of waves associated with the crystalline basement [83].

To assess the effect of curved-ray refraction in the underlying layer on the intensity of curved-path waves in sections close to the point of origin where the formulae of the ray-path method are inapplicable, computations have been performed for seven values of β from 0.000 10 to 0.000 26 m^{-1}, and for the following parameters of the medium: $H_v = 1880$ m, $v_{av} = 2600$ m s^{-1}, $v = 5000$ m s^{-1}, $f = 30$ Hz. The effect of curved-ray refraction on the intensity was computed for $l = 2500$ to 6500 m (Figure 72). While for $\beta = 0.000 10$ m^{-1} and $l = 5500$ m, the curved-path wave's intensity is 11 times that of the head wave, for $\beta = 0.000 20$ m^{-1} this ratio rises to 67. For $l = 3500$ m, the intensity of the curved-path wave is 7 times that of the head wave. For $l = 5000$ m, this ratio is equal to about 70, and for $l = 5500$ m to over 100.

To sum up, experimental data enabled the presence of a velocity gradient in a layer of great thickness to be established, and the theoretical results demonstrated the strong effect of the gradient on the intensity of the first arrival, usually interpreted as a head wave.

Considerable interest is attached to waves propagating in media with weak velocity differentiation, which are, in particular, typical of areas of inflexions formed by thick terrigeneous deposits or by very thin-layered marine deposits. It was assumed that the intensity of head waves in such media would be considerable [53]. Numerous VSP studies demonstrate that the head waves are not recorded in media with weak velocity differentiation. This can be convincingly demonstrated by observations in borehole N3 in Afipskaya, in the Krasnodar province, situated in an inflexion area. Here a vertical profile has been shot from eight points in the range 0 to 1500 m, in order to detect head waves. The first wave to be recorded on vertical profiles from all shotpoints (Figure 23) is the curved-path wave, with a great number of reflected waves recorded in subsequent arrivals. We have obtained similar results in other regions with weak velocity differentiation, namely, in several foothills and burried valley depressions (the Illi, Fergana, Naryn monoclines) formed by thick layers of coarsely fragmented terrigeneous deposits.

It is noteworthy that the effect of curved-ray refraction can be reliably established in observations along off-line vertical profiles as well as on surface observations practically everywhere. However, the velocity gradient does not always manifest itself in VSP along in-line profiles. This fact has already been established earlier.

COMPUTED DATA

The lower intensity of the head waves as compared with that of the transcritically reflected, as observed in experiment, is in good agreement with the theoretical results. To illustrate the point, we present data on the computed intensities of the head and the transcritically reflected waves corresponding to the case of a layer bounded by a half-space and of a velocity increase in the layer velocities in the overlying bed and in the layer (borehole N42). The computations* have been carried out using a surface impact source with observation points located on the surface behind the point of emergence of the head wave $l_0/2d < l/2d$, where l_0 is the distance to the point of origin of the head wave, d is the layer's thickness, and l is the distance from the shotpoint to the point of observation. The impulse is shown in Figure 73 (top).

The following parameters have been assumed in the computations:

$$\gamma = \frac{v_{S_1}}{v_{P_1}} = 0.394; \qquad \delta = \frac{v_{S_1}}{v_{S_2}} = 0.558; \qquad \Delta_0 = \frac{v_{S_2}}{v_{P_2}} = 0.5;$$

* The computations have been carried out at LOMI by I. S. Smirnova [114].

$$\sigma_0 = \frac{\rho_1}{\rho_2} = 1; \qquad \frac{d}{v_{P_1}} = 16.5,$$

where ρ is the density.

The results of the computations (Figure 73) demonstrate that at a distance $l/2d = 1.8029$, the head wave leaves the zone where it interferes with the reflected wave, and that the interval between the arrival times of the waves increases with the distance. The intensity of the head wave is 5 to 7 times less than that of the transcritically reflected. The intensity of the head wave declines drastically with the decrease in the layer's thickness. It remains less than the intensity of the transcritically reflected wave for all ratios of velocities of longitudinal waves in the first and second layers.

Consider the theoretical seismograms (Figure 73) calculated for various values of the velocity ratio v_{P_1}/v_{P_2}. At the distance $l/2d = 1.58$, when the head wave has completely separated from the transcritically reflected, its intensity is almost an order of magnitude less than that of the reflected wave. With a decrease in the velocity contrast, i.e., for media with a weak velocity differentiation ($v_{P_1}/v_{P_2} = 0.90$ and 0.99), this ratio experiences little change. The head wave remains weaker than the reflected. True, for certain structures of the media, observation may change those relationships.

Data concerning the situation of waves in media with weak velocity differentiation where in many cases the curved-path and the reflected waves should, probably, be regarded as head waves, are in need of a critical revision. Experiments should also be conducted to appraise the effect of curved-ray refraction on the intensity of waves interpreted as head waves.

Theoretical calculations of wave fields at great offset distances from the source, for horizontal layered media containing layers each of a thickness exceeding the wave length, using the ray-path approximation in accordance with a program drawn up for internal points of the medium, have demonstrated the applicability of models constructed from observations carried out in the proximity of the source [115]. As a result of this work, surface and VSP observations combined with theoretical results on the detailed velocity distribution enabled certain criteria for the determination of the nature of waves at great distances from the shotpoint to be formulated [115].

For head waves:

(1) A parallel course of overlapping traveltime curves, both for horizontal profiles and for VSP.

(2) The minimum of the vertical traveltime curve corresponds to the depth of the refracting interface, and its depth does not change with the distance.

(3) Asymmetrical shape of the vertical traveltime curve with respect to its minimum.

(4) Coincidence of observed and computed traveltime curves for head waves.

(5) The equality of v_{in} computed from the traveltime curves with the velocities in the refracting layer.

For reflected waves:

(1) With increasing offset distance from the shotpoint, the reflected wave approaches the first arrival event both on the surface profile and the VSP.

(2) The amplitude of the reflected wave exceeds that of the first arrival event.

(3) v_a increases with the distance from the source on the vertical profile.

(4) The vertical traveltime curves of the direct transmitted and the reflected waves intersect at the depth of the reflecting interface.

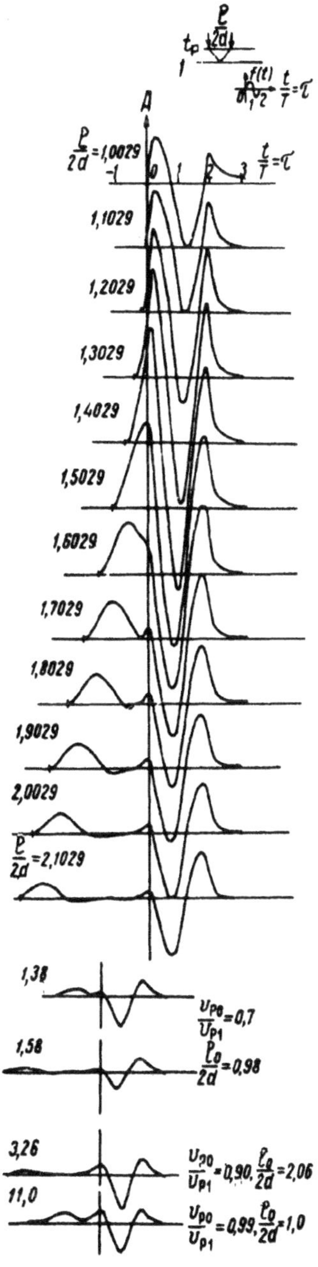

Fig. 73. Computed seismograms of head and transcritically-reflected waves [114].

It has also been demonstrated that the increase in the depth of the minimum on the traveltime curves of first arrival events may, in addition to curved-ray refraction, also be the result of interference with the wave reflected from the interface located below the refracting interface.

Beds of pure salt should be considered as the most homogeneous among the sedimentary deposits. A thick salt layer serves as a good model for the study of head waves because a zero, or a very small, increase in velocity with depth is typical of salt.

In the studies of waves from thick layers of pure rock salt (up to 1200 m) carried out using a wave length of $\lambda = 90$ m, waves have been discriminated from a comparatively heterogeneous thick layer that could be interpreted as head waves by virtue of their parameters. The computed and observed traveltime curves demonstrate the effect of the waves reflected from the low-velocity layer on the first arrival wave associated with the top of the salt. Unhappily, only the vertical oscillation components have been studied in the VSP refraction work described above, and this made the study of the relative intensity of the waves impossible. In addition, the effect of curved-ray refraction was not studied, and the results therefore need to be enlarged with additional data.

The results obtained in the study of the wave associated with the crystalline basement merit special attention [113]. They demonstrate that the properties of the waves with either a shallow (200–250 m) or a deep (400 m) basement are quite similar, and that in both cases, the wave cannot be regarded as a simple curved-path wave corresponding to a half-space with a linear law of velocity increase with depth. Inhomogeneities in the upper portion of the basement complicate the wave's kinematics, with local extrema appearing on the amplitude charts. Frequently, the same inhomogeneities can be used to explain the observed difference in the wave's velocities, and their variation with depth obtained from surface traveltime curves and from down-hole observations. In the first instance, we are dealing with the law of velocity variation vs depth averaged over a long base, and in the second, with the velocity measured at a point. Such quantities cannot coincide in conditions of fragmented heterogeneous media. It should also be kept in mind that the presence of a velocity gradient can be much easier established with the aid of waves, which are propagating mainly in directions close to the horizontal.

VSP data may be used to detect errors in the mapping of seismic interfaces from surface refraction method data, to establish the causes of such errors and to improve mapping accuracy. The method is based on the comparison of vertical traveltime curves obtained in VSP with those constructed from surface traveltime curves. The curves will coincide, if the earth model was chosen correctly. Differences between the traveltime curves enable more accurate estimates of the velocities to be obtained, and the interfaces established, taking account of intermediate reflections. The method is, in principle, applicable to different wave types, but is primarily of value for methods based on the recording of waves propagating in directions close to the horizontal. The inclusion of VSP data enables the accuracy of the construction to be substantially improved.

4. Effect of Upper Portion of the Geologic Section on Observations at Large Offset Distances

The effect of the upper portion of the section on the refraction method has been inadequately studied. However, the available VSP experience points to the fact that the upper portion of the section can exercise a very pronounced effect on the surface seismogram recorded at large offset distances from the source. This effect manifests itself in two ways. On the one hand, the seismograms are complicated by various unwanted waves, and, on the other hand, the natural seismic background noise associated with the upper portion of the section limits the useful sensitivity of instrumentation required for long-range observations.

Two groups of waves are normally typical of seismograms obtained at great offsets. The first group forms the early portion of the seismogram and consists of a large or small number of waves with velocities equal or close to the velocity of the first wave. The duration and the intensity of the waves in this group is not the same for all regions, depending substantially on the geologic structure of the upper portion of the section. In some regions, the intensity of such waves may exceed that of the first event. The second group of waves, arriving substantially later, is composed of high-intensity low-velocity waves that in some cases can be recorded up to very large distances (15–20 km). Low velocity values and rectilinear horizontal traveltime curves with origins practically at the shotpoint are proof that such waves do not penetrate to great depths and are associated with the upper portion of the section.

We shall consider the effect of the upper portion of the section on the basis of data observed in the Staro–Minskaya and the Kletsko–Pochtovaya areas.

Observations in Staro–Minskaya area

To study the effect of the upper portion of the section and to determine the origin of the waves recorded at the surface, we obtained a system of combined horizontal-vertical traveltime curves in the range 0 to 12.1 km.

The early portion of the seismogram obtained along the vertical profile at great depths and in the vicinity of the shotpoint is composed of a large number of downward waves (both longitudinal and converted) with intensities commensurate with that of the first event. Such waves originating from sharp interfaces in the upper portion of the section occupy the recorded interval up to 1.0–1.5 s in some cases. Part of the wave energy from deep interfaces returns to the surface in the form of multiple reflected waves. The remaining energy in the form of refracted (transmitted waves) follows the first arrival deep into the earth along a similar trajectory. Those reflected -refracted waves can be recorded and their parameters studied with the aid of down-hole observations at great depths, and at adequate distances from the source.

We were able to establish from observations along the vertical profile in the range 0 to 2.5 km that a large number of longitudinal and converted waves propagate into the earth (Figure 74). All such waves, upon reaching the basement (interface 5), are refracted and emerge at great distances in the form of the reflected-refracted waves in the original sequence.

Figure 74a is a schematic representation of the trajectories and the identification of waves recorded at short and long distances. Comparing the wave patterns at great depths we are entitled to state that the majority of the waves recorded in the initial portion of the record are combined reflected-refracted and refracted-reflected waves.

Low-velocity waves may be associated with interfaces located at comparatively shallow depths, if there is a velocity gradient present in the underlying layer. In this case, interference waves are formed at the interface whose kinematics almost coincide with

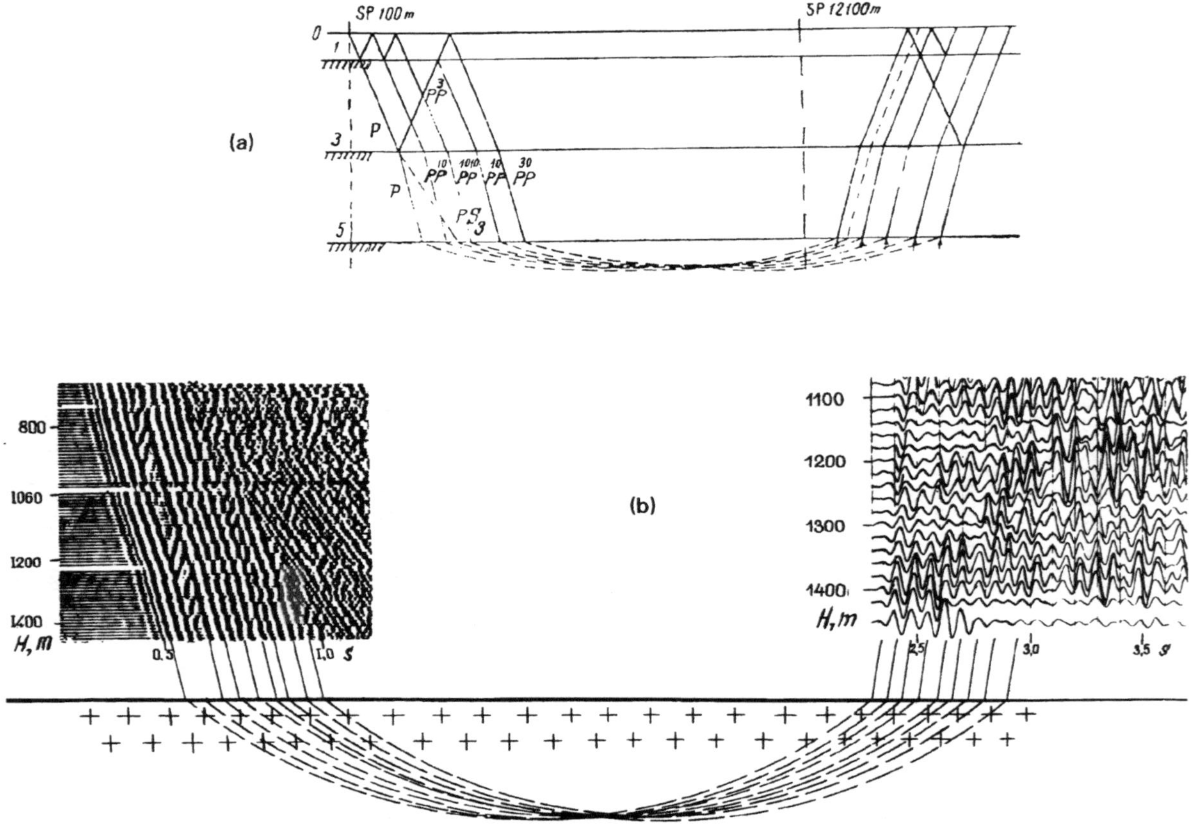

Fig. 74. Schematic representation of paths of multiples (a) and VSP seismograms (b) recorded in the vicinity of the explosive source (left) and at a distance of 11 km (b) (Kletsko–Pochtovaya area).

that of the head waves, but whose intensity is much higher and decays with the distance very slowly.

Such waves can be recorded up to very long distances as maximum- intensity waves on the seismograms. Possibly, they are sometimes interpreted as head waves. It should, however, be pointed out that in some cases low-velocity waves do not display such high intensities and decay relatively quickly. This may be the result of a zero or a very small velocity gradient in the upper portion of the section. The study of waves comprising the early portion of the record at great distances may serve as a still more vivid illustration of the effect exercised by the upper portion of the section. The early portion of the record on VSP seismograms obtained from *SP* 60 m, 200 m, and 600 m (Figure 75) is composed of a large number of downward waves with parallel line-ups associated with the upper portion of the section. The waves are of adequate intensities, and can be reliably correlated along the entire vertical profile. The most stable waves are those recorded in the early portion of the seismogram.

As the distance is increased (from 1020 to 2300 m), the wave curved in the carbonate deposit bed (H = 140-850 m, Figure 69) and returned to the surface begins to register first in individual sections of the vertical profile. Several curved-path multiples with the same time intervals between them as observed from near shotpoints stand out in the subsequent portion of the record. The early portion of the seismogram at great depths is also composed of downward waves, which reaching the crystalline basement are curved in it and return to the surface. Such waves are depicted on seismograms obtained from *SP* 7200 m, 10 900 m, and 31 700 m (see Figure 93). The multiples associated with the upper portion of the section 'duplicate' the wave field of the first event forming as a result a complex wave field with almost parallel line-ups observed in the early portion of the seismogram. Accordingly, at comparatively short ranges at which waves curved in carbonate deposits are recorded, and at long ranges where the waves are curved in the crystalline basement, the early portion of the record is mainly composed of waves associated with the upper portion of the section. In the studies of the wave field in the early portion of the seismogram at great offset distances from the shotpoint, much importance is attached to the high-intensity waves propagating with the same velocity as the first events. It has been demonstrated in repeated VSP and surface observations that the most important of these waves are the multiply reflected ones originating between the surface and the first reference interfaces. Such waves may even originate in the low-velocity zone, provided it is thick enough. Three waves with equal velocities can be distincly tracked after the first event in the time interval up to 1.2 s on the surface seismogram in Figure 76. It is clearly evident from the VSP seismogram that all these waves are the result of multiple reflection in the low-velocity zone, which in the case being considered is 80 m thick [116]. Such waves are recorded mainly in the early portion of the record, and because of that are unwanted in regional studies.

The presence of such waves has been long established. Their effect has been clearly demonstrated in deep seismic soundings in the Pacific [12]. The ocean is especially favourable for such observations, because of the presence of several sharp interfaces (the surface, the ocean bottom, the sea floor sediments) separated by 5 to 7 km of water with a low velocity of propagation of seismic waves. In such conditions the entire wave pattern in the early portion of the record is 'expanded' in time, and this makes for a good resolution of the multiples, which can be studied against a low-noise background. In

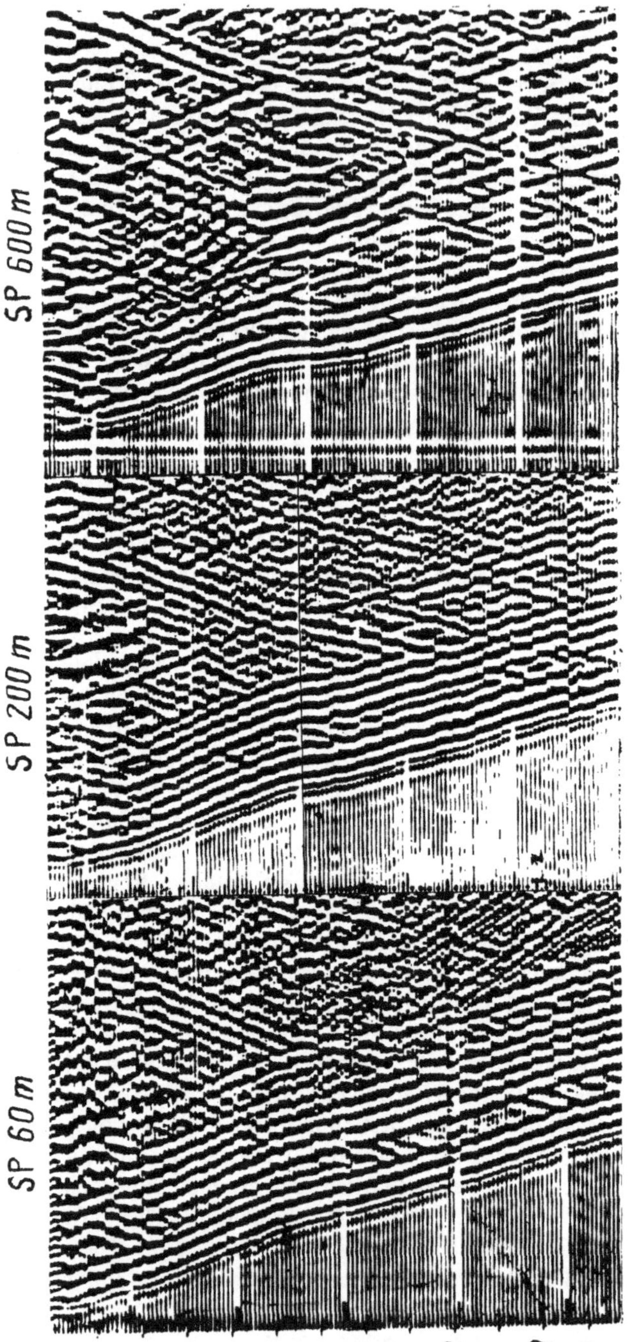

(a)

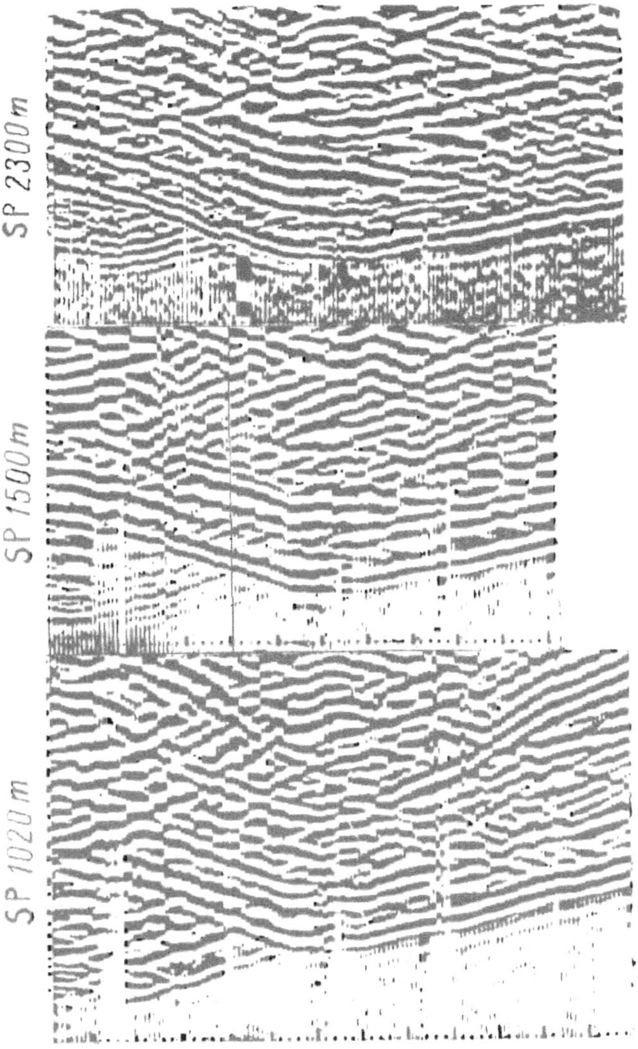

(b)

Fig. 75. Multiple reflected waves associated with the low-velocity zone. Surface (a) and down-hole (b) seismograms [116]

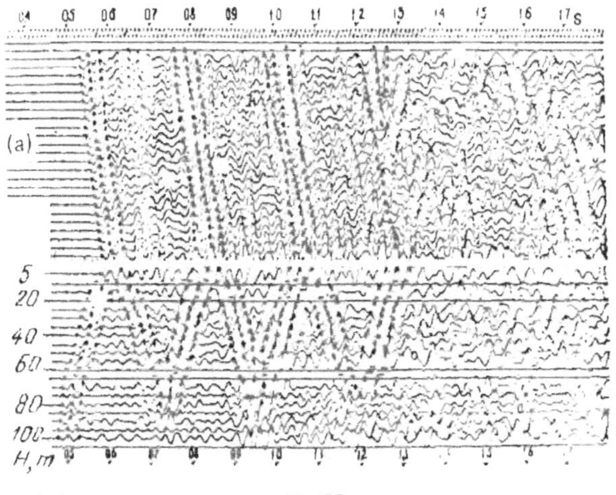

Fig. 76.

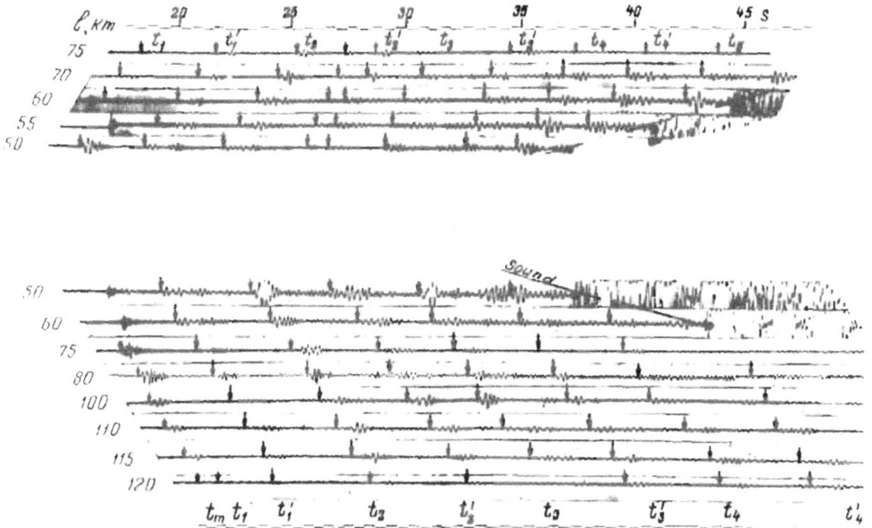

Fig. 77. Combined reflected-refracted and refracted-reflected waves registered in the Sea of Okhotsk [12].

such observations it has been possible to record in the later portion of the record combined reflected-refracted and refracted-reflected waves up to the 5th and 6th order of multiplicity (Figure 77).

The wave intensity diminishes very slowly with the increase in multiplicity and, as a rule, remains higher than that of the first arrival event. As the sea becomes more shallow, the wave pattern becomes gradually more complicated with the reflected-refracted waves drawing closer to the early portion of the record, until with no depth of water, the wave pattern becomes one with a complex early portion that is normally observed on land seismograms.

Summing up, we would like to emphasize that at large offset distances, the decisive effect on the wave field is excercised by the vertical velocity gradient. At the same time, the difference between wave kinematics on horizontal profiles in continuously-changing and homogeneous layered media is slight, and cannot always serve as a reliable criterion for determining the nature of waves. Experience demonstrates, in addition, that an important part in forming the early portion of the seismogram in a layered medium can be played by the transcritically reflected waves, which at large offsets come quite close to the early portion of the record, and interfere with the first curved-path wave. Sometimes, even the first event on the seismogram may be a complex one. This possibility explains the increase in the intensity of waves recorded after the first event sometimes observed on refraction-method seismograms. The problem of the relationship between the reflected and the refracted waves in the early portion of the record should be the subject of special studies. An essential part in the formation of the wave field in the early portion of the seismogram is played by the upper portion of the geologic section, with which many high-intensity combined multiples may be associated.

In order to appraise and to develop further the exploratory potential of the method of reflected waves, we shall have to study in detail the nature, and the kinematic and dynamic characteristics of waves that constitute the early portion of the seismogram. It should be pointed out that the longitudinal multiples observed at large offsets can serve as a source of information about the Earth. For example, in the polarization method of regional studies employing remote earthquakes [42], the longitudinal reflected multiples formed in the region of recording are processed and interpreted jointly with the converted transmitted waves, this enabling the authenticity of the geological results of regional studies to be improved.

CHAPTER 8

THE VSP METHOD FOR TRANSVERSE (MONOTYPE AND CONVERTED) WAVES

During the first stages of the development of the transverse reflected wave method, the principal advantages were associated with the prospects of improved resolution and accuracy of mapping geologic structures. In particular, this is required for the solution of problems such as low-dip structures, wedged layers etc. In recent years the principal advantages of the transverse wave method are being associated with the acquisition of additional information about the velocities and absorption coefficients of S-waves, such information being necessary for making predictions of the geologic section. In regional studies, converted transmitted waves also enable the accuracy of the studies to be improved as compared with methods utilizing longitudinal waves such as the correlational method of refracted waves and deep seismic sounding. This is even more true of methods employing remote earthquakes.

However, it is not always possible in practice to achieve the principal advantages. This perhaps explains why, despite the fact that the earliest attempts to develop the transverse (monotype and converted) reflected wave methods date back more than thirty years [117], and that in the last 15 to 20 years a great amount of work has been carried out, these methods have still not found much practical application. This is due to a large extent to certain specific difficulties mainly of methodological and technical origins. Until recently, the main difficulties of the transverse wave method have been associated with the energy source that limited the penetration. These difficulties could be largely overcome by the use of directional explosive sources. The major difficulties associated with converted, reflected and transmitted waves are due to the identification of the longitudinal and converted waves, corresponding to the same conversion interface, and with the stratigraphic correlation of the waves. The development of the method has been handicapped by inadequate information about the velocities of propagation of transverse waves.

The VSP method has been utilized to study monotype and converted transverse (reflected and transmitted) waves in regions with a great variety of seismological conditions. In the initial stages, the transverse waves were excited by special directional sources. Subsequently, conventional explosives or demolition cord were used. Our studies have demonstrated that:

(1) VSP enables the effectiveness and the directivity of a source to be studied objectively. The wave fields excited by directional surface impact or explosive sources are conspicuous by their simplicity. The direct transverse waves have adequate intensity and are trackable over long ranges.

(2) The polarization method of VSP has improved the chances of discrimination and tracking of various wave types, including the transverse (monotype and converted), and appreciably extended the possibilities of studying the conditions of excitation and propagation of the waves.

(3) Large amounts of VSP data obtained in a wide variety of seismological conditions

have demonstrated that high-intensity transverse* waves are, as a rule, also excited by explosions used for the longitudinal reflected wave method in seismic exploration. Such waves penetrate to great depths, are reflected and can be used together with longitudinal waves for the solution of geological problems [118].

(4) Converted reflected waves are quite common in the real earth. They have been identified from all interfaces at which longitudinal reflected waves originate and at all depths at which observations could be performed (over 4000 m), including waves from the top of the crystalline basement. In numerous cases, the converted reflected waves have been identified even at short distances from the shotpoint. The exploration potential of converted waves has been underestimated and not fully utilized. On the basis of the polarization method, the converted reflected waves can be widely used in combination with longitudinal reflected waves for the acquisition of information about the earth.

(5) Converted reflected waves propagating downwards display high intensity and stability in down-hole observations. They originate, as a rule, at all interfaces, and are persistent on the seismograms. The waves may be used as an independent source of information about the velocities of transverse waves.

(6) PM–VSP observations at long ranges (up to 63 km) have enabled the conditions for the formation of converted waves propagating upwards to be studied. Evaluating the exploration potential of the converted transmitted wave method in the platform region in the Volgograd province, we have demonstrated that the structure of the horizontal and the vertical components of the record is identical.

(7) Sharp interfaces in the upper portion of the section, when present, can exert a decisive influence on the wave field. In this case, a complex extensive wave train of downward longitudinal waves is formed in the vicinity of the shotpoint, with repeated and multiple converted waves appearing in the recording interval forming a complex interferential field. In such conditions, it is almost impossible to record simple waves on surface seismograms. At present, there are no reliable physical criteria for the identification of longitudinal and converted waves associated with deep interfaces to be applied in such conditions. The converted transmitted wave method may prove effective for studying the upper reference level (e.g., the interface at top of carbonate or salt deposits or top of the basement in the case of a terrigeneous overlying bed) with which the dominant *PS*-waves are associated.

(8) Similar difficulties in the determination of the nature of waves are encountered when the early part of seismograms from remote earthquakes are being processed. In the case of earthquakes, the situation is simplified by the circumstance that for deep-seated foci, the first longitudinal wave typically has a simple form. However, the decrease of velocity differentiation with depth substantially reduces the probability of the formation of high-intensity converted transmitted waves at deep interfaces in the Earth's crust. The problem is beyond the scope of this book and should be dealt with elsewhere.

(9) The principal trend in the perfection of methods of regional studies is the joint use of the longitudinal multiply reflected and the converted transmitted waves recorded in the early portion of the seismogram at large offset distances, as exemplified by the

* Here and below the term 'transverse' applies to a monotype wave.

polarization version of the correlational method of refracted waves, and by the polarization method of regional studies empoloying remote earthquakes [42].

(10) The transverse and the converted waves associated with the upper portion of the section can play an essential part in the formation of a complex wave field and obscure the correlation of useful waves on the seismogram, even for the simplest earth structures. In the presence of even one sharp interface in the upper portion of the section, a complex field is formed between it and the base of the low-velocity zone by converted reflected, transmitted and multiple waves, which, together with the longitudinal multiples, may be the reason for the absence of regular waves on surface seismograms. The results obtained in the study of transverse and converted waves make it possible and expedient to devise a general polarization method based on the utilization of different wave types [2].

Consider some examples of the application of VSP to the study of the transverse and the multiple waves. The problems of the excitation of S-waves have been discussed in Chapter 2.

1. Characteristics of Correlation of Transverse Waves

The correlation of transverse and converted waves in VSP is fraught with much more serious difficulties than that of longitudinal waves. As the result of the variations of direction of arrival of the waves at various points of the profile, the vertical components of transverse and converted waves may become commensurate with the vertical components of longitudinal waves, or may even exceed them, so that the intensity ratio will vary greatly along the profile. In the case of a homogeneous overlying bed, the angles of incidence of longitudinal waves on the conversion interface will be functions of the ratio l/H_v, and will exercise an appreciable effect on the conversion coefficients. This should be taken into account when the systems are being chosen for the recording of converted waves of various types. The intensity of the horizontal component of converted waves is higher in low-velocity layers, in which the direction of wave propagation approaches the vertical, and the refraction indices of transverse waves become greater.

The converted wave interferes with the longitudinal wave that is excited in sections of the vertical profile contiguous with the conversion interfaces. The extent of the zone of interference depends primarily on the ratio of the velocities of propagation of the longitudinal and the transverse waves, as well as on the relative disposition of the recording section and the shotpoint. At remote points from the interfaces, the disruption of the correlation of converted waves is the result of their interference with various wave types.

Certain difficulties with the correlation of transverse and of converted waves are also associated with particular polarization of such waves. As the result of the superposition of SV and SH-waves polarized in orthogonal directions, the trajectories of particle motion may approach the elliptical. The correlation of complexly polarized waves, and the timing of their arrivals, should best be executed from the records of the total oscillation vector. This is because in all cases when the maximum sensitivity axis of the geophone does not coincide with the major axis of the ellipse, there is a phase shift between the record of the maximum amplitude component and that of the total vector.

Correlation in accordance with the maximum amplitude, may produce a scatter of points on the traveltime curve, owing to errors in the timing of the wave's arrival which may be as great as a quarter of a period.

The correlation of transverse waves may be disrupted by low-velocity surface-type Stoneley waves excited at the interfaces and recorded on sections of the profile adjoining them. Such waves usually appear when the source is remote from the borehole being studied, and with increasing separation they originate at increasingly deeper interfaces. A rapid decline in the intensity with the distance from the appropriate interface is typical of such waves. In Figure 78 the P_{st} wave is the most intense in the later portion of the record. It is sharply distinguished by its low frequency and is recorded on both sides of the interface at $H_v = 1250$ m with which it is associated. Its vertical traveltime curves are parallel to the depth axis, and its propagation velocity is close to that of transverse waves.

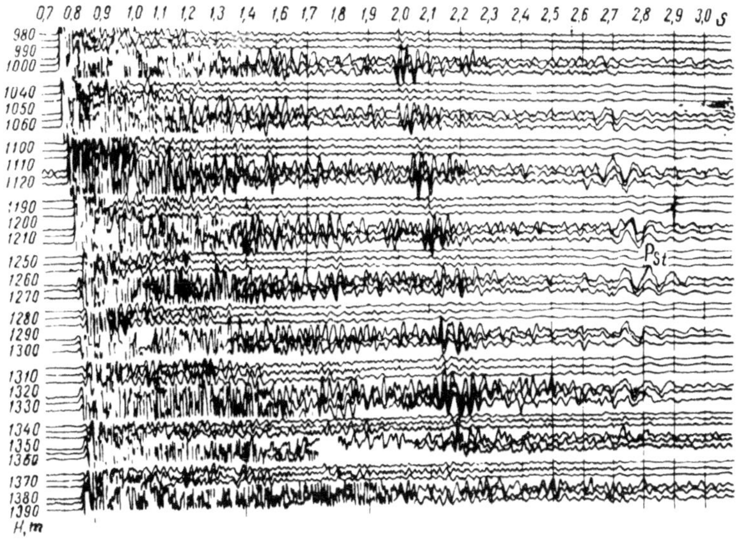

Fig. 78. Low-frequency waves of the surface type associated with interface 3 (H_v = 1250 m, shotpoint 1200 m, borehole N20 Staro–Minskaya). Records from a three-detector probe at two amplification levels are shown for each depth.

Converted transmitted waves display a high intensity, and their vertical component may even exceed that of the first arrival longitudinal wave. The variation of the intensity ratio of the longitudinal and converted transmitted waves with distance may clearly be seen from the seismograms of vertical components in the section of the profile intersecting the interfaces in Figure 79. The transmitted longitudinal and converted waves are recorded below the interfaces. At the distance l = 1580 m, the intensities of both waves are approximately equal, but at l = 2020 m the intensity of the transmitted converted waves has already risen sharply. The longitudinal refracted wave, because of its low intensity, can hardly be detected on the seismograms. The correlation of downward converted transmitted waves is normally only little disrupted by interference with the

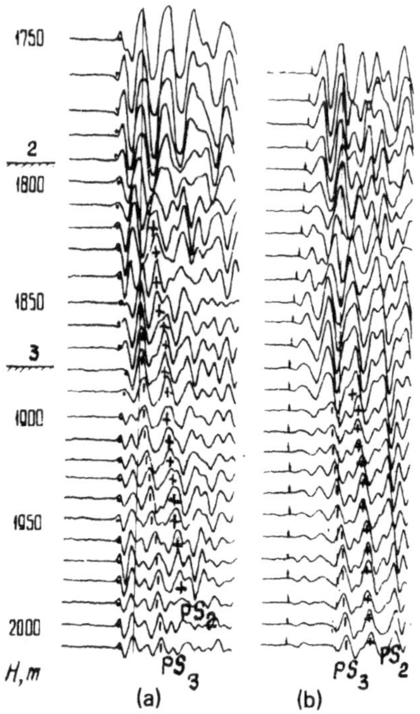

Fig. 79. Variations of intensity of the vertical component of longitudinal and converted waves for l = 1580 m (a) and l = 2020 m (b) (borehole N162 N. Stavropol').

longitudinal and the upward converted waves (mostly waves of low intensity reflected from lower interfaces).

Converted waves transmitted upwards may be recorded both at short and at long offsets. At short offset distances they are mostly associated with interfaces located in the upper portion of the section where the conditions for correlation are very unfavourable. On a vertical profile, such waves can usually be detected only in the early portion of the record at short distances from the interface (see Figure 29).

The principal unwanted waves obscuring the correlation of waves in high-velocity layers ($v > 5000$ m s^{-1}) are the longitudinal waves that may have high-intensity horizontal components. In such conditions, it is also very difficult to discriminate *P*- and *S*-waves accoring to their apparent velocities when comparatively short bases are used. Deviations from linear polarization owing to the interference between the longitudinal and the converted waves, serve as a criterion for detecting transverse waves. In low-velocity layers ($v < 3000$ m s^{-1}), the main difficulty in detecting the waves is the interference of the transverse converted waves with waves with parallel alignments, as well as interference with converted and longitudinal reflected waves.

Converted waves transmitted upwards, and originating at various interfaces in the upper portion of the section, display a high intensity. They are recorded as first arrivals on the *X*-component seismograms and allow the most reliable correlation. Apart from converted transmitted waves from deep interfaces, there are numerous transverse oscillations recorded in the later portion of the record, including waves produced from secondary longitudinal waves at shallow interfaces. It is often practically impossible to identify among such oscillations the converted transmitted waves from deep interfaces.

The presence of numerous different wave types greatly complicates their tracking. In this respect the polarization-positional correlation of waves is of special interest. By way of an illustration, consider the wave field in the upper portion of the section (Figure 80). Despite the fact that the observations have been carried out in the comparatively simple conditions of a horizontally layered medium, the wave field consists of a large number of various wave types and origins. The tracking component of the first longitudinal wave is given by the *P*-component whose orientation coincides with the direction of the displacement vector in the *P*-wave, and which changes along the profile. The longitudinal and the converted waves can be reliably tracked on the *Z*- and *P*-components. However, whereas the longitudinal waves reflected upwards can best be tracked on the dipping componentes with a zero azimuth, the longitudinal waves reflected downwards are trackable on components with an opposite azimuth (the top stereogram in Figure 80). The direct transverse waves S_1 and S_2 are best tracked on component 17, and the reflected upward and downward transverse waves on component 20. The converted *PS*-type waves reflected upward are trackable on the 20 and 12 components with a 180° azimuth, and the downward *PS*-multiples on the *P* and 7 components with the 0° azimuth. Converted longitudinal *SP*-waves are trackable on the components close to *Z* and 20. The examples cited above show how complex the wave field is in the upper portion of the section, and that it is possible to divide it into separate waves by studying various oscillation components in space.

2. Direct Transverse Waves

The study of transverse waves in boreholes began with *S*-wave well-surveys required for the interpretation of data obtained with the surface method of reflected transverse waves. Equipment and methods have been developed for three-component observations in boreholes, and an extensive volume of research on the dynamics of direct longitudinal and transverse waves has been carried out. However, the logging of transverse and longitudinal waves was intended to play an important although only an auxillary part – that of studying velocities of transverse waves required for the interpretation of surface observation data. At the same time there is no doubt that the use of VSP for the study of source conditions and the process of propagation of transverse waves could substantially facilitate and stimulate the development of the method. Direct *S*-waves have been studied for different methods and conditions of excitation.

THE FREQUENCY CHARACTERISTICS OF *P*- AND *S*-WAVES

Direct *S*-waves, as a rule, have a lower frequency spectrum than the *P*-waves from the same source. In the case of an impact source, the dominant frequencies of the *S*-waves are approximately 1.5 to 2.0 times lower than those of the *P*-waves. The differences in the frequency spectra of the *P* and *S*-waves are already detectable in the immediate vicinity of the source, where the transverse waves have only emerged from the interference region. The frequency spectrum changes little with depth. The differences between the frequency spectra of the *P*- and *S*-waves are determined by the source type,

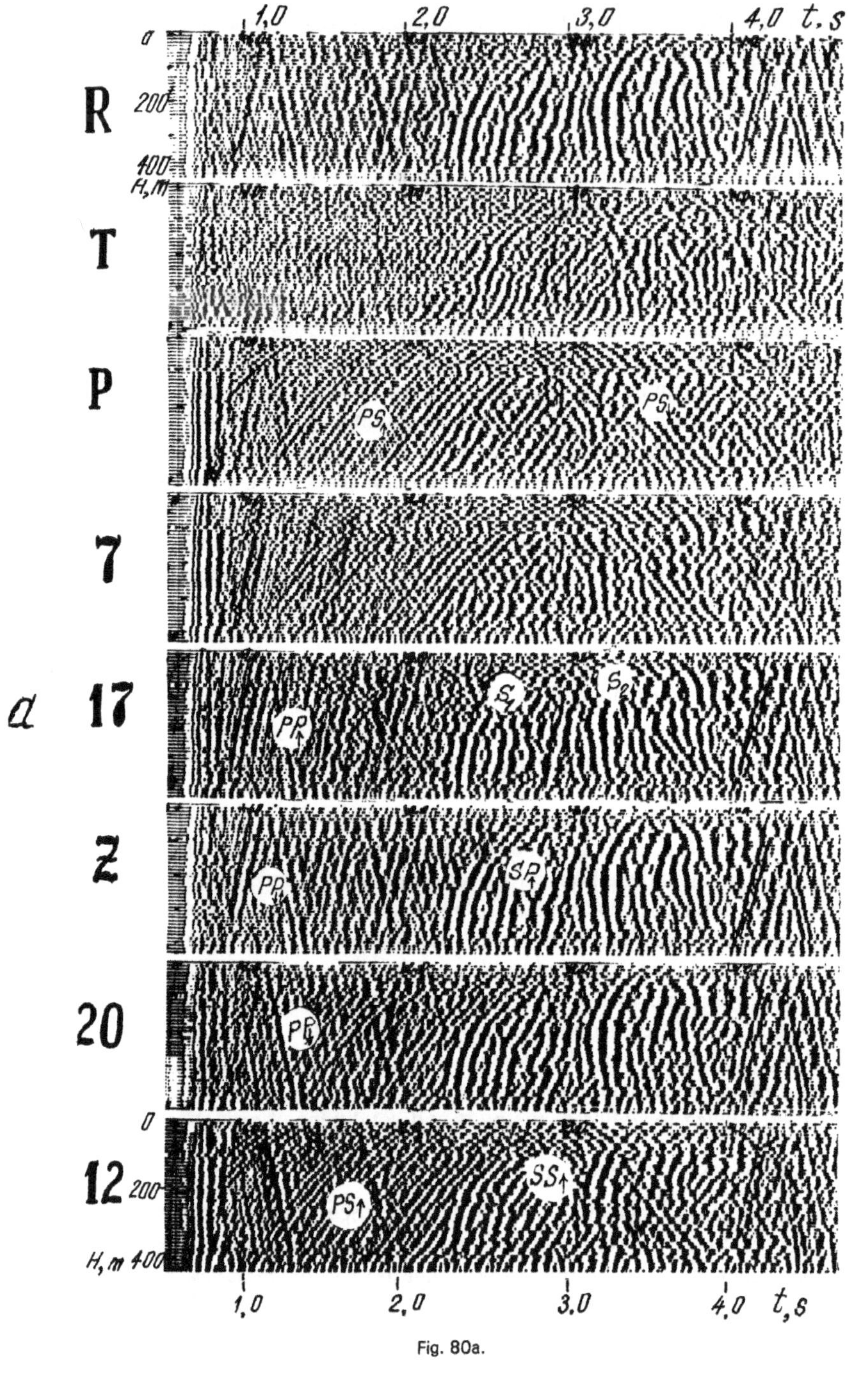
1,0
2,0
3,0
4,0 t, s
0
200
400
H, m
R
T
P
PS
PS
7
a
17
PP
S
S2
Z
PP
SP
20
PP
12
PS
SS
t, s

Fig. 80a.

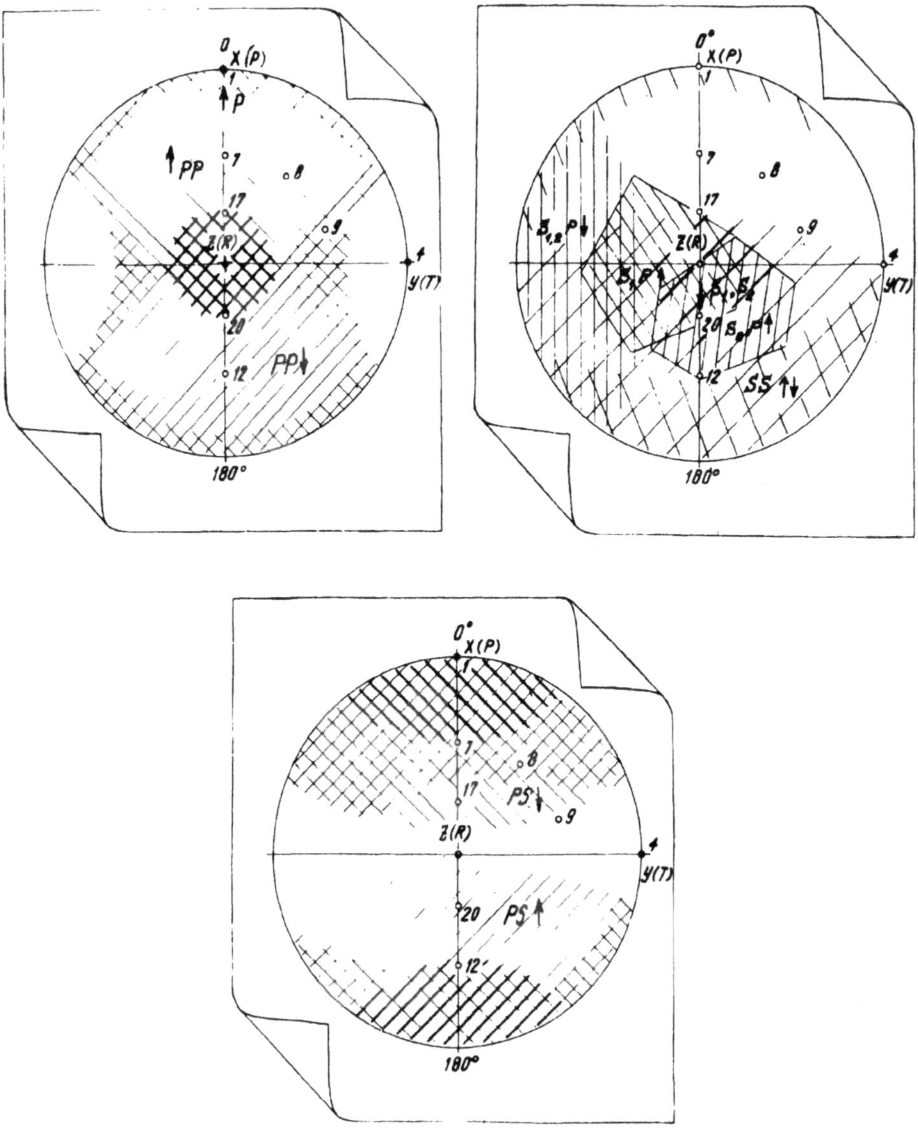

Fig. 80b.

Fig. 80a-b. Wave correlation in PM–VSP: (a) seismogram of fixed components; (b) regions of tracking components for waves of various types (borehole N137724–P, Kaluga).

by its mass, by the area of interaction and by the physical and mechanical properties of the ground at the point of interaction. In the case of constant source excitation conditions, there is usually an excellent reproducibility of the records.

In the experiments being discussed, the frequency spectrum of S-waves produced with an explosive source, practically coincides with that of S-waves produced by impact

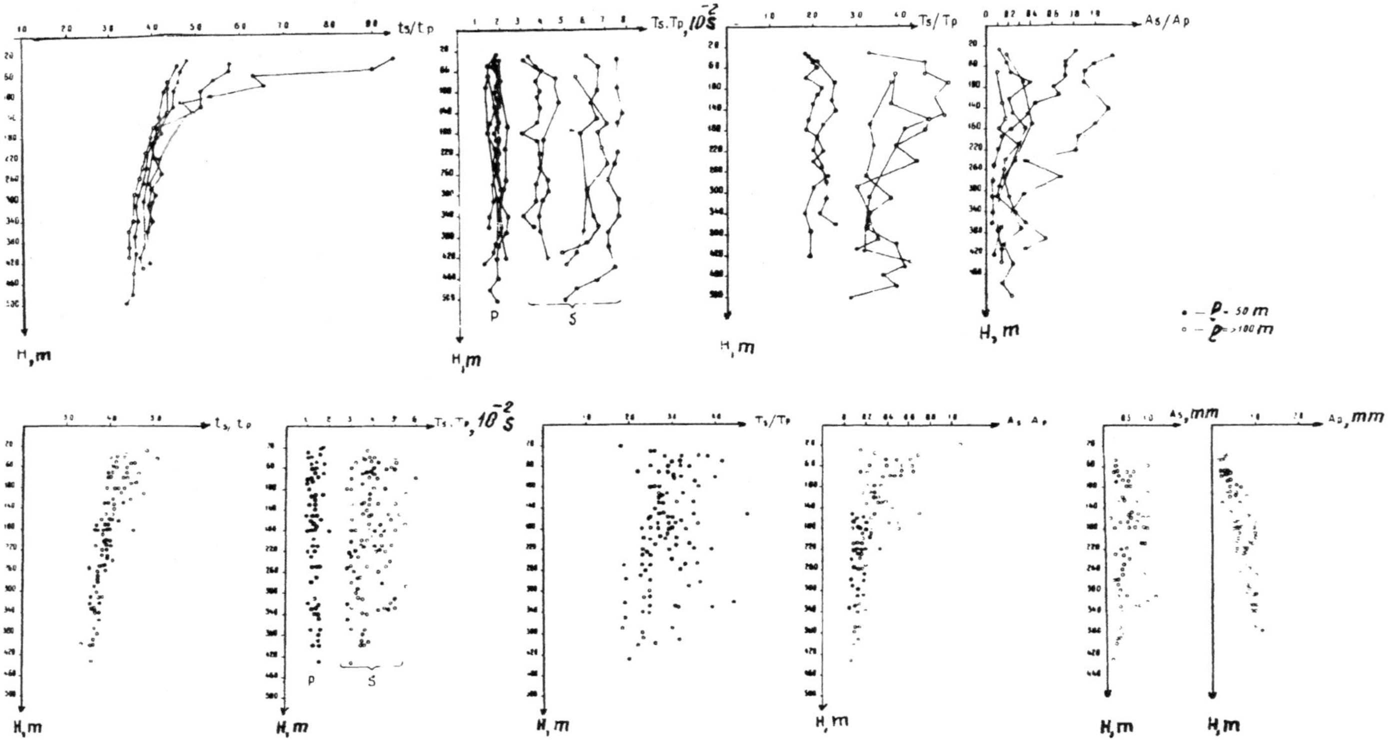

Fig. 81. Results of studies of parameters of direct *P*- and *S*-waves in two areas of Krasnodar province (courtesy Yu. D. Mirzoyan).

excitation. At the same time, the frequencies of longitudinal waves excited by an explosion are higher than in the case of impact excitation. In our experiments, the difference at shallow depths between the frequencies of transverse waves excited by explosions and by impacts may be due to the fact that the explosions have been performed at a distance of 110 m from the borehole, whereas the impacts were at 40 m.

The visible frequencies of the *S*-waves observed at the surface in the Novo–Dmitrovsk area are about 20 to 25 Hz, and experience practically no change when the depth of the charge varies in the range from 55 m to 15 m. Below the low-velocity zone, somewhat higher (35 to 45 Hz) frequencies of *S*-waves are observed when shots are fired below the *LVL*. The visible frequencies of *P*-waves shot at the base of *LVL* diminished from 55 to 60 Hz to about 38 to 40 Hz for shots in the *LVL*. The variation of *P*- and *S*-wave frequencies is determined by the rock's lithology, by the depth and the weight of the charge, and by other factors. The results of studies of the direct *P*- and *S*-waves in the upper portion of the section in two areas of the Krasnodar province are depicted in Figure 81. The frequency spectrum of the pulses originating at the source is decisively affected during propagation by absorption and by various phenomena taking place at geologic interfaces such as reflection, refraction and conversion.

A complex record form, determined by the superposition of transverse *PS*-type waves, whose conversion took place at sharp interfaces close to the source, is typical of *S*-waves excited by non-directional sources. Such waves display a high intensity and travel only slightly ahead of the monotype *S*-wave with which they interfere. In such a situation, the monotype *S*-wave can be discriminated by its lower frequency composition. It may also have a different polarization.

Polarization of Transverse Waves

The studies of polarization of transverse waves are of special interest. Our experiments demonstrate that the initial phases of the *S*-waves are, as a rule, linearly polarized. The linear polarization is distorted in the subsequent phases of the *S*-waves, due to the superposition of waves polarized in different directions. In anisotropic media, the transverse wave is observed to split into two waves, the *SV* and the *SH*, propagating with different velocities and characterized by different displacement directions. Different velocities of *SV*- and *SH*-waves have been observed for both impact and explosive source excitation. In axially symmetrical media, the *SH*-wave is polarized in the horizontal direction normal to the direction of propagation and is recorded on the *Y*-component. The *SV*-wave is polarized in the vertical plane and is recorded on the *Z*- and *X*-components. The shape of these waves on the seismograms is also different. The spectrum of *SV*-waves is usually higher than that of *SH*-waves.

It should be pointed out that the splitting of transverse waves on an in-line vertical profile does not always take place. In a horizontally layered medium, when the vertical borehole coincides in direction with the symmetry axis of the earth's elastic properties, the velocities of the *SH*- and *SV*-waves are equal, and they cannot be discriminated on the seismogram. The velocity ratio of the *SV*- and the *SH*-waves depends greatly on the ratio of elastic constants in an anisotropic medium.

Down-hole studies of the polarization of transverse waves have produced some information about the anisotropy of the Earth. In particular, in some regions we have been able to detect the exitence of anisotropy for *S*-waves in the absence of any anisotropy for P-waves. The experimental results obtained in a number of regions are in good agreement with those of theoretical calculations performed for transversely anisotropic media according to which: (a) the vertical velocity of the SH-wave must be equal to the horizontal velocity of the SV-wave; (b) the vertical and the horizontal velocities of the SV-waves are equal; (c) there may be a different degree of anisotropy for the P- and the S-waves. The polarization of S-waves depends substantially on the rock stresses which, with the exception of hydrostatic pressure, create anisotropy; the anisotropic characteristics of rock vary, and this changes the polarization of S-waves. The available observation data is inadequate for general conclusions to be drawn. The results presented here should demonstrate the feasibility of using the particle displacement directions for the study of the real earth and of the wave propagation in it. At present it is impossible to interpret quantitatively all the features of the polarization of transverse waves observed in experiment. However, the results obtained in PM–VSP stimulated theoretical studies of the process of propagation of seismic waves in anisotropic media, and it is to be expected that soon it will be possible to interpret the polarization parameters quantitatively in order to obtain detailed information about the earth, which is difficult to obtain exclusively from the analysis of kinematic characteristics of the waves.

Consider, by way of an example, the propagation of direct transverse waves in media with a pronounced velocity anisotropy for S-waves, for example, in the Maikop clays typical of well developed diapir tectonics (Taman). The velocities of S-waves in Maikop clays are four times lower than those of P-waves. Close to the shotpoint the S-waves constitute a complex integrated group of oscillations that interfere with surface waves (Figure 82). As the offset from the shotpoint is increased, the transverse waves diverge from the surface waves and split into two groups – the S_1 and the S_2. The waves of the S_1 group are linearly polarized in the horizontal direction at right angles to the plane of the rays (for the SH type), while the waves of the S_2 groups are polarized in the plane of rays (for the SV type). The horizontal and vertical traveltime curves of S-waves are proof of the existence of a vertical velocity gradient: the minimum on the vertical traveltime curve falls as the offset of the borehole from the shotpoint is increased. The existence of a velocity gradient results in the formation of multiple S-waves, and this produces an extensive group of S-oscillations. The difference between the recorded times of the S_1- and S_2-waves increases with the distance from the source reaching 0.5 s at $l = 1.5$ km. The ratio of S_1 and S_2 velocities is equal to 1.4, varying a little in different directions from the source. The intensity of the SV-wave group is higher, and the group is more persistent than the SH-group. The earliest clearly defined phases of the S_1- and the S_2-waves are linearly polarized. However, the polarization of subsequent phases is, as a rule, distorted by the superposition of waves of other types. Hence, two characteristics of the Earth have decisively influenced the pattern of the field of direct transverse waves in the example cited: a strong anisotropy of S-wave velocities that results in separate recording of the S_1- and the S_2-waves, and the vertical gradient of S-wave velocity that results in the formation of multiple S-waves. Such properties of the direct transverse waves exercise, as we are to learn below, a strong influence on the entire wave field observed both downhole and on the surface, and may also affect the exploration potential of seismic studies.

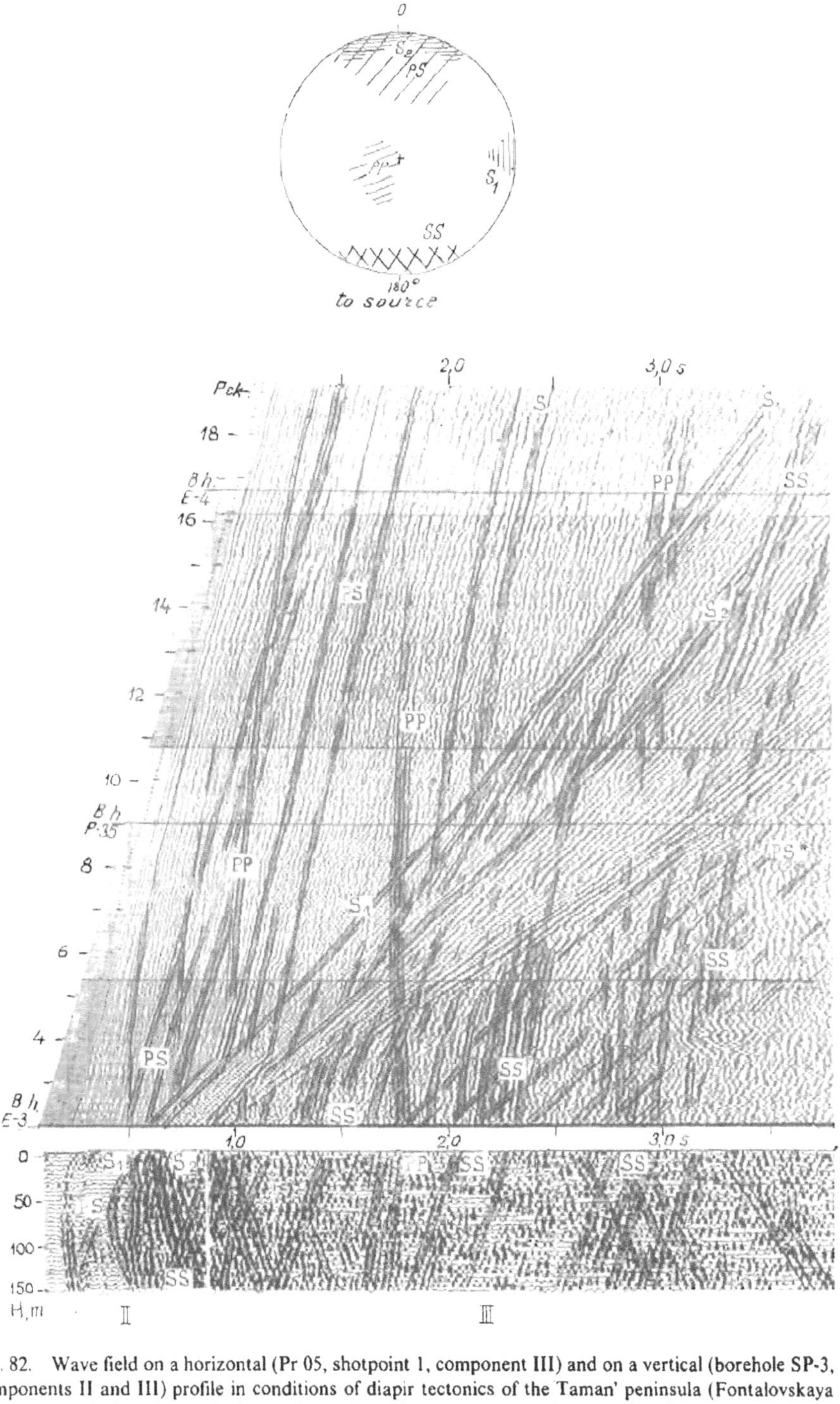

Fig. 82. Wave field on a horizontal (Pr 05, shotpoint 1, component III) and on a vertical (borehole SP-3, components II and III) profile in conditions of diapir tectonics of the Taman' peninsula (Fontalovskaya area). Stereogram depicts regions of tracking components of waves of various types.

Velocities of *P*- and *S*-waves

The ratio of the *P*- and *S*-wave velocities has been studied quite adequately, and we shall not dwell on it. VSP has confirmed that the velocity ratio depends greatly on the lithology and can vary in a wide range. In most regions it is about 1.5 to 2.0 for crystalline rock, 2.0 for coarsely fragemented terrigeneous deposits, and 3.5 to 4.0 in sandy clay beds. In individual areas, a much wider range of variation of the ratio is observed. For instance, in the Kudinovskaya area v_P/v_S varies from 5 to 6 in the uppermost portion of the section, to 2.0 in the Paleozoic bed at $H = 380$ m. Large values of the v_P/v_S ratio can also be encountered at great depths. For example, in the Staro–Minskaya area in the Krasnodar province the ratio of the average velocities v_P/v_S is equal to 3.3 down to the depth of 1300 m. Figure 83 depicts the average velocity curves for the longitudinal and transverse waves obtained in PM–VSP observations in twenty boreholes located in different areas in the Krasnodar province. The range of variation of the velocities will be seen to be very great. Thus, at the depth of 2000 m the average alues of the *P*-wave

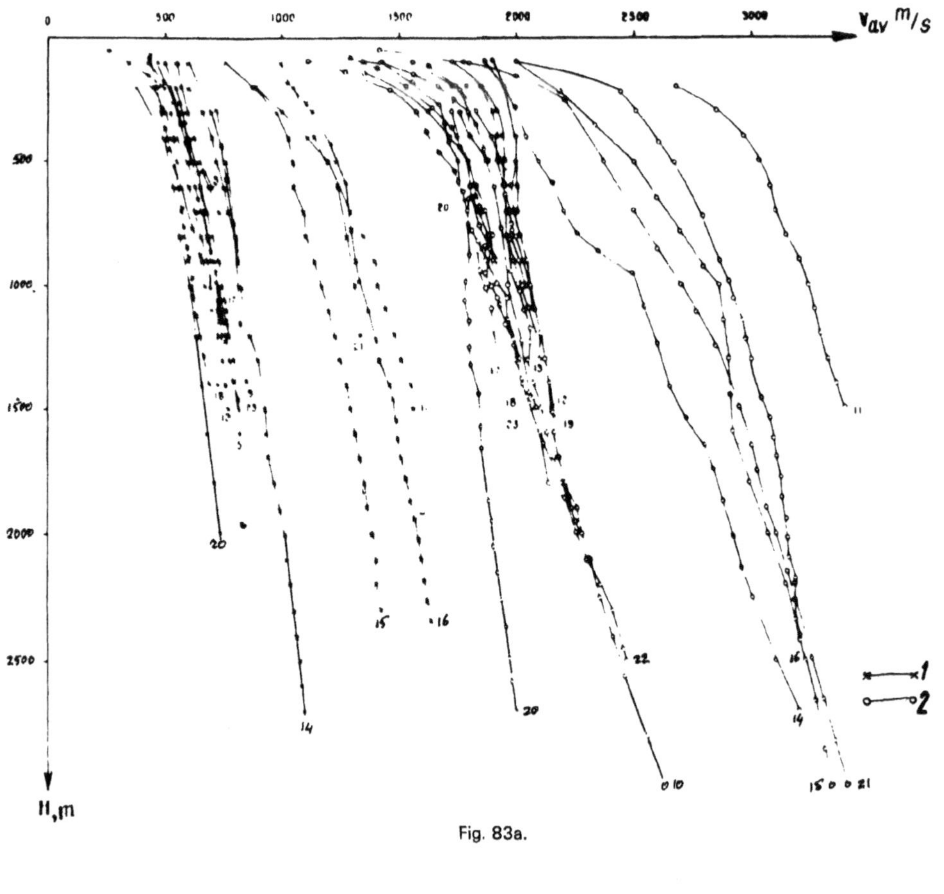

Fig. 83a.

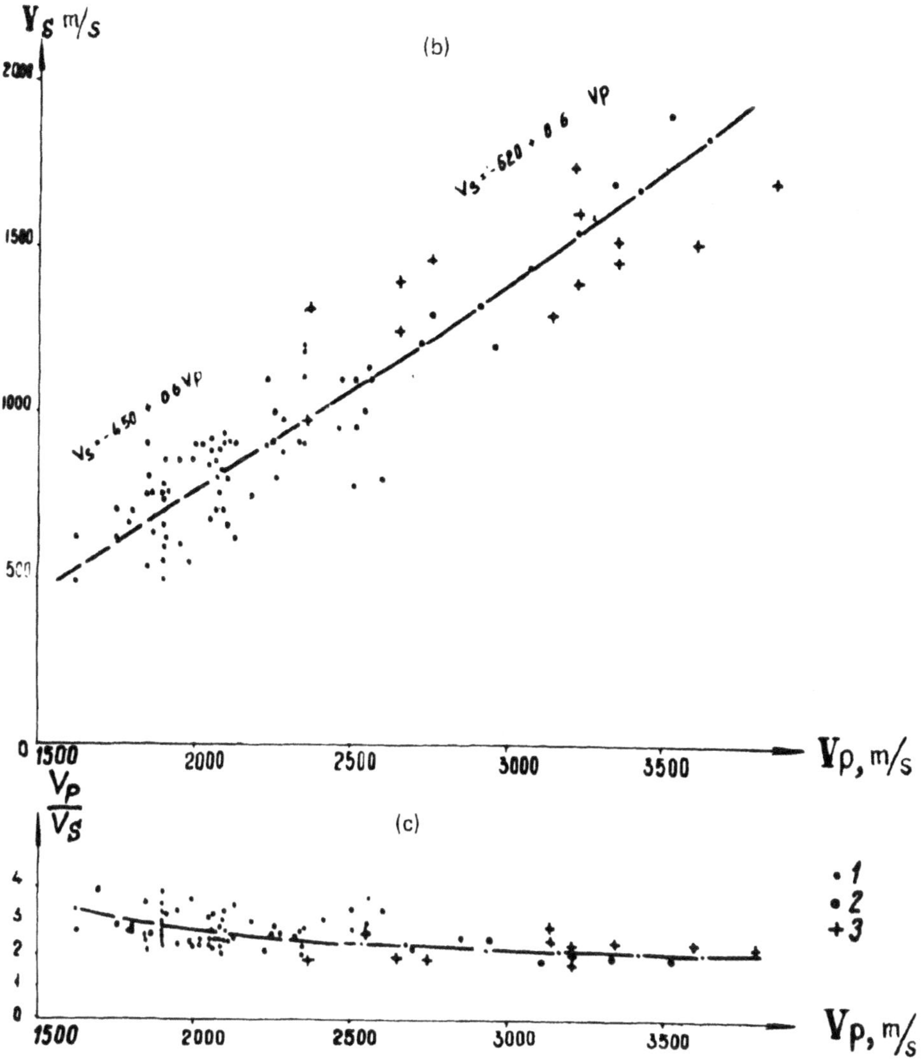

Fig. 83b-c.

Fig. 83a-c. Results of studies of velocities of *P*- and *S*-waves in twenty deep boreholes in the Krasnodar province [45]: (a) average velocity plots of *P*-(1) and *S*-(2) waves; (b) v_S vs v_P dependences; (c) v_P/v_S vs v_P dependence; 1 – Neogene–Oligocene deposits, 2 – Paleocene–Upper Chalk, 3 – Lower Chalk.

velocities vary in the range from 1800 to 3200 m s^{-1}, and of the *S*-waves, from 700 to 1600 m s^{-1}. The dependence of v_S on v_P and the dependence of the ratio v_P/v_S on v_P are also depicted in Figure 83. *S*-waves have been discriminated in all boreholes and tracked on records obtained from conventional explosives used in seismic exploration. It should be noted particularly that converted waves transmitted downwards may be used instead of direct S-waves to gain information about to velocities of S-waves. It has been demonstrated in numerous VSP studies that such waves associated with sharp marker beds in the upper portion of the geologic section display a high intensity, when shots are fired at comparatively short offset distances from the borehole, and propagate in directions close to the vertical.

THE POLARIZATION METHOD IN UPHOLE SHOOTING (PM–MSL)

Information about the P- and S-wave velocities in the shallowest and the most inhomogeneous portion of the section, in which the velocities can vary in a very wide range, is very important for the processing of PM data, and especially for the identification of various wave types. The method has been developed to obtain such information. In contrast to conventional uphole shooting, a three-component surface recording is carried out, and by means of polar correlation, the tracking components of S-waves (Figure 84b) are discriminated on the seismograms (Figure 84a) on which the arrivals of the S-wave can be reliably identified and used to determine velocity. PM uphole shooting carried out on numerous profiles has demonstrated that the P- and S-wave velocities as well as their ratio may vary along the profile. Figure 85, a depicts velocity plots for the P- and the S-waves, the ratio v_P/v_S and Poisson's coefficient. The velocities of the P-waves in the low-velocity zone vary both along the vertical and the horizontal from 500 to 1300 m s^{-1}. In underlying deposits the velocities of the P-waves are about 1500 to 2000 m s^{-1} and remain constant for the given area. The velocity field for the S-waves is simpler and stable both along the vertical and the horizontal. There is practically no low-velocity zone for S-waves. The ratio v_P/v_S varies most drastically in the low-velocity zone: from 1.6 to 7.0 (Figure 85b). In bedrock underlying the low-velocity zone, v_P/v_S is about 3 to 5, and depends greatly on lithology. A similar pronounced variation of these parameters along the profile must be taken into account in data processing.

The values of P and S-wave velocities obtained enabled the variation of the shear modulus (σ) of the rock constituting the upper portion of the section to be calculated (Figure 85a).

3. Transverse Reflected (Monotype and Converted) Waves

PM–VSP observations carried out in regions with different geologic structures have demonstrated that the reflected SS- and PS- waves abound in the wave field. They are recorded in various intervals of the seismogram, and may be very important for seismic exploration. It should be pointed out that, thanks to research being carried out during the last twenty years, the SS-waves have been extensively studied. Unfortunately, this cannot be said of the converted reflected waves. Comparatively small-scale research on the converted reflected waves [117] did not produce a method and was almost completely stopped. The main reason has been the difficulties in the determination of the nature of waves in surface observations, and most of all in their stratigraphic correlation. PM–VSP enabled such difficulties to be overcome, and the converted reflected waves to be used effectively. This is especially important, since the PS-waves promise a great exploration potential. PM–VSP experience has shown them to be well developed, to originate simultaneously with the longitudinal PP-waves at the same

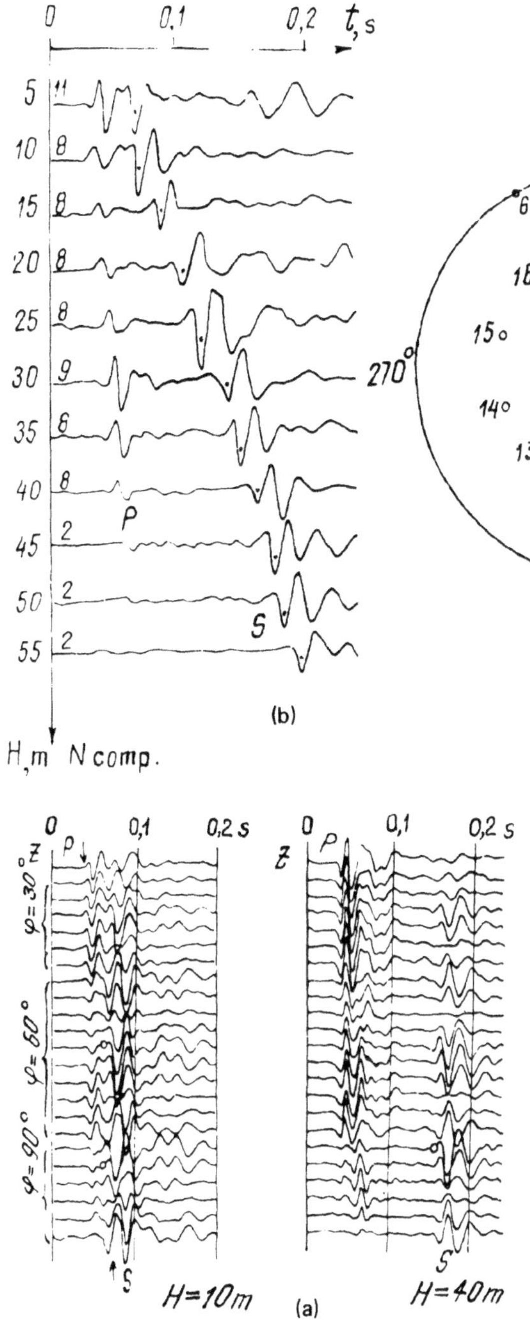

Fig. 84. PM uphole surveys: (a) polar seismograms; (b) seismograms of tracking components of an S-wave; (c) regions of tracking components of P- and S-waves.

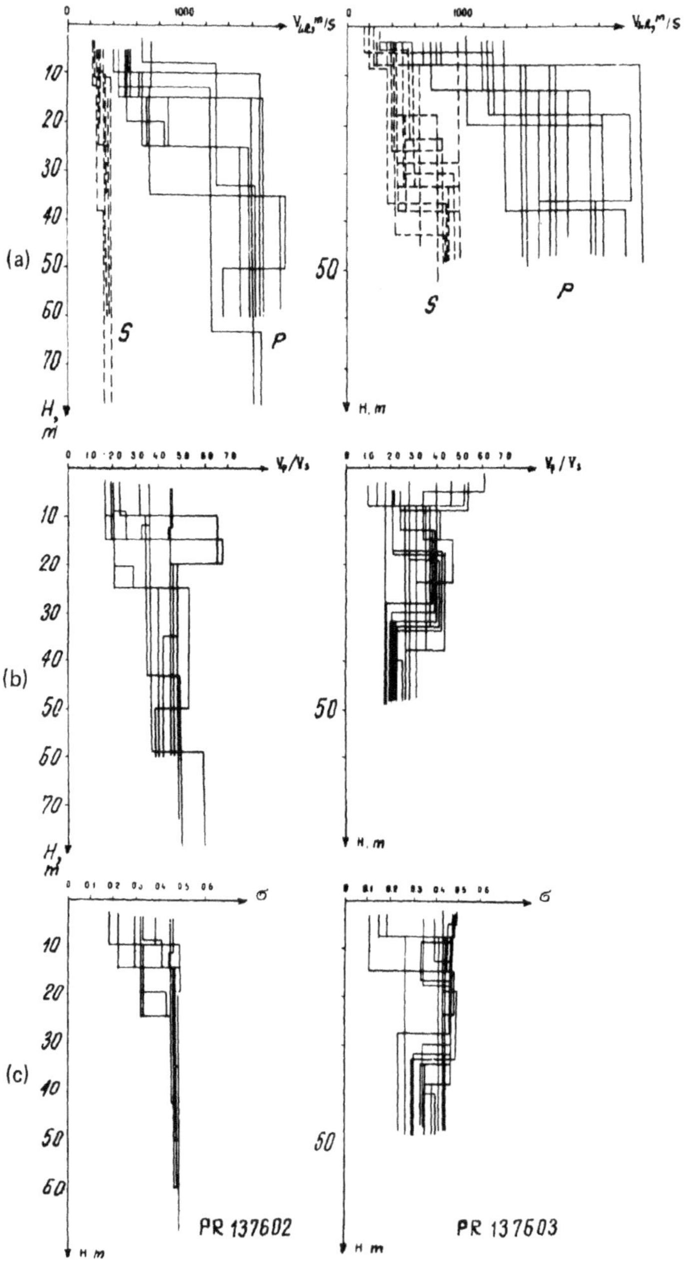

Fig. 85. PM uphole surveys: (a) $v_i(H)$; (b) $v_P/v_S(H)$; (c) $\sigma(H)$ for two profiles.

interfaces, and to facilitate the subdivision of the geologic section in terms of S-wave velocites. Moreover, the converted reflected waves can provide, in principle, the same information about the earth as the transverse monotype waves. Accordingly, the joint use of longitudinal and converted reflected waves can be very effective and easily realized.

Let us illustrate the potential of PM–VSP with the results of studies of the wave fields of reflected transverse and converted waves, obtained in some regions of different geologic structure: in conditions of diapir tectonics, greatly dislocated media and platforms. In this discussion we do not pretend to present detailed studies of the parameters of all waves, our only aim being to demonstrate the abundance of transverse waves in the recorded wave field produced by conventional explosive sources, the ability of PM–VSP to discriminate, track, identify and jointly employ such waves.

Waves in the presence of diapir tectonics in Taman

The complexity of the wave field in areas of diapir tectonics is the result, on the one hand, of the presence of a great number of various wave types polarized in different directions, and, on the other, of the variable conditions of superposition of such waves, which are greatly dependent on the relative disposition of the line of observations and the diapirs. All these factors are responsible for a very low effectiveness of seismic exploration. PM–VSP observations were carried out to study unwanted waves whose nature, despite a large amount of previous work, still remained vague, as well as to evaluate the possibility of discriminating waves reflected from interfaces in subdiapir deposits.

Waves associated with the upper portion of the section

A three-dimensional observation system was devised consisting of a network of orthogonal surface profiles at whose points of intersection vertical profiles were shot (Figure 14). PM–VSP seismograms display an abundance of transverse upward and downward waves, and a substantial difference between the records of various space components varying greatly with the offset distance from the source.

Transverse reflected SS-waves

These can be reliably tracked along vertical profiles (Figure 86). The SS-waves can be discriminated on various components and occupy a wide time interval. They feature a lower frequency spectrum than the S-waves, and in the majority of cases are polarized like the SV-oscillations. Propagating along directions close to the horizontal, the waves have an intense vertical component, and may act as unwanted waves when longitudinal waves are being tracked.

Converted PS-waves

Such waves are very well developed in diapir tectonic conditions and occupy a substantial part of the seismogram. The PS-waves may be both reflected and transmitted (Figure 86), depending on the relative disposition of the point of excitation and the conversion interface. PM–VSP enables their nature to be determined uniquely. PS-waves display a high intensity and are reliably trackable on PM–VSP seismograms from all shotpoints. The apparent velocity and the orientation of tracking components of PS-waves on the surface vary in accordance with the relative disposition of the

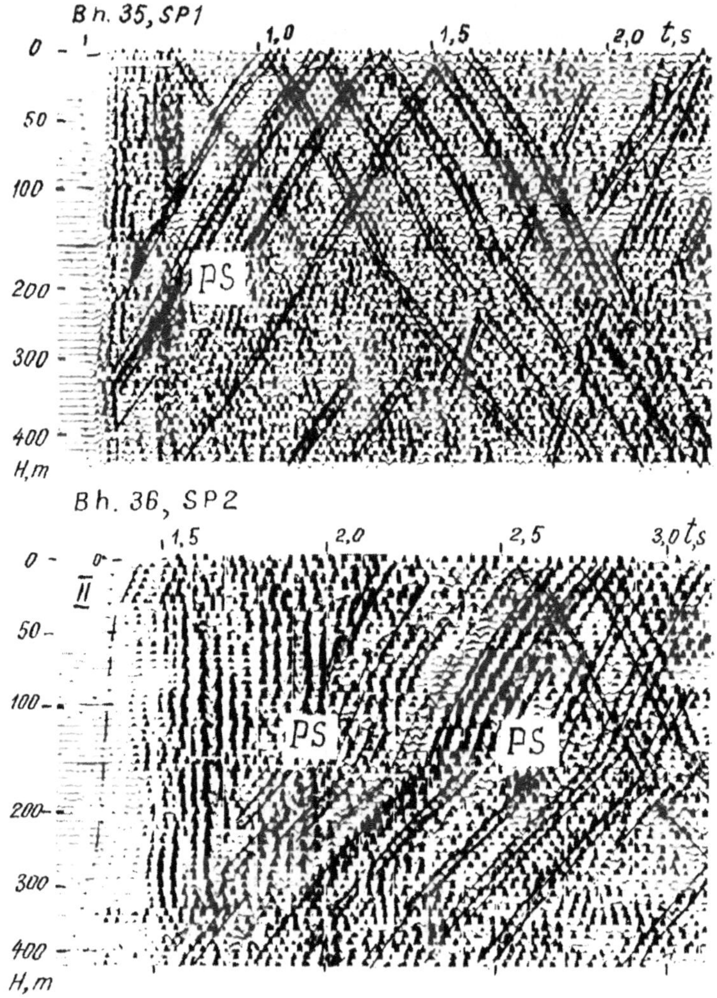

Fig. 86. PM–VSP seismograms of reflected *SS*- and converted *PS*-waves.

shotpoint and the conversion interface, and with the latter's orientation in space. The apparent velocity of *PS*-waves on vertical profiles is small and close to that of *S*-waves in the medium (about 500 m s^{-1}). *PS*-waves recorded outside the region of interference are polarized linearly like *SV*-oscillations.

Reflected *PP*-, *PS*-, and *SS*-waves associated with great depths were studied in borehole N5 in Taman. Figure 87 depicts the seismograms of components on which the transverse SS waves can best be discriminated. The tracking component for direct transverse waves is the component 10, whereas for the reflected transverse waves the appropriate components are 13, *R* and *X*. The *SS*-waves can be discriminated over the entire interval penetrated by the borehole. Comparing the vertical traveltime curves with

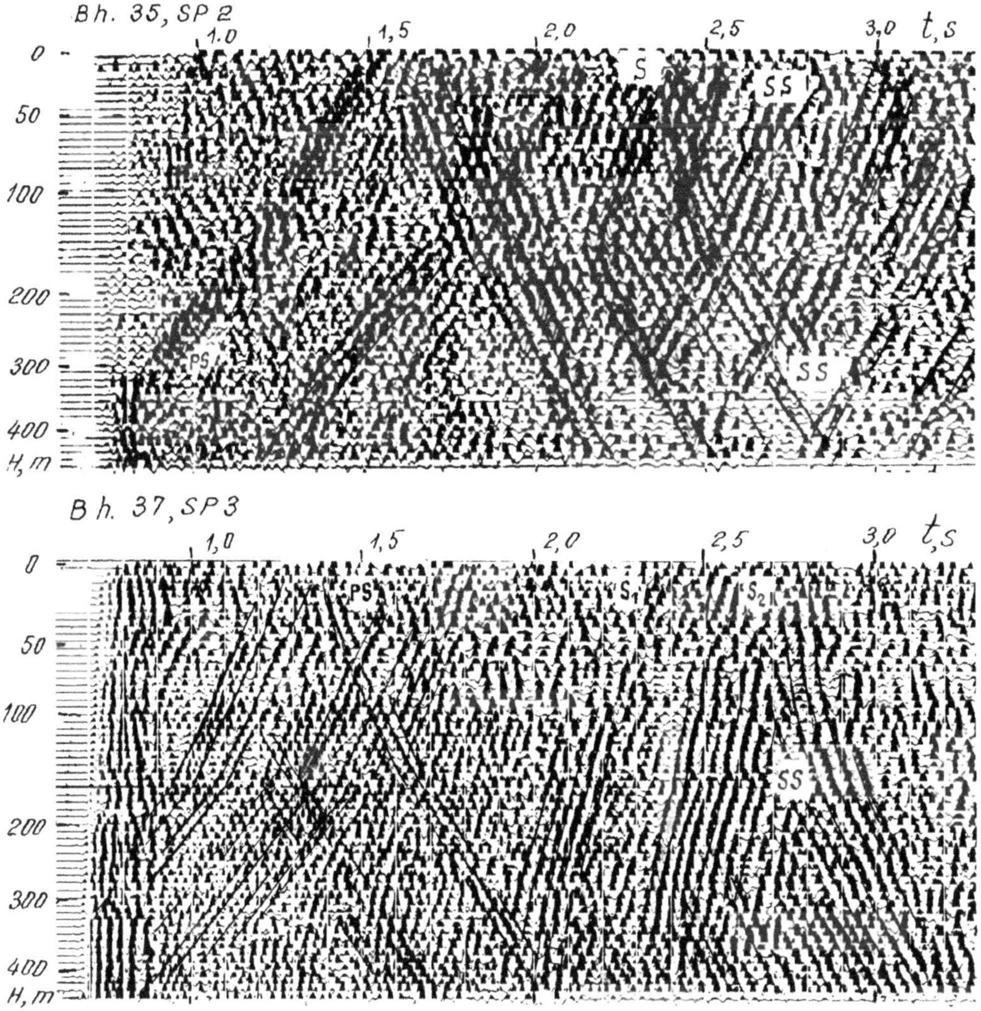

Fig. 86 (cont.).

the regions of existence of the PP_1-, PS-, and SS-waves, one can see a number of different wave types to be associated with the same interfaces.

Waves in the Presence of Dislocated Media

Let us illustrate the capabilities for discriminating and tracking various wave types with an example of PM–VSP observations made in a complex zone in the region of the Western Caucasian foothills, based on seismograms of fixed oscillation components (Figure 88a) whose orientation is shown on the stereonet (Figure 88b). The correlation of the first arrival event on the Z-component seismogram is disrupted in the upper portion of the section. The tracking component for this event is the P-component in the

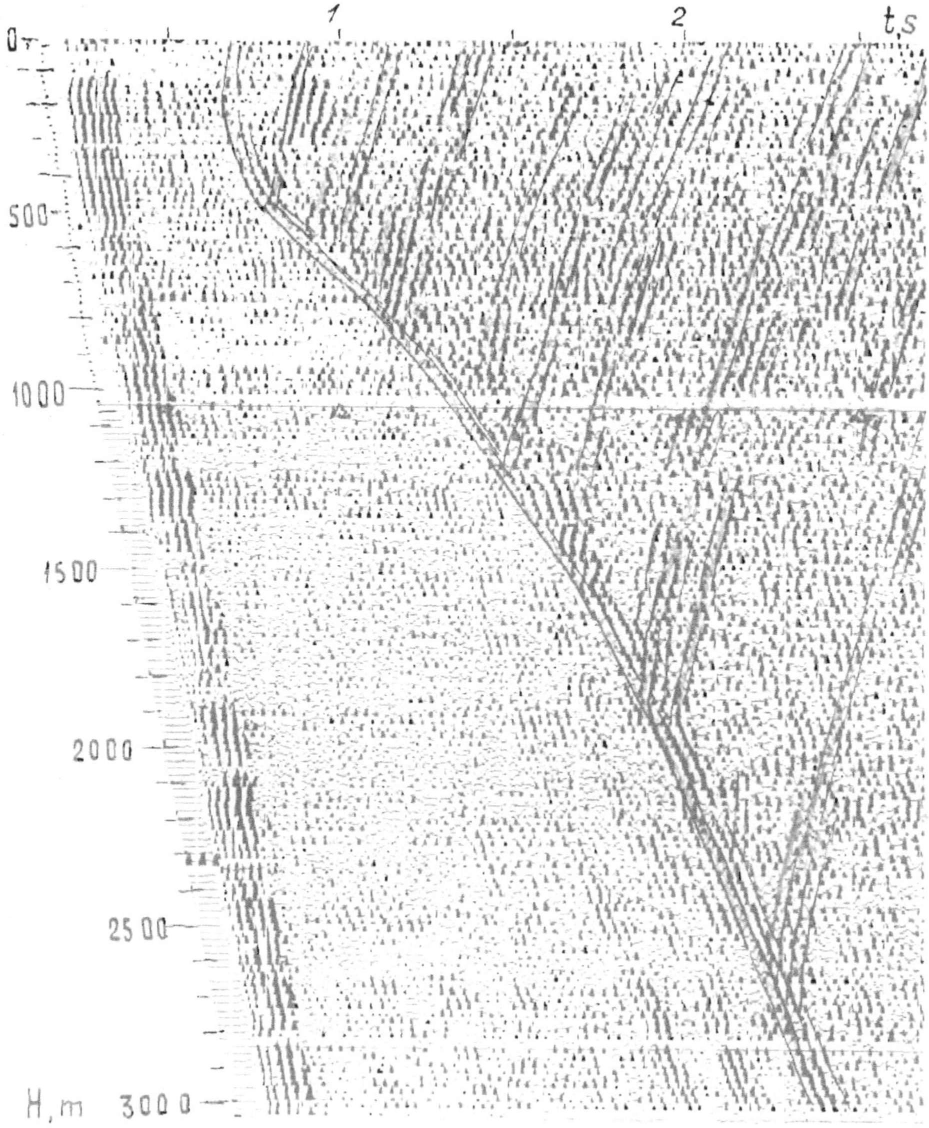

Fig. 87. PM–VSP seismograms of reflected transverse SS waves (borehole N5, Taman', shotpoint 1630 m).

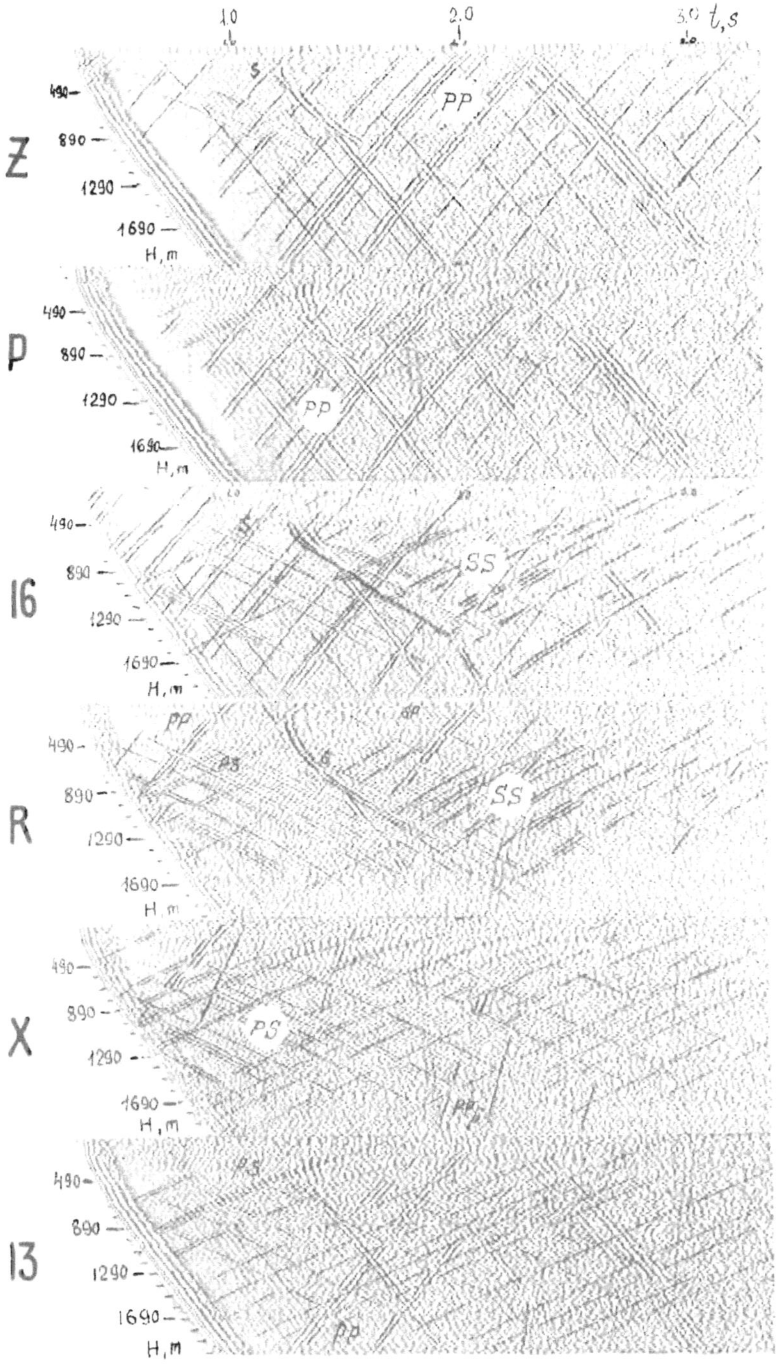

Fig. 88a.

Fig. 88a-b. PM–VSP seismograms of fixed components of oscillations (a) and regions of tracking components (b); l = 500 m (borehole N60, Kaluga).

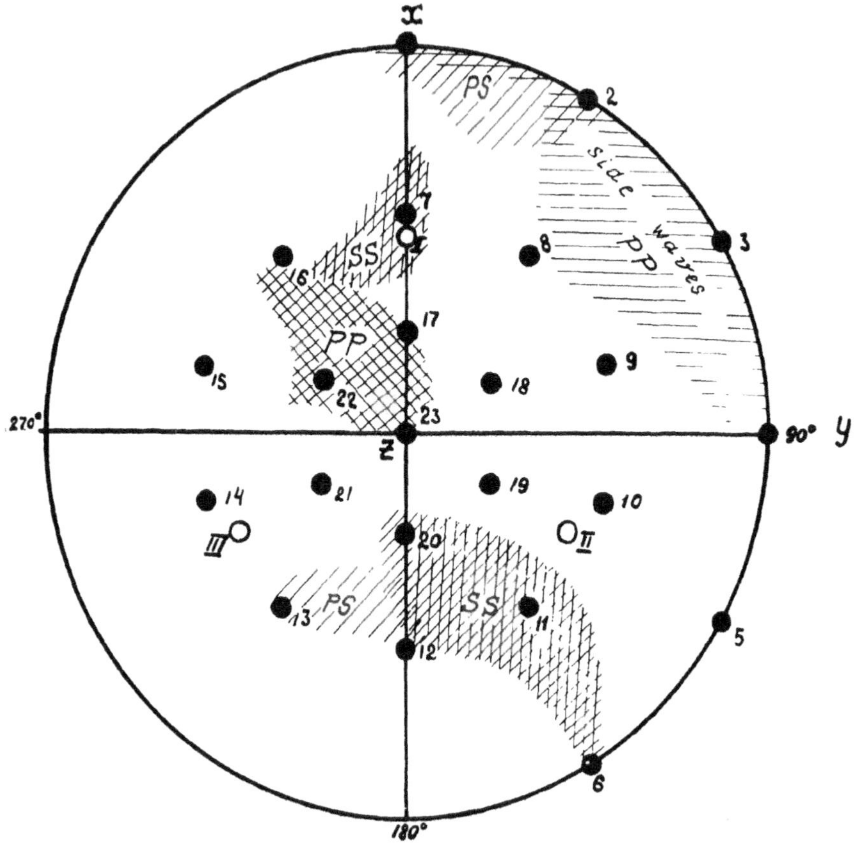

Fig. 88b.

local coordinate system, whose orientation at all points coincides with the displacement direction in the first wave. Naturally, on this component, the *P*-wave is trackable along the entire profile. The *P*-wave is also well trackable on the *Z*- and 13 components. Longitudinal reflected waves can also be reliably tracked on component 16. The direct transverse wave and the reflected *SS*-waves are best tracked on the *P*-component of the local coordinate system at right angles to the direction of *P* lying in the plane of rays. They are also trackable on component 16. The converted reflected waves are trackable on components 13 and *X*. The converted waves transmitted downwards are trackable on the *R* component. The tracking components will be seen from the stereogram to occupy a substantial region in the half-space. Thus, PM–VSP was able to discriminate waves of various types: longitudinal, converted and transverse produced by a conventional explosion. Surface PM observations enable these waves to be tracked on the surface, and the results to be processed and interpreted jointly with the down-hole observations.

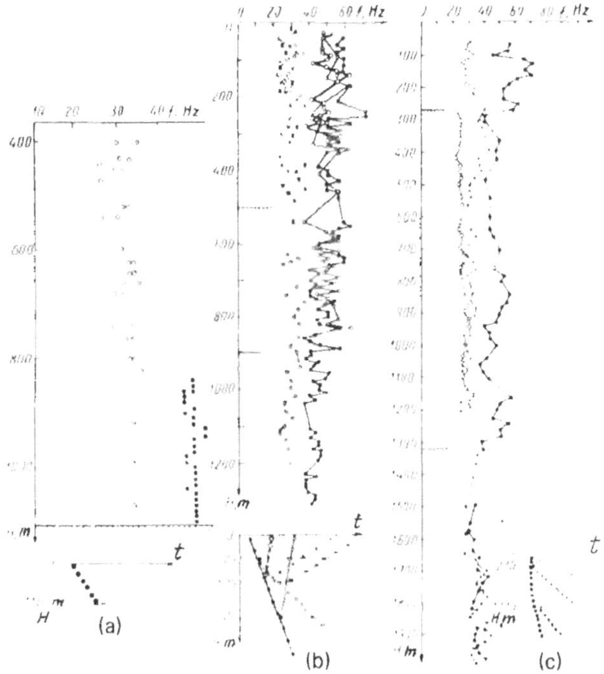

Fig. 89. Graphs of dominant frequencies of converted waves of various types: (a) shotpoint 1250 m (borehole N20 Staro-Minskaya); (b) shotpoint 1250 m (borehole N15 Chelbasskaya, Krasnodar province); (c) shotpoint 1775 m (borehole N42 Staro-Minskaya).

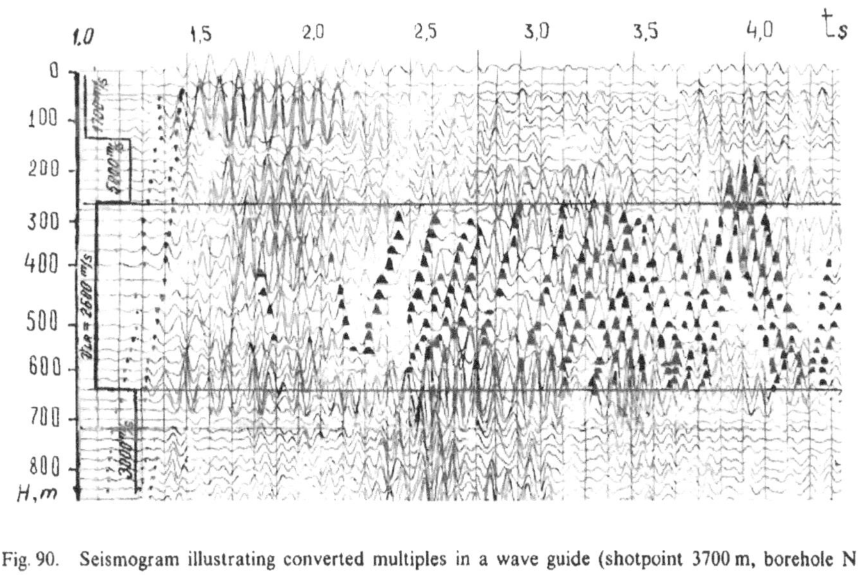

Fig. 90. Seismogram illustrating converted multiples in a wave guide (shotpoint 3700 m, borehole N1 Kletsko-Pochtovaya).

Converted waves in platform conditions

Consider converted reflected waves associated with two thin layers in a continuously changing medium (interfaces 1 and 3, Figure 46) recorded in the Staro–Minskaya area.

The dominant waves in the upper portion of the section are the converted reflected and the converted transmitted waves associated with the first interface at $H_v = 270$ m, which are the most intense on the surface seismograms. In addition, weak converted reflected multiples, originating between the bottom of the low-velocity zone and the first interface can also be discriminated.

Characteristics of the converted reflected wave from the second interface are stability and an extensive recorded interval. It is the most intense of all the converted waves recorded and is perfectly trackable on the vertical profiles from all shotpoints. The converted transmitted wave from this interface is the most reliable source of information about the velocities of transverse waves below the interface.

Frequency characteristics of converted waves

Converted reflected waves typically have lower frequency spectra than the longitudinal. Down-hole observations have demonstrated (Figure 89) that in the immediate proximity of the interfaces the spectra of converted waves contain much lower frequencies than those of the longitudinal. The variation of the frequency spectra of longitudinal and converted waves propagating between the interfaces is normally negligible. In this case, the loss of the high-frequency components in the converted waves is due to the conversion phenomenon in thin-layered conditions.

VSP has demonstrated that converted reflected waves are produced at most interfaces with which the longitudinal reflected waves are associated, and that high-intensity converted waves are associated with thin high-velocity layers. This is in accord with the experience gained in surface studies of converted reflected waves, and with theoretical computations. The converted waves, according to VSP data, are equally well-developed in thick-layered media, their frequency spectrum differing from that of longitudinal waves.

It is worth noting especially, that in low-velocity layers, the wave pattern may become quite complicated due to the appearance of numerous converted multiples. Such layers act as wave guides. VSP observations from different shotpoints enable the process of the formation of the wave field in a wave guide to be studied (Figure 90). However, phenomena in wave guides greatly complicate correlation of the converted waves. Longitudinal, transverse and converted waves from these layers discriminated in VSP, have subsequently been discriminated and tracked on surface observations with the aid of the *PM* common-depth-point method. Figure 91 depicts time cross-sections of recorded data for three wave types: PP_1, *PS* and *SS*. The seismic horizons are reliably associated with geologic interfaces.

It should specially be noted that in PM–VSP, one is able to observe high-intensity converted transmitted (*PS*) and reflected (*PSS*) waves that serve as an additional independent source of information about the absorption and the reflection properties of the gelologic section. Specificially, the variation of the amplitudes of the converted wave can be used to determine the absorption coefficient α_S, and the ratio of the amplitudes of the reflected (*PSS*) and transmitted (*PS*) waves can be used to determine the reflection coefficient (K_S). If K_S is known, α_S can be determined from the transmitted

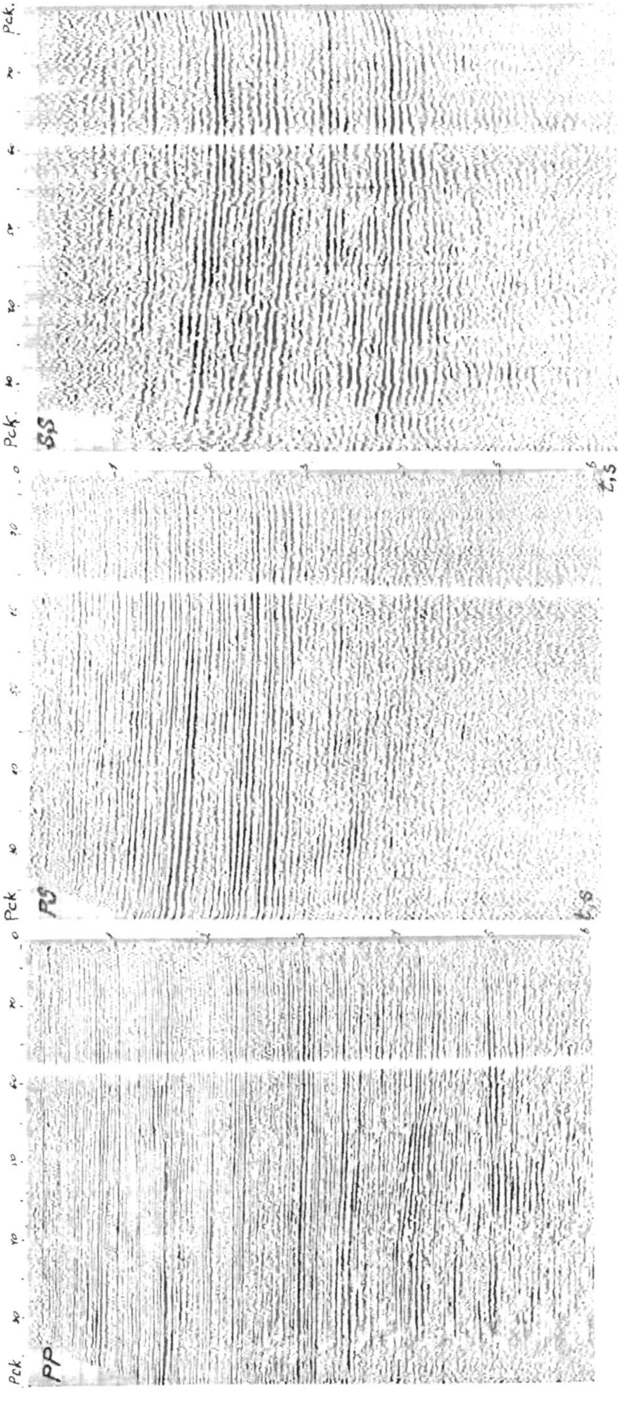

Fig. 91. Time cross-sections plotted for waves of different types in the course of surface PM common-depth-point observations (Severskaya area, Krasnodar province [43]).

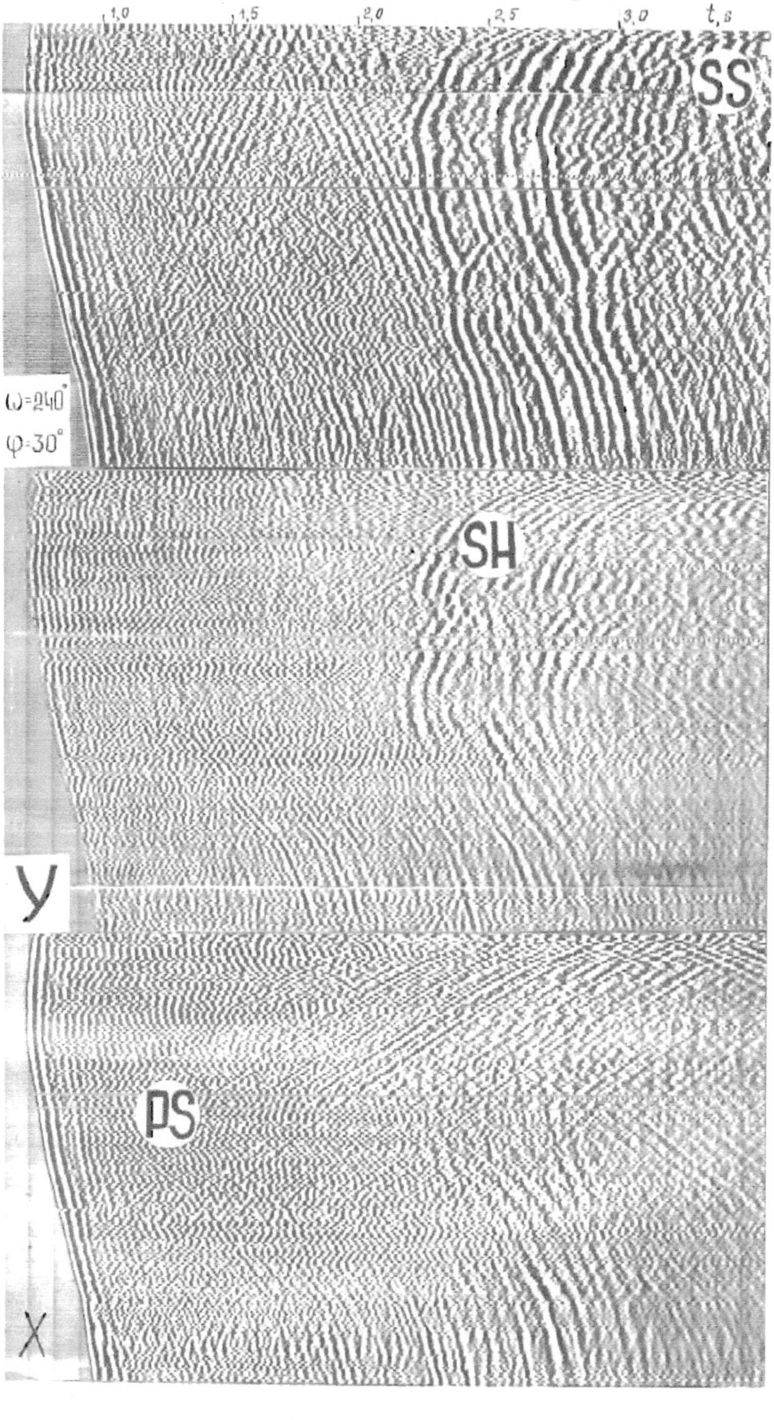

Fig. 92a.

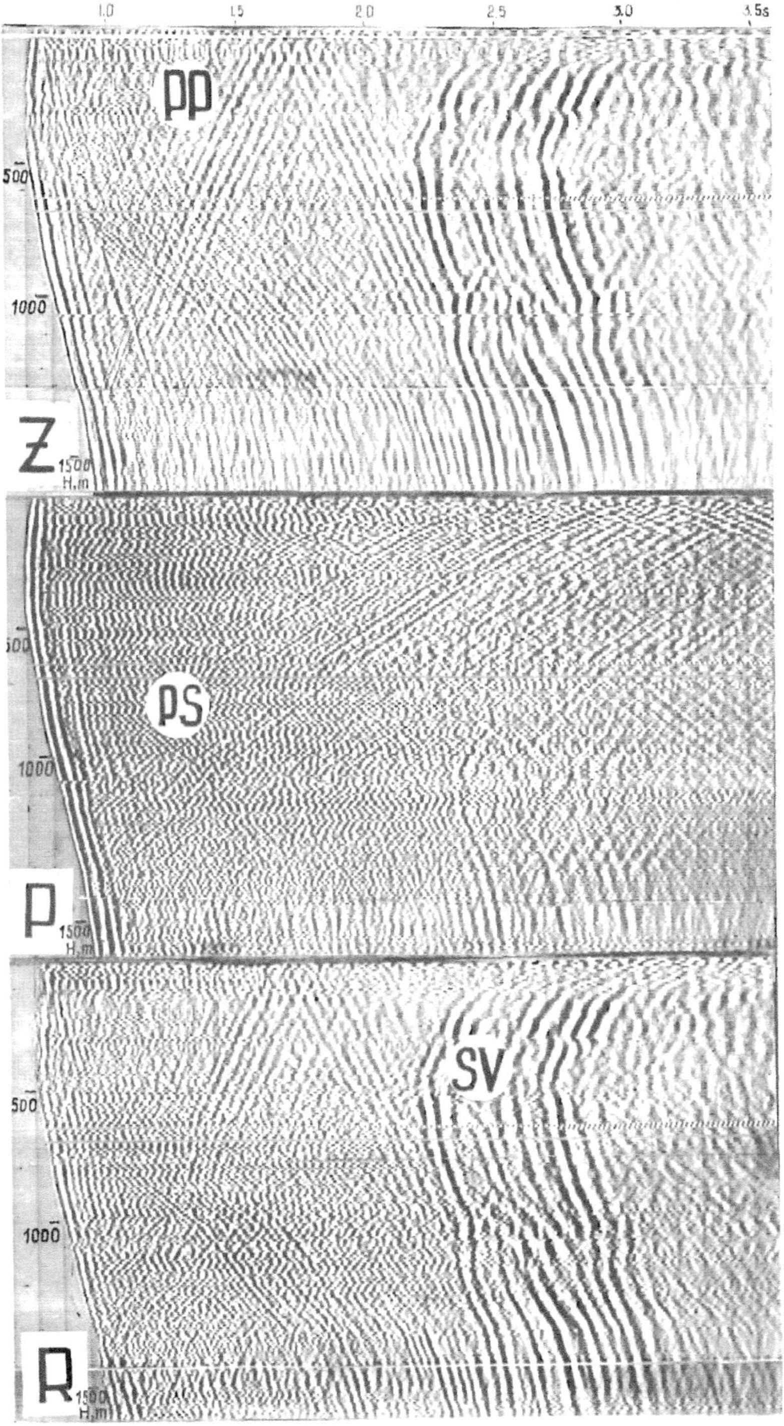
1.0
1.5
2.0
2.5
3.0
3.5s
PP
500
1000
Z
1500
H,m
PS
500
1000
P
1500
H,m
SV
500
1000
R
1500
H,m

Fig. 92a (cont.).

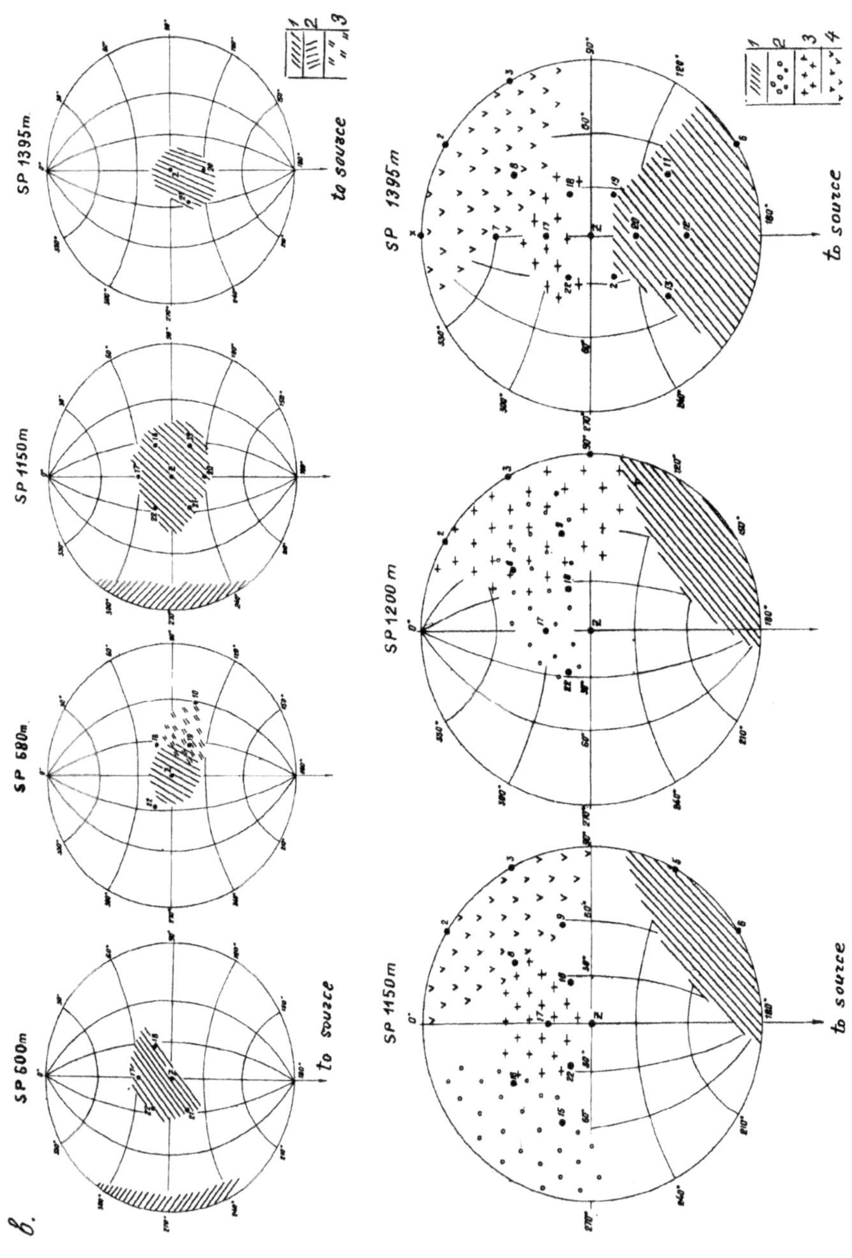
в.
SP 600m
SP 680m
SP 1150m
SP 1395m
to source
SP 1150m
SP 1200 m
SP 1395m
to source
to source

Fig. 92b.

and the reflected waves. The values thus obtained are approximate but effective. By way of illustration, Table VIII shows the values obtained in two boreholes in the Moscow syneclise [119]. The values of α_S obtained with downward waves will be seen from the table to exceed those obtained from the amplitudes of reflected waves by about 1.5 times, due to the fact that adequate account has not been taken of the transmission coefficients; a similar relationship also holds for longitudinal waves. The absorption coefficient for transverse waves is about 3 to 4 times that of the longitudinal. The reflection coefficient K_{PSS} is about 2 to 3 times that for the longitudinal waves, and this explains the commensurate intensities of such waves, despite their intense absorption.

The wave field observed in the Anastasiyevskaya–Troitzkaya area

A big gas and oil field is situated in this area. The wave field contains numerous waves, and PM–VSP has discriminated and tracked some different wave types (Figure 92) and plotted the regions of their tracking components. The converted reflected waves are recorded chiefly on the oblique components.

The converted transmitted waves originate mainly at the interfaces in the upper portion of the section with which the longitudinal reflected waves are associated. The polarization of such waves is complex and may be different for waves reflected by different interfaces. The polarization varies along the line of the profile.

The direct S-wave is recorded from all shotpoints. As offset increases, the shape of the recorded S-waves becomes more intricate, this being perhaps the result of the influence of the converted transmitted waves; the minimum on the traveltime curve corresponds to a medium with a velocity gradient $\beta = 0.001\ \mathrm{m}^{-1}$. At great distances, the waves separate into the SV- and the SH-types, which have different propagation velocities and typically different velocity gradients.

The SV-waves have higher intensities and are more stable than the SH-waves, and are characterised by a long wave train. The regions of tracking components of the S-waves depend on offset and for more remote sources approach the vertical. The transverse monotype waves are associated mainly with the interfaces in the upper portion of the section.

It is worth noting that in several cases the reflection and the conversion interfaces do not coincide.

4. Converted Waves Transmitted Upwards

Data obtained with the correlation method of refracted waves and based on the study of waves that travel a substantial part of their path in a direction close to the vertical are difficult to interpret. This has lead to a drastic increase in the extent of regional studies utilizing the method of converted transmitted waves, for both seismic explo-

---

Fig. 92a-b. PM–VSP seismograms (a) and regions of tracking components (b) of waves of various types (borehole N693 Anastasieyevsko–Troitzkaya, l = 1150 m). Regions of tracking components: *top*: 1 – transverse SV-wave; 2 – transverse SH-wave; 3 – reflected transverse waves; *bottom*: 1 – converted reflected waves; 2 – converted downward waves from the interface H = 300 m; 3 – same as 2, but H = 460 m; 4 – same as 2, but H = 520 m.

TABLE VIII

Depth interval in Tolbukhin borehole (m)	$\alpha_S \times 10^{-3}$ from transmitted downward *PS*-wave (m^{-1})		α_P/α_S from incident waves (*P* and *PS*)		$\alpha_S \times 10^{-3}$ from reflected waves (m^{-1})		α_P/α_S from reflected waves		Depth interval in Lezhsk borehole (m)	$\alpha_S \times 10^{-3}$ from transmitted downward *PS*-wave (m^{-1})	α_P/α_S from incident waves (*P* and *PS*)
	Distance from shotpoint to borehole being logged (m)									Distance from shotpoint to borehole 180 m	
	167	367	167	367	167	367	167	367			
850–1010	1.36	–	–	–	0.38	–	–	–	1860–2030	2.16	0.333
1060–1310	1.47	2.87	0.490	0.267	–	–	–	–	2030–2179	2.96	0.270
1310–1370	–	–	–	–	–	1.53	–	0.251	2190–2320	4.42	0.182
1380–1730	2.50	1.95	0.350	0.257	1.52	–	0.218	–	2330–2420	4.80	0.246
1500–1650	–	3.38	–	0.249	–	–	–	–	2420–2670	2.85	0.414
1760–1970	2.22	2.30	0.275	0.263	1.53	1.76	0.203	0.250	2680–2900	3.10	0.380
1970–2210	1.63	1.95	0.283	0.277	–	–	–	–	2900–3030	2.84	0.415
									3070–3169	3.83	0.240

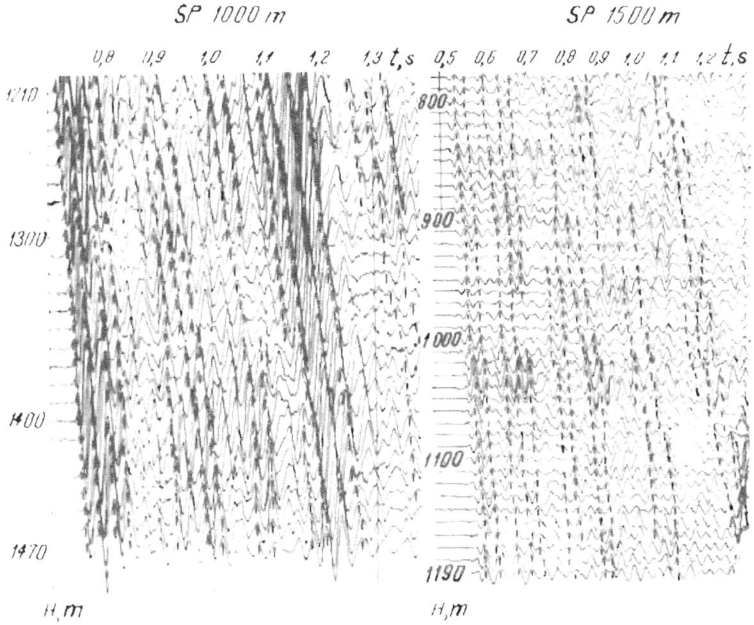
SP 1000 m
SP 1500 m
0,8 0,9 1,0 1,1 1,2 1,3 t, s
0,5 0,6 0,7 0,8 0,9 1,0 1,1 1,2 t, s
1210
1300
1400
1470
H, m
800
900
1000
1100
1190
H, m

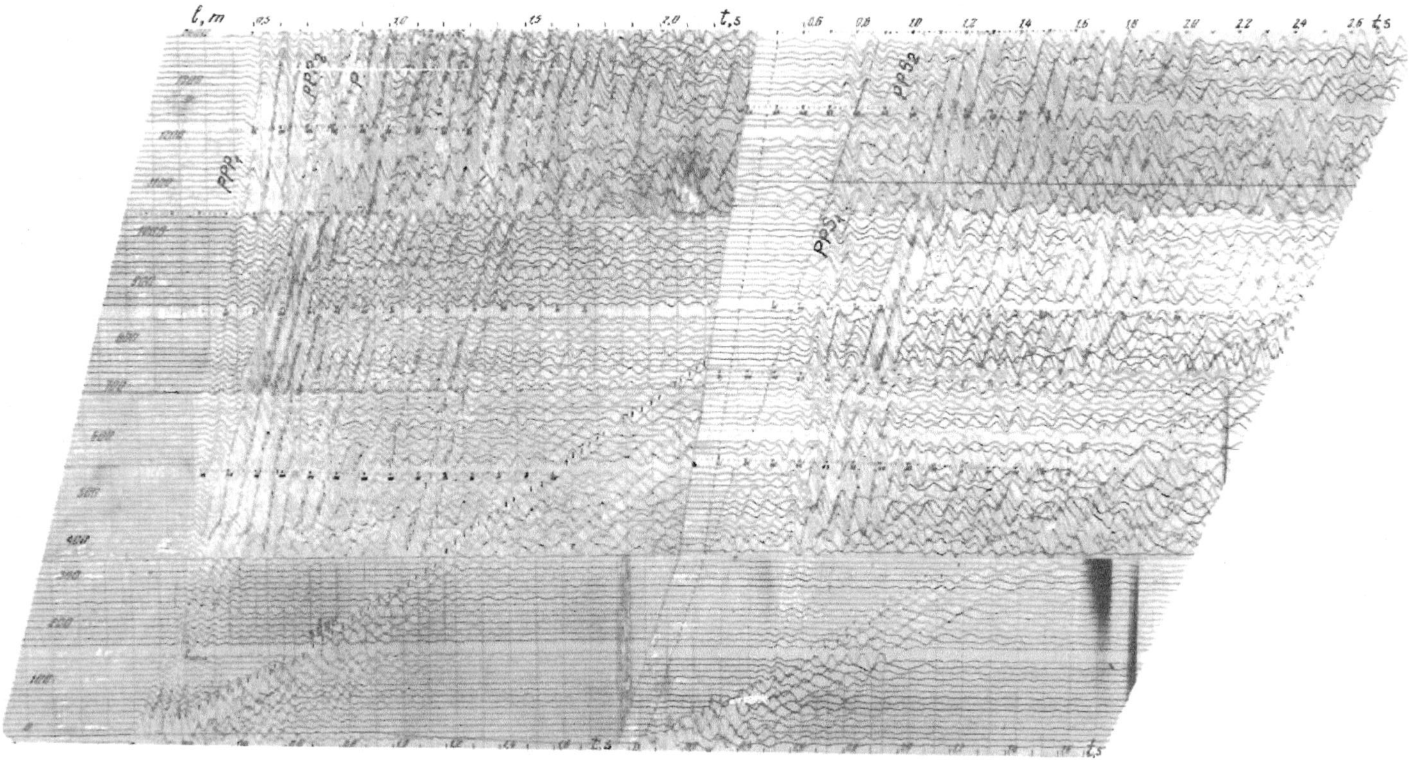

(a) (b)

Fig. 94. Surface seismograms of Z-component (a) and of X-component (b). Dotted line marks the arrival time of the first refracted PPP- and the direct P-longitudinal waves. Seismograms for l = 0–325 m have been recorded at a low amplification level.

ration employing explosions [120] and for seismology, where earthquakes act as wave sources [121]. An important part in the development of the method based on earthquakes was played by the mass-produced equipment 'Zemlya' designed in VNII Geofizika. Difficulties in the interpretation of converted transmitted waves are usually associated with the absence of physical criteria for the determination of the origin of the waves, and for their identification. This makes it expedient to employ VSP in order to appraise the exploration potential of the method for solving various problems. To this end, a study was made of conditions for the formation of converted transmitted waves at interfaces located at great depths, and the conditions and the facilities available for their recording and their identification on surface observations.

VSP boasts of the following advantages when used to study converted transmitted waves. For surface observations employing explosions, the source and the receiver are

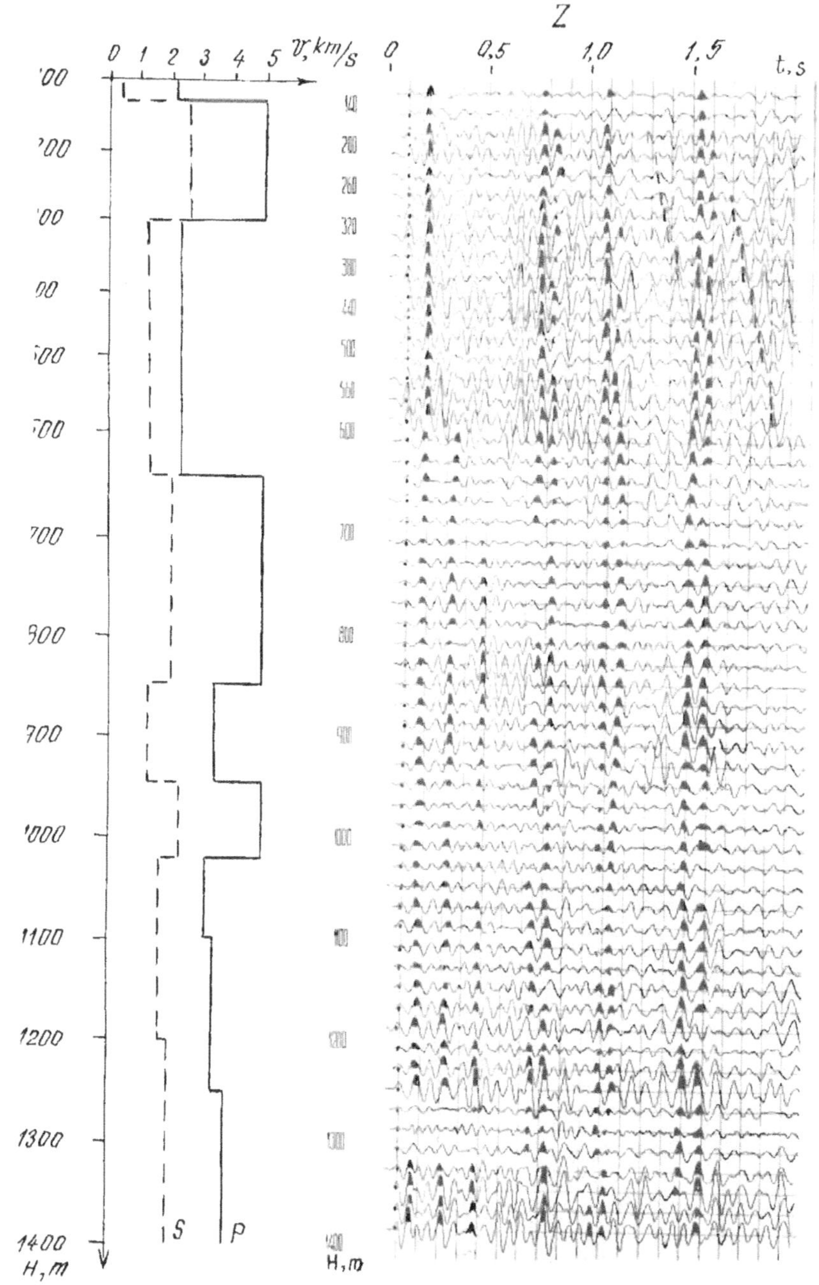

Fig. 95a.

Fig. 95a-c. Seismograms of Z- and of X-components (shotpoint 31700 m, borehole N1, Kletsko–Pochtovaya): (a) Z-component; (b) total X-component; (c) total X-component (1) and Z-component (2).

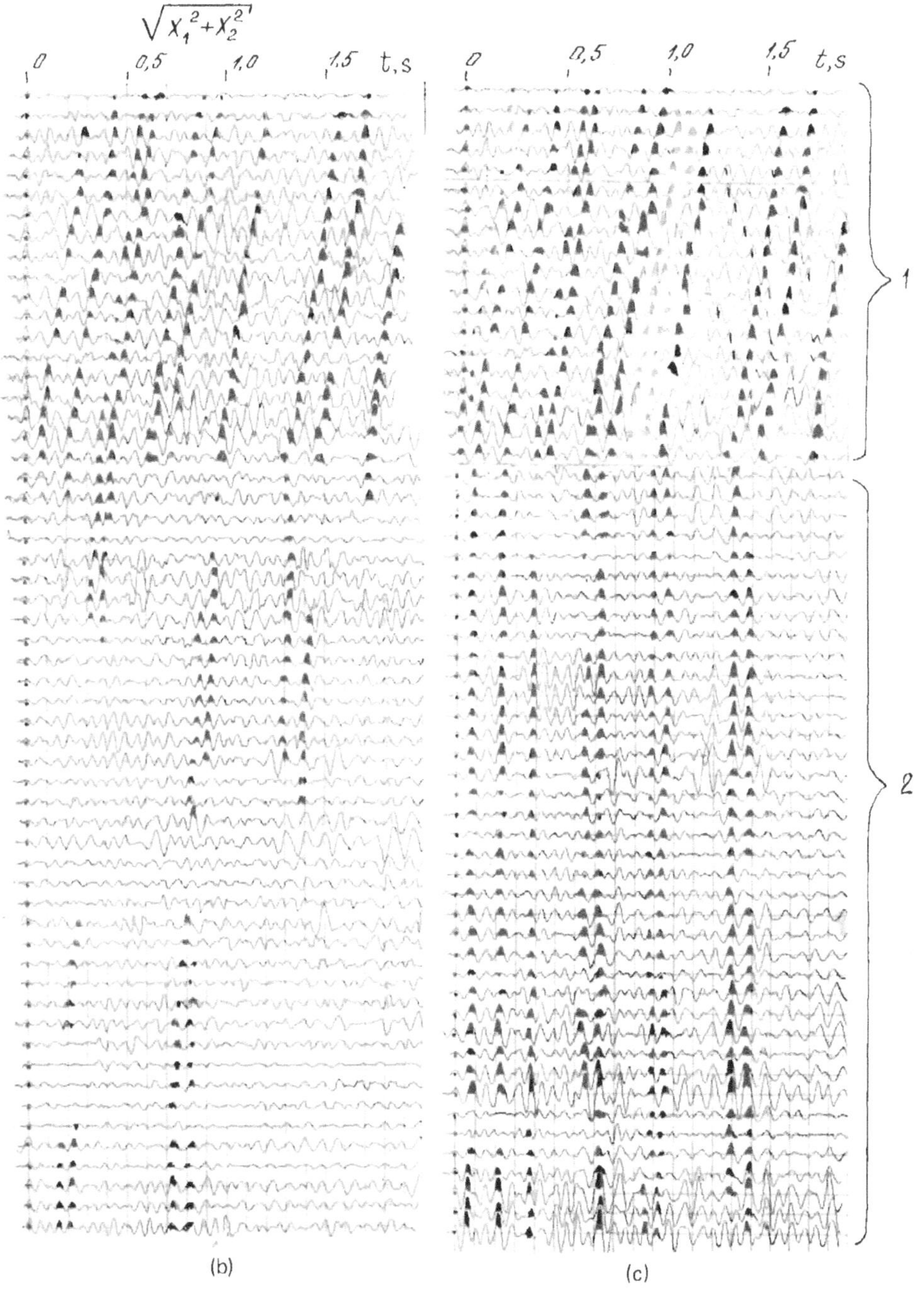

(b)

(c)

Fig. 95b-c.

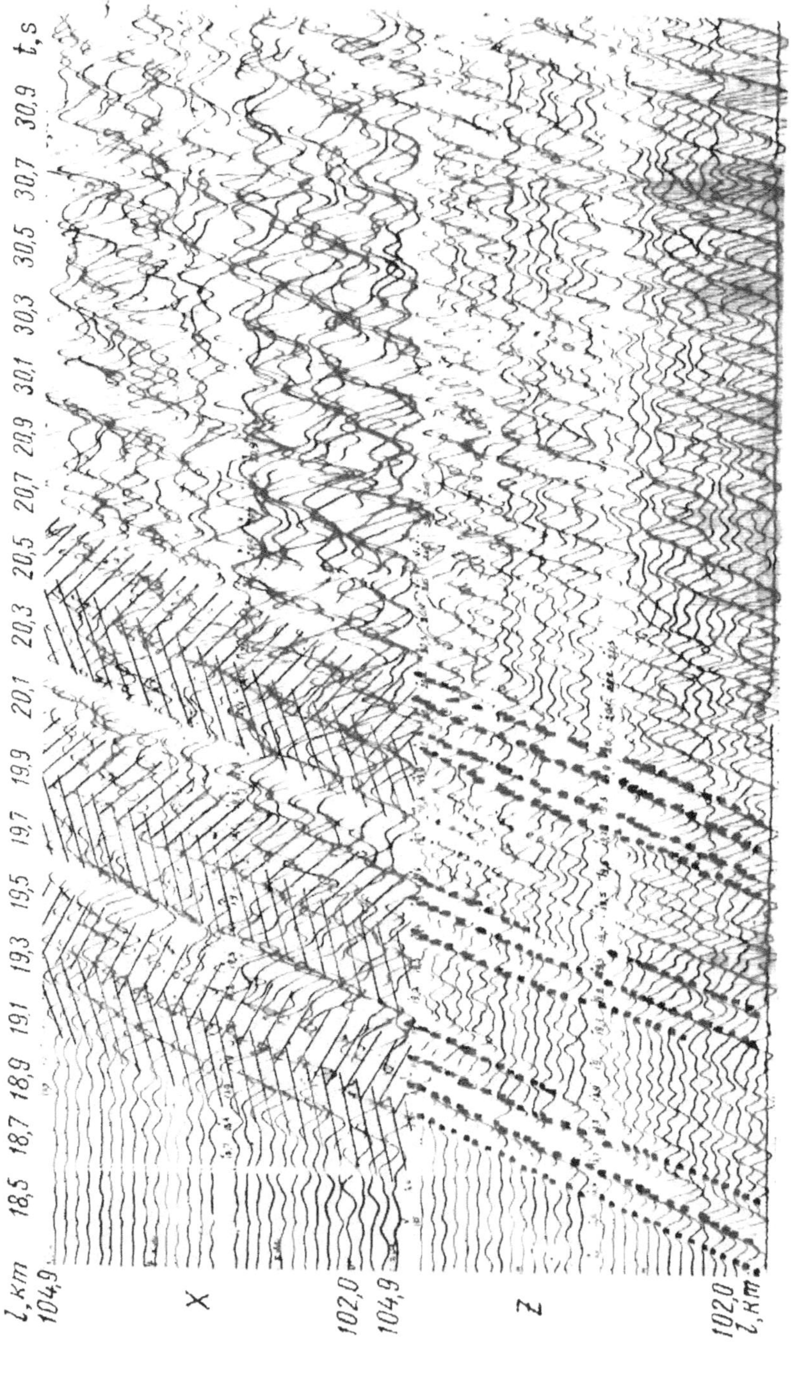

Fig. 96. Surface seismograms of *X*- and *Z*-components in the range of 102–104.9 km from shotpoint (Kletsko–Pochtovaya).

from interfaces located at different depths. Even one sharp interface in the upper portion of the section may be enough to explain fully the entire seismogram of the horizontal component. It will be seen from comparison of the surface Z- and X-component seismograms (Figure 94) that the record of the X-component (right) almost completely duplicates the Z-component record (left), being displaced in time by $\delta t = 0.25$ to 0.30 s. This time lag is due to conversion at the top of the Paleozoic at $H_v = 125$ m.

We shall now illustrate the formation of converted waves at the next deepest sharp interface at $H_v = 640$ m with observations on a vertical profile shot from a shotpoint located at a distance of 31 700 m from the borehole. The Z-component seismogram (Figure 95a)* shows numerous longitudinal waves propagating from below commensurate with the first longitudinal wave. Each longitudinal wave arriving at the interface will be seen from the seismogram (Figure 95c), composed of the vertical (2) and the total horizontal (1) components, to correspond to a converted transmitted wave above the interface, which can be reliably correlated along the vertical profile. All such waves, both longitudinal and converted, crossing the interface at a depth of 125 m, in turn give rise to a series of secondary waves subsequently recorded on surface seismograms. High-intensity converted transmitted waves originate at the sharpest interfaces, which are located, as a rule, in the upper portion of the section. The discrimination of converted waves associated with great depths on the vertical profile is effectively made with the aid of polarization-positional correlation, since in many cases the intensity of the horizontal component of the longitudinal waves may exceed that of the horizontal component of the converted waves. This applies to observations in high-velocity layers, and largely explains why the converted waves are difficult to track at great depths on the records of the horizontal components (Figure 95). The use of computerized controlled directional recording of the I kind [106] enabled the presence of converted waves to be established, the optimum components for discrimination to be determined, and seismograms of the tracking component, used subsequently for their correlation, to be drawn up.

In conditions of a complex interference wave field, the criteria for identifying P- and PS-waves (according to the form of the record, time lag, polarization) now in use, do not ensure reliable determination of the origin of waves associated with great depths. Even in the case of a comparatively simple geologic section of two or three sharp interfaces in the upper portion of the section, the interpretation of converted waves may be ambiguous. The situation is aggravated by the fact that, because of a more or less thin-layered structure of the earth, and of a comparatively low frequency of the waves being recorded, the converted transmitted waves belong, as a rule, to the interference type, and are characterized by records of a complex form. Figure 96 depicts surface seismograms illustrating the complexity of the wave field. The Z-component seismogram is of a comparatively simple type: a series of wave groups with comparable intensities are recorded in the early portion of the record. The time intervals on the X-component seismograms of converted transmitted waves associated with sharp interfaces at depths of 125 m, 640 m and 1470 m have been shaded. All converted transmitted waves of interest associated with the surface of the basement and with the series of deep interfaces in the sedimentary beds, are located exclusively in the region of complex interference.

* The first arrivals on the seismograms have been aligned to the line $t = 0$.

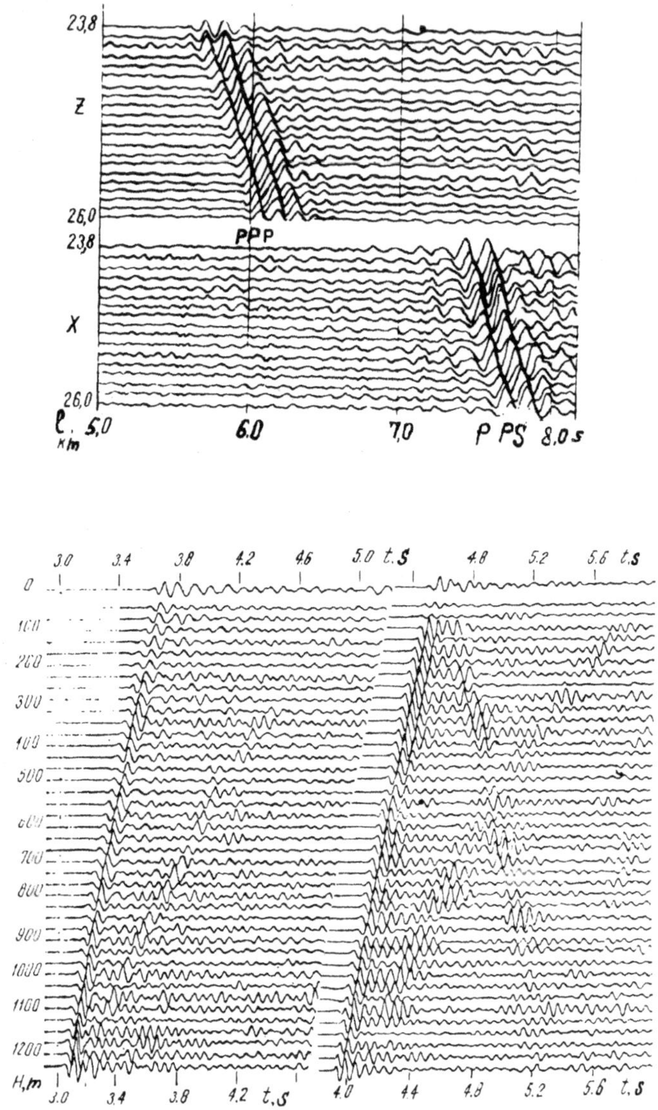

Fig. 97a.

Fig. 97a-b. Converted transmitted waves from the crystalline basement in case of a compact sedimentary bed: (a) in platform conditions for a shallow basement; *top*: surface seismograms, Q = 200 kg; *bottom*: VSP seismograms l = 12.1 km (left), l = 17.2 km (right) (courtesy E. Ya. Kutzenko); (b) inflection region, Iliy depression, Q = 900 kg (courtesy I. K. Pushkarev).

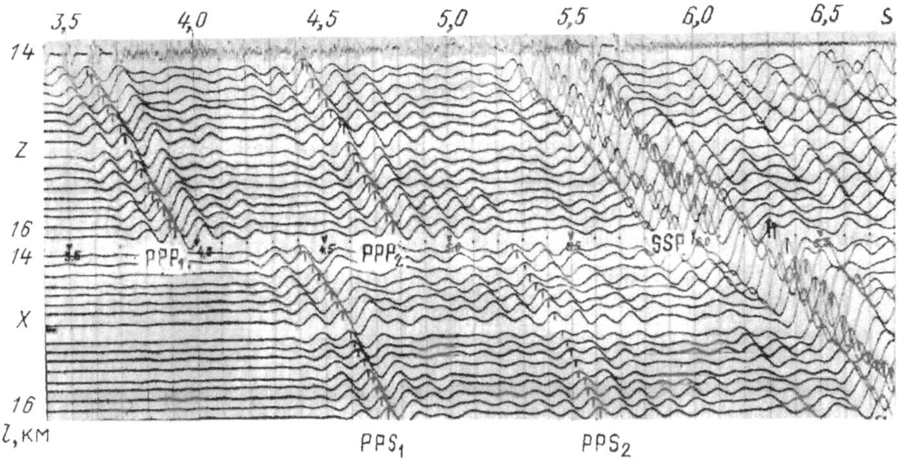

Fig. 97b.

This does not hold for converted waves recorded as first arrivals on the seismograms whose interpretation normally presents no difficulties. However, in the majority of cases, the first event is associated with upper interfaces and is of no geological interest. In some areas, the first sharp interface may be located at a depth of 4 to 6 km, e.g. in inflections where the basement is covered with a thick bed of weakly differentiated terrigeneous deposits. The converted wave associated with the interface may be a source of interest for exploration, and in such cases, the first interface can be very effectively studied by means of converted transmitted waves. Figure 97 depicts surface and VSP seismograms recorded in the platform area of the Krasnodar province, in conditions of a comparatively shallow crystalline basement (1500 m) and a weakly differentiated bed of terrigeneous deposits. The converted transmitted wave from the basement can be reliably discriminated on the VSP seismograms (Figure 97a, bottom) from shotpoints 12.1 km (left) and 17.2 km (right) even on the *Z*-components. A noteworthy feature is the simple first arrival wave whose form remains unaltered along the vertical profile. The absence of sharp refracting interfaces both in the region of source excitation and reception has produced a very simple wave field. By force of this, only the first longitudinal wave on the *Z*-component and the converted transmitted wave from the crystalline basement on the *X*-component register on the surface seismogram in Figure 97(a) top.

A similar picture is observed also for a deep basement, when the sedimentary beds consist of undifferentiated terrigeneous deposits. On the surface seismogram (Figure 97b), recorded in the Iliysk depression, containing a 4 km thickness of terrigeneous deposits, the *X*-component seismogram is seen to duplicate the *Z*-component seismogram although displaced in time.

In contrast to the case of a simple undifferentiated bed, Figure 98 depicts VSP seismograms recorded with thick sedimentary deposits and numerous interfaces. Seismograms obtained at comparatively short offset distances from the shotpoint (Figure 98a) show a complex wave field with numerous waves formed in the vicinity of

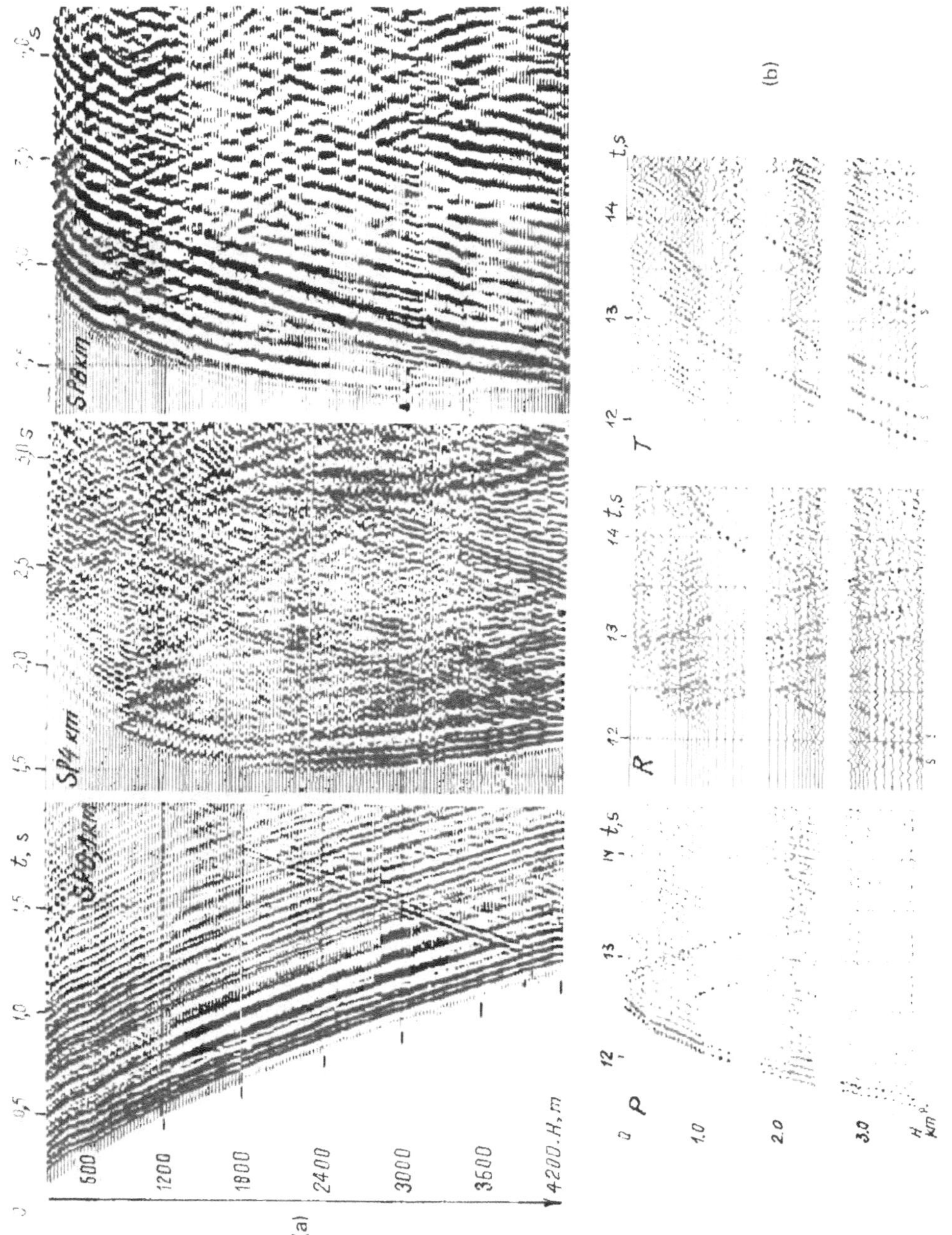

Fig. 98. VSP seismograms: (a) SP = 100 m, SP = 4000 m, SP = 8000 m, Z component seismograms; (b) seismograms of components of a local PRT coordinate system (borehole N2, Svetlosharinsk, l = 63 km [122]).

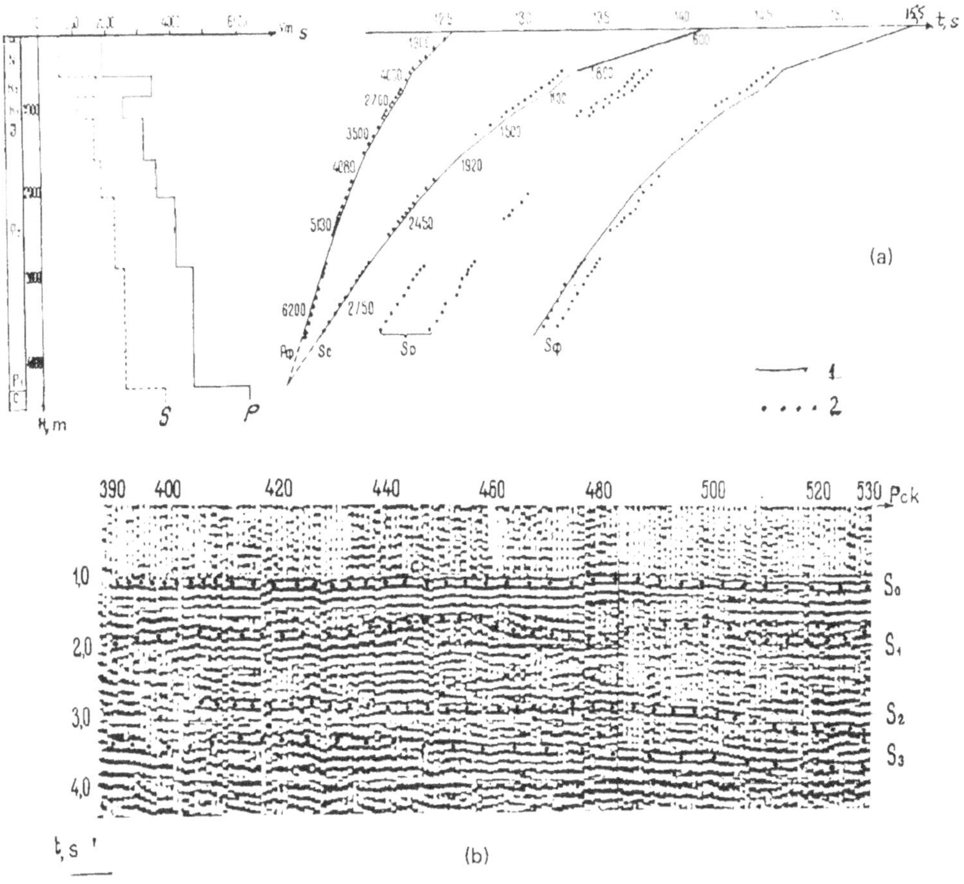

Fig. 99. Comparison between theoretical (1) and observed (2) traveltime curves of *P*- and *S*-waves (a); (b) time cross-section plotted from data on converted transmitted waves; S_{C2}, S_{P+T}, S_D, S_B – conversion interfaces associated with chalk, Perm–Triassic, Paleozoic deposits and with the surface of the basement, respectively (velocity cross-sections for *P*- and *S*-waves are shown on the right).

the shotpoint. A very complex wave field is seen on the PM–VSP seismograms recorded at a distance of 63 km from the shotpoint (Figure 98b). Only longitudinal waves stand out on the *Z*-component seismograms (Figure 98a). The converted transmitted waves can be discriminated only with the aid of polarization-positional correlation on the *R*-component of the local coordinate system, in the zero displacement plane of the *SP*-waves, which are the main unwanted waves that obstruct the discrimination of the converted waves. However, on the *R*-component too, their correlation is disrupted by downward longitudinal waves associated with the upper portion of the section. Subtracting the latter, we can track the converted transmitted waves associated with the interfaces in the sedimentary beds and with the surface of the basement. Theoretical and observed traveltime curves are in good agreement (Figure 99a). Those results enabled the time cross-section to be drawn for four conversion interfaces (Figure 99b).

The results of experimental studies of converted transmitted waves in the normal seismic frequency band are in good agreement with computed values. Figure 100a depicts synthetic seismograms of *Z*- and *X*-components calculated for the earth model shown in Figure 100b for the case of two original longitudinal wave pulses reproduced from the experimental seismograms recorded at depths of 1450 m (*SP* 10 900 m) and 1400 m (SP 31 700 m). An intricate interference pattern resulting from the superposition of converted transmitted waves from various interfaces, is observed on the *X*-component seismograms. Unambiguous identification of the different phases of the longitudinal and converted waves is almost impossible in such conditions. The data obtained cast doubt on the use of the method of converted transmitted waves produced by explosive energy sources, for the purpose of studying interfaces located at great depths, in the presence of sharp interfaces in the upper portion of the section. Further studies of the conditions for the formation and the physics of propagation of deep converted transmitted waves, and of criteria for their identification, are required.

In recent years earthquakes have been used increasingly for regional studies. In such studies, converted transmitted waves originating at interfaces below the geophone locations are employed to obtain information about the earth's structure. Earthquakes possess substantial advantages over explosives and enable information about deep interfaces in the Earth's crust and the mantle to be obtained. However, the determination of the origin of waves associated with the upper portion of the section using the method of converted waves produced by earthquakes, is fraught with difficulties.

Results obtained in PM–VSP have suggested the need to develop a polarization method for seismic regional studies. The use of polarization-positional correlation enables the early portion of the record to be studied in more detail, and results of great practical interest to be obtained. Regular longitudinal waves constituting the early portion of earthquake records (up to 15 s), and recorded on the *Z*-component, have been subdivided into two groups. The first group is composed of waves with intensities close to that of the first event. Their number and arrangement is not the same for different earthquakes. The second group includes waves of substantially lower intensity than the first, however, the waves of the second group from different earthquakes are recorded in the same time intervals. The waves of the first group are associated with the area of the source; waves of the second group are associated with the region of registration and are multiples formed between the surface and interfaces within the Earth's crust. The second group of waves are of interest and may be used to study the structure of the Earth's crust. The discrimination and tracking of such waves could be improved by observing their tracking components in the zero-displacement plane of unwanted (longitudinal) waves. It has been demonstrated that the converted waves may have a simpler form than that of the first longitudinal wave.

Repeated converted waves recorded in the time interval 9 to 15 s have also been discriminated. The train of repeated waves is similar in shape to that of primary converted waves. These waves can be classified according to their arrival times, record shape and polarization, as belonging to converted waves originating at the same interfaces as the primary converted waves, but associated with longitudinal multiply reflected waves.

These studies were the pre-requisites for the development of the polarization method of regional studies employing earthquakes [42], and are based on the joint use of various

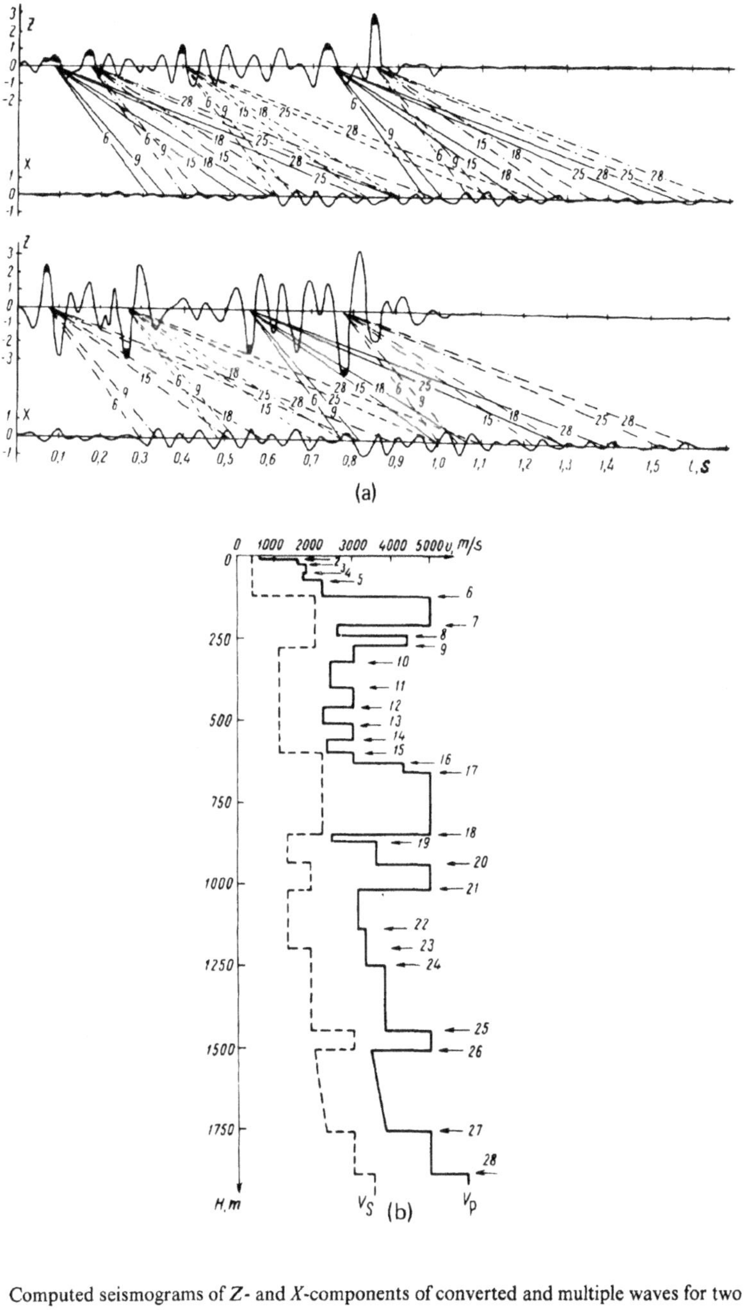

Fig. 100. Computed seismograms of Z- and X-components of converted and multiple waves for two wave trains of different shape impinging on the top of the basement at an angle of 70° (a) and velocity model of the Earth (b). Numbers on lines denote the ordinals of interfaces with which the converted waves are associated.

wave types. Signal stacking by grouping records from various earthquakes, has served to suppress waves associated with earthquake sources, and to discriminate the stable part of the wave field associated with the region of reception. This refers to the recording of seismograms of the common receiver point and the common point of source excitation, and to their use in processing by methods similar to the common-depth-point technique used in seismic exploration.

Thereby, the development of PM–VSP has appreciably broadened the scope of the study of transverse and converted waves. It is now possible to use jointly different wave types, which are abundant in the gelologic section, to study the earth.

PART THREE

VSP AND INVESTIGATION OF THE REGION SURROUNDING THE BOREHOLE

The potentialities of VSP are not limited to its applications for improving the effectiveness of the principal surface seismic methods. The advantages accruing from downhole recording of waves can also be exploited for direct studies of the region surrounding and between boreholes. The majority of borehole geophysical methods such as electric and sonic measurements, provide information about the borehole walls in a radius of investigation of approximately 1.5 to 2 m from the borehole axis. Only seismic observations are capable of substantially augmenting the scope of subsurface studies to include the region surrounding the borehole up to several hundred meters, and in some cases even 1.5 to 2.0 km, from the borehole shaft, as well as to depths exceeding those penetrated by the borehole. This region may to the first approximation be represented by a cylinder or a truncated cone with the borehole as its axis.

In the course of the years, studies of the region surrounding the borehole have become of increasing interest. This is explained by the fact that the increase in drilling depth and the drastic increases in drilling costs have been acoompanied by a noticeable reduction in the effectiveness of exploration, and by a growth in the relative number of unproductive boreholes. In the period 1971–75, in the whole of the USSR, on the average three-quarters of all exploration boreholes proved unproductive, and more than half of the structures explored turned out to be barren. Therefore the reduction of drilling in the zone beyond the limits of the producing fields, and in barren structures, would have a great effect for the economy.

A general trend to achieve this goal, and to improve the effectiveness of geologic exploration, lies in a more complete utilization of the potentialities of seismic observations during the stages of exploration and exploitation of the fields. Until recently seismic methods have been employed mainly in the exploration and the drilling planning stages. This enabled the effectiveness of geologic exploration to be greatly increased. Suffice it to say that over 90% of the increase in the oil and gas reserves has been obtained in fields where drilling plans were based on seismic methods. However, the effectiveness of surface seismic studies falls with the increase in exploration depth, mainly as the result of an increasing level of irregular and high-velocity noise, and the reduction in the size of structures.

At the same time, due to the development of VSP and subsequently of PM–VSP, recent years have witnessed a far-reaching improvement in the technology of seismic observations in boreholes that has greatly contributed towards a drastic extension of the exploration potential of studying the region surrounding the borehole. Reflected waves observed in VSP have greatly improved the effectiveness of solving structural problems. With the development of PM–VSP that has made the extensive use of wave dynamics possible, there appeared new possibilities for the study of the physical properties of the geologic section, and for the solution of lithologic and stratigraphic problems. By making use of the wave polarization parameters, it became possible to obtain independent and more detailed information about the Earth.

The newly-acquired potentialities for the study of the region surrounding the borehole have provided a foundation for the application of seismic studies not only during the stages of the search for and the selection of structures for drilling, but also during such important stages of the geologic exploration process as exploration, production drilling and exploitation of the fields. Such studies enable the fundamentals of a new trend in seismic exploration (production seismics, or seismics of the region surrounding the borehole) to be formulated [51]. Production seismics is used in areas where a network of deep boreholes is being developed; it is intimately associated with drilling. Production seismics includes VSP, which is combined with surface and sub-LVL observations, and is amalgamated with other geophysical (sonic, electric, neutron, etc.) methods of research.

The importance and the value of such studies lie in the fact that by proposing a more reasonable arrangement of boreholes during the appraisal stage, the amount of costly deep drilling required for the development and production drilling of the field may be cut. In addition, the deposits contained in the region surrounding the borehole may be studied, and primary information about potential reserves may be obtained together with data required for the selection of optimum systems for working the deposits. During the exploitation stage they enable the deposits to be controlled, the effectiveness of methods of acting on the layer to be evaluated, and the increase in output to be planned [123, 124]. The effect of such work on the national economy does not require any comment.

In order to solve the problems enumerated above, the extent of research carried out in every borehole should be greatly broadened, with the goal of gaining more detailed and more accurate information about the region surrounding and separating boreholes. This is well worth the costs, since the only alternative to seismic methods is drilling.

In studying the region surrounding the borehole, two main groups of problems can be identified: the structural and litho-stratigraphic. Each of these groups include numerous more specific problems. For the solution of a variety of problems, VSP makes use both of various wave types and various wave paramters. Whereas the kinematics of reflected waves is chiefly used for the solution of structural problems, litho-stratigraphic problems are solved mainly by the extensive use of wave dynamics.

Part Three of the book discusses the application of VSP to production seismics for the study of the region surrounding the borehole. The attention is focused on the study of geologic structure and the physical properties, and on the solution of lithologic and stratigraphic problems. The use of the polarization of oscillations to obtain independent and more detailed information about the earth has been made the subject of special discussion. Not all aspects of the study of the region surrounding the borehole have attained an equally advanced level, and this was bound to limit the scope of the discussion.

CHAPTER 9

CERTAIN ASPECTS OF THE DETERMINATION OF VELOCITIES FROM VSP DATA

Seismic observations in boreholes have always served and continue to serve as the main source of information about velocities. In well velocity surveys, the in-line vertical traveltime curve of the first arrival is normally used to determine the velocity. There are several methods of processing this data, which have been extensively discussed in the literature and will not be considered here. We will consider only specific aspects of velocity determinations that are typical of VSP.

The wave fields observed in VSP establish the existence in the earth of numerous thin reflecting interfaces and layers, which cannot be practically resolved from conventional well velocity data on account of its poor depth resolution. Accordingly, it is important to utilize VSP data in order to obtain more detailed information about the velocity distribution, the more so since VSP is almost the perfect substitute for the traditional velocity survey. The principal advantage of VSP is the opportunity to use not only the first arrivals to determine the apparent, interval and average velocities, but also the waves of various origin and type recorded in the later portion of the record. In this chapter attention is focused on this issue.

1. Determination of Velocities from the Traveltime Curves of First Arrivals

A large amount of VSP observations are carried out along off-line profiles. To determine a velocity distribution from an off-line traveltime curve, the latter is reduced to the vertical one. The corrections needed for such a reduction are determined from very simplified ideas about the geology with a constant increment law of variation of velocity with depth. However, the real earth is a thin layered continuously-variable medium, and the velocities in it vary continuously. Accordingly, the means for the determination of the velocity distribution from an off-line traveltime curve without a preliminary reduction to an in-line traveltime curve is of practical interest. The utilization of off-line traveltime curves merits attention, because data obtained this way is usually less distorted by unwanted waves of various types (casing, cable, etc.) than in the case of observations along in-line profiles. In the latter case, such interference makes the study of the upper portion of the section impossible. True, polarization-positional correlation in PM–VSP enables the interference to be suppressed, and the first arrival of the direct wave to be discriminated. Consider one of the methods of determining the curve of velocity variation with depth $v(H)$ from an off-line traveltime curve [125].*

* Note that $v(H)$ can also be determined from the displacement direction variation curve (see Chapter 12).

The vertical traveltime curve of a curved ray wave can be represented by the equation

$$t(H) = \int_0^H \frac{d\zeta}{v(\zeta)\,[1 - v^2(\zeta)]\left[\dfrac{1}{v^2(H)} - \left(\dfrac{dt}{dH}\right)^2\right]^{1/2}} .$$

For the function sought $v(H)$, this is a non-linear integral Volterra equation of the first kind [125]. Since its solution is difficult, $v(H)$ can be found by means of successive estimates. The velocity value at the first point ($H = O$) determined from the formula

$$v(0) = \frac{1}{\left(\dfrac{dt}{dH}\right)_{H=0}}$$

serves to determine the value at the second point, which corresponds to the interval of observations along the vertical profile ($H = \Delta H$). The value $t_1(\Delta H)$ is found numerically after $v(O)$ has been substituted as a first approximation for the value of velocity at point $H(\Delta H)$. Next the value $t_1(\Delta H)$ is compared with $t(\Delta H)$ known from experiment. If $t_1(\Delta H) \neq t(\Delta H)$, by successively changing the value $v(\Delta H)$, the calculated and the observed values of the time are made to correspond with the accuracy desired. The value of $v(H)$ obtained at the second point is assumed as the first approximation, and the value at the following point is determined. Such a procedure is maintained for all the points of the vertical profile. Such a method does not lead to any accumulation of errors, since they are continuously compensated from point to point. The method is applicable for a series of off-line vertical traveltime curves of direct waves with a positive apparent velocity, i.e., in case the rays intersect the borehole axis before they curve upwards.

The method has been computerized and tested on models. Figure 101 depicts the results of the determination of $v(H)$ for two models: one with a continuous function of depth (left) and the other with a constant-increment function.

Traveltime curves for offsets of 100, 500, and 1000 m have been calculated for both models, and velocity curves have been determined from them. The curves obtained will be seen from Figure 101 to almost completely coincide with the original curves.

The apparent velocity of waves in the layer above the interface in the case of a plane dipping interface with a homogeneous overlying bed can be determined from the traveltime curve of the *head wave* (for $l > l_{cr}$, off-line profile). From the traveltime curve equation (Chapter 4, Equation (9)) [3] we obtain

$$v_a = -\frac{v_i}{\cos(i_{cr} \pm \varphi)},$$

i.e., v_a is a function of the angle of incidence φ.

In the case of a horizontal interface $\varphi = 0$

$$v_a = \frac{v_1 - v_2}{\sqrt{v_2^2 - v_1^2}} = -\frac{v_1}{\cos i_{cr}} .$$

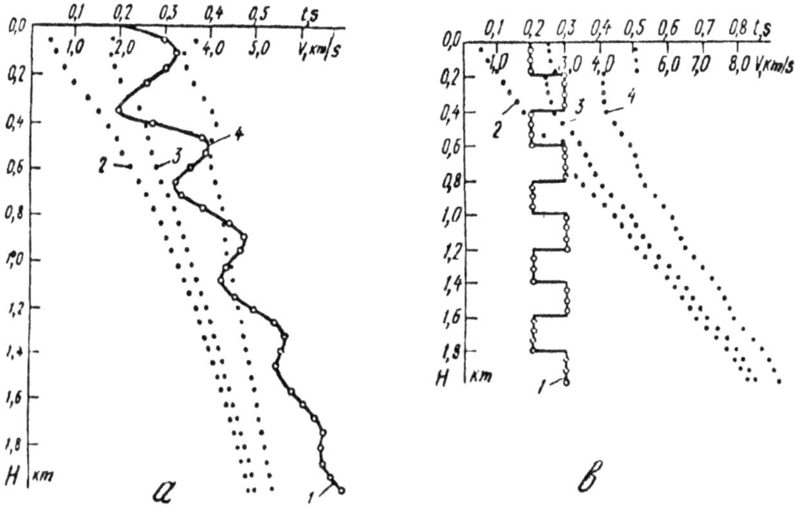

Fig. 101. Examples of determination of $v(H)$ from off-line vertical traveltime curves in case of a continuous (a) and a constant-increment (b) law of variation of velocity with depth [157]: 1 – specified $v(H)$ function: 2, 3, 4 – theoretical traveltime curves for shotpoints for l = 100, 500, 1000 m. Circles indicate the velocities obtained.

The former expression can be presented in the form

$$v_a = \frac{v v_{\text{in}}}{\cos \varphi \sqrt{v_{\text{in}}^2 \pm v^2} \pm v \sin \varphi},$$

where the plus sign corresponds to the shotpoint located down-dip, and the minus sign to one located up-dip.

The interfacial velocity is determined as follows:

$$v_{in} = \frac{2l}{\sqrt{(\sum \Delta T)^2 + l^2 \Delta v_a^2}},$$

where

$$\sum \Delta T = t_{0+} - \Delta t_+ + t_{0-} - \Delta t_-,$$

Δt is the difference in the arrival times of the waves at the surface ($H = H_0$) and at the interface ($H = H_v$), and t_0 is the time of the first arrival at the point of intersection of the vertical profile with the interface.

The velocity in the top layer is

$$v = \frac{2 v_{\text{in}} \cos \varphi}{\sqrt{\left(\frac{1}{v_{a+}} + \frac{1}{v_{a-}}\right)^2 v_{\text{in}}^2 + 4 \cos^2 \varphi}}.$$

In a multi-layered medium the traveltime curve takes the form (Chapter IV, Equation (26)). Making use of the traveltime curve of the head wave and the value of velocity in the refracting layer $v_n = v_{in}$, one is able to calculate the velocity in the m-th layer

$$v_m = \frac{1}{\sqrt{\frac{\cos^2 i_{mn}}{v_m^2} + \frac{1}{v_n^2}}} .$$

The relative error in the value v_m thus determined is

$$\delta v_m = \left(\frac{v_m}{v_{in}}\right) \delta v_{in} ,$$

where δv_{in} is the relative error in the value of the velocity along the interface.

In the case of a thick layer, the value of the velocity in it can be found from the traveltime curve of the transmitted wave, since the slope of the asymptote of the traveltime curve is equal to $1/v_n$.

One of the modifications of well velocity surveys by means of refracted waves is based on the studies of interval and average velocities from data on refracted waves.

The apparent velocity of transmitted waves is determined from the expression [5]

$$v_a = \frac{v_2}{\sqrt{1 - p^2 v_2^2} \cos \varphi - p v_2 \sin \varphi} .$$

From the analysis of the values of apparent velocity, which is of considerable interest in the study of velocity parameters of the Earth, it follows that $v_a = v_l = v_2$ for $i_2 = -\varphi$, or

$$\varphi = \arc\tan \frac{p v_2}{\sqrt{1 - p^2 v_2^2}} ,$$

$$v_a = \infty \quad \text{for} \quad i_2 = \frac{\pi}{2} - \varphi ,$$

or

$$\varphi = \arc\tan \frac{\sqrt{1 - p^2 v_2^2}}{p v_2} .$$

Accordingly, when determining velocities v_l from in-line vertical traveltime curves of transmitted waves, one should take into account the dip of the interface, otherwise excessive values may be obtained. In some cases when $v_2 < v_1$, a layer may be omitted.

When solving various seismic exploration problems, one often has to evaluate the precision with which the arrival times of waves are determined. Basically, it is determined by the precision of timing of the arrival of a wave on the seismogram, and this depends on the signal intensity. The arrival will be observed the earlier the higher the signal amplitude; an increase in the intensity reduces the arrival time by perhaps 3 to 5 ms. However, in practice this does not change the recorded times of extrema. Due to velocity dispersion, absorption and the inhomogeneities of the real earth, the front of the pulse becomes less pronounced as the pulse propagates and loses its high-frequency components. Accordingly, the velocity determined from arrival times may differ from that determined from the positions of the extrema, since the latter correspond to the velocity of propagation of harmonic oscillations with a frequency equal to the dominant frequency of the pulse. Normally this difference does not exceed 1 to 2%.

When highly absorptive or sharply inhomogeneous media are being studied, the problem can be solved rigorously be constructing phase spectra of the pulses (for example, raising the frequency from 20 to 100 Hz increases the inverval velocity by about 10%). In this case, it is possible to determine the magnitude of the velocity and its variation with frequency, and the results are independent of the pulse intensity and the variation of its shape with distance. If the pulse shape remains practically unchanged with the distance, it is possible to find the velocity vs frequency dependence by constructing vertical traveltime curves obtained with different filters using oscillation extrema (the first, the second) for timing on copies of the record obtained with narrow band-pass filtering.

It was established as the result of the analysis [5] of errors in visual timing of the first arrivals, and the accuracy of determination of the difference in arrival times of the seismic signal at various points in the geologic medium (as observed in boreholes) that, should the position of the signal in time be defined as the position of the maximum slope of the pulse front, namely at $0.5A_0$, when measuring the propagation velocity of oscillations over intervals shorter than the wave length, it would be permissible to neglect the inelastic absorption of energy in the Earth, and the variation of signal shape associated with it.

The flattening of the traveltime curves resulting from the loss of high-frequency components may reduce the velocities measured in VSP by as much as 2% [126]. The application of deconvolution and the introduction of dynamic corrections enables the effect of absorption on the pulse shape to be eliminated.

2. Determination of Apparent and Layer Velocities from Traveltime Curves of Reflected Waves

The apparent velocity of a reflected wave at each point H of the vertical profile can be determined from the traveltime equation of secondary waves including the longitudinal reflection [14]

$$v_a = -v_1 \frac{\sqrt{(2H_H \cos\varphi - H)^2 + (2H_H \sin\varphi - l)^2}}{2H_H \cos\varphi - H}$$

$$= -v_1 \frac{\sqrt{l^2 + H^2 + 4H_H(H_v - H)\cos\varphi}}{2H_H \cos\varphi - H}.$$

If the borehole is tilted with a constant Zenith angle α, the value v_a along the borehole axis is equal to

$$v_a = \frac{v^2 t\sqrt{k^2+1}}{(k^2+1)z - 2H_H(k\sin\varphi + \cos\varphi) + kl},$$

where $k = \tan\alpha$.

The relative error in v_a, if the borehole's tilt is not taken into account, is

$$\frac{\Delta v_a}{v_a} = 1 - \frac{k(kz + l - 2H_H \sin \varphi)}{(z - 2H_H \cos \varphi)\sqrt{k^2 - 1}} - \frac{1}{\sqrt{k^2 - 1}},$$

where H_H is the depth of the interface along the normal from the shotpoint, and H_v is the vertical depth. This error vanishes at the surface and increases smoothly with depth. The maximum values correspond to the point of intersection of the borehole with the reflecting interface, and for high tilts of the borehole can be quite large (up to 30% or more) resulting even in a change of sign of the apparent velocity. The effect of a tilted borehole on v_a should be taken account of when the type of reflected wave is determined from the value of v_a. Such boreholes are commonly encountered in mining exploration.

The apparent velocity of a reflected wave of any multiplicity is a quasi-periodic function and at each point of the profile is determined as follows [5]:

$$v_a = \pm v_1 \sqrt{1 + \left(\frac{2H_H \sin^2 k\varphi - l \sin \varphi}{H_H \sin 2k\varphi \pm H \sin \phi}\right)^2}.$$

The apparent velocity of a wave singly reflected downwards is:

$$v_a = v_1 \sqrt{\frac{(2H_H \cos \varphi + H)^2 + (2H_H \sin \varphi - l)^2}{2H_H \cos \varphi + H}}.$$

It is always positive, irrespective of the slope of the principal interface. For small angles of dip φ, the apparent velocity is close to the velocity of the direct wave. For $\varphi = 0$, the apparent velocity diminishes with the increase in reflection coefficient K, and in the limit it becomes equal to the layer velocity.

The value of v_a for transmitted, primary and multiple reflected waves in multilayered horizontal media is related to v_l by means of the following approximate equality:

$$v_a = v_l \sqrt{1 + \frac{l^2 v_l^2}{\left(2 \sum_{i=1}^{n} d_i v_i\right)^2}}.$$

In VSP the magnitude of v_a may be used to identify the wave type. From the expression for v_a, one can find the equation for v_l

$$v_l = \frac{1}{\sqrt{\left(\frac{dt}{dH}\right)^2 + \left(\frac{dt}{dl}\right)^2}}.$$

Note that in the interpretation of VSP data, v_l is very important for the identification of various wave types recorded in observations along off-line profiles. The latter formula may serve to find v_l both for the first arrivals and for the reflected waves of any multiciplicity and to partition the section according to v_l. In the case of continuously-changing media, this formula remains valid, provided a function describing the law of variation of the velocity with the depth can be substituted for v_l The dependence of v_a

on v_l is determined by the expression

$$v_a = v_l \sqrt{\frac{l^2}{(2H_v - H)^2 + 1}},$$

i.e., the number $l/(2H_v - H)$ determines the difference between v_a and v_l: the closer the ray is to the vertical in the case of great depths, or of waves of high multiplicity, the smaller is this difference.

For fixed geophone depth surveys, traveltime curves of the longitudinal reflected wave (Chapter 4, Equation (17)) enable the layer velocities for the proportion of the section not penetrated by the borehole to be determined. Applying the constant-difference method to the traveltime equation for $\varphi = 0$, we obtain the effective velocity down to any interface

$$v_{\text{ef}} = \sqrt{2l_1 \frac{\Delta l}{t_2^2 - t_1^2}},$$

where l_1 is the distance from the borehole to the nearest shotpoint, and Δl is the separation between the shotpoints.

Hence the layer velocity is

$$v_l = \sqrt{\frac{v_{\text{ef}\cdot b}^2\, t_{0b} - v_{\text{ef}\cdot t}^2\, t_{0t}}{t_{0b} - t_{0t}}},$$

where $v_{\text{ef}\cdot b}$, $v_{\text{ef}\cdot t}$, t_{0b}, t_{0t} are effective velocities and vertical traveltimes to the bottom and the top of the layer. v_l can also be computed from the values of apparent velocities at points A and B equidistant from the borehole:

$$v_l = \sqrt{\frac{l}{\Delta t_0}(v_{aB} - v_{aA})}.$$

Traveltime curves corresponding to various fixed geophone depth surveys enable independent values of v_l to be repeatedly obtained for subsequent averaging.

Determination of the interval velocity and thickness of a layer not penetrated by the borehole from the reflected wave [33] traveltime curve [28] at a fixed geophone level. The method is based on the selection of approximate models of the Earth and the problem is fomulated as follows. Given a model of a horizontally-layered medium (d_i, v_i) containing n layers above the layer not penetrated by the borehole, and m layers above the level where reflected waves are recorded, and in addition, the traveltime curve of the wave reflected from the bottom of the layer of unknown thickness d_j and interval velocity v_j, determine the paramters d_j and v_j, taking into account the refraction of seismic rays in the overlying beds.

A relationship connecting the velocity sought with the quantities that describe the path and the traveltime of the ray in a medium of known parameters has been derived from the traveltime equation (Chapter 4, Equation (28)):

$$v_j = \sqrt{\frac{l_k - l_i(p)}{[T_k - t_i(p)]p}},$$

where l_k is the distance of the kth source point from the borehole, T_k is the recorded time of the reflected wave at depth d_m, and p is the ray parameter.

The thickness of the layer that corresponds to such a velocity is found from the equation:

$$d_j = \frac{v_j}{2}\,[T_k - t_i(p_k)]\,\sqrt{1 - p_k^2 v_j^2}\,.$$

By testing the values of v_j on a computer, the minimum mean-square deviation of the values of d_j for various source positions is obtained:

$$\sigma d_j = \sqrt{\frac{\sum_{k=1}^{n} (d_{jk} - d_j)^2}{N - 1}}\,,$$

where N is the number of shotpoints.

This method is limited by the assumption of a horizontally-layered earth.

Experience of the determination of layer velocities from VSP records has been gained with the aid of the *visual method* [17]. The essence of the method is in the transformation of sections of VSP records into two-dimensional spectra, and in the computation of interval velocities v_l from the angle α the spectral lines make with the frequency axis. Use is made of the fact that in VSP the value of v_a of down- and upward waves in a horizontally-layered medium is approximately equal to v_l:

$$v_l = \frac{\Delta HT}{l \tan \alpha}\,,$$

where ΔH is the observation interval; T is the number of millimetres corresponding to 1 s on the VSP seismogram; l is the distance in mm between the traces of the VSP seismogram from which the microcopy for visual processing has been produced. The accuracy of determination of v_l is influenced by errors in the determination of α, by the layer thickness and by the extent of the time interval and its position on the time axis (this is essential in the case of an off-line profile). The velocities v_l for S-waves are determined more precisely with this method than v_l for P-wave, since the values of α are smaller for S-waves. Processing experience shows the minimum layer thickness that can be reliably determined from actual VSP seismograms using this method to be 150 to 200 m, because of the presence of various types of interference and irregular static shifts between seismic traces.

3. Determination of Average Velocities for P- and S-waves from Vertical Traveltime Curves of Reflected Monotype Waves*

In the case of a plane dipping interface with a homogeneous overlying bed, the average velocity in the section penetrated by the borehole is determined from the formula

$$v_{\text{av}\,i}^2 = \frac{l^2 + (2H_v - H)H}{t^2 - 2t[H_v - H]\dfrac{dt}{dH}}\,,$$

* The method of determining v_{av} for P- and S-waves from the traveltime curves of monotype and converted waves has been devised by Yu. D. Mirzoyan.

where $v_{\text{av}\,i}$ is the average velocity along the surface-reflecting interface-observation point path, H_v is the depth of the reflecting interface, H is the depth of the point of observation and φ is the dip of the interface. The formula does not contain φ, since dt/dH can be obtained from traveltime curves.

To determine average velocity in the section below total depth of the borehole, one should calculate the average velocity for the path surface-reflecting interface-observation point for different points of the vertical profile from the following formula

$$v_{\text{av}\,i}^2 = \frac{v_a^2 + \sqrt{v_a^4 - \dfrac{4l^2}{t^2 v_a^2}}}{2}.$$

The value of this velocity $v_{\text{av}\,i}$ for various points of the profile is used to find the depth of the reflecting interface H_{vi}:

$$H_{v_i} = \frac{v_{\text{av}\,i}^2 t_i^2 - (l^2 + H^2)^2}{2\left(v_{\text{av}\,i}^2 t \dfrac{1}{v_a} + H\right)} + H,$$

Next, $V_{\text{av}\,i}$ and v_{vi} are determined for each point of the medium. They can be used to obtain information about the velocities of wave propagation below the bottom of a deep borehole. Since H_{vi} is determined for the same reflecting interface, the values obtained at various points of the profile yield the following average value

$$\overline{H}_{vi} = \frac{n}{\sum_1^n \dfrac{1}{H_{vi}}}.$$

The average value of H_v serves to determine the average velocity in the section down to the reflecting interface

$$v_i = \frac{2\overline{H}_{vi} v_{\text{av}\,i}}{H_i \dfrac{v_{\text{av}\,i}}{v_{Hi}} + 2H_{vi} - H_i},$$

where v_i is the average velocity in the section from the surface to the reflecting interface and v_{Hi} is the average velocity in the section from the surface to the observation point.

The values of H_{vi} from observations at various points of the vertical traveltime curve will coincide only, if the velocities v_i have been determined correctly. Accordingly, in the same way as in [33], values of v_i are determined that yield the minimum mean-square deviation of the values of H_{vi} obtained from various points of the vertical traveltime curve

$$\sigma_{Hin} = \sqrt{\frac{\sum_1^n (H_{v\,in} - \overline{H}_{vn})^2}{n-1}},$$

where n is the number of observation points. The values of v_i, H_{vi} obtained are used to

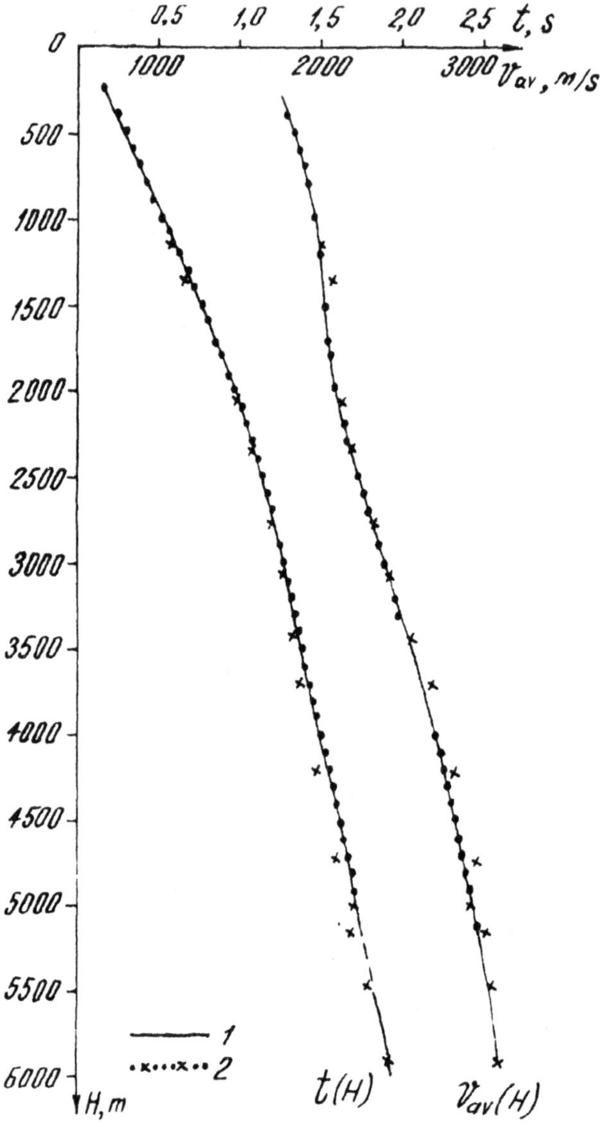

Fig. 102. Determination of velocities from VSP data in the geoligic section not penetrated by the borehole (borehole N11 Yubileinaya, courtesy Yu. D. Mirzoyan): 1 – computed, 2 – observed from the first arrivals (crosses), from reflections (circles).

extrapolate the vertical traveltime curve $t(H_v)$, or the curve $v(H_v)$, beneath the total depths of a deep borehole. In the case of angles of dip of reflecting interfaces equal to or greater than 15°, the dip must be taken into account, for which purpose iterative methods are employed.

When processing a system of several overlapping vertical profiles, the following equations resulting from the joint solution of such traveltime equations can be used:

$$v_{\mathrm{av}\,i}^2 = \frac{l_2^2 - l_1^2}{t_{2i}^2 - t_{1i}^2}; \quad H_{vi} = \frac{H_{vi} + \sqrt{\dfrac{l_2^2 t_{1i}^2 - l_1^2 t_{2i}^2}{t_{2i}^2 - t_{1i}^2}}}{2};$$

$$\sin \varphi = \frac{v_{\mathrm{av}\,i}^2 (t_{2i}^2 - t_{1i}^2) - (l_2^2 - l_1^2)}{2(H_{vi} - H_i)(l_2 - l_1)},$$

where l_1 and l_2 are distances of the shotpoints from the borehole, t_{1i} and t_{2i} are the recorded times of reflected waves associated with the same reflecting interface at the same level H_i. Figure 102 depicts the plots $t = f(H)$ and $v = f(H)$ constructed from the first arrivals of the direct P wave (down to the depth 4980 m) and from velocities determined from the traveltime curves of longitudinal reflected waves (down to 5900 m).

4. Determination of v_S from Vertical Traveltime Curves of Converted (Reflected and Transmitted) Waves

In the case of a homogeneous overlying bed and a horizontal conversion interface, the traveltime equation of a converted wave and its derivative is of the form

$$H = H_v \mp \frac{(l - H_v \tan i_0)\sqrt{1 - \gamma^2 \sin^2 i_0}}{\gamma \sin i_0};$$

$$t = \frac{H_v}{v_P \cos i_0} \pm \frac{H_v - H}{v_S \sqrt{1 - \gamma^2 \sin^2 i_0}};$$

$$\frac{\mathrm{d}t}{\mathrm{d}H} = \frac{1}{v_{a\,PS}} = \mp \frac{\sqrt{1 - \gamma^2 \sin^2 i_0}}{\gamma v_P};$$

where v_P is the velocity of propagation of the longitudinal waves to the interface; v_S is the velocity of propagation of transverse waves between the interface and the point of reception; $\gamma = v_S v_P$; i_0 is the angle of emergence from the source; H is the observation depth; H_v is the depth of the reflecting interface measured along the vertical (beneath the top of the borehole). The upper sign corresponds to the converted reflected, and the lower to the converted transmitted waves.

The determination of the velocities v_S involves the solution of systems of three equations with three unknowns (v_S, H_v, and i_0). Eliminating the parameter γ from the system, we obtain two equations

$$H = H_v \mp \frac{(l - H_v \tan i_0)\, v_P}{v_{a\,PS} \sin^2 i_0};$$

$$H_v = \frac{t \pm H \dfrac{v_P^2 + v_{a\,PS}^2 \sin^2 i_0}{v_{a\,PS}\, v_P^2}}{\dfrac{1}{v_P \cos i_0} \pm \dfrac{v_P^2 + v_{a\,PS}^2 \sin^2 i_0}{v_{a\,PS}\, v_P^2}};$$

whose solution determines the angle i_0. The values of i_0 are used to determine the depths H_v from the formula, and subsequently to determine average velocities in the section between conversion interface and the reception point. When the values of H_v and i_p are known, the velocities v_S are determined from the relationship

$$v_{Si} = \pm \frac{H_v - H}{\left(t - \dfrac{H}{v_P \cos i_0}\right) \dfrac{v_P}{\sqrt{v_P^2 + v_{a\,PS}^2 \sin^2 i_0}}}.$$

The velocities obtained with the aid of this formula will vary in accordance with the position of the reception point with respect to the conversion interface. The interval velocities ($v_{Si\,\text{int}}$) between the observation points, and the transit times of the S-waves from the interface to the reception point, are determined from the equations

$$t_i = \pm \frac{H_v - H}{v_{Si}}$$

$$v_{Si\,\text{int}} = \mp \frac{H_{i+1} - H_i}{t_{i+1} - t_i}.$$

The presence of several PS-waves in the interval being studied enables the accuracy and the reliability of the determination of arrival times and velocities of transverse waves to be improved. When jointly interpreted, the expressions for v_{Si} and Δt_{Si} are of the form

$$v_{Si\,\text{int}} = \mp \frac{H_{i+1} - H_i}{\dfrac{1}{m} \sum_{l=1}^{m} (t_{i+1,l} - t_{i,l})},$$

$$\Delta t_{Si} = \frac{1}{m} \sum_{l=1}^{m} (t_{i+1,l} - t_{i,l}),$$

where Δt_{Si} is the increment of transit time of an S-wave between adjacent reception points i and $i+1$, l is the wave's index, and m is the number of PS-waves. The coincidence of the vertical traveltime curves $t_S(H)$ constructed from the direct S- and the converted PS-waves serve as a criterion in the evaluation of precision with which the velocities v_S were determined using the method described above. The maximum difference in the interval velocities was 2 to 4% and in the times 1 to 2%. Errors in the velocities v_P exercise an essential effect on the precision with which the velocities v_S are determined. It was established as the result of evaluation of the effect of the dip of the conversion interface on the precision of the determination of v_S that for dips of 15 to 20°, the velocities of transverse waves v_S are determined with a precision of 5 to 8%. In the case of dipping reflecting interfaces, the problem may be solved with the aid of the method of interations carried on until the differences in the values of v_S of the successive cycles falls below a specified threshold value σv_S, i.e., until

$$|v_{S\,\text{pres}} - v_{S\,\text{suc}}| \leq \sigma v_S$$

5. Determination of Anisotropy

Both the direct and the reflected waves can be used in VSP for the study of anisotropy.

The vertical traveltime curve of the direct wave in a homogeneous transversely-anisotropic medium with a vertical axis of anisotropy and a wave surface in the shape of a rotational ellipsoid is described with the aid of the equation [86]:

$$t^2 = \frac{l^2}{v_X^2} + \frac{H^2}{v_Z^2},$$

where v_X and v_Z are velocities in the horizontal and the vertical directions. The velocity along the vertical will be equal to

$$v_Z^2 = \frac{H_2^2 - H_1^2}{t^2(H_2) - t^2(H_1)}.$$

For an in-line profile, $l = 0$, when observations are made at two points H_1 and H_2

$$v_Z = \frac{H_2 - H_1}{t(H_2) - t(H_1)}.$$

For observations at a single point of waves from several shotpoints, the velocity along the horizontal is described by the expression

$$v_X^2 = \frac{l_2^2 - l_1^2}{t^2(l_2) - t^2(l_1)}.$$

For an anisotropic medium with a tilted anisotropy axis, the traveltime curve of the direct wave is given by the formula

$$t^2 = \frac{l^2 + H^2}{v_{\varepsilon \pm \alpha}^2},$$

where ε and α are the angles the anisotropy axis and the ray, respectively, make with the vertical. Hence, $\tan \varepsilon = l/H$. The angle the ray makes with the anisotropy axis is $\varepsilon \pm \alpha$. The traveltime equation may be represented by the equality

$$t^2 = \frac{H^2}{v_Z^2} + \frac{l^2}{v_X^2} \pm \frac{Hl(A^2 - 1)\tan \alpha}{A^2 v_\perp^2},$$

where

$$A = \frac{v_X}{v_Z} = \frac{v_\parallel}{v_\perp}$$

is the anisotropy coefficient, and $v_\parallel$, $v_\perp$ are the velocities in the direction of and at right angles to the anosotropy axis. The traveltime curves can be used to obtain the values of the velocities. In the direction of the ray that makes an angle $\varepsilon \pm \alpha$ with the anisotropy axis

$$v_{\varepsilon \pm \alpha} = \frac{A v_\perp}{\cos(\varepsilon \pm \alpha)\,[A^2 + \tan^2(\varepsilon \pm \alpha)]^{1/2}}.$$

In the direction z

$$v_z^2 = \frac{H_2^2 - H_1^2}{t^2(H_2) - t^2(H_1)}\left(1 \pm \frac{2l(A^2 - 1)\tan\alpha}{(H_1 + H_2)(A^2 + \tan^2\alpha)}\right).$$

In the direction x

$$v_x^2 = \frac{l_2^2 - l_1^2}{t^2(l_2) - t^2(l_1)}\left(1 \pm \frac{2H(A^2 - 1)\tan\alpha}{(l_1 - l_2)\cos\alpha(A^2 + \tan^2\alpha + 1)}\right).$$

The velocities obtained make it possible to determine the direction of the anisotropy axis, the velocity in the direction of the axis and the magnitude of the anisotropy coefficient. The vertical traveltime curve of the reflected wave in the case of an anisotropic medium located on a horizontal reflecting interface is described by the equation

$$t^2 = \frac{(2H_v - H)^2}{v_z^2} + \frac{l^2}{v_x^2}.$$

The velocities in the vertical and the horizontal directions can also be determined from the traveltime curve of the reflected wave:

$$v_z^2 = \frac{(2H_v - H_2)^2 - (2H_v - H_1)^2}{t^2(H_2) - t^2(H_1)},$$

$$v_x^2 = \frac{l_2^2 - l_1^2}{t^2(l_2) - t^2(l_1)}.$$

In studies of the anisotropy of layer velocity in the Moscow syncline, VSP data established an anisotropy of $A = 1.07 \pm 0.03$ in a rock unit consisting of carbonate deposits with the inclusion of terrigeneous interlayers. The cause of the anisotropy was attributed to a thinly-layered structure [27].

Sonic logging and VSP have been used to study the anisotropy of longitudinal waves associated with a pile of high-velocity layers between 1200 and 1400 m. The anisotropy coefficient was equal to 1.18 in the depth interval 1230 to 1390 m. Velocity anisotropy may also manifest itself in the curvature of exploratory boreholes in directions corresponding to those of minimum velocity. In this case the anisotropy of the rock mass is explained chiefly by a field of natural stresses.

6. Constructing Velocity Cross-Sections from VSP Data and Sonic Logs

The presence of a large number of reflected waves in the later portion of a VSP seismogram enables additional and independent information about the Earth to be gained, and a finer sectioning of interval velocities, by means of observations in the seismic frequency band, to be achieved.

There are several methods in existence for the determination of velocities from the traveltime curves of reflected and converted waves. Consider a method based on the joint use of down- and upward waves in the later portion of the record [79]. This method employs the determination of the slopes of the traveltime curves by means of summation

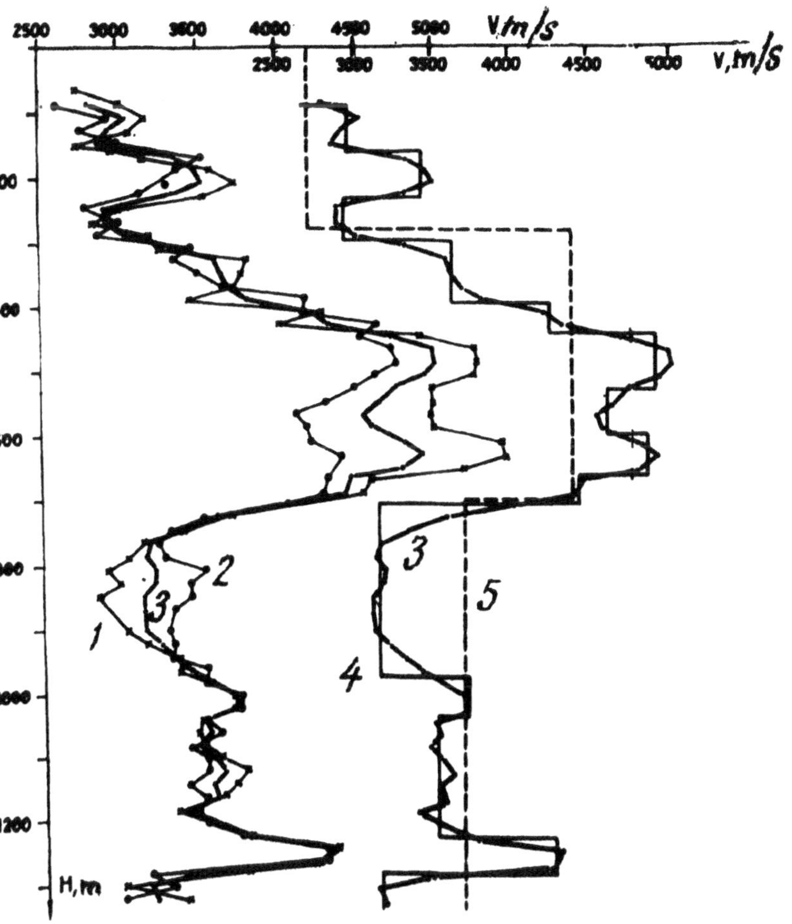

Fig. 103. Detailed subdivision of a section using waves recorded in the subsequent portion of the record [79].

at different times. To this end, after the amplitudes of all traces have been equalised, and the waves have been separated into the down and the up, one has to determine the pseudocorrelation function whose maximum value corresponds to the most probable value of the apparent velocity on the basis chosen. This procedure is carried out on a computer, and numerous values of velocities determined from the down- and the upward waves for each interval are subsequently averaged. The results of processing VSP data obtained in borehole N31 (Vsevolod–Vilven area) for $l = 200$ m are depicted in Figure 103. In the course of processing, seven traces have been summed with increments $\Delta t = 0.004$ s. The plot shows the velocity curves corresponding to the downwave (curve 1) and the upwave (curve 2) to be approximately equal. The averaged velocity curves obtained from VSP data (3, 4 – blocked version of the curve 3) are distinguished from the uphole survey curve (5) by substantially finer details: it shows up to 14 layers and agrees with the resistivity logging curve. The plot of layer velocities obtained may be used for the subsequent reduction of the off-line traveltime curve of direct waves to the vertical, which is used to determine the curve of layer velocities in the method described above.

COMBINING SONIC LOGGING (SL) AND VSP

A further trend in the direction of improving the resolution of the velocity section is the joint use of data obtained with the aid of VSP and ultrasonic logging. Such a combination makes it possible to differentiate comparatively thick layers identified in VSP into separate thin interlayers. The joint processing of sonic and seismic observations is automatic and performed on a computer. Such joint data processing is exemplified by results obtained in the borehole N18 in Kolganovskaya (Figure 104). Figure 104 depicts a vertical traveltime curve of the direct wave and a sonic log from which two graphs of layer velocity have been constructed displaying a great difference in resolution. The velocity curve obtained [127] makes it possible to compute the reflection coefficients shown in the figure on the right. The combination of sonic logging and VSP has not, as yet, been developed to a sufficiently advanced stage.

When interpreting the results of combined seismic and sonic observations in boreholes and comparing the velocities of elastic waves measured at different frequencies, one should keep in mind the difference existing between the methods of measuring velocities in seismic and sonic logging. In sonic logging, a much greater influence is exercised on the velocity being measured by the zone surrounding the borehole, whose elastic parameters have been altered by drilling. The value of v_{SL} drops in a fractured zone (by as much as 20 to 30%), which is followed by the region of the borehole wall, where the velocity is greater.

Depending on the extent of penetration of the waves into the borehole wall, v_{VSP} can be either greater or smaller than v_{SL}.

In addition, a change in the velocity with time after the borehole has been drilled has been observed. Too low a velocity may also be the result of a weak signal, in which case only the subsequent phases are recorded. On the other hand, in well geophone surveys the waves which propagate in the unperturbed rock mass, can be affected by dispersion of the phase velocity, which may be the result of the absorption properties of the Earth (body dispersion) as well as wave interference in heterogeneous thin-layered media. Each type of dispersion may have a different value, with the total dispersion being dependent on their signs. Accordingly, it would be interesting to study both kinds of dispersion separately at different frequencies, and subsequently to consider the combined dispersion. One deficiency of the method of the separate studies of dispersion in boreholes is the comparatively narrow frequency band: 10 to 100 Hz for well geophone surveys and 10 to 25 KHz for sonic logging, and in this respect the joint appraisal of traveltime curves obtained with the aid of sonic logging and well geophone surveys whose frequencies differ by three orders of magnitude is preferable. Here one should not fail to take into account the whole variety of dispersion types and their signs, not excluding the possibility of 'compensation mechanisms'.

It has been established as the result of studies of the effects of absorption, and of the phase velocity dispersion associated with it and on the magnitude of the average and interval velocities obtained with the seismic and the sonic logging methods, that the values of the average and interval velocities measured in seismic logging, in the case of a homogeneous isotropic medium in the presence of absorption are not the same and do not depend on the separation between the source of excitation and the point of observation.

Comparing the experimental traveltime curves obtained in sonic and in seismic logging for unsaturated rock in uncased boreholes, we have discovered a velocity dispersion lying within the limits of 3 to 8 % due to measurement at different frequencies. The theoretical dispersion for the above-mentioned frequency ratio in the presence of

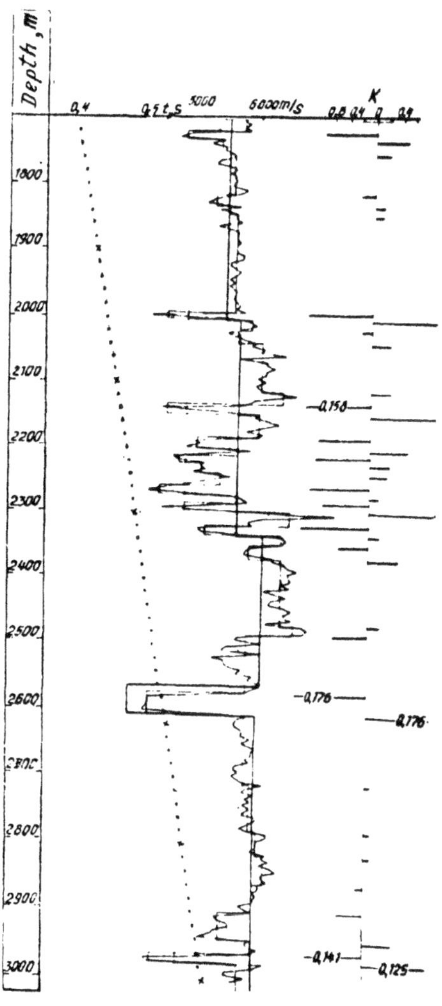

Fig. 104. Comparison between the results of processing of sonic and seismic logs (borehole N18, Kolganovsk area [127]).

wide ranging variations of attenuation typical of barren sedimentary rock does not exeed 7%. The velocity dispersion in carbonate reservoirs (porous media saturated with oil) measured in combined seismic and sonic studies in an uncased borehole amounts to 18 to 25%, rising for the same kind of rock in a cased borehole to 29 to 33%. The discrepancy between the traveltime curves obtained in sonic and in seismic logging,for the range of depths considered here, and the dispersion of the phase velocity (of the order of 30%), are apparently associated with a high concentration of cracks and with upward migration of oil and thermal mud. The difference may be due to the difference in frequencies in thin-layered media: for a specific geologic section, the velocities of high-frequency waves are 5 to 10% higher than those of low-frequency waves. As a result, insufficient accuracy of the quantities being compared makes it impossible to establish clear relationships in the velocity ratios.

The interval and layer velocities measured using seismic and sonic logging methods differ even for the same observation basis. The sonic velocities often prove to be somewhat higher than the seismic. The differences in layer velocities measured in sonic and seismic logging are shown in Table IX [128].

TABLE IX

Rock	v_{SL} (m s^{-1})	v_{VSP} (m s^{-1})	$\Delta\bar{v}$ (m s^{-1})	$K\% = \frac{v_{SL} - v_{VSP}}{v_{VSP}} \cdot 100\%$
Dense clay	3300	3000	330	10.0
Clay in low-velocity zone (moist)	1500	1600	− 100	6.2
Sandstone	3800	3630	170	4.9
Limestone	5360	5330	30	0.6
Dolomite	6030	6100	− 70	1.1
Anhydrite	5920	5700	220	3.8
Pure salt	4600	4650	− 50	1.1

The minimum discrepancies are observed in high-velocity rock, and the maximum in terrigeneous low-velocity rock. Extra-high values of velocities obtained from sonic logging data can be explained mainly by the presence of a zone of high stress around the borehole formed in the course of drilling. When observation data on several boreholes in one location are available, it is possible to determine correction coefficients for the transition from the velocities in the sonic frequency band to those in the seismic exploration band.

It should be emphasized that the value of sonic logging for seismic exploration is not limited to the establishement of empirical relationships between v_{VSP} and v_{SL}. The development of production seismics increases the resolution of studies based on the use of wave dynamics which imparts a special significance to sonic logging, so that it becomes an integral part of the general scheme of seismic exploration. Sonic logging is needed for the computation of synthetic seismograms, for the inversion of seismic traces to pseudo-sonic logs and for constructing models of the earth and its deposits. VSP data may serve as the connecting link between sonic logging and surface seismics.

7. Determination of Velocity and Density from Production Geophysical Data

An extensive combination of VSP and sonic logs is, to a large measure, retarded by an inadequate volume of sonic logging. This can be surmounted, if a detailed velocity distribution is computed from data obtained by production geophysics in boreholes.

The opportunity to gain information about the velocity section in those numerous boreholes in which sonic logging is not performed is very important for practical purposes. In this respect, the computation of velocity curves from production geophysical data is of some interest. Computational procedures have been developed and widely used, particularly in studies aimed at constructing effective seismic models. They depend on the correlation of the electric and elastic properties of rock. The computation of velocity is based on Faust's empirical formula [129]:

$$v_P(Z) = \gamma \sqrt[6]{Z_i \rho_i(Z_i)},$$

where

$$\gamma = \frac{\bar{v}_P}{a},$$

$$a = \frac{\sum\limits_{i=1} \sqrt{Z_i \rho_i(Z_i) h_{bi}}}{h_n}$$

and

$$h_n = \sum h_{li},$$

$v_P(Z)$ is the velocity of longitudinal waves at the depth $Z_i(m)$ of the ith layer; $\bar{v}_P$ is the average interval velocity in m s^{-1} of longitudinal waves in the rock computed from sonic logging or VSP data; ρ_i is the measured value of the relative resistivity of the layer or of the rock at the depth Z_i in Ohm · m; h_{li} is the thickness of the layer in m; h_n is the thickness of the rock bed in metres.

Using data from a combination of the apparent resistivity method, sonic logging, well velocity surveys and VSP, the effectiveness of the computation of $v_P(Z)$ functions from apparent resistivity data has been checked. Comparison of the velocity section obtained from curves computed using Faust's formula with the section obtained by sonic logging in the Belgorod district of the Kursk Magnetic Anomaly [129], has shown satisfactory agreement between velocities in the upper portion (chalk, marl, sands, clays) but large discrepancies (up to 2000 m s^{-1} and more) in limestones and basement rocks. Such discrepancies can apparently be explained by the fact that during computations of v_P from the VSP and well geophone survey traveltime curves, rocks of different origin and different elastic properties in a relatively thick depth interval are regarded as a single rock unit.

The next stage in improving the accuracy of the $v_P(Z)$ curve computed from borehole seismic data involves the use of sonic logging data to compute average interval velocities. The most accurate velocity section was found to be closest to the sonic over the entire depth range. Discrepancies are smallest in low-velocity rocks in the upper portion of

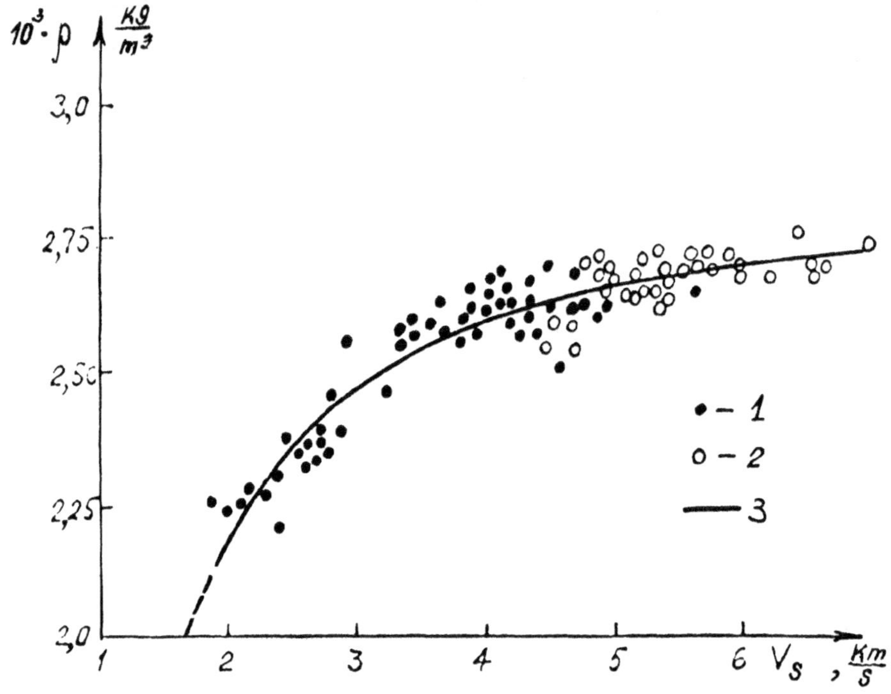

Fig. 105. Plot of density of sedimentary rock vs layer velocity v_1 [130]. 1 – for terrigeneous rock; 2 – for carbonate rock; 3 – line of regression of ρ on v_1.

the section (8 to 11%), and highest in chalks (16%). The results obtained from the computation of velocity curves from apparent resistivity curves, and the comparison with sonic logging data, are proof of their feasibility in the study of the velocity sections for the interpretation of seismic observation data.

In addition to resitivity logging data, other kinds of production geophysical data are being used nowadays to study the velocity section, such as natural gamma-ray and other radioactive logging. At present, the relative improvement in the quality of neutron and natural gamma-ray logging data apparently makes neutron data the optimum for the computation of velocities from all the production geophysical data used for that purpose. The next in order of merit is natural gamma-ray and resistivity logging data.

Determination of Density

Of great importance in the study of the reflecting properties of the geologic section are the rock densities. The experience gained from information about the velocity section in the study of its density parameters is of essential importance. The determination of density from rock sample measurements gives a wrong picture of the geologic section, and the process itself is very labour-intensive. Besides, the use of seismic wave velocities for this purpose enables information to be obtained about densities in portions of the section not penetrated by the borehole. For some areas in the Dnepr-Donetz depression, a parabolic correlational relationship between the density and propagation velocity of elastic waves has been established. It is described by the equation:

$$\rho = 1.4661 + 0.4212v - 0.0463v^2,$$

where ρ is the density in g cm^{-3}, and v_P is the velocity of longitudinal waves in km s^{-1}.

For the Western and the Central Caucasian foothills region, the relationship is described by two equations: for values of v_P between 1 and 4 km s^{-1} $\rho = 1.8 + 1.18v$, and for values of v_P from 4 to 6.3 km s^{-1}

$$\rho = 2.52 + 0.12(v - 4).$$

For sedimentary rocks in the Pechora and the Upper Pechora depressions, the relationship between density and velocity in the terrigeneous and carbonate rocks (Figure 105) is expressed by a quadratic equation in $1/v_P$ (the regression equation) [130]:

$$\rho = 2.89 - \frac{0.9919}{v_P} - \frac{0.8244}{v_P^2}.$$

The average error in density determined using this equation is equal to $\pm$ 0.0329 g cm^{-3}, which is low enough to permit the application of the density values obtained for all practical purposes.

The relationships cited above are not universal and may turn out to be different in other areas. They enable the densities to be calculated for the entire geologic section, their spatial variation to be determined, and density maps of various complexes to be drawn. By combining borehole seismic and gravimetric logging data it will be possible to obtain a still more detailed subdivision of the velocity section.

CHAPTER 10

SOLUTION OF STRUCTURAL PROBLEMS

Observations made in boreholes for the purpose of solving structural problems were initiated as far back as the thirties using a borehole geophone method to study the flanks of salt domes. This method involves the positioning of the borehole geophone below the refracting interface and recording the waves arriving from shotpoints located on the surface. In the case of a high-dip interface with a sharp velocity differentiation, processing of the observed data with the aid of the time field enables the refracting interface in the section contiguous with the borehole to be mapped. However, the method has not become widespread, because it utilizes only the first arrivals of the transmitted waves.

The development of VSP has substantially enhanced the potentialities for structural delineation. The reason is that in VSP not only the transmitted wave first arrivals are used for the purpose, but reflected waves as well, and this in turn enables not only the interfaces intersected by the borehole, but also those located below the bottom of the borehole to be studied. The advantages of the transition from the transmitted to the reflected waves for the solution of structural problems using observations along a vertical profile can be compared to those of a similar transition, which took place in the thirties, and which made seismic studies the principal geophysical exploration method.

The work of Yu. B. Demidenko, V. A. Teplitzkiy, G. E. Rudenko, L. L. Khudzinskiy played an important part in the development of VSP as applied to solving structural problems using the method of reflected waves, VSPRM [28–30, 33, 99–102].

The development of PM–VSP not only improved the conditions for the correlation of longitudinal waves, but also made it possible to utilize reflected waves of various types, thus greatly enhancing the potentialities of the VSP reflection method in the study of structure in the region surrounding the borehole.

Until recently the VSP reflection method has been used most frequently when despite quite complex observation systems and processing methods, the effectiveness of discriminating useful waves on surface seismic data has remained inadequate, because, for example, of high levels of unwanted signal, particularly multiples, a highly disturbed upper portion of the section, and adverse surface conditions. Down-hole observations below the principal marker reflecting interfaces in such conditions may make possible the discrimination and tracking of reflected waves.

A point deserving special mention is the effectiveness of the VSP reflection method used to study complex folded high-dip media in which case the reflected waves in certain sections of the profile do not arrive at the surface at all. Such a situation is encoutered mainly in mining areas, but is also feasible in oil-producing areas, for example, in the study of the flanks of salt domes, reefs, diapirs. In such cases, by using combined horizontal (surface and fixed geophone depth) and vertical profiling one is able to track some sections of the interface on a surface or a fixed geophone level profile, and others (high-dip) – on a vertical profile. Moreover, the VSP reflection method may be desirable on account of the presence of a thick bed of gravel causing severe energy absorption, as well as in case of very rough terrain, which makes surface observations difficult and the grouping of geophones and/or sources impossible. The VSP reflection method can be recommended also when it is required to improve the resolution. It can play a special part in the deep drilling process in ascertaining the structural position of the borehole being drilled for the purpose of orienting further drilling. If deep boreholes are already available, the cost of additional studies will be negligible, and this will make VSP reflection studies advisable (in combination with other production-geophysical methods), especially where surface methods yield no positive results.

In the studies of localised sections contiguous with the borehole, the VSP reflection method may be important by itself, but in the majority of cases it is combined with surface observations. VSPRM plays a very important role in applications during the exploration and exploitation stage of a field as an indispensable part of the production seismics.

The VSP reflection method envisages the recording of reflections both on the vertical and the fixed geophone level profiles. Systems based on observations along fixed geophone level profiles became known as 'the inverted traveltime curve method' [34, 88]. Naturally, in both cases we are dealing with VSP modifications differing only in the mutual disposition of the sources and the receivers. Since in fixed level observations recording is performed as a rule not at a point, but in a section of the vertical profile, such methods constitute in effect a continuous combination of observations along the vertical and the fixed level profiles.

The combination of the VSP reflection method and surface common-depth-point observations is of great importance for the discrimination of waves reflected from deep interfaces in the presence of high-intensity multiples, and in the case of deep multiple producing interfaces, for example, in the search for sub-salt structures.

1. Observation Procedures in the VSP Reflection Method (VSPRM)

There is a reflection point on the interface, which corresponds to every point of a vertical traveltime curve, and a section of the interface, which corresponds to a section of the traveltime curve. The length of this section is determined by the formula [88]:

$$\Delta L = L_2 - L_1 = l\left(\frac{H_v - H_2}{2H_v - H_2} - \frac{H_v - H_1}{2H_v - H_1}\right)\cos\varphi,$$

where H_v is the depth of the interface in the borehole measured along the vertical; H_1, H_2 are the depths of recordings; φ is the angle of dip of the reflecting interface; l is the distance from the shotpoint to the borehole; L_i are the distances along the interface for $\varphi = 0$.

$$\Delta L = \frac{l \cdot H_v \cdot \Delta H}{(2H_v - H_2)(2H_v - H_1)}.$$

For an in-line profile,($\Delta L = 0$) all reflections arrive from one point. ΔL increases with l up to a certain value depending on the recording interval along the vertical (H_2–H_1) and on the depth of the reflecting interface: the greater is H_v the less is the increment ΔL. If $H_2 = 0$ (surface observations),

$$\Delta L = l\left(\frac{1}{2} - \frac{H_v - H_1}{2H_v - H_1}\right)\cos\varphi.$$

If $L_1 = 0$ (the borehole intersects a reflecting horizon)

$$H_v = H_1; \qquad \Delta L = L_2; \qquad \Delta L = \frac{l}{2}\cos\varphi.$$

If $l = 0$, the wave cannot be tracked up to the surface, but only up to the recording depth, H_2. For interfaces located below the bottom of the borehole, tracking becomes possible starting from a definite distance from the borehole

$$L_1 = l\left(\frac{H_v - H_1}{2H_v - H_1}\right)\cos\varphi.$$

Specifying the necessary distance along the interface, one can use the above equations to find l – the required distance to the shotpoint.

By utilizing the multiples, it is possible to extend the subsurface coverage. In this respect, the downward reflected multiples from the bottom of the low-velocity zone, which have a high intensity and can be reliably tracked along the profile, may be of major interest. The lengths of the interface illuminated by the multiples reflected downwards are the continuations of the lengths illuminated by the waves reflected upwards. The total length of the interfaces covered may be approximately 1.5 to 2.0 times greater than the lengths illuminated only by waves reflected upwards. The total length of the section of the interface illuminated by the waves reflected both upwards and downwards depends greatly on the dip, offset and depth of the interface, and in some cases may exceed the length of the vertical traveltime curve being studied.

For observations at fixed geophone level profiles, the geophones are located in boreholes at fixed depths, and the shotpoints are arranged on the surface (along the desired lines of the profiles). The transit times are related to the shotpoints in accordance with the reciprocity principle. This produces a profile located at a specified depth. The trackable section of the interface is given by the equation [88].

$$L = \frac{(l \pm H\tan\varphi)(H_v - H)\cos\varphi}{2H_v - H \pm \tan\varphi}.$$

For $\varphi = 0$,

$$L = \frac{l(H_v - H)}{2H_v - H}.$$

The trackable section of an interface becomes shorter with increasing depth of the interface.

The observation procedure with the VSPRM as well as the methods of discrimination and tracking of reflected waves, is similar to that used in VSP. VSP data is of essential importance for drawing up plans of the VSPRM. They are used to study the wave field, to choose the depth intervals for fixed geophone level studies and source excitation conditions. Both the up- and the downward reflected waves are studied in the VSP reflection method, and observations performed on a certain vertical basis enable traditional methods of interferential reception to be used, and the waves of both types to be discriminated. The most important pre-condition for the correlation of waves to be possible, is the repeatability of source excitation conditions and the shape of the direct wave. Unfortunately, this cannot always be achieved, especially in fixed geophone level observations along a section of the profile where the low-velocity zone has a variable structure. The choice of depth for the fixed geophone observations is based on seismologic conditions. In the presence of high-intensity multiples, recording is per-

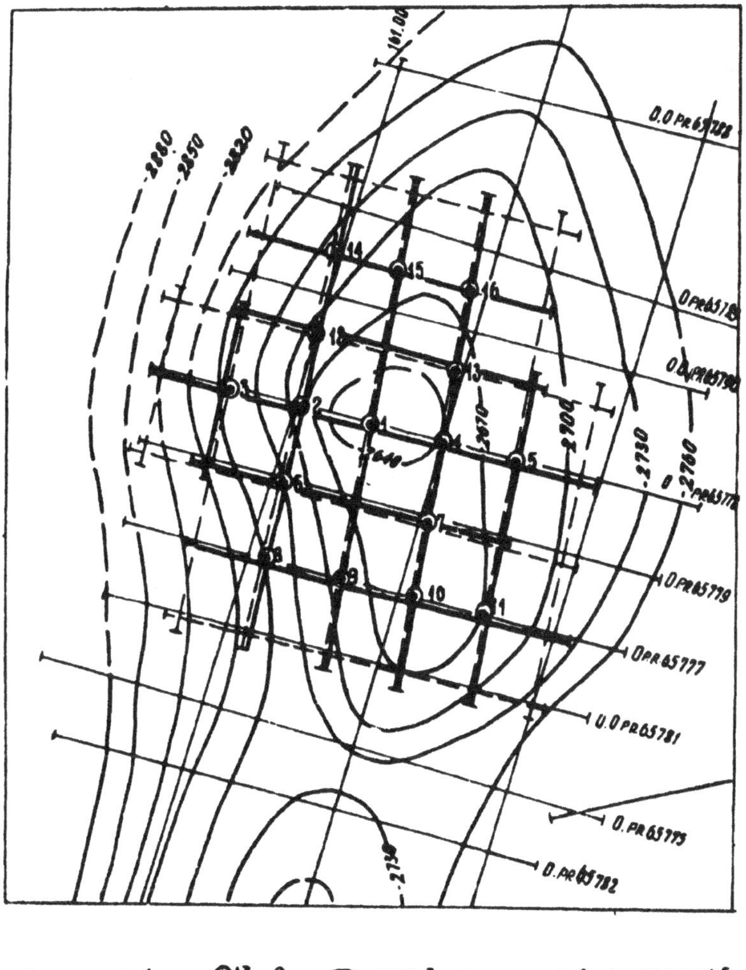

Fig. 106. Observation system and structural map for the method of reflecting horizons (Tangikuduk [88]): 1 – reflection method profiles; 2 – observation holes; 3 – contours of the reflecting horizon close to the salts surface; 4 – off-line method-of-reflecting-horizons (VSP-reflection method) profiles; 5 – in-line method-of-reflecting-horizons (VSP-reflection method) profiles.

formed beneath the principal interfaces, which generate multiples outside the zone of interference of the downward and the reflected waves. When the intensity of the multiples is not too high, observations in shallow boreholes penetrating the low-velocity zone may even improve the quality of the data and the resolution of the record. Such observations are, on the one hand, much less costly, and on the other, more easily standardized and thus enable complex spatial observation systems to be shot.

The use of converted transmitted waves, in addition to the longitudinal, facilitates the study of the structural arrangement of interfaces intersected by the borehole. The effectiveness of combined studies of such waves has been proved in mapping the flanks of salt domes.

Observation systems. The VSP reflection method may be put to work both in a series of boreholes and in an isolated one. In the former case, it provides for continuous tracking of the waves and the construction of extensive cross-sections. Systems of anti-parallel traveltime curves with overlap are being employed for this purpose, since there are no related points that could be used to combine observation systems, and continuous correlation can only be achieved by means of overlapping traveltime curves. Systems of fixed geophone level observations are determined by the goals of the studies. Linear systems are shot to obtain complete anti-parallel curves with continuous tracking of reflecting interfaces and the necessary overlap. For areal observations, systems of in- and off-line profiles are shot. Figure 106 depicts an areal system in which fixed geophone level observervations were shot on the Tagikuduk structure (Eastern Turkmenia) and a structural map drawn as the result of the VSP reflection studies. With such a system by recording seismograms in a single line of boreholes from shot points located at various offsets in different directions one is able to obtain a series of interrelated cross-sections. This may reduce the amount of drilling of observation boreholes, increase the coverage and the productivity.

Multiple overlap systems are set up by recording in 3 or 4 boreholes the waves arriving from the same shotpoints, or by carrying out in a single borehole at different depths, observations of the waves from different shot points. The magnitude of the overlap is

$$\Delta l = H \frac{l \pm (H_v - H) \sin 2\varphi}{2(H_v - H) \cos^2 \varphi + H},$$

where + refers to down-dip, and − to up-dip. The maximum length of the traveltime curve depends on the conditions for correlation of the waves. It is determined experimentally, and, whenever possible, must not exceed the depth of the probe. For continuous mapping of the interface, the separation between the observation points (the boreholes) must be less than the maximum length of the traveltime curve.

The separation between shotpoints (Δl) is determined by the angle of dip of the reflecting interface, by the velocities in the overlying medium, by the periods of oscillations, by the depth of recording, by the terrain and by the low-velocity zone. In general

$$\Delta l = \frac{vT(2H_v - H)}{4(H_v - H) \sin \varphi}.$$

In platform regions, the separation between shotpoints lies within the range of 100 to 200 m, and for high-dip interfaces drops to 50 m. Continuous tracking of the interfaces is possible, provided special structural boreholes and traveltime curves of adequate length are available. An essential drawback to fixed geophone level observations is the need to drill observation boreholes.

In the majority of cases, structural problems are solved with the aid of VSP observations in the vicinity of marker beds in isolated exploration and production boreholes. In- and off-line observation systems are shot in the process. Source excitation points are arranged on the surface along radial lines or in a regular network that enables an element of the reflecting interface in the section intersected by the borehole to be mapped and the interface coordinates to be determined. This facilitates the orientation of subsequent drilling.

Fixed geophone level observations should be extensively combined with observations along a vertical profile (VSP) for the solution of structural problems.

The type of observation system and the relationship between the amount of data recorded along the vertical and along the fixed geophone profiles are determined in each particular case in accordance with the geological problems to be solved and the seismological conditions. Observations should be limited to the vertical profile where the terrain makes the establishment of a line of shotpoints impossible, or to deep boreholes, where a long vertical profile can be shot, and a corresponding section of the interface mapped. The relationship between the fixed geophone and VSP observations is determined from considerations of the complexity of the deep geologic structure and the surface conditions. In complex surface conditions, it is usually difficult to arrange shotpoints along profiles, and accordingly observations along the vertical profile begin to play a prominent part. The same is true in the case of a deep complex structure, and in particular, in cases of intricately shaped high-dip reflecting interfaces with disjointed faults, when reflected waves are best recorded at sufficiently large intervals of the vertical profile. The use of three-dimensional observation systems along the vertical and the horizontal (surface and fixed geophone level) profiles, facilitates comprehensive studies of the rocks around the borehole. By combining the vertical and the fixed geophone level profiling in VSP reflection observations, it is possible to obtain phase correlation of the observed data and to improve the validity of structural mapping.

2. Processing details

Specific outputs of processing data obtained in the VSP reflection method consist of the construction of cross-sections in time and depth. These outputs are preceded by preliminary stages of processing, which include the traditional procedures of discrimination and tracking the waves, determining their nature, and the computation and introduction of kinematic corrections. Determination of the nature of waves may present difficulties in high-dip interface conditions, because the differences between the kinematics of the monotype and the converted reflected waves may not be adequate for and unambiguous determination of the nature of the waves. In this respect, joint studies of the kinematics and particle displacement directions with the aid of PM–VSP may prove of great help. When it is possible to track waves arriving from the same shotpoint along the horizontal and the vertical profiles, the velocity of the waves can be determined from the isochron field or from the components of the field's gradient. In good correlation conditions, such measurement may be made in comparatively shallow boreholes. Considerable difficulties are encountered in the identification of similar waves recorded along the vertical profile from different shotpoints. They can be identified with the aid of horizontal (surface or fixed geophone level) profiles shot along a line passing through the shotpoints. Kinematic corrections are computed from well-known formulae, however, the most precise data is obtained with the aid of VSP. The magnitude of the kinematic correction is determined from the relationship $\Delta t_k = t - t_{ost}$, and can be computed approximately, taking account of the recording depth, from the expression

$$\Delta t_k = \frac{l^2}{2t_{ost} v_{av}^2},$$

where v_{av} is the average velocity of the reflected wave from the top of the borehole to the point of observation.

In the process of computation of kinematic corrections, the velocity will vary as a function of the depth of the observation point. It can be computed from the formula

$$v_{av} = \frac{2H_v - H}{\dfrac{H_v}{v_{av}} - t_v}.$$

Various methods may be used to compute static corrections Δt_{st}. In the first method, known data about the velocity in the near-surface zone are used to reduce all observations to one datum, which is chosen to be near and somewhat below the bottoms of the shotholes. In this case

$$\Delta t_{st} = \frac{h_1 - h_2}{v_l};$$

where h_1 is the elevation at the top of the shotpoint with respect to that of the observation borehole; h_2 is the depth of the explosion, v_l is the velocity in the interval between the datum and the bottom of the shallowest shothole. More exhaustive data are obtained if use is made of the times of first arrivals of the direct wave recorded in the borehole or along the surface next to the borehole being logged. Normally, in VSP reflection studies, the corrections are computed with the aid of three methods, and used as initial corrections to be adjusted subsequently in the course of construction of the time cross-sections. The latter are convenient, since they enable VSP data to be represented in a form similar to those of surface seismic exploration, thus substantially simplifying the correlation and comparison of data obtained in surface and downhole observations.

The time cross-section is constructed repeatedly with different variants of static corrections. The adjustment of static and kinematic corrections is usually executed on the basis of several (3 to 5) easily trackable reflecting horizons. The improvement in correlation of all reflecting interfaces serves as a criterion that the corrections have been introduced properly. In recent years algorithms have been devised for the construction of time cross-sections with kinematic corrections for every trace [131]. Time cross-sections enable the correlation of waves to be improved, and allow coherency methods such as summation and fan filtering etc, which are hardly applicable to the original seismograms because of the broken-line nature of vertical traveltime curves, to be used more effectively. Various types of controlled record summation are used to improve the signal-to-noise ratio. In the case of observations along vertical profiles, summation from common shot points is preferable, but summation from common reception points is preferable in the case of observations along fixed geophone level profiles. By combining both types of controlled summation, one is able to guarantee reliable determination of the kinematic parameters of the waves and to improve the quality of the time cross-sections.

The combined processing of time cross-sections obtained from various shotpoints presents great opportunities for improving the effectiveness of VSP observations [78]. A drawback to time cross-sections is the non-linear relationship between the time and the depth scales; another is the failure to take account of migration. Time cross-sections are transformed into depth cross-sections by means of computations involving the recording of times of reflected waves. This is conveniently done from the first marker interface down to which the average velocity is known. To construct a depth cross-section, use is normally made of the time field method or the selction of traveltime curves. It should be pointed out that the accuracy with which interfaces can be constructed with the aid of the VSP reflection method depends on the dip of any intermediate refracting interfaces on the presence of lateral velocity gradients, and the deviation of the borehole from the vertical. Typical of depth cross-sections is the unequal overlapping of sections of the reflecting interfaces located at different depths, this being a general case in seismic exploration: if adjoining sections of deep interfaces overlap within a definite interval, shallow interfaces are not continuously trackable.

Specialized processing routes have been set up for VSP reflection method data that enable all the processing procedures from the discrimination of waves to the construction of depth cross-sections to be performed. Certain aspects of the observation and processing procedures have been developed to a more advanced stage specifically for the purpose of structural mapping; errors in mapping interfaces and the factors responsible for them, as well as methods of construction (the time field and the gradient methods), including computerized ones, have been evaluated.

3. The Polarization Method of a Common Depth Point in VSP

Multiple overlap observation systems may be developed to improve the conditions for discrimination and tracking reflected waves, in the same way as it is done in surface observations.*

A specific feature of the common depth point method in VSP, as distinct from surface observations, is that a change in the location of the shotpoints changes the direction of particle displacement and, by force of it, the relative intensity of the vertical or, in general, any fixed-in-space oscillation component being measured. In such conditions the summation of records of the full vector or the tracking components, instead of the records of fixed components, improves the signal-to-noise ratio and enables weak deep reflections to be detected against a background of regular unwanted waves. The polarization common depth point method can be comparatively easily executed on fixed geophone level profiles. Methodological difficulties are associated with the formation

* This trend has been developed by Yu. D. Mirzoyan, and the results cited belong to him.

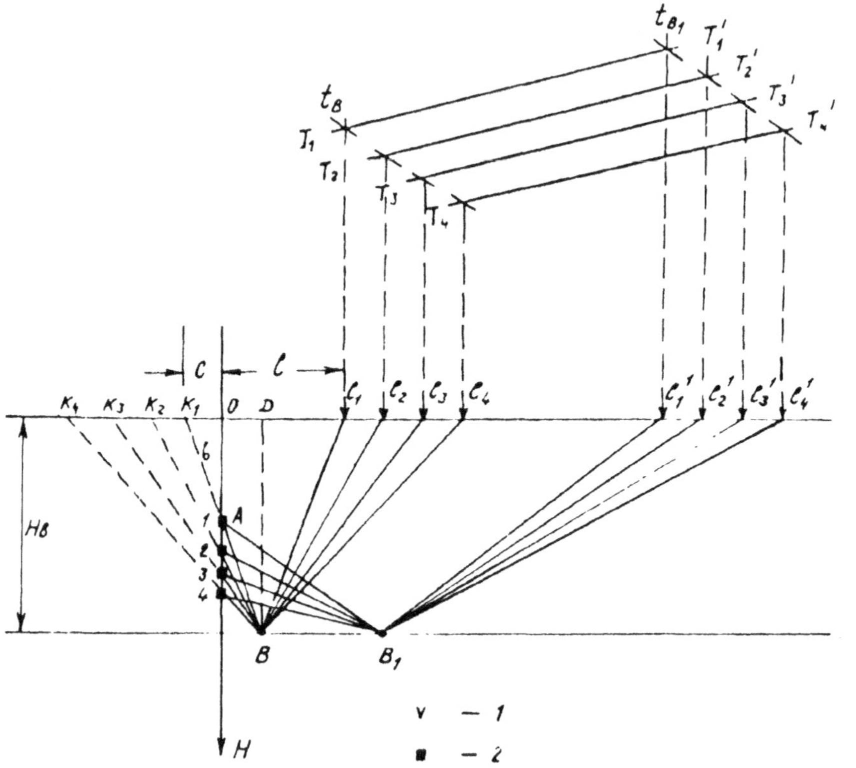

Fig. 107. Illustrating application of the common-depth-point method on fixed geophone level profiles; 1 – source excitation points; 2 – reception points.

of reflected waves corresponding to the common depth point. Consider the recording on a vertical profile of waves reflected from a common depth point, assuming a horizontal interface (Figure 107). Let $l_1 \ldots l_i$ be the coordinates of the shotpoints, and $K_1 \ldots K_i$ those of the points of emergence of the corresponding rays on the surface. The time the wave arrives at point K_1 is determined as follows

$$t_{(l)} = \tau_{(l,\,H)} + \Delta\tau_{v(l,\,H)},$$

where $\tau_{(l,\,H)}$ is the common-depth-point VSP traveltime and $\Delta\tau_{v(l,\,H)}$ is the transit time of the ray from the reception point in the borehole to the surface at point K_1. The position of point K_1 on the surface is determined by the expression

$$\tilde{l} = l + c$$

The projection of the common depth point on the surface profile is

$$L = \frac{l + c}{2}$$

where c is the distance from the point of emergence of the ray on the surface to the borehole entrance. In such conditions, c and $\Delta\tau_{v(l, H)}$ are the unknowns. The quantity c can be found from the triangles $K_1\ OA_1$ and K_1BD (Figure 107):

$$c = \frac{Hl}{2H_v - H},$$

where $H_v = H_{\mathrm{CDP}}$ is the depth to the reflecting interface. The correction $\Delta\tau_{v(l, H)}$ can be found from the expression

$$\Delta\tau_{v(l, H)} = \frac{\sqrt{c^2 + H^2}}{v} = \frac{H}{v}\sqrt{1 + \frac{l^2}{(2H_v - H)^2}},$$

Substituting the values of l, C, $\Delta\tau_{v(l, H)}$, one can write down the traveltime equation of the reflected wave in VSP–CDP as the difference between the transit times of the wave to a point on the surface profile and from the down-hole observation point to the surface:

$$\tau_{(l, H)} = \sqrt{\frac{4H^2_{\mathrm{CDP}}}{v^2} + \frac{\tilde{l}^2}{v^2}} - \frac{H}{v}\sqrt{1 + \frac{l^2}{(2H_{\mathrm{CDP}} - H)^2}},$$

$$\tilde{l} = \frac{2H_{\mathrm{CDP}}\, l}{2H_{\mathrm{CDP}} - H}.$$

After appropriate transformations, we obtain the equation of the in-line common depth point traveltime curve for a homogeneous medium in VSP–CDP

$$\tau^{\cdot}_{(l, H)} = \sqrt{\frac{(2H_{\mathrm{CDP}} - H)^2}{v^2} + \frac{l^2}{v^2}}, \qquad L = \frac{H_{\mathrm{CDP}}\, l}{2H_{\mathrm{CDP}} - H}.$$

Since

$$\frac{2H_{\mathrm{CDP}} - H}{v} = \tau_{\mathrm{ost}(l, H)},$$

we obtain

$$\tau_{(l, H)} = \sqrt{\tau_{\mathrm{ost}} + \frac{l^2}{v^2}}, \qquad L = \left(1 + \frac{H}{v\tau_{\mathrm{ost}}}\right)\frac{l}{2}.$$

τ_{ost} is the common depth point traveltime curve's 'zero' time that takes account of the depth of recording. The former equation means that the common depth point traveltime curve in VSP–CDP is a hyperbola. To transform the line-up into τ_{ost}, one should choose appropriate values of the velocity(v_{CDP}), which can be obtained by controlled summation of common depth point seismograms over a set of hyperbolae, with a definite value of v_{CDP} corresponding to each value of τ_{ost}. If at every point (l, H, t) of the common depth point seismogram corrections for normal increments are introduced, (in the base of a horizontal interface – kinematic corrections), we reduce the recorded time of the channel with the coordinates l, H to the time τ_{ost}:

$$\tau_{\mathrm{ost}} = \sqrt{\tau^2_{\mathrm{ost}(l, H)} + \frac{l^2}{v^2}} - \Delta\tau_{k(l, H)}.$$

The magnitude of the kinematic correction for various wave types (*PP*, *PS*, and *SS*) can be computed approximately with the aid of the equation

$$\Delta\tau_{k(l,\,H)} \approx \frac{l^2}{2v^2\,\tau_{\mathrm{ost}(l,\,H)}}. \tag{34}$$

The origin, and the type of waves can be reliably determined with the aid of the PM–VSP common-depth-point method, and this makes it possible, in addition to solving exploration problems, to study the displacement directions and the intensities of various wave types, to appraise their contribution to the total energy of the wave field and to make a realistic choice of a field observation system.

Such observation systems should provide for reliable discrimination and tracking of various wave types (*PP*, *PS*, *SS*, etc.). When devising such an observation system, as in the case of surface seismic exploration, one will have to choose: overlap multiplicity, separation between shotpoints and the maximum length of the traveltime curve at which the necessary time lag of an unwanted wave, e.g., a multiple, sets in. In addition, the interval of observations on the vertical profile, shotpoint number and the channels whose records belong to the common depth point must be determined. The overlap multiplicity depends on the level of the unwanted waves. In PM–VSP–CDP, such estimates for various wave types (*PP*, *PS*, *SS*) are obtained from PM–VSP observations. The maximum length of the traveltime curve (the summation interval) is obtained, after corrections to the useful wave have been introduced, from the condition that the difference between the transit times of the unwanted wave across the interval be equal to 1.5 periods (30 to 50 ms), when its suppression takes place. For quantitative evaluation of the maximum length of the traveltime curve, one may use the equation for the kinematic correction or the time-lag functions of the multiples with respect to the singly-reflected wave obtained from PM–VSP data. There are certain specific points about the choice of the sampling rate, and the separations between the shotpoints in PM–VSP–CDP even in the case of waves of the same type. The distance Δl for two traveltime curves obtained at adjacent depths H_1 and H_2 is determined from the expression

$$\Delta l = l\left[1 - \frac{(H_v - H_1)(2H_v - H_2)}{(2H_v - H_1)(H_v - H_2)}\right].$$

Summation over the common depth point can be performed only for one interface, and Δl changes with l. Accordingly, when studying closely-spaced interfaces, one can choose a medium summation sampling rate and vary it for different time intervals, i.e., obtain several time cross-sections that jointly provide information on the geologic section being studied. However, on the whole, such limitations are not of essential importance, since frequently only deep interfaces need to be studied. In VSP, marked differences in the kinematics of singly- and multiply-reflected waves begin to be noticed sooner when depth is increased than when the summation interval is extended. For this reason when observations are made at different depths, an effective summation can already be achieved with short traveltime curves, with the maximum separations between the points of observation remaining constant. Such a procedure will be especially effective in the case of multipoint (18–24 points) recording on a vertical profile. The conditions the sampling rate must satisfy on a vertical profile are similar to those in surface seismic exploration, a reliable correlation of various wave types being one of them. These conditions are normally met in studies of deep horizons that are of major interest in exploration.

The procedure for computerized processing of PM–VSP common depth point data differs from that employed in surface seismic exploration in that the numbers of channels and traces used to form the traveltime curves of the common depth point for the interfaces (reflections) being studied must be calculated before the summation i.e. the points of registration must be determined. Tabulated data or charts computed on the basis of information about the velocity model of the Earth may be used for this purpose. The VSP common depth point traveltime equation in the case of a dipping interface is obtained in a similar way by introducing two corrections to the surface common depth point traveltime curve: the wave's transit time from the reception point on the vertical profile to the surface in the direction of the seismic ray and the normal increment (kinematic correction).

Various types of controlled summation may be used in data processing. Consider one of the possible variants of controlled summation using constant velocities, $v_{\mathrm{CDP\ const.}}$ The controlled summation of polarization common depth point data is performed over the fixed X, Y, Z components with a wide set of velocities constant in time. The algorithm can be written as follows

$$S_j(\alpha, \tau_{\mathrm{ost}}) = \sum_{1}^{N} Y_i(\tau_{\mathrm{ost}} + \beta),$$

where S_j is the summation, Y_i are the observed values of the times $i = 1, 2, \ldots, N$ is the number of traces to be summed;

$$\alpha = v_{\mathrm{const.}}, \qquad \beta = \Delta\tau_{ki}(v_{j\ \mathrm{const.}})\,;$$

$\Delta\tau_{ki}$ is computed from Equation (34). The controlled summation seismograms are used to obtain effective velocities, $v_{\mathrm{CDP\ const.}}$, of various wave types (*PP*, *PS*, *SS*) and to determine their nature (multiples, diffracted, etc.). The set of velocities, $v_{\mathrm{CDP\ const.}}$, is selected so as to guarantee uniform variation of the time-lag function over the summation interval. This variation depends chiefly on the separation, as stipulated by the equation

$$\delta\Delta\tau \approx \frac{l^2 \Delta v}{\tau_{\mathrm{ost}} v^3}\,.$$

Normally, $\delta\Delta\tau$ is put at 5 to 10 ms. The summation common depth point time cross-sections thus obtained can be reduced both to a definite fixed level (at the recording depth H) and to the surface. In the latter case, a correction for the reception point equal to the wave's transit time along the vertical from the reception point to the datum line is introduced into τ_{ost} at each depth point.

The application of the PM–VSP common depth point method can guarantee high accuracy and validity of structural mapping, and the most comprehensive studies of the wave field and the physical parameters of the geologic section in the vicinity of the borehole.

The potential of VSP is very important for the study of tectonic faults around and between boreholes. The advantages of VSP in the studies of fracture zones stem from a sharper difference between the waves associated with geologic interfaces and fracture zones, as well as from the opportunity to map high-dip interfaces with the aid of VSP. The experience gained in fracture zones in oil, ore and coal geology confirms these advantages, although the nature of the fracture zones is not the same in each case. For instance, the experience gained in the discrimination of buried fractures in oil geology has demonstrated that waves coming from the fractures are characterized by low and unstable intensities, and are hardly trackable along the profile. In mining geology however the fractures are usually regarded as strong reflecting interfaces, this being the result of a substantial drop in the velocity of the waves in the zone itself as compared with the velocity of the rock containing the ores.

4. Experience Gained in the Application of VSP to the Solution of Structural Problems

It has been demonstrated in numerous observations that the VSP reflection method is very effective as a tool for studying the rocks around the borehole as well as in those cases when the tracking of reflected waves on the surface becomes difficult: (a) in the study of high-dip interfaces; (b) in the presence of high-intensity multiples associated with the upper portion of the geologic section; (c) in the presence of fractured structures that create complex surface conditions. Up to the present, a large amount of experience in the application of the VSP reflection method to the solution of structural problems in oil, ore and coal geology has been gained.

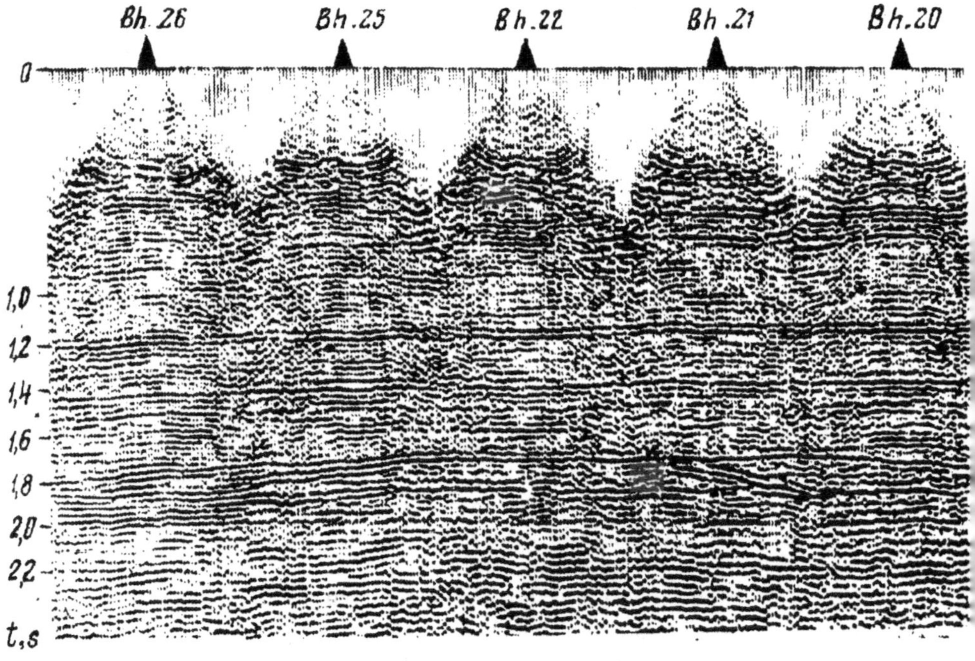

Fig. 108. Studies of the top of a fold under unfavourable surface conditions. VSP-reflection method cross-section (Gagarin area[88]).

In seismic exploration for oil, VSPRM studies have been carried out in various seismic and geologic conditions both in platform and in geosyncline regions. The VSP reflection method has been used to explore rock mass of reef origin in sub-salt Jurassic deposits (Uzbekistan and Turkmenia), chalk deposits beneath basalt (Armenia), in complex structural conditions in the chalk and Jurassic horizons, caused by the presence of wash-outs and angular faults (the Karakum desert), of extensive low foothills with a typically heteregeneous structure (Tadjikistan), in the presence of high-intensity unwanted waves of the multiple type associated with the upper portion of the geologic section (Perm, the Kama river basin, Moscow syneclise) as well as in conditions of highly developed salt-dome tectonics in the presence of an extensive network of tectonic faults, and of individual thinning horizons and complexes (North Caspian and Dnepr–Donetz depressions). Let us illustrate how the VSP reflection method has been applied to the arch of the Gagarin structure in the presence of extremely rugged terrain, with a low-velocity zone of greatly and sharply variable thickness (up to 100 m), and high-intensity unwanted waves (Figure 108). In such conditions it is not possible to track deep waves on surface seismograms, although on the flanks of the structure they are easily trackable. VSP reflection observations made it is possible to track horizons in the arch part of the structure [88].

The effectiveness of the VSP reflection method in the studies of reefs and especially sub-salt deposits has been demonstrated in observations. The VSP reflection method facilitates optimum planning of drilling, thereby reducing the number of wells to be drilled. Figure 109c shows a VSP reflection cross-section of the Zeverdinsk reef.

With an adequate observation system, the VSP reflection method makes it possible not only to construct the cross-section, but also to map the structure. The structural map

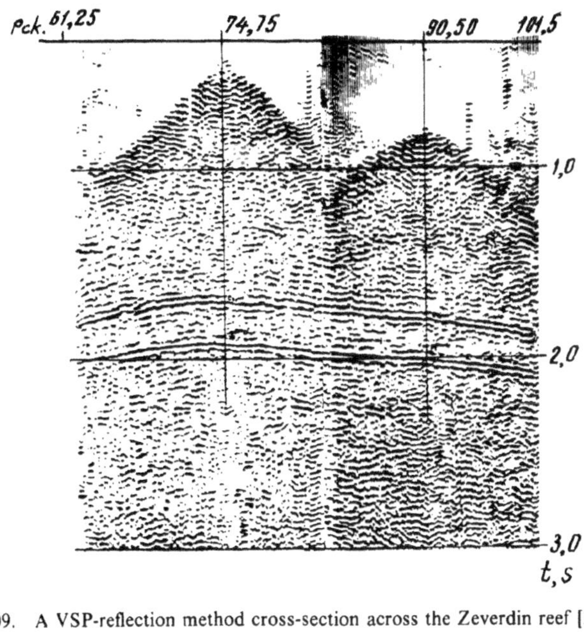

Fig. 109. A VSP-reflection method cross-section across the Zeverdin reef [36].

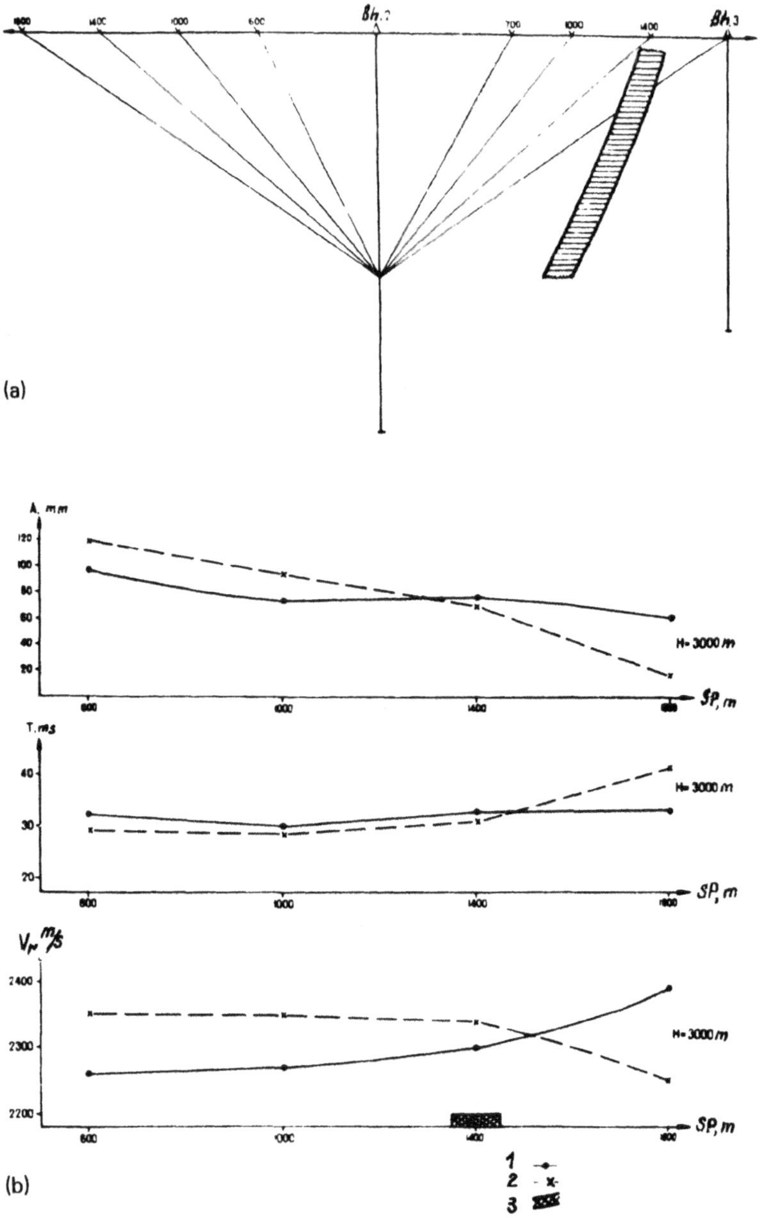
1400
1000
600
700
1000
1400
Bh.3
(a)
A, mm
120
100
80
60
40
20
H=3000 m
SP, m
600
1000
1400
T, ms
40
30
20
H=3000 m
SP, m
600
1000
1400
2400
2300
2200
H=3000 m
SP, m
600
1400
(b)
1
2
3

Fig. 110a-b.

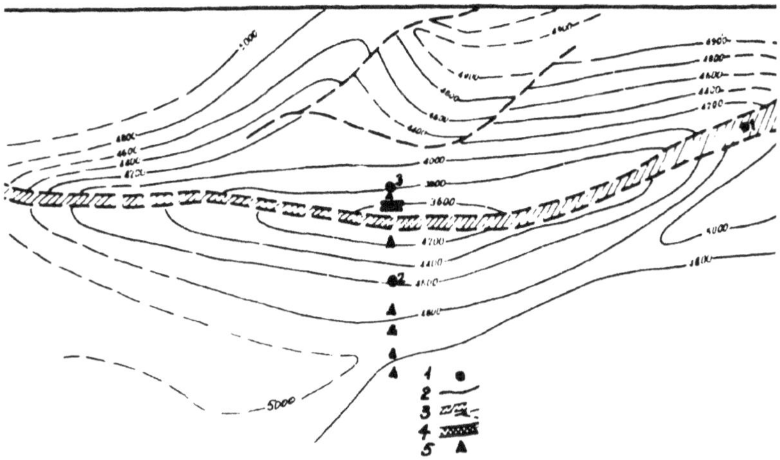

Fig. 110c.

Fig. 110a-c. Determination of the exact position of a fault. Slyusarev area (courtesy A. N. Amirov): (a) arrangement of VSP shotpoints in borehole N2; ray-path schematics and tectonic fault in the cross-section; (b) parameters of the direct wave as observed from southern 1 and northern 2 shotpoints; 3 – position of tectonic fault at the depth of 1000 m; (c) a fragment of structural map along reflecting horizon II: 1 – deep exploration holes; 2 – contours along reflecting horizon; 3 – tectonic fault from surface common-depth-point-observation data; 4 – tectonic fault at the depth 1000 m from VSP data; 5 – VSP shotpoints.

of the Tangikuduk field, shown in Figure 106 [88], may serve as an example of results obtained in such areal observations.

In oil geology, the study of tectonic faults is of great importance. Consider an example of the application of VSP to obtain a more accurate position of a fault.

In borehole N3 (Figure 110) a gas deposit has been found whose dimensions are governed by the fault. The tectonic fault is located between the boreholes N2 and N3 and has been detected by surface observations. It has been reliably traced on the structural map of the reflecting interface at a depth of 4000 m. To obtain a more accurate position of the fault, VSP with a sampling rate of 20 m has been executed in borehole N2. Eight source offset points have been positioned along a linear profile crossing the structure and passing through both boreholes. Four shotpoints were located to the South of borehole N2 and four more between the boreholes (Figure 110)). The ray velocities, the amplitudes and the periods of the direct wave have been determined from the records obtained. By comparing these wave parameters for the southern shotpoints, from which they have passed through the geologic section where no faults are contained, with those of the northern shotpoints, some of which were located in the region of the fault, it has been possible to locate the fault at a depth of about 100 m on the plan and on the cross-seciton.

It is especially noteworthy that the experience gained shows comparatively shallow structural boreholes to be suitable for the VSP reflection method. This not only substantially cuts the costs, but also enables more complete observation systems required for the solution of specific geological problems to be developed.

(a)

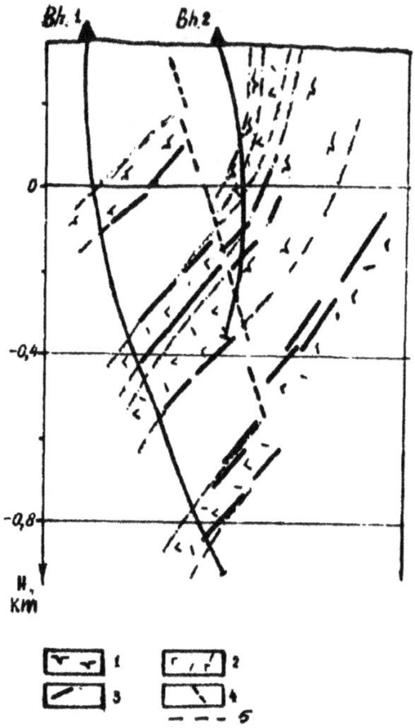

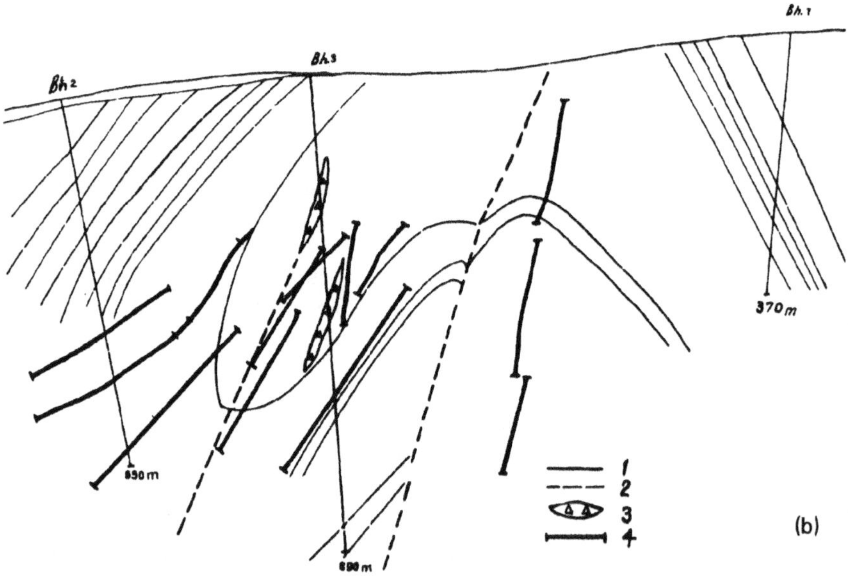

(b)

Fig. 111. Geologic cross-sections constructed from VSP reflection data. (a) Pechenga region [132]: 1 – peridotites, 2 – gabbro-diabases, 3 – rock contacts, 4 – reflecting horizons corresponding to rock contacts, 5 – reflecting horizons corresponding to zones of tectonic faults; (b) Karatan: 1 – lithologic interfaces constructed from drilling data; 2 – tectonic faults; 3 – karst cavities; 4 – reflecting interfaces from PM–VSP data.

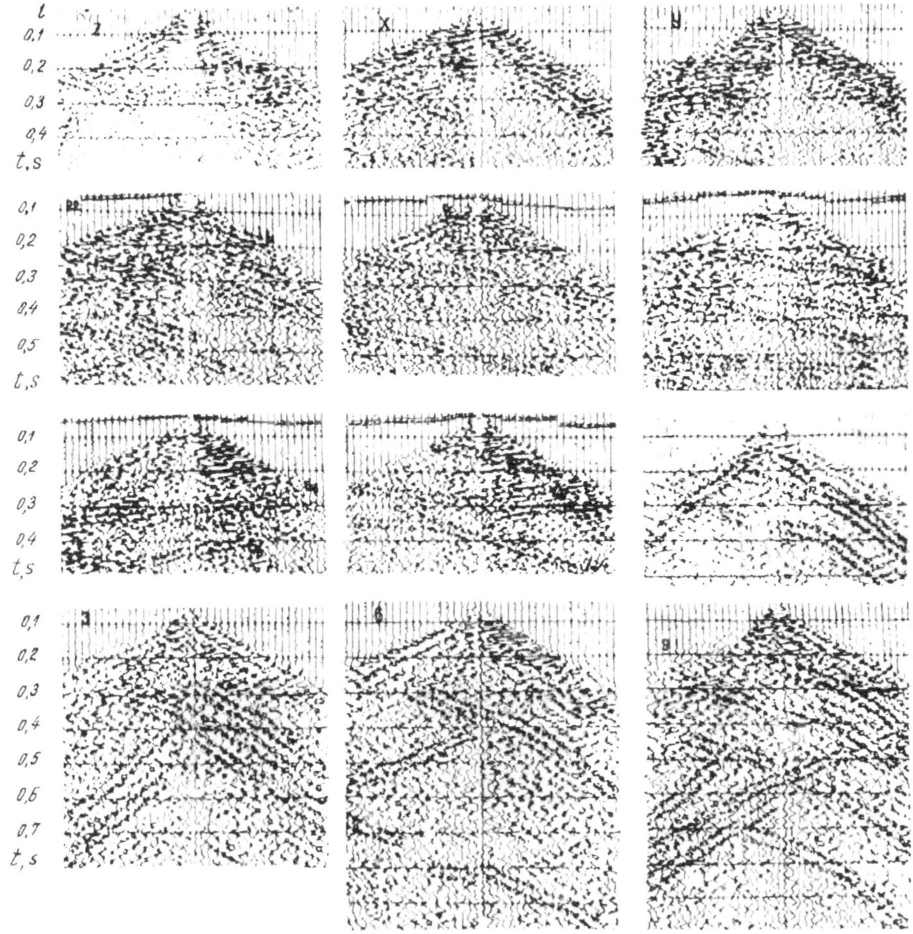

Fig. 112. Seismograms of fixed components in level observations ($H = 120$ m, borehole N3123, mining field Zhairem, courtesy T. G. Chastnaya)

In mining geology, the exploration potential of the VSP reflection method is associated mainly with studies of greatly dislocated media. The method enables mapping of high-dip interfaces from which refracted waves cannot be recorded on the surface. When ore deposits are explored, despite comparatively small separations between the boreholes, a reliable correlation of the geologic section is impossible, because of a complex structure. Accordingly, if one is able to map horizons, the validity of the sections can be appreciably improved. The availability of numerous prospecting holes typical of mining areas favours the development of the necessary observation systems, and is a major reason why the VSP reflection method is widely used in mining geology.

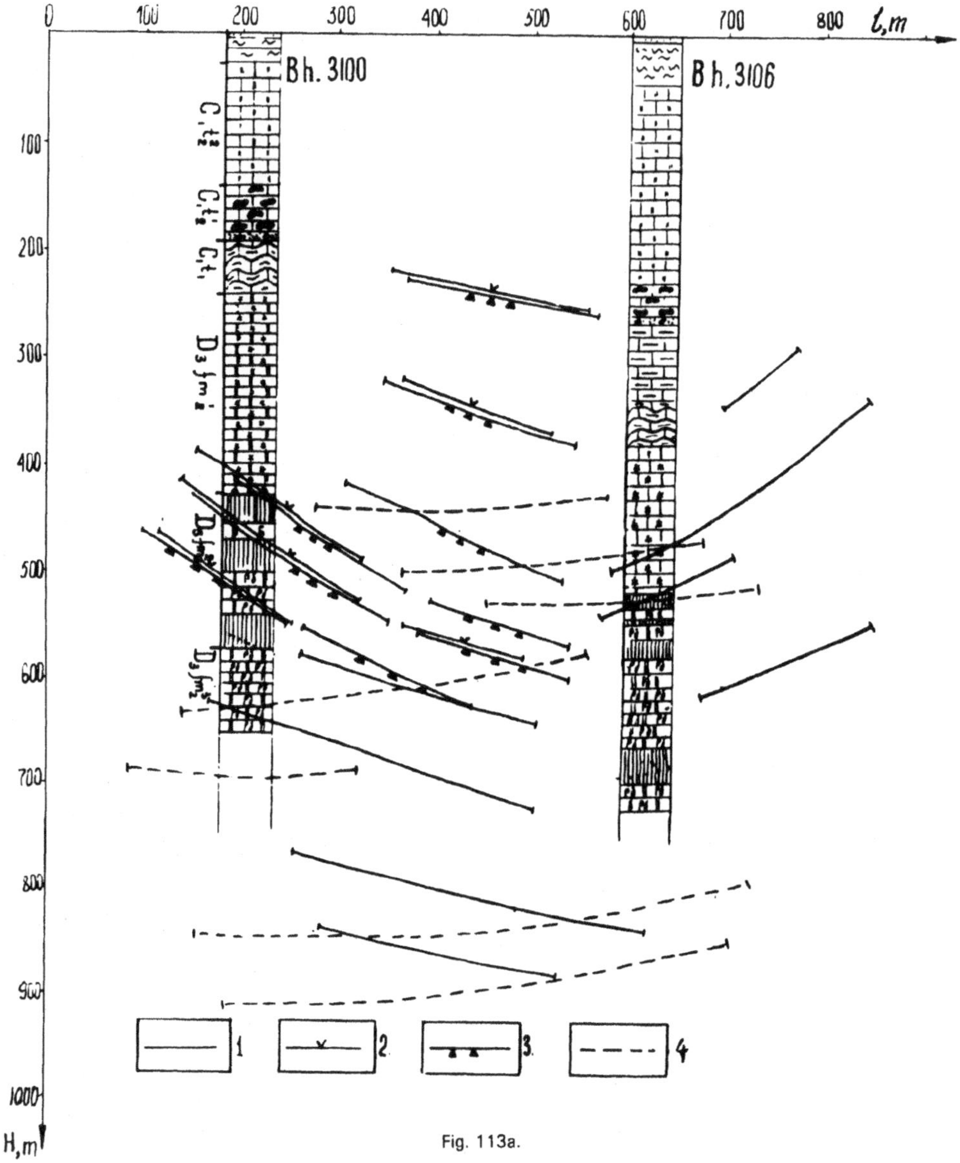

Fig. 113a.

Fig. 113a-b. Examples of cross-sections constructed with the aid of PM–VSP and PM reflection methods. (a) cross-section from surface and level observation data: 1 – interfaces established with *PP*-waves (from up-hole data); 2 – interfaces established with *PP*-waves (VSP); 3 – interfaces established with *PS* and *SS*-waves (VSP); 4 – side waves. (b) cross-section from VSP reflection data (Khibin mountains): 1 – reflecting horizons, 2 – alkaline contacts, 3 – petrologic differences of alkaline rock, 4 – tectonic fault [105].

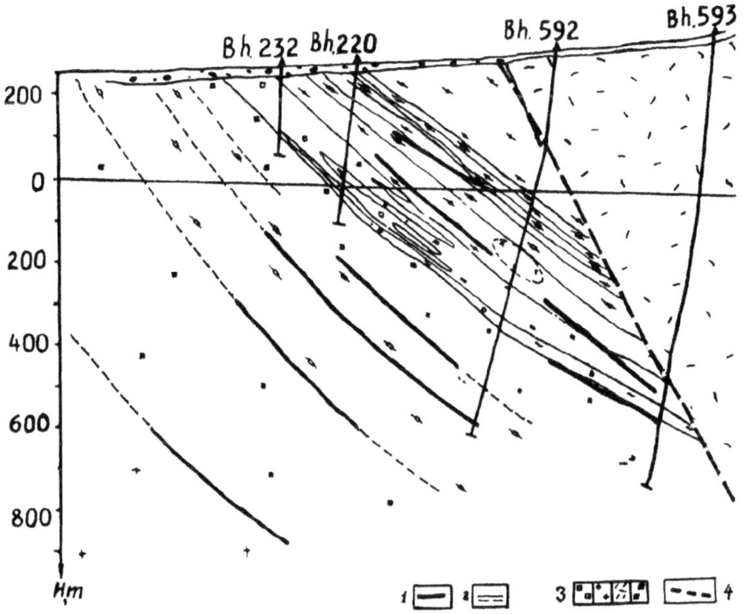

Fig. 113b.

To illustrate the point, we would like to cite a cross-section obtained from VSP reflection data in the Pechenga district (Figure 111a). The VSP reflection method enabled high-dip interfaces corresponding to zones of tectonic faults intersecting a monocline bed to be tracked. The sections for the boreholes NN 1 and 2, separated by a distance of 200 m, could not be compared, because of the presence of the fractures. The VSP reflection cross-section constructed in the borehole N1 explained the sharp difference in the structure as a result of the presence of a fault between the boreholes.

Figure 111b depicts a cross-section constructed from VSP observations in several boreholes in the Zhairem field.

The development of PM–VSP substantially enhances the opportunities for the solution of structural problems. This is the result not only of improvement of the conditions of correlation of longitudinal waves, but also of the possibility of jointly using waves of various types (transverse and converted) together with the polarization of oscillations to gain independent information about the position of interfaces in space. To illustrate the point, we would like to cite the example of the application of the VSP reflection method in the Zhairem ore field. Figure 112 depicts seismograms of fixed components obtained on a fixed level profile. The regular waves are practically untrackable on the Z-component seismograms. At the same time, various wave types propagating in various directions are well identifiable and trackable on seismograms of components corresponding to different orientations in space. An example of a cross-section constructed from combined surface and fixed level observations is presented in Figure 113b. Such a combination substantially improves the prospects for applying seismic exploration in studies of complex greatly dislocated media, and in particular of ore fields. Figure 113a depicts a cross-section constructed in the region of the Khibin alkali mass from VSP reflection observations, utilizing reflected longitudinal, converted and transverse waves.

In the Northern Ural bauxite region, VSP made possible the mapping of the South–Cherepukhov step fault, with a total throw of up to 150 m, which splits the Devonian (Figure 114a) [113]. The traveltime curve of the *PP*-wave reflected from the fault will be seen from the figure to differ appreciably from the traveltime curves of waves coresponding to the bauxite interfaces (*B*) and to the bottom of the Devonian deposits (*D*) (Figure 114b). VSP observations enabled the waves from the fault zone (Figure 114c) and from the interfaces at the bottom of the Devonian deposits (Figure 114d) to be identified. The correlation of waves dislocated by the step fault is disrupted. The waves corresponding to the bauxite horizon are trackable chiefly from the elevated block. In the sunken block, the waves from the bauxite horizon are practically untrackable. In the middle portion of the low-velocity zone, whose total thickness amounts to 70 m, the velocity discontinuity is equal to 2.0 from 2.5 km s^{-1}, the velocities in the carbonate deposits both above and below the fault zone are equal to $v_P = 6.1 \pm 0.37$ km s^{-1}.

In the VSP reflection method the dip (in the range of 0 to 10°) is determined with an error of fractions of a degree. The error increases in the case of interfaces located below the bottom of the borehole, and for angles of dip of ~3° may be as high as 1°. The accuracy can be improved, if velocity heterogeneities of the section are taken into account.

It should be specially emphasized that in ore regions with a typically very complex structure an optimum combination of observations along fixed geophone level and horizontal profiles is of paramount importance. Observation systems should be designed in each particular case according to the geologic structure and to the problem being solved.

Numerous VSP studies of polymetallic fields in Central Kazakhstan, copper-nickel sulphide fields in the Norilsk region, bauxite fields of the Northern Ural, ore fields of the Kola peninsula, etc, have demonstrated the effectiveness of production seismics as an instrument for solving prospecting and exploration problems. This refers to the detailed studies of the rocks surrounding and separating boreholes, including mapping interfaces in ore-containing complexes, establishing the boundaries of ore bodies, detecting and tracing tectonic faults, fragmentation zones, etc. In connection with the current tendency of prospecting for deep ore bodies, a detailed study of structural maps of interfaces at great depths assumes special importance. Production seismics substantially raises the importance of seismic studies in mining geology and improves the effectiveness of geologic exploration in mining regions. It should be pointed out that new opportunities for stratigraphic correlation of seismic horizons enable the validity of the results of surface seismic observations in the studies of complex ore regions to be enhanced.

Vast VSP reflection experience has also been accumulated in coal-fields, which as a rule are distinguished by a complex distribution of coal seams and highly developed tectonics. In coal fields, to increase the seismic detail, high-frequency (100 to 500 Hz) observations with a sampling interval of 2 to 3 m are performed, and this enables tectonic faults with a throw of 3 to 5 m to be detected and traced. Numerous VSP

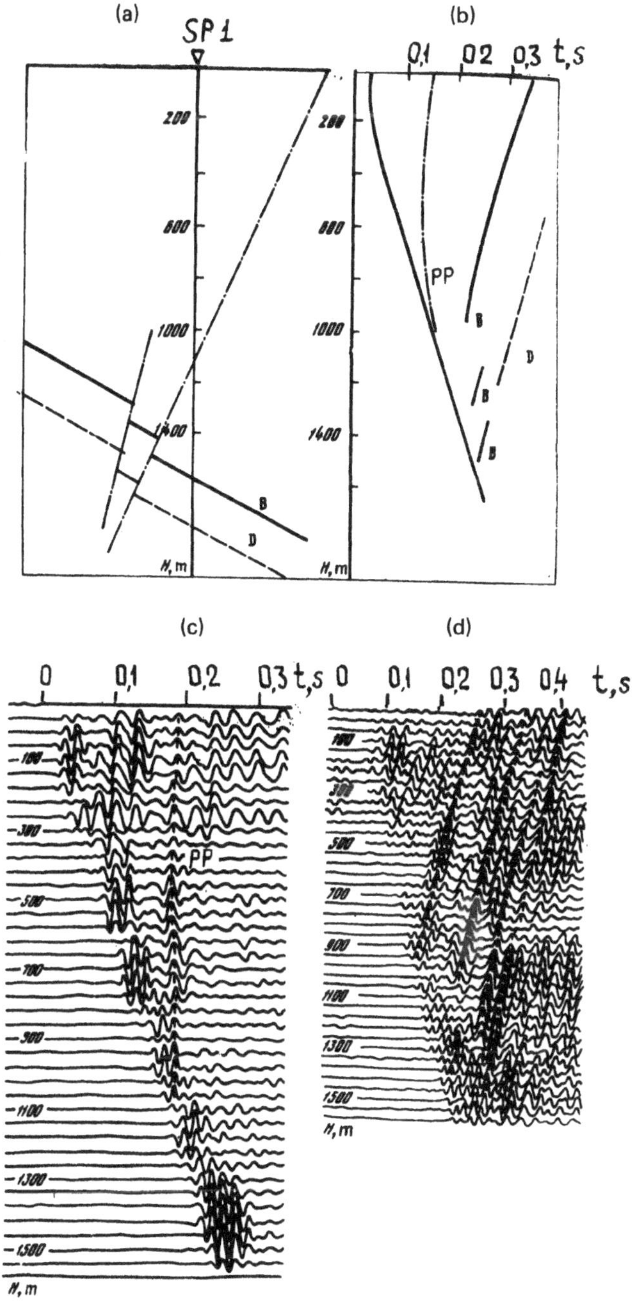

Fig. 114. The effect of a dislocation on the wave field (borehole N5707 [133]): (a) geologic cross-section; (b) theoretical vertical traveltime curves; (c) *PP*-wave reflected from dislocation plane; (d) waves reflected from interfaces at the bottom of Devonian deposits.

(a)

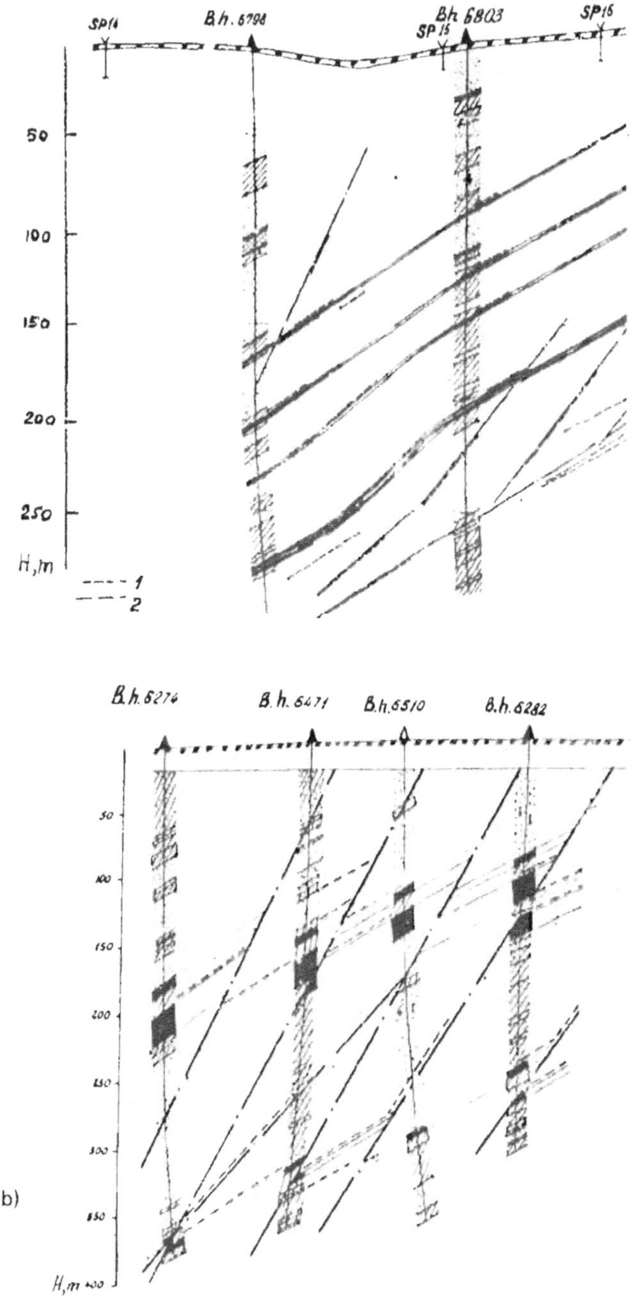

(b)

Fig. 115. Examples of application of the VSP reflection method in coal fields [61]. Seismologic cross-section for mines 'Yuzhnaya' (a) and 'Severnaya' (b), Kuzbass: 1 – reflecting horizons; 2 – lines of tectonic faults.

reflection studies in Kuzbass [60] (fifty coal prospecting holes located in eight mining fields) confirmed the effectiveness of the VSP reflection method as an instrument for detailed studies of the tectonic structure of coal fields, for the detection and mapping of low displacement fractural faults in coal seams, and for the detection of heterogeneities in the rocks around the boreholes in various stratigraphic conditions. High-frequency (50 to 500 Hz) VSP reflection observations enabled longitudinal, converted and multiple waves reflected from all coal seams of 0.5 m thickness or more, at distances up to 200 m from the observation hole, to be detected and tracked. Tectonic faults on VSP reflection seismograms are identified by a discontinuity in correlation, and by a relative displacement of the alignments of waves reflected from the coal seams as well as by waves reflected from the plane of discontinuity.

Tectonic faults with a throw of 2 to 3 m or more, that could not be detected by drilling, were detected in VSP reflection observations [60]. With the existing network of prospecting holes, only tectonic faults with a throw no less than 10 to 15 m could be detected by drilling. Faults with a small throw that have been detected in the coal seams can be traced in the rocks surrounding and separating boreholes. Coal seams can be mapped up to a distance of 200 m in all directions from the borehole. this enables the geologic-sections of neighbouring boreholes to be reliably identified. Figure 115 depicts the results of the application of the VSP reflection method in two mining fields in Kuzbass. The coal seams penetrated by and separating the boreholes will be seen to be intersected by a series of tectonic faults which by mapping with the aid of the VSP reflection method have made possible the identification of the geologic section in neighbouring boreholes.

Some experience in the application of PM–VSP to the detection of coal seams between the boreholes has also been gained. In the Moscow coal field PM–VSP enabled karst manifestations to be detected, and their dimensions to be determined, with the aim of appraising whether the coal fiels were suitable for fully mechanized production technology, as well as the coal seams to be identified from the geologic sections in neighbouring boreholes. Primary and channel waves propagating in lignite seams, which act as wave guides, have been used for such investigations. By observing the intensity of seismic oscillations, it has been possible to detect virgin coal seams characterized by interferential space-polarized pseudo-Rayleigh-type waves of high intensity, with group velocities of 400 to 600 m s^{-1} and frequencies lying in the 60 to 100 Hz band. The maximum intensity is observed when both the receiver and the source are located in the coal seam. Any dislocation of the coal seam, or a reduction in its thickness, results in a drastic reduction in intensity, and a simultaneous rise in the velocity and the frequency of interferential waves. To sum up, the experience gained in VSP reflection studies of the rocks surrounding the borehole in oil-producing, ore- and coal-mining regions is proof of its effectiveness, and justifies recommending VSP for large-scale use in production work. The VSP reflection method is to be extensively combined with surface reflection and common-depth-point method. This facilitates more exhaustive studies of the geologic section and enables the tracking distance of horizons in the presence of high-dip interfaces to be increased. Such combinations enable areal observation systems to be developed.

CHAPTER 11

THE STUDY OF THE PHYSICAL PROPERTIES OF ROCKS SURROUNDING THE BOREHOLE

Of great importance in the study of the region around the borehole, in addition to mapping structures, is the study of the physical properties of rock which is essential for determining the geologic section, and particularly hydrocarbon and solid mineral deposits, as well as establishing physical foundations for assessing the section from surface observations. To this end, it is first of all necessary to study the effect of various heterogeneities on the parameters of the wave field. Such effects should be capable of making detectable any deposits of oil and gas penetrated by the borehole, which were not detected in the testing of the borehole (such instances have occured), and should also help in outlining the deposits, in studying their extent and geometry as well as in controlling them in the process of exploitation.

The basic factors governing the application of seismic exploration at the exploration and exploitation stages of a field, are associated with studies of the physical properties of rock, and there is no doubt that in the coming years this trend will rapidly develop.

The determination of the physical properties of rock is to a large extent founded on the use of the dynamic properties of waves in addition to their kinematic properties. This sets special standards for the quality of field data that substantially affects the potential and the accuracy of such determinations. The conditions that field data will have to satisfy depend on the problems to be solved. Here we can discuss some general requirements only.

1. Conditions the Initial Data Must Satisfy

For dynamic processing one needs VSP data that provides stable information about the characteristics of the shape and intensity of the waves, primarily of the direct wave. This is especially important in the study of absorption.

The following may serve as criteria for the quality of the data:

– the facility to obtain records in a frequency band and with a dynamic range free whenever possible of various regular and irregular unwanted waves;

– the reproducibility of the record (of both amplitude and frequency) when repeated observations are made at a point;

– the reproducibility of conditions governing the variation of amplitudes and of spectra, when repeated observations are made in the same borehole (along the vertical profile) of waves excited from the same shotpoint;

– the validity of the data available for statistical processing.

The quality of the data is influenced by a number of factors associated both with the conditions of excitation and of reception.

Among the conditions of excitation, constancy and control are the chief factors. This is true in the first instance of excitation by explosives. With non-explosive excitation, it is easier to guarantee constancy, because the earth is less affected in this case. With explosives, it is possible to achieve the required reproducibility for a great number of shots only if the excitation conditions are kept strictly constant. In particular, by exploding comparatively small charges at the same depth each time, and aligned in the borehole, it is possible to keep the excitation conditions constant and to carry out numerous explosions.

In addition to the constancy of such traditional parameters as the intensity and the spectral composition, the constancy of source directivity becomes a factor of essential importance for the solution of certain problems. In such cases directional excitation sources are usually employed.

To eliminate the effect of changing excitation conditions, particularly when absorption is to be determined, the pulses of the direct wave observed at great depths are frequently normalized with the aid of records of the same wave obtained at a comparatively shallow depth relatively close to the source. Such a normalization does not always improve the quality of the data, because the effect of various source excitation conditions on the direct wave near the source may be different from that at a great depth (as has been pointed out earlier in Chapter 5). One can prevent this by placing the monitor geophone at a depth where the shape of the direct wave has already stabilized. Sometimes better results can be obtained by using records from monitor geophones to reject low-quality data.

Reception conditions should guarantee undistorted recording in the specified dynamic range and frequency band at every point. To this end, the instrumentation should be calibrated, and the effects of the temperature and pressure, which vary along the vertical profile, should be eliminated. Recording should be performed in the frequency band corresponding to the problem being solved. For instance, to study absorption and its frequency dependence, a frequency band approximately from 5 to 500 Hz and a dynamic range up to 100 dB should be available.

Downhole measurement technology should exclude various mechanical parasitic resonances associated with the design of the borehole tool itself, its clamping to the borehole wall, and the cementation between the casing strings and between the casing and the earth, which may change from point to point along the vertical profile. The extension of the frequency band of the waves being studied necessitates a substantial increase in the force with which the borehole tool is clamped to the borehole wall. The possible effect of the cable remaining on the drum should also not be overlooked. The inductive reactance of the cable changes with its length, and this can affect the frequency response of the recording channel and thereby distort the results of the measurement. Such an effect can be assessed by recording calibrated electric pulses. All these effects merit special attention, because in several instances attempts have been made to determine physical parameters with the aid of wave dynamics, using data known to be low-grade (e.g., from records of a sliding-clamp or bowspring tool, or even worse, from a tool suspended by the cable, etc.), with completely erroneous results being the outcome.

When solving certain problems, e.g., when studying the mechanism of absorption, it is advisable to make special checks of the coupling conditions at every depth point, e.g., clamping reliability and casing cementation quality.

The observation system should require observations not at disconnected points, but at ones allowing a continuous correlated tracking of a wave, yielding an adequate amount of data for statistical processing. Specifically, it is in many cases advisable to carry out independent repeated observations.

2. Absorption of Seismic Energy

As the seismic pulse propagates, its amplitude decreases, and its shape changes. In homogeneous media the decline of the amplitudes is due to geometric spreading of the wave front, as well as to the absorption of energy in a medium that is not ideally elastic resulting in an irreversible transformation of the energy of elastic vibrations into heat. However, the real earth is normally inhomogeneous. In such an earth, other factors that appreciably influence the seismic pulse, in addition to geometric spreading and absorption, are reflection and refraction at the interfaces of layered media as well as scattering of the energy by small heterogeneities. In thin-layered media, part of the energy lost in reflection from the interfaces is returned in the form of multiple reflected waves propagating in the same direction. Accordingly, the intensity of a wave depends essentially on the conditions of superposition of it and the multiples, which could both enhance or weaken it. In individual cases, the intensity of such combined waves may exceed that of the transmitted wave.

Among the factors enumerated above that affect the seismic pulse, absorption is of special interest. Research into absorption is of great practical and theoretical importance.

The practical importance follows from the fact that the absorption coefficient (α) is independent of the propagation velocity of seismic waves and, like the velocity, is a parameter that characterizes the properties of a rock such as lithology, porosity, density, and the presence of fractures and inclusions etc. Recently, absorption began to attract greater attention after it had been established that hydrocarbon deposits exercise a strong influence on the intensity of the waves, after discovering waves with anomalous amplitudes on some seismograms. At present, the absorption coefficient is one of the chief parameters in determining the geologic section. Besides, absorption enables the intensities of various wave types to be evaluated during the course of propagation, this being necessary for the interpretation of complex wave fields observed both on the surface and down-hole.

The theoretical importance is associated with studies of the absorption mechanism itself. The two trends (the applied and theoretical) require data of fundamentally different precision.

Studies of absorption in VSP

Absorption has been studied both in the laboratory and in natural conditions, with the aid of surface and downhole observaitons. In the laboratory, the absorptive properties of hard rocks have been studied on samples of natural and synthetic materials. However, major importance is attached to the experiments in actual media, primarily in conditions of their natural occurence. The volume of such work grows progressively from year to year.

Several methods have been proposed for the determination of absorption from surface observations based on the use of different wave types (reflected and refracted). Such methods are well known, and we shall not dwell on them.

The advantages of determining absorption from down-hole observations stem from the fact that down-hole records are to a considerable degree free from the distorting influence of the uper portion of the geologic section and the low-velocity zone, as well as from the ability to study the changes in the spectra of the same wave at various points along the vertical profile. The direct waves recorded against a quiet background along the line of the vertical profile are of special interest from this point of view. It should be pointed out that down-hole observations to determine absorption were undertaken long before VSP was born. As a result of such observations, Ricker [92] demonstrated in 1958 that the absorption in homogeneous shales had some correspondence to absorption in an elastic viscous medium. It is true, these results have not been substantiated by any subsequent more accurate observations or by spectral analysis in the course of data processing [96]. Jolly [6] and Levin and Lynn [7] have studied the spectra of the direct wave pulses, but because of changing conditions of source excitaiton and coupling between the geophones and the borehole walls, were unable to determine the absorption. Their absorption studies in the upper portion of the section gave in certain cases, values that were too low, due to, on the one hand, imperfect experimental procedure, and on the other, the complexity and heterogeneity of the earth.

With the development of VSP and the general improvement in procedure and instrumentation of seismic observations in deep boreholes, it became possible to develop methods of studying absorption.

The dependence of the absorption factor on frequency $\alpha(f)$ can be determined in VSP with both vertical and oblique ray paths. Technically, the first method is simpler, however, when probes with a small number of channels are used, it is difficult to keep the source excitation conditions sufficiently constant. By using multichannel equipment, it has been generally possible to surmount those difficulties and improve the accuracy of the results. Observations with oblique ray paths require simultaneous recording in two boreholes, and for this reason they are carried out at shallow depths. Since the vibrations are excited by a single explosion, it is in principle possible to obtain more accurate data. However, to this end one needs to be well informed of the geometry of the rays, and this is not always possible in practice, because such knowledge requires a very detailed study of the velocity profile. For this reason the majority of observations aimed at studying absorption are carried out with vertical ray paths.

ON THE SHAPE AND STRUCTURE OF THE DIRECT WAVE PULSE

The most reliable data on absorption can be obtained from studies of the spectra of the direct wave. When measuring absorption, one should take account of the strong influence on a direct wave of the upper portion of the geologic section, which manifests itself primarily by addition of longitudinal multiple reflected waves on the first arrival. Such a superposition usually produces a stable interference pattern changing little with depth and relatively easily studied to be either eliminated or taken into account. It is much more difficult to take into account the effect on the direct wave of scattered energy associated with various heterogeneities of the geologic section as well as the energy of secondary body waves formed as the result of transformation of surface waves. In such cases, it is better to carry out observations in conditions of a comparatively uniform upper portion of the cross-section. Moreover, when working with explosions, it is desirable to lower the charge to a depth at which the ghosts cease to interfere with the direct wave. In all cases the structure of the upper portion of the section should be thoroughly studied in advance.

The majority of the factors listed above are well-known and usually more or less taken account of in VSP observations. The development of PM–VSP created new opportunities mainly for the study of the structure of pulses, by means of the polar correlation and controlled directional reception (CDR–I). It has been demonstrated as the result of numerous PM–VSP studies in a great variety of seismologic condtions (see Chapter 5), that even the simplest and the shortest direct wave pulses on the record of the Z-component may consist of several events. In addition to pulses polarized in the same sense as the longitudinal wave, pulses polarized like the transverse wave can also be present. Phase shifts between the pulses may amount to fractions of a period. The shape of the direct wave is determined by the conditions of superposition of the pulses, such conditions being liable to change. Direct wave pulses recorded in various seismologic conditions are depicted in Figures 30 to 33, 35.

In order to study the structure of the pulse and to eliminate the effect on it of various factors, the recording of only vertical oscillation components should be rejected in favour of three-component PM–VSP observations, and the full vector should be studied, even when observations are carried out along a vertical profile.

The methodological and technical difficulties involved in the measurement of pure absorption make it expedient to subdivide the problem of studying the laws governing the decline in amplitudes into two problems: the determination of the total decline of amplitudes (effective absorption) and the determination of pure (inelastic) frequency-dependent absorption. Unfortunately, not always a clear distinction is drawn between the data being obtained. This might be the reason why there is still no reliable information on the frequency dependence of the coefficient of absorption, the data often being contradictory. We whall consider here only some aspects of the procedure for determining absorption.

In principle, to determine the absorption coefficient as a function of frequency for various rock types constituting the geologic section, one should compare the spectra of the same wave in a wide frequency band at several points of a homogeneous bed. This may be comparatively easily accomplished in a homogeneous bed of great thickness, in which the decline of the amplitudes with distance is the result solely of absorption of energy and of spreading of the wave front. In inhomogeneous media of small thickness, it is very difficult to separate pure absorption from the above mentioned factors affecting the seismic pulse. In addition to technical problems, fundamental difficulties stem from the necessity to eliminate the effects on the total absorption of geometric spreading and the passage through thin-layered media, as well as scattering by small heterogeneities. To evaluate such losses, the usual procedure is to apply digital modelling in which synthetic seismograms are computed from sonic logging data. The wave front spreading function can be computed also from interval velocities. The following factors exercise a considerable influence on the accuracy of determination of spreading: the accuracy with which the velocity has been determined, especially at relatively shallow depths, where geometric spreading greatly affects the attenuation of energy, the velocity anisotropy and the interface dip.

By comparing field and synthetic VSP seismograms, it is possible to establish the correspondence between the real earth and the theoretical model. Despite improvements in methodology, the experimental results of the measurement of inelastic absorption are very contradictory and frequently agree neither with the wave fields observed nor with theoretical computations. One of the possible causes of this lack of succes could be the failure to take account of the complex structure of the direct wave pulse, whose mechanism of formation in layered media has been studied inadequately. In addition to multiple reflected waves formed in thin-layered media, a strong influence is exercised on the direct wave by the upper portion of the section. In practice, this influence has also not been studied at all.

EFFECTIVE ATTENUATION

The experimental values obtained in the real earth characterize not abosorption, but total decline of the amplitudes and can be regarded as 'effective attenuation'. At the same time, for certain practical purposes in the majority of cases (e.g., for studying the wave field, constructing models of the real earth, determining the intensities of various wave types and especially multiple reflected waves, selecting optimum recording parameters) it is enough to know only the effective values of attenuation that characterize the total decline of the amplitudes, and the change in the shape of the record, corresponding to the filtering properties of the geologic section. The coefficient of effective attenuation usually exceeds that of pure absorption, since it also includes energy lost in the passage through thin layers and scattered by small heterogeneities [134]. In the conditions of a homogeneous bed, the amplitudes should decline monotonously. In the real earth, there is a spread in the amplitudes associated with the variation in the rock properties, which substantially exceed the errors of measurement in the case of high-quality observation data. Usually the sharpest changes in the smooth course of the amplitude curve are associated with the superposition of waves in sections of the profile, which intersect the interfaces.

Numerous studies of effective attenuation have been carried out in conjunction with the work on the effective seismic model method [16–18]. They have demonstated that the nature of the decline of the amplitudes greatly depends on the structure of the geologic section and is not the same over different intervals.

By way of an example of the procedure of determining effective attenuation, consider the observation results in the Anastasiyevskaya–Troitzkaya area obtained by the 'Krasnodarneftegeofizika' trust (Figure 116). The observations were carried out on an in-line vertical profile with the aid of a three-point probe containing Z-component geophones with each probe clamped independently. By lowering the charge to a depth of 68 m, it was possible to separate the ghost reflection by 60 ms, and to record the direct wave outside the zone of interference. The use of small charges (five detonators) enabled a wide-band pulse to be produced, and the excitation conditions to be kept constant during observations along the entire vertical profile (Figure 116a).

The records of the down-hole monitor geophone and the instruments of the three-point probe have been used to compute amplitude spectra of the direct wave in the time interval during which practically its entire energy arrives. The major portion of the energy of the direct wave measured by the down-hole instruments is concentrated in the frequency band from 20 to 120 Hz. Within this band, the spectrum is respresented by a smooth curve. The amplitude variation curves in the depth interval 400 to 1570 m have been constructed separately for each instrument of the probe, thus enabling the effects

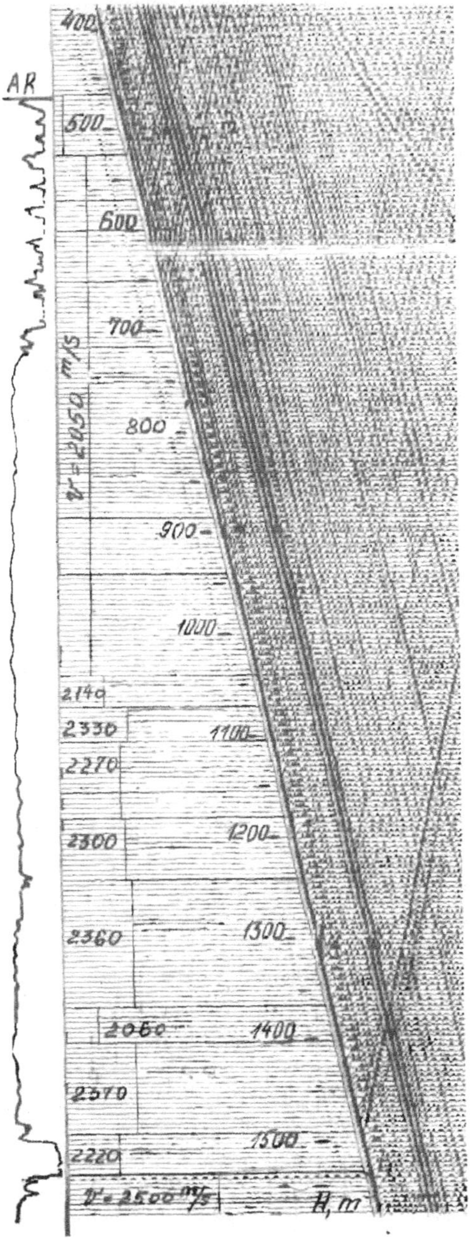

Fig. 116a.

Fig. 116a-c. Determination of effective attenuation in borehole N693, Anastasiyevsko–Troitzkaya, l = 100 m, charge of 5 detonators, H = 68 m: (a) seismogram of Z-component, geologic column and production geophysical data; (b) plots of variation of amplitudes of various spectral components for two geophones (1, 2) as well as of amplitudes of geophone 1 uncorrected from control geophone (3) data; (c) plot of effective attenuation for three geophones (1, 2, 3) for 40, 60, 80, 100 Hz components.

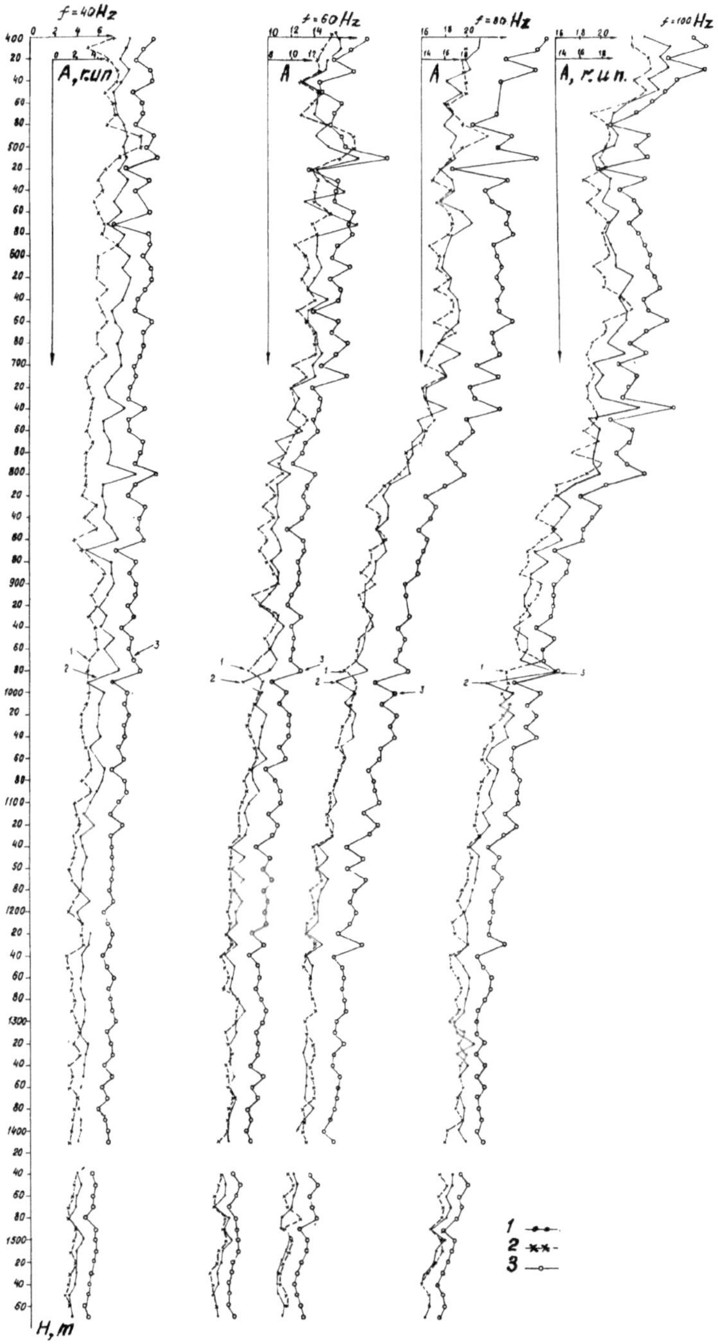

Fig. 116b.

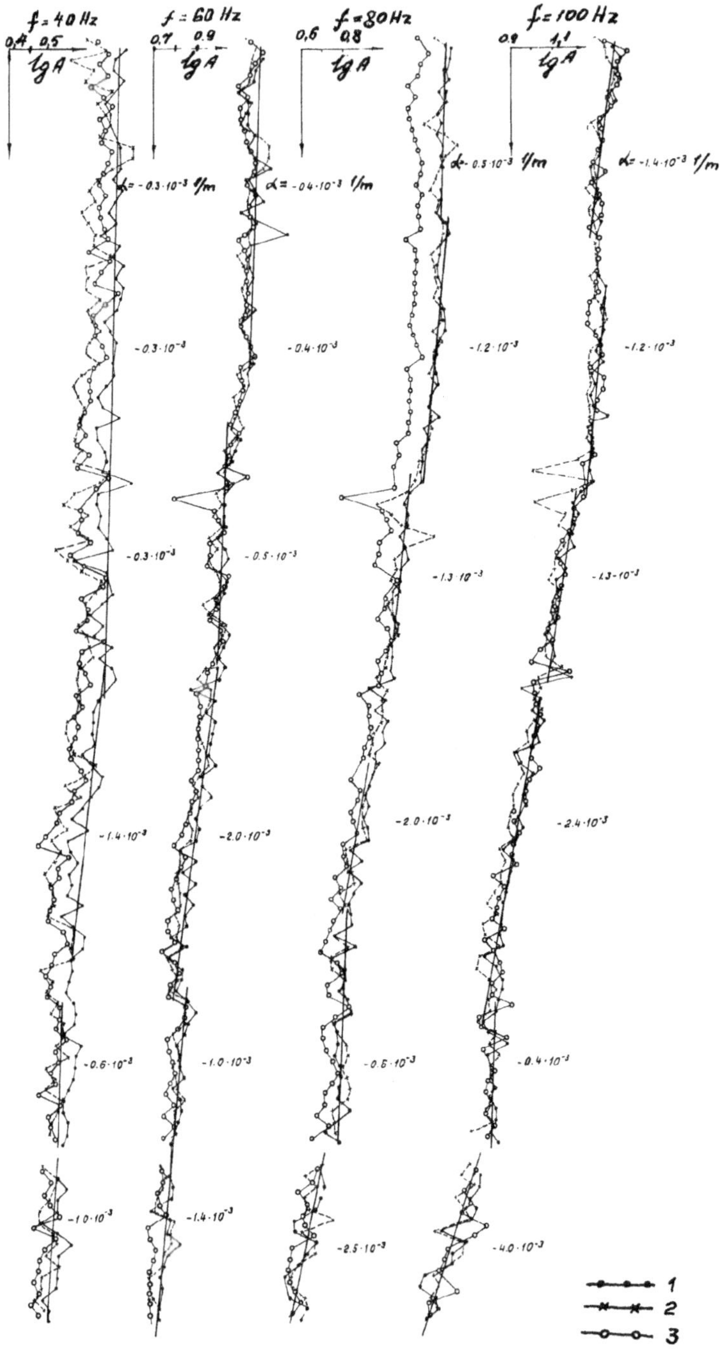
f = 40 Hz
0,4 0,5
lg A
f = 60 Hz
0,7 0,9
lg A
f = 80 Hz
0,6 0,8
lg A
f = 100 Hz
0,9 1,1
lg A
α = -0.3·10⁻³ 1/m
α = -0.4·10⁻³ 1/m
α = -0.5·10⁻³ 1/m
α = -1.4·10⁻³ 1/m
-0.3·10⁻³
-0.4·10⁻³
-1.2·10⁻³
-1.2·10⁻³
-0.3·10⁻³
-0.5·10⁻³
-1.3·10⁻³
-1.3·10⁻³
-1.4·10⁻³
-2.0·10⁻³
-2.0·10⁻³
-2.4·10⁻³
-0.6·10⁻³
-1.0·10⁻³
-0.6·10⁻³
-0.4·10⁻³
-1.0·10⁻³
-1.4·10⁻³
-2.5·10⁻³
-4.0·10⁻³
1
2
3

Fig. 116c.

TABLE X

Borehole 693

Depth interval (m)	Effective attenuation (1/m)	Rock
400–800	1.3×10^{-3}	Water-saturated sand-clay deposits
800–1020	1.6×10^{-3}	Clay bed
1020–1200	2.0×10^{-3}	Clays with sandstone inclusions
1200–1400	0.5×10^{-3}	Pure clays
1450–1550	3.3×10^{-3}	Gas-saturated sandstone

Borehole 1641

Depth interval (m)	Effective attenuation (1/m)	Rock
700–970	0.5×10^{-3}	Clay bed
970–1150	3.5×10^{-3}	Clays with small sandstone content
1200–1450	1.5×10^{-3}	Pontic clays
1450–1550	6.7×10^{-3}	Gas-saturated portion of the sandstone
1600–1750	1.0×10^{-3}	Clays with sandstone interlayers

of a possible inequality of the probe's channels to be reduced. The curves for some fixed frequency components (Figure 116b) show the rate of decline of the amplitudes to grow with frequency. In addition, by comparing the amplitude curves of the first instrument corrected for changing source excitation conditions (1) with the uncorrected (2), one can see how effective such correction is.

In individual depth intervals where the amplitudes decline monotonously, the effective absorption (attenuation) coefficient was determined from the formula

$$\bar{\alpha} = \frac{\lg A_{i2} - \lg A_{i1}}{\Delta H \lg e},$$

where A_{i2} and A_{i1} are the amplitudes at the lower and upper interfaces of a layer ΔH thick. The quantity $\bar{\alpha}$ is determined from the slope of the curve showing the variation of the logarithm of amplitude with depth after geometric spreading has been approximately taken into account. Variations of the logarithm of amplitude for each frequency component in individual sections have been averaged with the least-square method (Figure 116b). The spread in the amplitudes observed can be explained by the fact that variations in source excitation conditions have not been adequately taken into account, as well as by the interference of waves in a thin-layered medium.

Several zones with different attenuation coefficients can be identified on $\lg A = f(H)$ plots constructed separately for the three instruments of the probe for each frequency component 40, 60, 80, and 100 Hz. The values of $\bar{\alpha}$ thus determined are tabulated in Table X. The values obtained generally correspond to the attenuation of saturated clay and sandstone. Transition from sandstones to clays affects the attenuation coefficient.

TABLE XI

Area	Observation method	Depth interval (m)	$\bar{\alpha}$, 1/m from direct wave	$\bar{\alpha}'$, (1/m from reflected wave)	Number of measure-ments
N. Kamashi	UhS*	0–10	1.0×10^{-1}	–	1
Pamuk	VCD**	0–20	–	$(1.7 \pm 0.3) \times 10^{-2}$	3
N. Kamashi	UhS	15–60	$(2.7 \pm 0.8) \times 10^{-2}$	–	5
Pamuk	VCD	20–70	–	$(2.5 \pm 1.5)10^{-3}$	3
Gumbulak	VCD	20–90	2	3×10^{-3}	1
Pamuk	VSP	500–1500	$(9.2 \pm 3.9)10^{-4}$	$(3.3 \pm 1.3)10^{-3}$	9, 11
N. Kamashi	VSP	700–1400	$(1.6 \pm 0.2)10^{-3}$	–	2
N. Kamashi	VSP	900–1400	–	$(1.2 \pm 0.5)10^{-3}$	2
Urtabulak	VSP	500–1500	8×10^{-4}	6×10^{-4}	2
Pamuk	VSP	1500–3000	$(6.7 \pm 2.3)10^{-4}$	$(3.7 \pm 1.0)10^{-4}$	8, 7
N. Kamashi	VSP	1500–2200	$(7.2 \pm 1.7)10^{-4}$	$(4.5 \pm 1.2)10^{-4}$	4, 4
Urtabulak	VSP	1500–3000	5×10^{-4}	4×10^{-4}	2

* Uphole survey (UhS).
** Variation of charge depth (VCD).

A point deserving special attention is the anomalously high attenuation measured in both boreholes in the depth range 1450–1550 m that corresponds to a gas-saturated sandstone. Here the value of $\bar{\alpha}$ exceeds from five to seven times the corresponding value in the overlying bed and five times the values of $\bar{\alpha}$ in the underlying rock. Within this layer, the spectrum narrows sharply on account of the attenuation of high frequencies. In contrast to other invervals of the section, the frequency dependence of the attenuation coefficient is, in this case, non-linear. The most intense wave is that reflected from the top of a gas-saturated layer, as can be easily seen from the seismogram. This layer can also be detected with the aid of polarizaton (see Chapter 12).

The effective attenuation coefficient can be determined from VSP data not only with the aid of the direct wave, but also with the aid of the reflected and even multiple reflected waves. Since a reflected wave, in contrast to the direct downward wave, is an upgoing wave, the velocity will have a reverse effect on its nature, and the decline of the amplitudes along the reflected ray path will not be the same as in the case of a downgoing wave. Table XI depicts the values of the effective attenuation determined with the aid of the direct ($\bar{\alpha}$) and the reflected ($\bar{\alpha}'$) wave [17].

The values of $\bar{\alpha}$ at the larger depths (from 500 to 3000 m) will be seen from the table to exceed those of $\bar{\alpha}'$ by 1.5 to 2 times.

Experiment shows the values of $\bar{\alpha}$ determined from the decline of amplitudes along an in-line vertical profile for the direct ($\bar{\alpha}$), the singly-reflected ($\bar{\alpha}'$) and the doubly-reflected ($\bar{\alpha}''$) waves to differ most in the upper portion of the section, which is usually represented by terrigeneous deposits. In recent years methods have been devised of determining $\bar{\alpha}''$ from amplitude ratios of the doubly- and singly-reflected waves measured near the points of intersection of their vertical traveltime curves [18]. The values of $\bar{\alpha}$, $\bar{\alpha}'$, and $\bar{\alpha}''$ give an idea of the geologic section and of the principal factors affecting the observed relationship between the intensities of the singly- and multiply-reflected waves.

A change in the velocity is always accompanied by a change in attenuation, but it has not been possible to establish a unique relationship between them. Sectioning of the section according to attenuation is of essential importance, especially in media with weak velocity differentiation.

Effects of absorption

The chief effects of absorption are to substantially increase the duration of the pulse and to reduce the amplitudes of the extrema, this reduction being different for different extrema. With increasing distance, the pulse becomes less clearly defined, assumes a smoother shape (the attenuation drops). The amplitude extrema are displaced in the direction of longer times, resulting in a change of the ratio of the areas of the pulse either side of the ordinate of its middle point. Should the area of the pulse be divided into two parts S_1 and S_2 by the ordinate of the middle point, the ratio of the areas S_1/S_2 will diminish with increasing absorption, i.e., absorption brings about a disproportional change in the areas of both halves. On actual seismograms, the anomalies of shape may be obscured by interference.

Experimental data obtained in surface reflection observatons indicate a regular increase in interval velocities and a decline in α. The ratio S_1/S_2 decreases slowly with depth. The quantitative characteristics of the effects of earth filtering on the shape of the pulse can be obtained from the variation of the dominant frequencies, the absorption coefficient with time, the average amplitudes and the ratio S_1/S_2 [97]. The ratio S_1/S_2 depends on the frequency composition of the oscillations. In the initial portion of the record, in the case of a narrow band, S_1/S_2 is smaller.

Determination of the reflecting and absorbing properties for transverse waves

In recent years, the extent of VSP studies of the dynamic characteristics of the coefficients of absorption and reflection of transverse waves has been growing, prompted by the development of the polarization method and others based on the combined use of longitudinal and transverse waves. Such studies are of special importance for the problem of direct prospecting for oil and gas.

Not only the first direct transverse waves are used to determine absorption parameters, but downgoing converted transmitted and reflected SV-waves excited as the result of a conventional non-directional explosion [119] as well.

The development of PM–VSP creates great opportunities in this respect. However, given certain assumptions, i.e., of a horizontal layered medium and an in-line profile, it is sufficient to study only one horizontal component of the SV-wave in the plane of the rays. The method used for this purpose is similar to that used for P-waves: α_s is determined from the amplitude plots of the downgoing converted wave $A_{PS}(H)$, and from the dependence of the ratio of the incident and the reflected waves $A_{PPS}/(A_{SS}(H)$ on the depth

$$\alpha_{si} = \frac{1}{\Delta H_i}\left(\ln\frac{A_1}{A_i} + \ln\frac{L_1}{L_i}\right),$$

where i is the number of the observation point within the processing interval; ΔH_i is the distance between the first and the ith points; A_1/A_i is the amplitude ratio and L_1/L_i is the relative geometric spread between the first and the ith points computed with the aid of formulae from the method of rays. α_{si} is determined from the ratio $A_{PSS}/A_{SS}(H)$ for each fixed depth with the aid of the formula

$$\alpha_{si} = \frac{1}{2\Delta H_i}\left[\ln K_S + \ln \frac{A_{PS}}{A_{PSS}} + \ln \frac{L_{PS}}{L_{PSS}}\right],$$

where ΔH_i is the distance from the observation point to the reflecting interface where the PSS wave has been formed, A_{PS}/A_{PSS} is the ratio of the amplitudes at the depth H_i; L_{PS}/L_{PSS} is the relative geometric spreading of the PS and PSS wave-fronts computed down to the depth H_i. The quantity K_S has been read from the A_{PSS}/A_{PS} plot in the immediate vicinity of the reflecting interface. It is required for the computation of α_{si} and is interesting by itself as an estimate of the coefficient of reflection.

This method has been tried in the Moscow syneclese, The values of the coefficients of absortpion of waves are presented in Table IX [133]. It follows from the table that the coefficients of absorption determined from downgoing waves exceed those found from A_{PSS}/A_{SS} by on the average 1.5 times. Such a discrepancy may be attributed to the fact that the transmission coefficient of thin layers not discovered from well velocity surveys data, has not been adequately taken into account.

The effective amplitude coefficients of transverse waves α_S are greater than the corresponding values for the longitudinal. The values of α_P/α_S vary in the range 0.24 to 0.27, i.e., the absorption of S-waves in the real earth is 3 to 4 times greater than that of the P-waves.

The coefficient of reflection of the converted PSS waves is, on average, 2 to 3 times greater than that of the longitudinal waves. Thus, for $H = 1480$ m for the top of the middle Ordovician $K_S = 0.66$ and $K_P = 0.27$. This, in particular, is a possible explanation why, despite great absorption of the S-waves, the intensities of the PSS and the PP waves are comparable.

Dispersion

The variation of the shape of a seismic pulse is determined by the frequency depedence of both the absorption and the phase velocity. An absorbing medium displays dispersion. This dispersion should be distinguished from the dispersion caused by the non-ideal elasticity in homogeneous and isotropic media. It should be pointed out that the absorption parameters determined from experiment are strongly frequency-dependent. The absorption coefficients in the sonic band may exceed the corresponding values measured in the seismic band 5 to 6 times. Such a discrepancy is explained mainly by damage of the borehole walls in the process of drilling.

Determining inelastic absorption

Despite the fact that digital recording makes quantitative processing of records possible with undistorted amplitudes in a wide frequency band, it is difficult to determine absorption associated with non-ideal elasticity in the real earth. It is still more difficult to determine the frequency dependence of the absorption coefficient required for the study of the mechanism of inelastic absorption. However, the absorbing properties of the real earth have been studied much more thoroughly than dispersion. Only the joint studies of absorption and dispersion will make it possible to develop the theory of the absorption of waves in a non-ideally elastic medium, and to obtain agreement between the theoretical and the observed data. Such a joint study of the coefficients of absorption $\alpha(f)$ and velocity dispersion $C(f)$ was carried out on mountain rock samples in laboratory conditions, in natural bedding conditions, and on man-made materials. Here are some results.

V. V. Zhadin [135] obtained from observations of direct transverse waves in boreholes a linear $\alpha(f)$ dependence and an absorption decrement of 0.25 in the frequency band from 25 to 100 Hz for sandstones, but failed to detect practically any dispersion. Similar results were obtained earlier for longitudinal waves in sandstones in the 100 to 500 Hz frequency band. P. Wuenschel, [136] processing the experimental data of McDonal [196] obtained in clay shales in the frequency band from 25 to 500 Hz, established a linear $\alpha(f)$ dependence and demonstrated that the phase velocities increase with frequency by 2% from 2205 m s^{-1} for 25 Hz to 2251 m s^{-1} for 400 Hz, i.e., there is an anomalous dispersion of phase velocities. He also demonstrated the possibility of computing the dispersion of waves from experimental data with the aid of the well-known frequency dependence of the coefficient of absorption. Such calculations have been possible thanks to W. Futterman [137], who derived a formula connecting dispersion with absorption, when the parameters of wave propagation and pulse form are known.

$\alpha(f)$ and $C(f)$ have also been studied using Rayleigh waves produced in the course of the excitation of stationary oscillations by electromagnetic radiators, as well as waves excited by explosions. E. V Karus [138] obtained an almost linear $\alpha(f)$ dependence in the frequency band from 80 to 5000 Hz for various rock types with for example 0.25 for iron hornfels, 0.12 for sylvanites, 0.66 for loams and 0.48 for glacier ice.

Many measurements of $\alpha(f)$ and $C(f)$ have been carried out on samples of metals and plastics. L. I. Bokanenko [139] studied the absorption and dispersion of the transverse wave in two media: in a poor absorber-plexiglass ($\gamma = 0.064$), and in a good absorber-vinyl plastic ($\gamma = 0.14$). As the result of meticulous experiments, the author demonstrated that the absorption coefficient increases linearly with frequency, and velocity dispersion is of an anomalous nature, with $C(f)$ growing with an increase in the absorption decrement in a fixed frequency band (12 to 140 kHz). In the band from 14 to 140 kHz, $C(f)$ in plexiglass increases by 1.7%, and in vinyl plastics, where γ is 22 times greater than in plexiglass, $C(f)$ increases by 3.2%.

When studying dispersion with the aid of VSP, the phase velocities can be determined by the direct method, utilizing the spectra of the direct wave recorded along the vertical profile. The phase spectra may be used to construct vertical phase traveltime curves, and the latter – to determine the phase velocity.

Thus, absorption is a factor exercising a decisive influence on the pulse shape. Data on the magnitude of dispersion and on its connection with the absorption coefficient for rock, are very scanty and frequently contradictory. Still, most authors presume that, if it exists at all, dispersion in the case of body waves is quite small (not more than several percent for a 10 to 12 fold change in frequency) and of anomalous type. As to quantitative relationship of dispersion with the absorption coefficient or with the absorption decrement, there is practically very little information.

Absorption causes the frequency composition of a pulse to change in the course of its propagation: its front becomes still smoother, the pulse itself expands, and its extrema are displaced, the displacement being the greater the higher the extremum's number. Because of this, the velocities measured from the first arrivals and from subsequent phases may not coincide. The velocities in greatly absorbing media (e.g., in the upper portion of the section) can be determined from the phase spectra of the pulses, thus enabling both the magnitude of the velocity and its frequency dependence to be determined simultaneously. In the stratigraphic correlation of reflected waves on the basis of VSP data, a change in the pulse shape may eventually result in errors in determining the depth of reflecting interfaces. Phase shifts in the seismic recording channel at various frequencies should also not be ignored. The use of inverse filtering practically eliminates the effect of absorption on the shape of the record and on the determination of velocity.

3. Reflecting Properties of Horizons

The reflecting properties, determined by the values of the reflection coefficient (K) and its frequency dependence, are one of the fundamental properties of the section.

The relative intensity, and the shape of reflected and multiple waves close to interfaces, are determined by the characteristics of the incident wave, by the values of the coefficient of reflection (K) at the interfaces, and by its frequency dependence $K(f)$. The latter, as is well known, depends in turn on the velocity ratio, the densities, and the wave's angle of incidence on the interface. If the reflecting interface serves as a contact between two thick homogeneous layers, the reflection coefficient's argument is equal to zero or π at angles of incidence less than the critical, and is independent of the frequency, and the reflected wave is a replica of the incident. In a thin-layered medium the coefficient of reflection is frequency-dependent, and the reflected wave will be different in shape from the incident [54]. Knowledge of the $K(f)$ dependence, gained directly from field observations, enables the intensities of waves of various multiplicities to be evaluated.

The reflecting and absorbing properties of a geologic section, constitute the initial data required for the calculation of the intensities of various wave types both on the surface and within the earth.

Various methods of determining K have been devised based both on surface and down-hole observations [140]. The surface observations are unable to provide sufficiently accurate data about the reflection coefficients. VSP enables $\alpha(f)$ to be determined from observations at different points of the profile, and K to be determined subsequently by comparing the direct and the reflected wave observed near a reflecting interface. K is determined from the records of the direct and the reflected wave at points located near the interfaces, at such distances from them that the direct wave has already left the area of interference with the reflected wave. The absorption, which is determined with a large error, has, in this case, a much smaller effect on K than when it is determined from surface observations. Moreover, in observations near the interface, use can be made of the real incident and reflected waves to determine K, this being not always possible in surface observations.

To illustrate the point, we shall consider the results of the determination of the coefficient of reflection of waves as a function of the frequency for two angles of

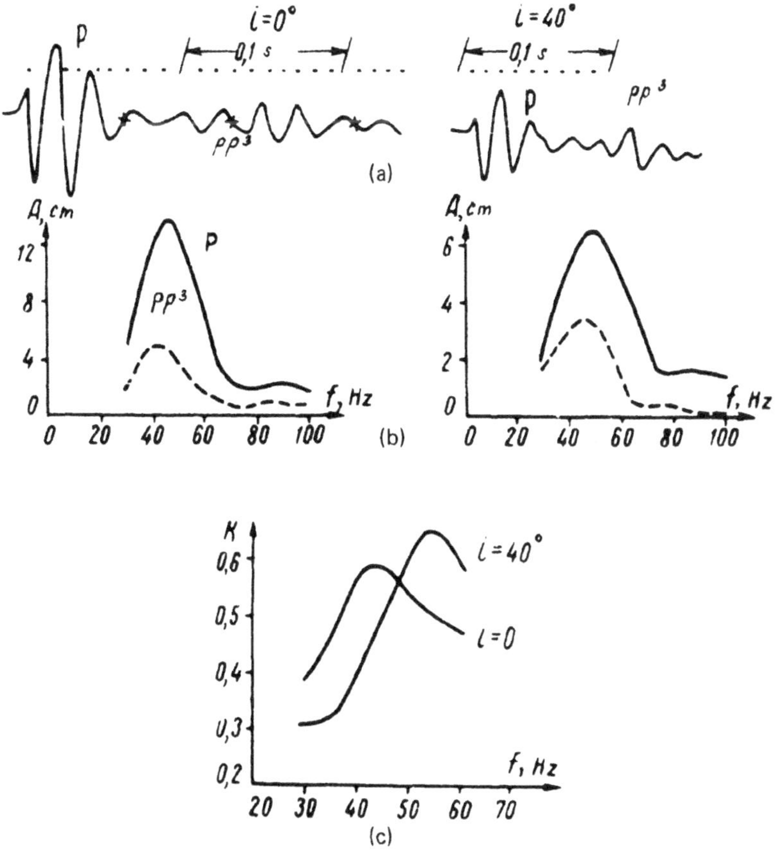

Fig. 117. Records of direct and reflected waves for normal (shotpoint 100 m) and oblique (shotpoint 1200 m) incidence (a), spectra of the waves (b), and spectra of reflection coefficient (c) (borehole N2 Staro–Minskaya).

incidence, from a stack of thin layers. The records of waves from two shotpoints at 100 m and 1200 m (Figure 117), obtained at the depths of 700 to 800 m, 900 to 920 m and 1080 to 1100 m have been used to determine $K(f)$. The angles of incidence corresponding to these shotpoints for the depth of 1100 m are equal to 0 and 40°, respectively. As the angle of incidence increases, the maximum value of the reflection coefficient K shifts towards higher frequencies, from 43 Hz at 0° to 54 Hz at 40°. The values of the reflection coefficient at the extrema are equal to 0.58 and 0.65, respectively. It will be seen from the shape of the spectrum of the reflection coefficient that the reflecting interface is not a 'thick' layer.

A method for determining K from records of the direct and reflected waves not at a point, but over the entire tracked interval of the reflected wave, and from the record of the direct wave in the region of the reflecting interface, has also been devised. It should be emphasized that in VSP the determination of K, as well as other parameters based on the dynamics of waves recorded in the later portion of the record, is fraught with great difficulties stemming from interference phenomena and from the intricasies of the discrimination of the reflected wave in a pure form. To discriminate waves in the least distorted form, conventional methods are used, including the subtraction of waves and the establishment of the precise form of the discriminated wave, as well as various forms of frequency selection, especially inverse filtering. Such procedures are sufficiently effective, this being corroborated by the stability of the incident wave shape along the vertical profile.

TABLE XII

	Salts	Anhydrites
Density (g cm^{-3})	2.13 ± 0.02	2.95 ± 0.05
Velocity (m s^{-1})	4450 ± 200	6000 ± 200
Acoustic impedance (g cm^{-2} s^{-1})	9478.5 ± 435	17580 ± 658

To illustrate the potentialities of VSP in studies of the reflecting properties of the geologic section, we present here the results of the determination of the coefficient of reflection at the anhydrite-salt interface of the upper Jurassic in the region of the Dengizmul elevation in western Uzbekistan [141]. The following values of K have been obtained from VSP data:

upper anhydrites – upper salts	0.15 ± 0.03
upper salts – middle anhydrites	– 0.22 ± 0.04
middle anhydrites – lower salts	0.22 ± 0.04
lower salts – lower anhydrites	– 0.23 + 004

The physical properties of the rocks shown in table XII enabled the reflection coefficient K to be calculated as 0.306 ± 0.034.

A comparison between the theoretical and the observed values of K brings out a systematic reduction by 25 to 50% of the values of K observed in VSP as compared with the theoretical. The authors think it may be due to the gradual nature of the transition from the salt to the anhydrites resulting from the presence of shallow interlayers of salt in the anhydrides and of interlayers of anhydrides in salts close to the interface. This allowed interfaces with a gradual transition to be identified from VSP data.

Of special interest are the studies of the reflecting properties of the section below the bottom of the borehole.

To predict the section below the bottom of the borehole use can also be made of the intensity of reflected waves associated with the interfaces located ahead. Such studies are based on the solution of the reciprocal dynamic problem and are in essence very similar to the widely-used method that received the name of pseudo-sonic logging in Soviet literature. Putting this method to work in VSP has fundamental advantages over surface observations, consisting of a substantially more effective suppression of the multiples and consequently offering the possibility of determining the intensity of singly-reflected waves. For an example of such an analysis of the wave field, see Chapter 3.

4. Determination of Elastic Parameters

The elastic parameters of an isotropic medium depend on a number of factors: on the structure and the composition of rock, its bedding depth, its degree of deformation, which as a set can characterize the rock and its local variations. The elastic properties are determined by one of the pairs of constants: Young's modulus (E) and Poisson's coefficient (σ), by Lamé constants (μ and π), by the velocities of the longitudinal, v_P, and the transverse v_S, waves connected with one another through the following well-known expressions:

$$v_P = \sqrt{\frac{\lambda + 2\mu}{\rho}} = \sqrt{\frac{E(1-\sigma)}{\rho(1+\sigma)(1-2\sigma)}},$$

$$v_S = \sqrt{\frac{\mu}{\rho}} = \sqrt{\frac{E}{\rho}\frac{1}{2(1+\sigma)}},$$

whence

$$E = \frac{\rho v_S^2(3v_P^2 - 4v_S^2)}{v_P^2 - v_S^2}, \qquad \sigma = \frac{v_P^2 - 2v_S^2}{2(v_P^2 - v_S^2)},$$

where ρ is the density of rock.

PM–VSP is able to study the velocities of both the longitudinal and the transverse waves simultaneously. The values of the density may be obtained by direct measurements on samples or from the results of gamma-gamma logging. In the absence of such experimental data, information about the densities may be obtained from the correlational dependence between velocity and density (see Chapter 11). PM–VSP data provides information about the elastic parameters of rock and enables the geologic column to be sectioned in accordance with such parameters.

The results of studies of the physical properties of Eocene-Paleocene sand-clay deposits in the Kaluzhskaya area in Western Ciscaucasian region are depicted in Figure 118. The section is represented by alternating layers of clays, sands, sandstones. There is evidence of several layers from 200 m to 500 m thick associated with various lithologic and stratigraphic complexes. The section is somewhat more differentiated with respect to the S-wave velocities than with respect to the P-waves. The ratio v_P/v_S, even at depths down to 1700 m, is equal to about 2.5 due to the properties of Maikop clays. Higher values of Young's and shear moduli correspond to the maximum values of the parameters v_P/v_S and σ. The lithology of the rock exercises a decisive influence on its elastic properties. A noteworthy point is the sharp increase in Young's modulus at the interface between the Chokrak and the Maikop deposits at H $\sim$ 1750 m where there is no observable drastic change in the density.

Table XIII [26] shows the results of the determination of elastic parameters of rock in the Uzen section of Southern Mangyshlak. The wave velocities have been determined with the aid of downward transmitted waves observed on off-line profiles. The average formation densities have been obtained from gamma-gamma logs. The errors in the values of Young's modulus were as high as 4 to 5% and of Poisson's coefficient from 1 to 8%. The errors in both parameters increased with the increase in $\gamma = v_S/v_P$. The

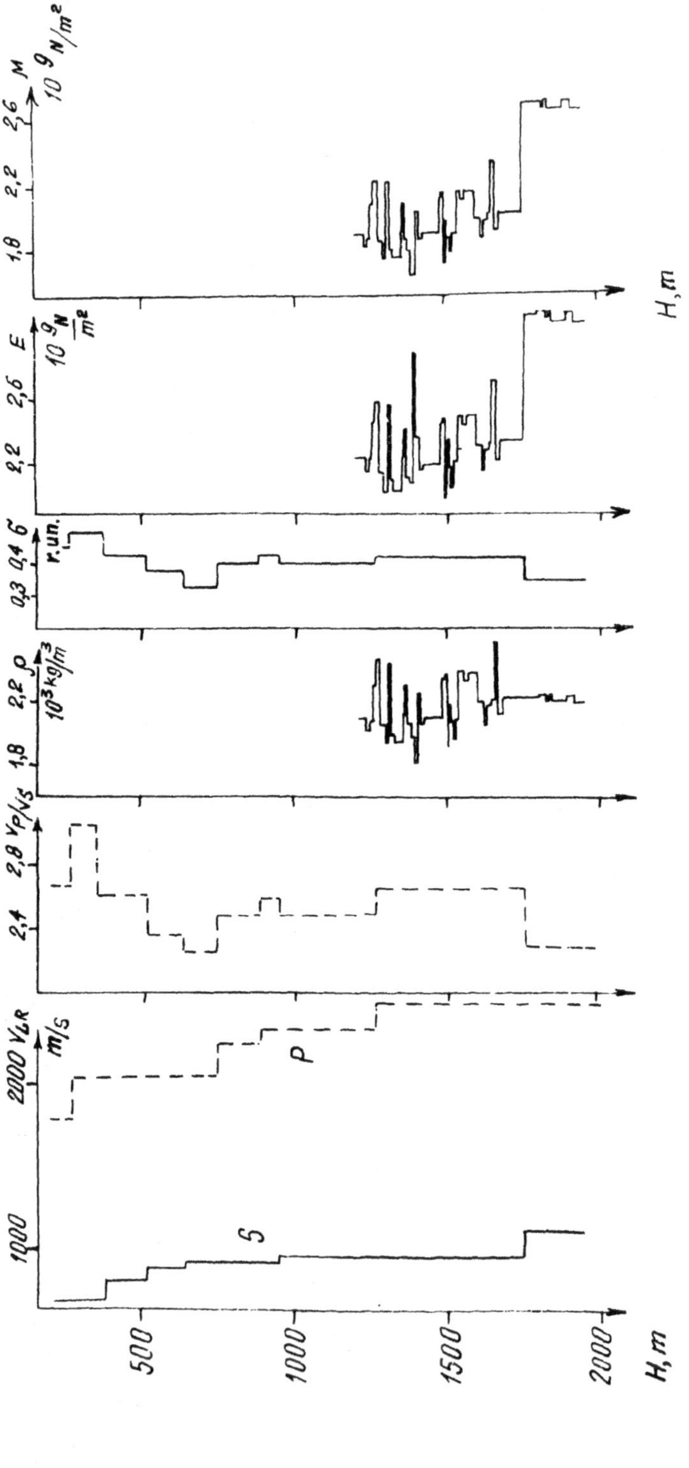

Fig. 118. Studies of physical properties of a cross-section. Kaluzhskaya area, (Western Ciscaucasian region) [45].

TABLE XIII*

Rock age	Depth of velocity interface (m)	Velocities v_S (m s^{-1})	Velocities v_P (m s^{-1})	$\gamma = \frac{v_S}{v_P}$	Poisson coefficient σ	Density ρ (g cm^{-3})	Young's modulus $E \times 10^5$ (g cm^{-2})
Paleocene*	70	730	2900	0.25	0.47		
Santonian*	180	680	1800	0.38	0.42	1.86	0.25
Turonian							
Cenomanian	275						
	360	770	2300	0.33	0.44	1.96	0.34
	430	770	2000	0.38	0.42	2.07	0.35
	575	930	2250	0.41	0.40	2.40	0.55
Albian	680	930	2430	0.38	0.42	2.49	0.63
	790	1000	2430	0.41	0.40	2.55	0.73
	850	1130	2900	0.39	0.41	2.37	0.87
Aptian	920	1130	2400	0.47	0.36	2.35	0.86
Oxfordian	1055	1440	3100	0.46	0.37	2.43	1.40
Oxfordian + Calliovian		1440	2780	0.52	0.31	2.42	1.37

* Studied at the depth of 1150 m.

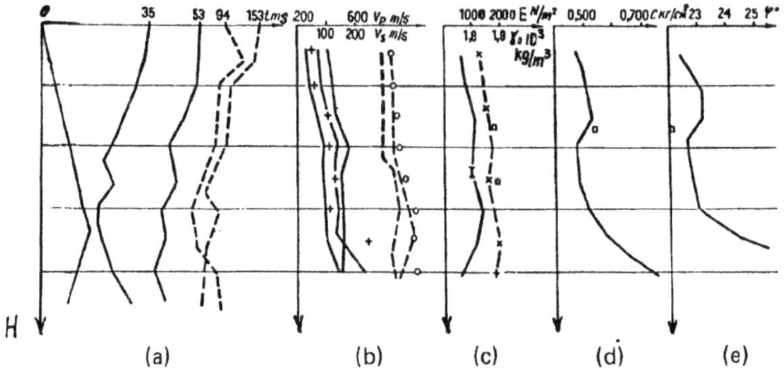

Fig. 119. Results of studies of mechanical properties of soils with the VSP method (borehole N43 Bulanash [62]): (a) traveltime curves of longitudinal (solid lines) and of transverse (dotted lines) waves; (b) velocity cross-sections obtained with P- and S-waves; (c) density γ_0 and deformation modulus E), dash shows the value of deformation modulus; (d) cohesion c; (d) angle of internal friction, squares show the results of determination of γ, C, and φ from laboratory tests on core samples from borehole N43.

magnitude of Poisson's coefficient will be seen from the table to diminish with the depth, and that of Young's modulus – to grow. By comparing similar data for rock of identical age, but located at different depths, we are able to demonstrate the dependence of the physical parameters of rock on depth. Such a dependence may be due to various factors (pressure, material composition etc), however, to establish the causes, we shall have to accumulate new data that may serve as an additional source of information about the Earth. The data obtained warrants the opinion that the study of elastic parameters may become an effective additional means of assessing the geologic section, specifically for the study of rock lithology, its porosity, ease of drilling and the search for reservoirs.

Determining the Physical and Chemical Properties of Soil

In engineering geological boreholes, VSP is executed to determine the physical and mechanical properties of soils required from the vertical traveltime curves. Figure 119 [45] depicts the results of the determination of the physical and mechanical porperties of soils.

5. VSP and Forecasting the Geologic Section

The task of forecasting the geologic section includes an extensive complex of problems, which can be subdivided into two groups – the geologic and the technological. The geologic problems include:

– forecasting lithology and the genesis of sedimentary rock; sectioning the geologic column into lithologic and stratigraphic complexes;

– discovering structural and stratigraphic traps with good prospects for oil and gas; evaluating the properties of the rock and the composition of fluid in reservoirs;

– establishing the boundaries of deposits, sectioning and mapping production deposits.

The technological problems include those associated with the design and drilling of the borehole itself, to make forecasts about the ease of drilling and to select the optimum drilling procedure, to establish the zones of abnormally high formation pressure and to determine the pressure, to map the bottom of the borehole in space and to control directional drilling, the latter point being of special importance for drilling deviated wells, as well as in sundry emergencies.

The list of problems of this sort could be continued, and it will grow as research progresses. Their solution depends essentially on the opportunities existing for the study of kinematic and dynamic parameters of various wave types aimed at utilizing them to determine the physical properties of rock consituting the geologic section. Such opportunities are limited by a number of factors: by the complexity of the real earth, by the comparatively small effect exercised by a deposit, by inadequate resolution of the methods and the purely technical quality of the data obtained. Given such conditions, we cannot regard the physical and the geologic foundations of the problem of forecasting the section as one that has been adequately studied, despite the sharp rise in recent years of the volume of studies associated with it. In the author's opinion, one of the main reasons for such a situation is that a large percentage of such studies are performed with the aid of surface observations whose potential for the study of wave dynamics is, as has already been repeatedly pointed out, extremely limited.

In this respect VSP, which makes it possible to study waves in the immediate vicinity of the interfaces associated with them, and their variation in the course of propagation through various geologic complexes, presents a special interest. Such studies executed on a modern technical level, serve as a source of information that makes it possible to appraise the potential of forecasting the geologic section and to place such forecasts on a sound physical basis. There can be no doubt that all the effects of heterogeneities, specifically, of deposits, on the wave field will be felt much more acutely in down-hole observations than in surface observations, and that only the most pronounced effects can be detected on the surface. Two principal trends can be seen to emerge from the VSP approach to the problem of forecasting the section: (a) the study of the characteristics of the section and of their influence on the parameters of seismic waves; (b) direct prospecting for deposits in the vicinity of a borehole. The first trend is particularly important for establishing physical foundations for forecasting the section from surface observation data.

The use of seismic exploration for direct prospecting for oil and gas is based on the quantitative differences between the kinematic and the dynamic parameters of seismic waves that have passed through water- and oil-saturated rock or of waves reflected from it.

THE EFFECT OF OIL AND GAS DEPOSITS ON THE MAGNITUDE OF SEISMIC VELOCITIES

Theoretical, model and field studies carried out by many authors have demonstrated that the velocity of propagation of longitudinal waves (v_P) in oil- and gas-rich reservoirs is lower than that in the water-rich. However, in this case numerical estimates yield widely different values. The velocity v_P in oil-saturated reservoirs is by 6 to 10% lower than in water water-saturated. The presence of a deposit results also in a reduction of effective velocities, however, not more than by 1.0 to 1.5% [143]. Sonic logging data also points to a substantial decrease in v_P in oil-saturated and especially in gas-saturated deposits. The presence of a gas condensate deposit results in a regular decrease in v_P by 3.5 to 5.8% in boreholes located inside its bounds compared with the boreholes drilled outside.

Differences in the velocities of longitudinal waves in oil- and water-saturated formations of 15%, as measured by low-frequency sonic logging, and of 10 to 25%, as measured by high-frequency sonic logging, have been discovered, such differences in the case of water- and gas-saturated formations amounting to 25 to 30%. The decrease in velocity associated with an increase in the volume of pores or cracks is one of the criteria for forecasting the presence of reservoirs. At the same time saturation with oil and gas has practically no effect on the propagation velocity of transverse waves (v_S).

The velocity dispersion obtained in the course of combined seismic and sonic studies as the result of measurement at different frequencies in carbonate reservoirs (oil-saturated porous media) in an uncased borehole amounted to 18 to 25%, the value rising in cased boreholes for the same kind of rock to 29 to 33%. The divergence of the traveltime curves obtained with sonic and seismic measurements and the dispersion of phase velocity (of the order of 30%) are evidently the result of the presence of a large number of cracks and of the ensuing upward migration of oil and thermal water. Greater dispersion in a cased borehole (as compared with that in an uncased one) in the case of oil-saturated rock can be explained by the deformation of the invaded or altered zone in a cased borehole and by greater absorption of the acoustic signal. The difference may be influenced by the difference in the frequencies (in thin-layered media); in cement the velocities of high-frequency waves are 5 to 10% higher than those of low-frequency waves. The consequent low accuracy of determination of the quantities being compared makes it impossible to establish definite relationships governing the velocity ratios.

The Effect of a Deposit on the Absorption of Seismic Waves

It has been established as the result of analysis of field surface observation data in various regions that there is an increased atttenuation of seismic waves in the area of a deposit. However, in the majority of cases the validity and reliability of the data have been inadequate for final conclusions. This is the result of a number of difficulties of a methodological and a technical nature encountered in the course of the processing of records obtained on the surface. Recording waves down-hole, one is able to improve the accuracy of observations and eliminate a number of distorting factors.

The attenuation of seismic waves in a reservoir and in the overlying bed with the reservoir having different fluids (gas, oil-gas-water) has been studied at various stages of VSP development. Despite meticulous experiments with a rigorous control of the instrumentation and the source excitation conditions, it has not been possible to establish a definite effect that the deposit has on the total absorption of seismic waves. Recent research in the same Anastasiyevskaya–Troitzkaya field, as we have already seen (Figure 116), was successful in reliably establishing increased absorption. It should be pointed out that increased absorption of the energy of seismic waves is normally not just limited to the depth of the deposit itself, but spreads over a considerable interval above the deposit, where secondary deformation of rock is a typical phenomenon. Such anomalous effects depend greatly on the deposit's depth and on the zone of secondary deformation above the deposit. The absorption increases drastically with an increase in the porosity. The deposit markedly affects the absorption of S waves. The dynamic characteristics of waves are several times more sensitive to variations in porosity than are the kinematic.

THE EFFECTS OF A DEPOSIT ON THE REFLECTING PROPERTIES OF A LAYER

The variation of intensity of the waves reflected from the top of a gas-saturated reservoir is the result of the variation of velocity and bulk density governed by the kind of fluid filling the pores: the gas-saturated portion of a reservoir has a lower acoustic impedance than the water-saturated. Even the replacement of a small fraction of water by gas results in an appreciable increase in the reflection coefficient. Data is also available in support of the thesis that it is possible to record waves reflected from water-oil and gas-oil contacts, on account of the difference in acoustic properties of gas-, oil-, and water-saturated formations in terrigeneous and carbonate deposits. PM–VSP observations carried out in the Anastasiyevskaya–Troitzkaya field have confirmed that in gas-saturated rock the velocity drops sharply, absorption increases, the coefficient of reflection from the top of the deposit increases, and polarization parameters associated with this layer become anomalous. Work carried out in recent years has confirmed that it is possible to establish the boundaries of a deposit in the vicinity of the borehole and to discover a deposit not discovered in the drilling. VSP observations made with a wide dynamic range and a wide frequency band enable wave dynamics to be used more extensively for forecasting the geologic section, and the amount of detail established in the studies to be substantially increased. Such forecasts may be of great importance for the choise of a rational deep-drilling technology and, on account of the high costs of deep drilling, may result in great savings.

As an example of the influence exercised by a gas deposit on the wave field, we would like to present observation data obtained in the Golitzyn area (Black Sea). It is worth pointing out that sea conditions make it fairly easy to achieve good reproducibility of source excitation conditions, and in this respect are favourable for research. Observations were carried out in borehole N4 located in the centre of the gas deposits area (Figure 120). Source positions were located at equal distances, but in different directions from the borehole both inside and outside the deposit area. The gas deposit in Maikop sediments was located at a depth of approximately 500 m, its thickness being about 50 m. The effective absorption coefficient determined from the first longitudinal wave was 2 to 5 times greater in gas-saturated portions of the reservoir than below the deposit (Figure 120). For this reason the amplitudes of the first longitudinal wave arriving from the shot-point located above the deposit (shotpoint 2000 m East) were about twice as low as those from shotpoints located outside the deposit. The velocities above the deposit and inside it were lower than in the surrounding rocks (Figure 120). The intensity of transverse waves that cross the deposit (converted transmitted waves associated with shallow interfaces contained in the upper portion of the section) were substantially lower than those of similar waves that did not cross it. This is evident from the seismograms of waves from shotpoints 2000 m East and 2000 m South (Figure 120). The strong influence exercised by the deposit on the intensity of waves can be illustrated with the aid of the seismograms obtained from remote shotpoints (4000 m) (Figure 120). The converted transmitted waves associated with interfaces located at a depth of about 2000 m decay almost completely in the gas-saturated portions of the section. The data cited is unfortunately inadequate for a quantitative evaluation of all the parameters,

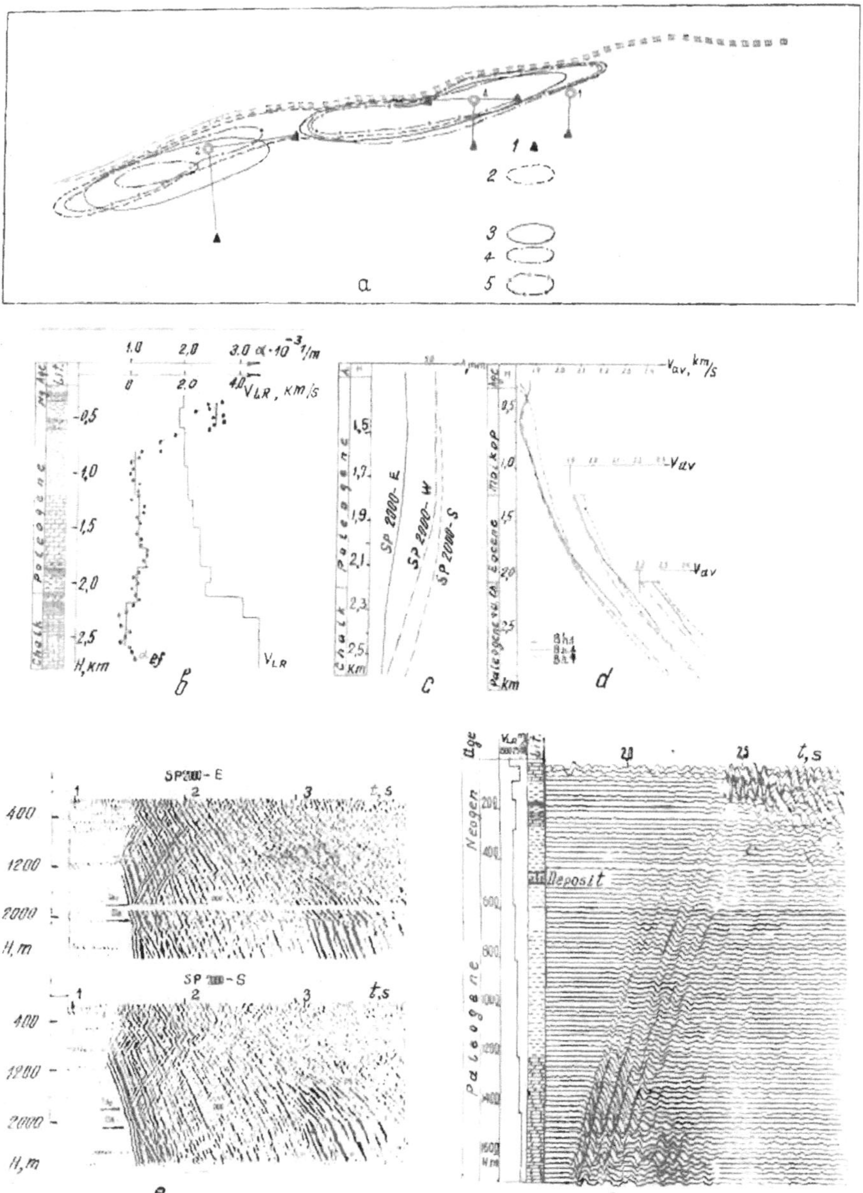

Fig. 120. Effect of a gas deposit on some parameters of the wave field Borehole N4 Golitzyn (courtesy A. N. Amirov): (a) schematic representation of disposition of shotpoints and of boreholes: 1 – shotpoint; 2 – outlines of the Golitzyn structure along reflecting horizon IIIa, predicted outlines of gas deposits in Maikop deposits (3, 4) and in Paleocene (5); b, c, d – plots of v_s, α_{ef}, A, v_{av} vs depth; e – VSP seismograms from shotpoints located below the deposit (shotpoint 2000E) and outside it (shotpoint 200S); f – seismogram from shotpoint 4000 m. Converted transmitted waves cross the deposit from below.

because observations were carried out with a *Z*-component geophone. Nevertheless, even such data helped to establish precise boundaries of the deposit.

We shall consider the use of VSP in forecasting the outline of a gas deposit using the example of the South Sivash borehole. Before VSP observations were initiated, the gas deposit had been discoverd from the results of drilling in borehole N1 (Figure 121). The deposit is located in sediments belonging to the Maikop series at a depth of about 2200 m. The reservoir was made up of clay-sand-siltstone rocks and had a porosity of about 30%. The thickness of the productive layer was about 10 m. Interlayers of clay 5 to 15 m thick served as a cap for the deposit. The deposit was not detected in boreholes NN 3, 4, 5. No anomalies of the 'bright spot' type were detected from data of surface common-depth-point observations (*Z*-component recordings), and velocity variations were minimal and not significant for the detection of the deposit. The absorption coefficient (α_{ef}) determined from VSP data provided valuable information. In the region of the deposit and the diffusion zone that accompanies it, α_{ef} increases 2 to 3 times and in the deposit itself 5 to 6 times. At the frequency of 30 Hz α_{ef} is equal to from 0.002 to 0.003 l/m in the diffusion zone and in the above lying mass – 0.001 l/m (Figure 121). The thickness of the diffusion layer is considerably greater than the thickness of the deposit itself, and the anomalies observed are due to secondary metamorphosis of the geologic section. It is an interesting fact that absorption anomalies have been detected in surface common-depth-point observations from changes in the spectrum. However, they have not been established reliably, because of strong unwanted waves, which cannot be adequately suppressed by the common-depth-point system.

To delineate the deposit VSP observations with *Z*-component geophones were carried out in two boreholes: N2 and N3. The observations in borehole N2 involved shots from 8 locations at distances from 700 m to 3300 m to the north, north–east, east and south–east (Figure 121). Recording was performed in the depth range 2400 to 1400 m with a depth sampling interval of 20 m. Observations in borehole N3 were carried out from two shotpoints at distances of 1500 m and 2500 m (Figure 121) in the depth range 3000 to 100 m with depth increments of 20 m. The shots were fired below the low-velocity zone. Source excitation conditions were rigorously controlled at each shotpoint with the aid of geophones specially installed for that purpose.

The average velocities in the boreholes that penetrated the deposit (NN 1, 2, 3) were lower than in boreholes outside the boundaries of the deposit (Figure 121). However, the factor of major importance was the variation of amplitudes of waves emerging from shotpoints located in different directions. The wave parameters from shotpoints located to the north of borehole N2 were regarded as anomalous, and the wave amplitudes from other shotpoints were compared with them. The shallow depth of the borehole greatly limited the forecasting chances, since the zone surrounding the bottom of the borehole could only be illuminated by direct waves within only a short distance of the shaft (Figure 121). To illuminate the section at great depths, we had to resort to reflected waves, which were markedly distorted by downgoing waves. The VSP data provided evidence of regions of abnormally high absorption of seismic energy ((7) in Figure 121(a)) and low absorption (9), thus enabling the contour of the deposit to be established. The latter coincided neither with the contour of the gas deposit expected from geologic data (6) nor with the predicted line of reservoir thinning (5). It is worth noting that, because of the shallow depth of borehole N2, the region (8) was not sampled.

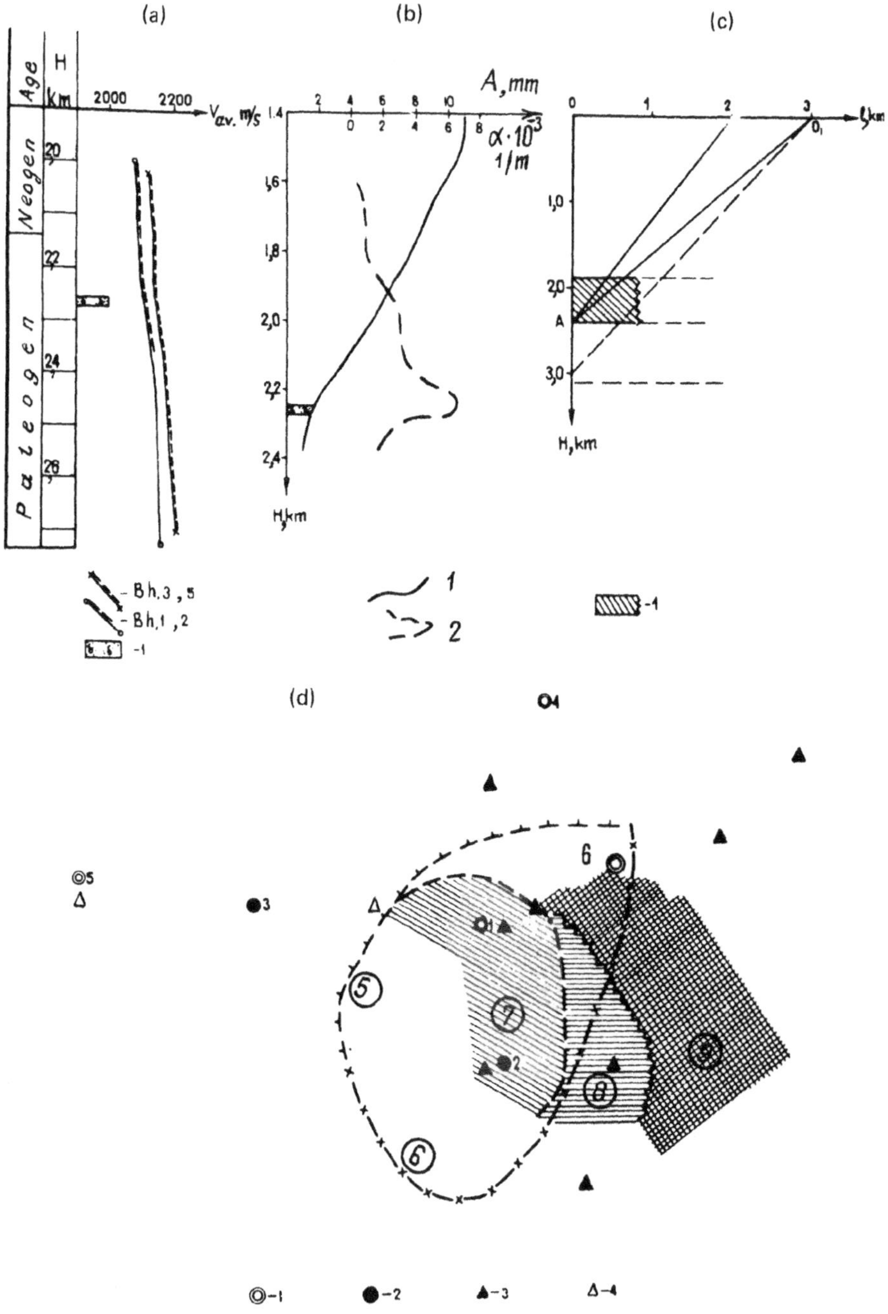
(a)
(b)
(c)
Age
H km
$V_{av.}$ m/s
2000
2200
Neogen
Paleogen
20
22
24
26
- Bh.3, 5
- Bh.1, 2
-1
A, mm
$\alpha \cdot 10^{-3}$ 1/m
1.4
1,6
1,8
2,0
2,2
2,4
H, km
1
2
l, km
O_1
A
1,0
2,0
3,0
H, km
-1
(d)
5
6
7
8
9
1
2
3
4
◎-1
●-2
▲-3
Δ-4

Fig. 121a-d.

The correctness of the VSP forecast of the contour has since been substantiated. Specifically, borehole N6 drilled inside the bounds of the predicted contour (9) failed to strike the deposit.

Zones of abnormal pressure

Experience has also been gained in applying VSP to forecasting zones with abnormally high layer (pore) pressure. Such forecasts are required for the mechanical design of boreholes to ensure against surprise blow-outs and other accidents in the course of drilling over-pressured rocks, for increasing the drilling rate and for cutting drilling costs. At present, the seismic method is thought to be the most promising method of forecasting such zones before drilling, in which case the depth and the pressure will be determined. To provide physical foundations for such a method, VSP data obtained from twenty-three boreholes in the Eastern Ciscaucasian area have been analyzed [45]. For a normal increase in density with depth, rock velocities also increase. The decrease in interval velocities indicates a reduction in the density on account of the high pressure in the pores. Comparing the curves of the variation of interval velocity with depth in conditions of normal and abnormally high pressure, we have demonstrated that it was possible to detect zones of over-pressure and were able to appraise its magnitude. It should be remembered that structure maps will be distorted, if in interpreting data the velocity variations associated with the presence of a zone of abnormally high pore pressure are ignored.

It should, however, be kept in mind that quantitative estimates of the effects of the deposits on wave field parameters are not always in good agreement with each another. This is primarily due to the complexity of the problem that necessitates special experiments and requires experimental data of appropriate quality. Unfortunately, such conditions are not always met, and sometimes results obtained from data clearly unsuitable for the solution of the problem are published. PM–VSP data provides the most objective estimates of the effects of a deposit on the wave field. In our opinion, it is necessary to continue the studies of the effects of deposits, combining VSP with sonic logging. Of special interest may be the joint studies of the effects of deposits on the longitudinal and the transverse waves. The down-hole observations should be combined with surface data, in order to find out how the effects of a deposit established in the borehole will manifest themselves on the surface, and hence to appraise the chances of detecting a deposit from surface observations as well as of mapping the internal and external contours of an oil-gas deposit. In order to study the structure of the deposit itself, one will have to increase the amount of detail established in the study of the geologic section, using for this purpose, records with frequencies up to 150 Hz. Of great importance may be the transformation of seismic traces into acoustic rigidity curves, that enable thin layers to be tracked and their variation to be observed over an area.

← Fig. 121. Predicting outlines of a deposit from VSP data, South Sivash area, Crimea courtesy A. N. Amirov): (a) plots of average velocities, 1 – gas-saturated layer; (b) plots of amplitudes (1) and of α_{ef} (2); (c) ray-path schematics, 1 – gas-saturated layer; (*d*) schematic diagram showing absorption properties of the cross-section: 1 – deep exploration borehole, 2 – hole where VSP observations were carried out, 3 – shotpoint for VSP in borehole N2, 4 – shotpoint for VSP in borehole N3, 5 – predicted line of reservoir thinning, 6 – contour of the gas deposit as expected from geologic data, 7 – region of abnormally high absorption of seismic energy, 8 – unsampled region, 9 – region of low absorption of seismic energy.

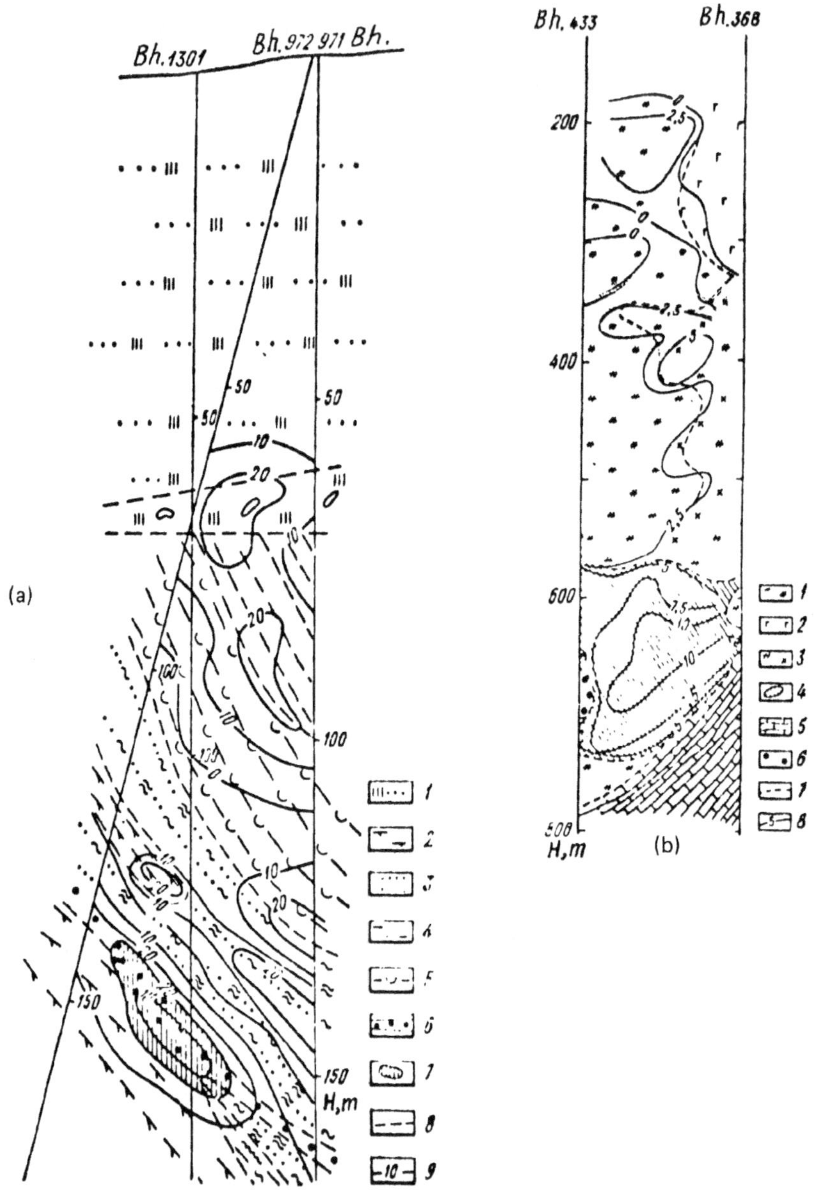

Fig. 122. Examples of application of inter-borehole measurements [144]: (a) copper-zinc fields, 1 – loess, 2 – tufoids, 3 – quartz-sericitic- and sericitic-chlorite-quartz schists, 4 – carbonate-chlorite-clay shale, 5 – chlorite, 6 – streaky impregnated ore, 7 – ore body according to inter-borehole measurements, 8 – geologic interfaces, 9 – lines of equal attenuation; (b) magnetite field, 1 – medium- and fine-grained diorite, 2 – plagioclaseous little fissured porfirite, 3 – succession of diorite and plagioclaseous porfirites, 4 – magnetite ore, 5 – massive fine-grained limestone, 6 – greatly fissured albite metasomatic granite, 7 – geologic interfaces, 8 – equal attenuation lines.

6. Measurements Between Boreholes

Inter-borehole sampling provides a good method for the study of the region separating the boreholes. Such studies have, as yet, not become widespread, but the available experience suggests good prospects. Such a technique is used most extensively in mining geology, a tone signal being employed for this purpose. The differentiation of rock according to the absorption coefficient at fixed frequencies in the audio band enables heterogeneities to be located, ore bodies to be detected by measuring various elastic properties, and their morphology to be studied. The method has been used in the chromite fields of Kazakhstan where magnetic and electric methods are useless, because the respective properties of the ore bodies do not differ from those of the ultrabasic rock containing them. At a frequency of about 3.3 kHz, chromite bodies were detected when the separation between the boreholes was up to 200 m, this exceeding almost 4 times the normal separation between boreholes employed in exploration. In studies of copper pyrite fields, the predictable range in the virgin rock mass, without faults, was as high as 300 to 400 m and in the zone containing faults 100 to 150 m. In a magnetite field, the range exceeded 200 m.

As a result of studies of the zone separating the boreholes NN971 and 972 in a copper-zinc field (Figure 122) we have established the presence of a bulge of copper-zinc ores, which was subsequently confirmed by borehole N1301. In a magnetite field (Figure 122), the morphology of ore bodies between the boreholes was determined [142]. The creation of stable high-frequency borehole sources will facilitate rapid development of the methodology of inter-borehole measurements and enable the principal potential of such observations to be put into effect. It is advisable to combine inter-borehole work with the VSP reflection method on both vertical and horizontal (surface and fixed geophone level) profiles.

The experience already available demonstrates that VSP provides the most exhaustive information about the deposit such as its geometry, intactness, pressure, lithologic and stratigraphic characteristics, such information being required to obtain an initial idea about the possible reserves of the field and to select the best pattern of working it.

CHAPTER 12

THE POLARIZATION OF SEISMIC WAVES AND ITS POTENTIAL FOR STUDYING THE ROCKS SURROUNDING THE BOREHOLE

The idea of employing wave polarization in seismic exploration as an additional parameter of the wave field is due to G. A. Gamburtzev, Member of the Academy of Sciences of the USSR, who in the fifties had already initiated studies of polarization, which resulted in the creation of the azimuthal method of seismic studies [9]. Strong distortion introduced by the low-velocity zone and by the upper portion of the geologic section, however, made the transition from polarization parameters at a point to the parameters of a wave in space impossible. Experimental surface and down-hole studies of polarization [15] have demonstrated the expediency of using the polarization in surface observations chiefly for wave discrimination, and this formed the basis of the polarization-positional correlation method [2], a specific feature of the polarization method.

In early VSP studies [15], polarization was used for the analysis of the wave field and to obtain direct information about the Earth. However, because of the specific difficulties associated with three-component observations in deep boreholes on the one hand, and inadequate theoretical knowledge of the polarization of seismic waves in the real earth on the other, the latter application lagged in its development somewhat behind the first. The use of polarization parameters in VSP to obtain additional information about the earth is the subject of special chapters in monographs [2, 14].

The trajectories of particle motion, when observed down-hole, are less liable to distortion due to the influence of the low-velocity zone, and this substantially broadens the scope of their employment in studying the zone around the borehole. The primary reason for this is that polarization is a very sensitive space-time characteristic of the wave field. Besides, it has been established in numerous studies of polarizaiton in boreholes that, in constrast to such dynamic parameters of the direct wave as shape and intensity, which are greatly dependent on source excitation conditions (see Chapter 3 and 5), the polarization parameters depend very little on excitation conditions, and are determined chiefly by the Earth's structure.

The reason why the direct wave is of special interest for the study of the region surrounding the borehole, is that it is recorded against a quiet background without interference from other waves over substantial intervals of the vertical profile, thus facilitating quantitative interpretation of polarization parameters.

Polarization parameters are described by the variation in time of the full vectors of particle motion in the earth and are characteristic of the trajectory described by a particle drawn into the vibration process. For a linearly polarized oscillation, such parameters are the direction of displacement, defined by two angles (φ in the vertical and ω in the horizontal planes), and the vector's modulus. For an elliptically polarized oscillation, such parameters are: the orientation in space of the plane of polarization described by the direction of its normal (φ_n, ω_n), the shape of the ellipse described by the ratio of it's semi-axes ($\varepsilon = a/A$ or $\varepsilon = A/(A + a)$, and by the orientation of the ellipse in the polarization plane, described by the direction of its major axis (φ_A, ω_A). Sometimes, it is convenient to study the parameters of a trajectory with the aid of their projections onto the vertical and the horizontal planes. Analyzing polarization parameters in the time and frequency domains, one is able to obtain more complete information about the Earth, and, in particular, to appraise the effects of inhomogeneities of different sizes.

Various polarization parameters may be used to gain information about the Earth. The connection between the polarization parameters and the standard features of the Earth was established early in VSP studies [15]. Since the polarization method was created, its use to obtain direct information about the Earth has experienced a boom [2, 145]. However, it should be pointed out that the potentialities of polarization have yet to be realised. Further studies of both a theoretical and an experimental nature are needed. Of special interest are experimental studies of the polarization of pulses in the real earth.

1. On the Polarization of Seismic Pulses

Numerous papers deal with seismic pulses, but in them, attention is paid chiefly to the shape of the pulse and to the effects of various factors on it. Very little is known about the polarization of seismic pulses even though it may carry a large amount of information. In contrast to the shape of the pulse, polarization depends substantially less on the source exctiation conditions, than on the Earth's structure [2, 14, 15]. In accordance with the zero approximation of the method of ray-paths, the first body wave pulses in isotropic media are usually assumed to be linearly polarized. However, there are still no completely convincing data, that determine the extent to which the real earth 'endures' the zeroth approximation of the method of rays-paths. Also, it has not been verified whether linear polarization is retained in the first and the subsequent phases of direct waves in the real earth.

In connection with the introduction of PM–VSP and the development of methods based on the use of polarization, the study of the polarization and the structure of pulses of direct body waves in the real earth, and the effects of various factors on them, assumes a special importance. This is of interest also, because the polarization of oscillations is a very sensitive parameter of the wave field, and in the first approximation of the ray path method it already becomes necessary to take account of additional transverse types in expressions for longitudinal waves, and additional longitudinal types in expressions for transverse waves.

The presence of vibrations 'transverse' to the principal displacement direction in subsequent phases of *P*- and *S*-wave pulses near the wave front should result in the violation of linear polarization in the subsequent phases of *P*- and *S*-waves. PM VSP enables the polarization of the pulses of the first *P*- and *S*-waves to be studied. The most reliable data can be obtained for *P*-wave pulses, which are recorded first on the seismograms against a quiet background, and have not yet been greatly distorted by the superposition of various wave types formed by the heterogeneities of the section. As to the *S*-wave pulses, their polarization is more sensitive to the nature of the geologic section than that of longitudinal waves. However, owing to difficulties resulting from the fact that transverse waves are recorded in the later portion of the record in conditions of superposition of various wave types, the polarization of *S*-waves has been studied less extensively than that of *P*-waves. This section deals with the results of studies of polarization of direct longitudinal waves generated at the surface and recorded in boreholes. We have studied pulses recorded in various geologic conditions, for example, high-velocity greatly metamorphosed media typical of mining regions, and of sedimentary rocks characteristic of oil areas.

Detailed studies of polarization have been carried out with the aid of polar correlation using polar seismograms obtained with a polarization analyzer [2, 14] simultaneously with the field recording of explosions. Methods of determining polarization parameters are discussed in detail in [2]. We would like to remind the reader that the criteria for the discrimination of a linearly-polarized oscillation with polar correlation [2, 9] includes a vertical alignment of the polar seismogram, conservation of the shape of the record, and a cosine law of amplitude variation. The presence of a phase shift on the traces of a polar seismogram results in a curved alignment, this being one of the criteria for non-linear polarization. It should be pointed out that the reason why the study of the polarization of pulses using the phase shifts and the shape of the alignment on polar seismograms is so effective, is that the traditional seismic exploration instrumentation is very sensitive to phase shifts. The study of phase shifts permits such fine detail to be established that could not be realized using amplitudes. Changes in the shape of the record are established on the basis of the variation of the ratio of amplitudes of adjacent extrema on various traces of a polar seismogram.

The term 'simple oscillation at a point' means a linearly polarized oscillation, which retains its direction of polarization over the entire recorded interval.

In the study of the polarization of the direct wave, of special interest is the polarization of simple short pulses. Among numerous such pulses recorded in PM–VSP, are sometimes those whose polarization remains linear during the whole recorded time of the wave. Figure 123, depicts three-component(a) and polar(b) seismograms of simple oscillations whose first two extrema are linearly polarized.

However, in the overwhelming majority of cases, direct waves in the real earth are not linearly polarized. The complexity of the polarization and of the structure of the direct longitudinal waves have been demonstrated earlier, when the potential of the polarization method was being described [2]. The polarization, as a rule, becomes less linear as the number of the extrema grows, and furthermore, the polarization of the first extremum is quite frequently observed to be already distorted. The complexity of the shape and the polarization of the direct wave are clearly visible on CDR–I seismograms (Figure 123 c) whose components are located in the first pulse P zero displacement plane. The first pulse P is not recorded on such seismograms, and because of this the pulses P_1 and P_2 forming part of the observed direct wave are clearly detectable. The first and the second extrema on the seismogram 1 (Figure 123, right) are linearly polarized, however, the shape of the oscillation at the third extremum is distorted, as is clearly visible on the record of the 21-st component where the first extrema are reduced to zero, and pulse P_2 with clearly-defined onsets is recorded. On seismograms 2, 3 the linear polarization of the first extremum has already been distorted and this manifests itself in phase shifts on the polar seismogram components. Similar examples of distortion of linear polarization are visible on seismograms in Figures 37 to 40 in Chapter 5.

A detailed analysis of voluminous data has demonstrated that in the study of the polarization of the direct wave, a large number of factors associated with the conditions

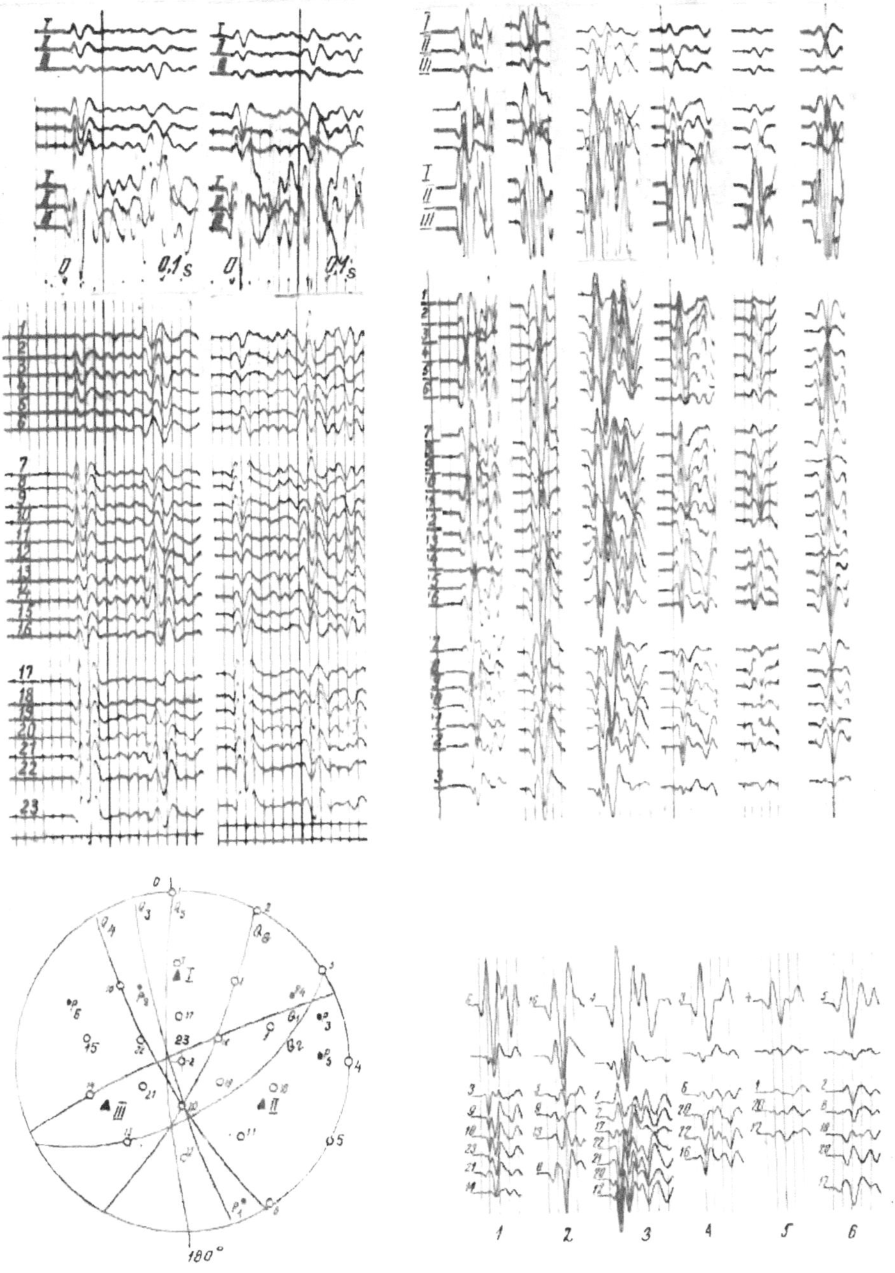

Fig. 123. PM–VSP seismograms illustrating direct wave wavelets. *Left*: seismograms of two wavelets first two periods of which are linearly-polarized. *Right*: seismograms of six non-linearly-polarized wavelets: (a) seismogram of components of the original I, II, III three-component coordinate system at three gain levels; (b) overall polar seismograms [1–23]; (c) planar CDR–I seismograms. Orientation of all seismogram components, displacement directions in the *P*-wave (P_1 – P_6) and zero-displacement planes (Q_1 – Q_6) shown on stereogram (d).

of source excitation, with recording techniques, and with geologic structure in the vicinity of the point of recording have to be investigated.

The polarization of oscillations is determined primarily by the structure of the direct wave, by the parameters of the pulses constituting it and by conditions of their separation (see Chpater 5). In cases where the pulses constituting the wave are very weak as compared with the first pulse, the linearity of the direct wave is usually not distorted (Figure 123, right, seismograms 1, 4, 5). The polarization of the first extrema is also determined by the magnitude of the phase shifts between the first arrival and the P_1 and the P_2 pulses. If Δt is large enough, the first phases can retain a linear polarization. Moreover, when an in-phase superposition takes place, linear polarization is not distorted, with the displacement direction changing only slightly, which is established chiefly by a change in the amplitude ratio of neighbouring extrema.

The nature of polarization of oscillations, is decisively influenced by the characteristics of the geologic structure in the vicinity of the point of observation. A noteworthy point is the association of particular sections of the cross-section with disruption of the linear polarization condition. There is evidence of an interconnection between the nature and the complexity of the polarization of a direct wave and the complexity of the seismogram as a whole. The latter is determined by the complexity of the geologic section itself.

Non-linear polarization of the first extremum is more often observed in the case of an oblique direction of arrival. For instance, the incidence of non-linearly polarized oscillations in observations at a depth of 260 m is much greater that at 760 m (Figure 39e).

Moreover, the number of points where the polarization of the first extremum of the direct wave is distorted is not the same for different directions of the profile, or for different shotpoints, which may also be the result of the characteristics of the geologic structure.

A strong dependence of the nature of the polarization of the first arrival pulse on the frequency spectrum of the direct wave is also observed. As dominant frequencies rise, the polarizaton, as a rule, becomes more complex. The reason is that progressively finer inhomogeneities begin to affect the nature of polarization. For each type of rock, there is a limiting frequency above which the polarization remains complex. This limiting frequency, determined by the structure of the rock itself, can serve as a parameter that characterizes the rock. Naturally, this parameter should be measured in thick homogeneous beds. In this respect it is similar to absorption.

Several factors may exercise a strong influence on measured polarization parameters. These include the conditions of installation of the geophone, the rigidity of the borehole tool-borehole wall contact, the vibrations of casing strings, parasitic vibrations of the tool itself etc. Experience shows that measurement techniques can greatly distort the polarization, and rigorous control of the quality of the data obtained is necessary.

It would appear expedient to subdivide the research on polarization into two stages. During the first stage only particle displacement directions should be studied and utilized, disregarding any elliptical motion. During the second stage, the parameters studied and utilized should include the nature of polarization, the shape of the trajectory, the orientation of the polarization plane etc. Until recently, the studies have been chiefly limited in scope to the model of linearly polarized oscillations. However, the experience gained justifies the exptectation that further studies of polarization parameters will result in their more comprehensive use within the non-linear polarization model to obtain detailed information about the geologic section.

The present chapter deals with the polarization studied first within the framework of the linear polarization model, and subsequently within the framework of the elliptical polarization model.

2. Studying the Direction of Particle Displacement Along a Vertical Profile

Generally the directions of arrival of waves as well as the directions of particle displacement, when observed downhole, vary from point to point. In homogeneous and isotropic media, these directions on the whole correspond to the directions of propagation. The same holds also for homogeneous isotropic layered media where there is no superposition of waves. Given such conditions, the laws of variation of particle displacement directions can be interpreted quantitatively to obtain independent and additional information about the heterogeneities of the earth, about the velocity distribution, and about the elements determining the position of the interfaces in the section contiguous with the borehole.

Certain laws of the variation of the direction of particle motion in isotropic media. Consider first the variation of directions of particle motion for longitudinal waves in an isotropic medium as a function of the relative disposition of points of excitation and reception. In the general case of an inhomogeneous medium with non-horizontal layers, these directions may have different azimuths. In the simplest case of an axially-symmetrical isotropic medium, the discussion can be limited to the vertical plane only. Such cases are not often encountered in practice, but still they are of some interest, because they can be used to approximate certain laws of variation of velocity with depth actually observed. In this respect they can be regarded as a kind of standard.

A constant-velocity medium. The particle displacement directions for $\beta = 0$ for a specific depth are functions only of source offset and are independent of the velocity. It will be seen from the curves of the directions of motion for three offset values (Figure 124a, line 5) that for each offset the curve approaches the vertical as depth increases.

A continuously-changing velocity medium. Consider a medium in which velocity increases linearly with depth with the coefficient β assuming values from $-0.0025\ \mathrm{m}^{-1}$ to $0.00505\ \mathrm{m}^{-1}$. In Figure 124a there are particle displacement direction curves to correspond to seven values of the velocity gradient. In media with a positive gradient, $\beta > 0$ (lines 1 to 4), the particle displacement direction curves in the upper portion of the profile are contained in the sector $\varphi = 0$–$90°$. As the depth is increased, the curves shift to the sector $\varphi = 90$–$180°$. For $\beta < 0$, the direct wave begins to be recorded only from a definite depth depending on β and offset. With the increase in offset for constant $\beta > 0$, the direction curves shift to the left. The shape of the curves changes little. The family of curves just described can be regarded as a chart for easily and rapidly determining the coefficient of velocity increase β. To this end, one needs only to have a section of the curve corresponding to horizontal displacement, which can be reliably identified on the record of the vertical component by the reversal of signs of arrivals.

A layered medium. Consider the case of a medium made up of two layers with a horizontal interface; in the first layer the velocity varies linearly and in the second layer it is constant. There is a velocity jump at the interface between the layers (Figure 124b). The particle motion direction curves have the following characteristics.

(1) Down to the interface the particle motion direction curves correspond to particle motion of curved wave path in the first layer.

(2) The curve experiences an abrupt discontinuity at the interface (a jump in the

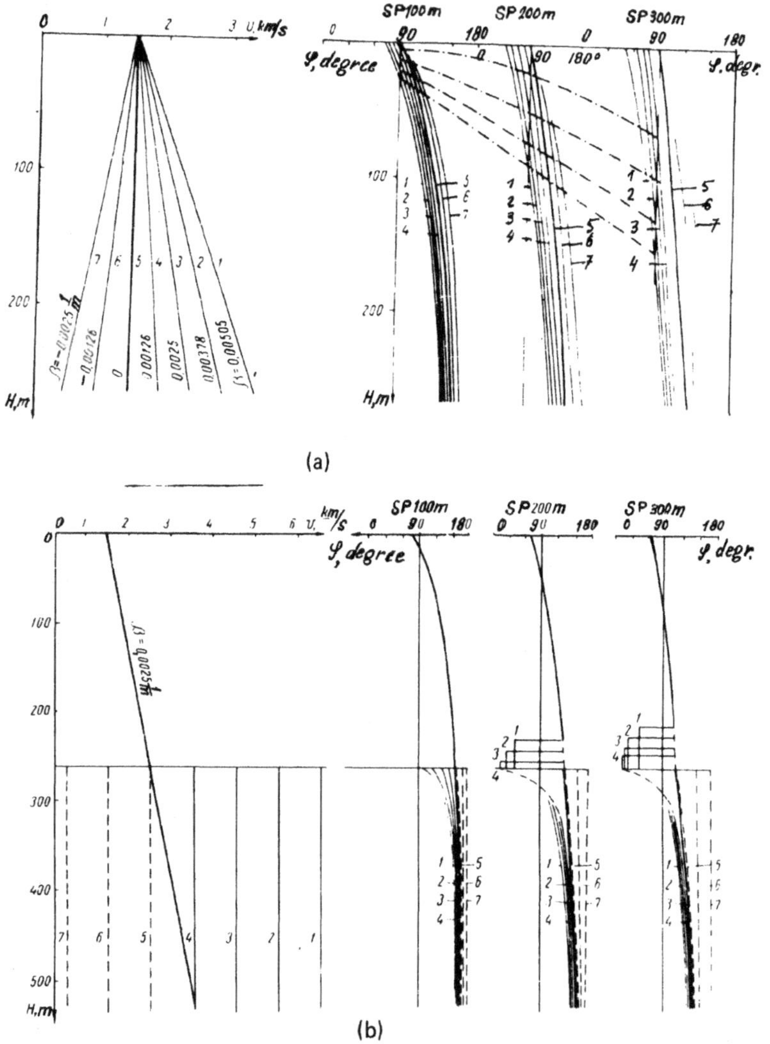

Fig. 124a-b.

Fig. 124a-c. Plots of particle displacement directions: (a) for media with a linear velocity vs depth variation (line of change of signs of horizontal displacements shown by dash-and-dot line); (b) for media with a velocity jump (the representation for head waves is a schematic one); (c) for media with a velocity gradient jump.

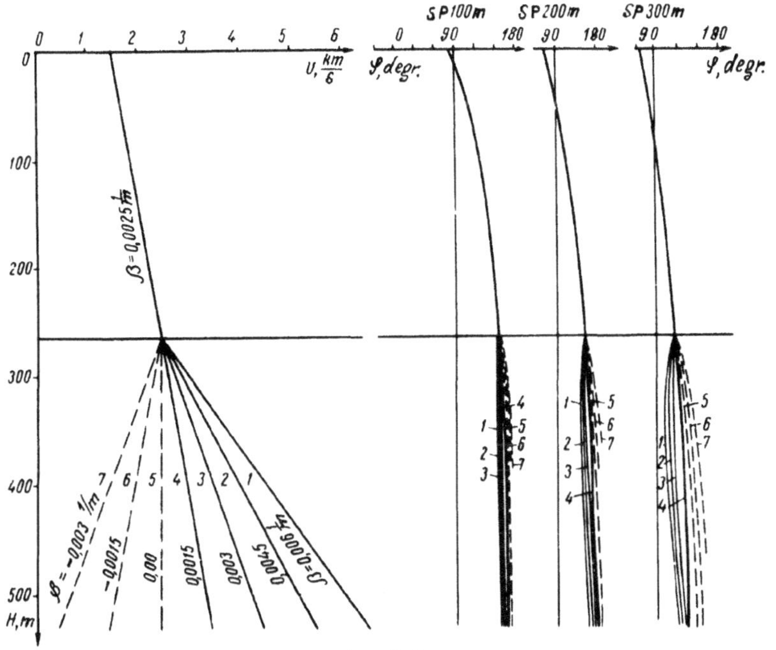

Fig. 124c.

displacement directions). The magnitude of the jump depends on the ratio of the velocities at the interface between the upper and the lower layer v_2/v_1. The ratio of these velocities can be determined from the magnitude of the change of displacement directions and a known displacement direction in one of the layers. As the velocity increases, the curves shift towards greater values of φ, and *vice-versa*.

(3) For head waves, the curve of particle motion direction lies in the sector $\varphi = 0–90°$, is a function of the velocity ratio v_2/v_1 and does not change with depth. At the point of interference of the direct and head waves, the curve experiences a discontinuity and makes a jump from the sector $\varphi = 90–180°$ to $\varphi = 0–90°$. The transition from a head wave to a transmitted one is accompanied by distortion of linear polarization, and the curves are indicated schematically by a dotted line (Figure 124b).

In the case of a change of the velocity gradient (Figure 124c) no discontinuity in the curve is observed at the interface. The discontinuity is experienced by the derivative of the curve. The transition to the medium with a greater gradient results in the curves being displaced to the left, a transition to the medium with a smaller gradient causing them to be displaced to the right. Such displacements grow with the increase in offset and the magnitude of the gradient change. As offset increases, the curve may either partially or completely move to the sector $\varphi = 0–90°$.

On the polarization of waves in anisotropic media. In contrast to isotropic media, the particle displacement direction of longitudinal waves in anisotropic media may not coincide with the direction of propagation, and greatly depends on the parameters of the anisotropy, in particular on the orientation of the direction of propagation of waves with respect to the axis of anisotropy. The theory of propagation of seismic waves in anisotropic media [146] developed in recent years has extended the means for studying the displacement field. This is quite important, because difficulties have been encountered in experimental studies of polarization, in particular those associated with the discrimination between the effects of anisotropy and those of heterogeneities of the Earth.

In an anisotropic medium, a longitudinal P-wave and two transverse waves, S_1 and S_2, can propagate independently. At no point on the longitudinal wave front, do the vectors of the normal $\mathbf{v}_n$, and the ray velocity $\mathbf{v}_r$ and the particle displacement $\mathbf{u}_P$ coincide. The displacement vectors $\mathbf{u}_{S_1}$ and $\mathbf{u}_{S_2}$ do not lie in planes orthoganal to the ray or the normal to the wave front ($\mathbf{v}_r$ or $\mathbf{v}_n$). The vectors $\mathbf{v}_r$, $\mathbf{v}_n$, and $\mathbf{u}_P$ will coincide only in the simplest case of a transversely-anisotropic model, the displacement in the S-wave being orthogonal to the anisotropy axis, i.e. lying in the plane of anisotropy. For other directions, the directions of the vectors usually do not coincide. However, the vectors $\mathbf{v}_r$, $\mathbf{v}_n$ and $\mathbf{u}$ for the P- and S_1-waves always lie in one plane passing through the anisotropy axis. For the S_2-wave, vectors $\mathbf{v}_r$ and $\mathbf{v}_n$ remain in the same plane, but the vector $\mathbf{u}_{S_2}$ lies in the anisotropy plane and is therefore always orthogonal to the vectors $\mathbf{v}_r$, $\mathbf{v}_n$. The vectors $\mathbf{u}$ and $\mathbf{v}_n$ remain parallel to each other along a ray path. The vector $\mathbf{u}_{S_2}$ is orthogonal to the vectors $\mathbf{u}_P$ and $\mathbf{u}_{S_1}$. The vector $\mathbf{u}_{S_1}$ will be orthogonal to $\mathbf{u}_P$ only if the normals to the P- and S-wave fronts coincide. We shall not dwell upon the general aspects of the theory of propagation of waves in anisotropic media, advising the reader instead to turn to appropriate literature [146], but to illustrate the point, only cite displacement directions for certain of the most elementary cases*.

Consider particle displacement directions for the P_1-, S_1-, and S_2-waves refracted at a horizontal interface between a two-layered medium and recorded immediately beneath the interface, with the upper isotropic layer having $v_P = 2.5$ km s^{-1}, $v_S = 1.5$ km s^{-1} and being 1.5 km thick. The lower half-space is anisotropic (the velocity profiles are depicted in Figure 125a), the anisotropy axis making an angle of 20° with the vertical, and the angle between the direction pointing to the source and the anisotropy axis being equal to 45°. For the sake of convenience and clarity, all plotting has been made on stereo nets.

Q_r is a plane passing through the anisotropy axis (point 0) and the directions of $\mathbf{v}_r$ (point 4), $\mathbf{v}_n$ (point 3); $Q_{P(1,2)}$ is a plane passing through the normals to the incident (point 1) and the refracted (point 2) wave fronts.

Consider the displacement directions for the P-, S_2-, and S_1-waves recorded at a point (Figure 125, top).

In the case of an isotropic medium, the normal for the incident wave and the ray path coincide in direction (point 5). The same is true of the refracted wave (point 6). Both directions lie in the same plane Q_{P_1} drawn through the source, the receiver and the normal to the interface (in the example being considered it is vertical).

In the case of an anisotropic medium, the directions of the normal and of the ray path of the incident wave (point 1) and of the normal to the refracted wave (point 2) deviate from the plane Q_{P_1} and lie in the plane of incidence Q_{P2}. The displacement vector of the refracted wave (point 3) and the ray of the refracted wave (point 4) for the P- and

* B. M. Kashatan has carried out the calculations, and V. A. Gokhblit has plotted the graphs.

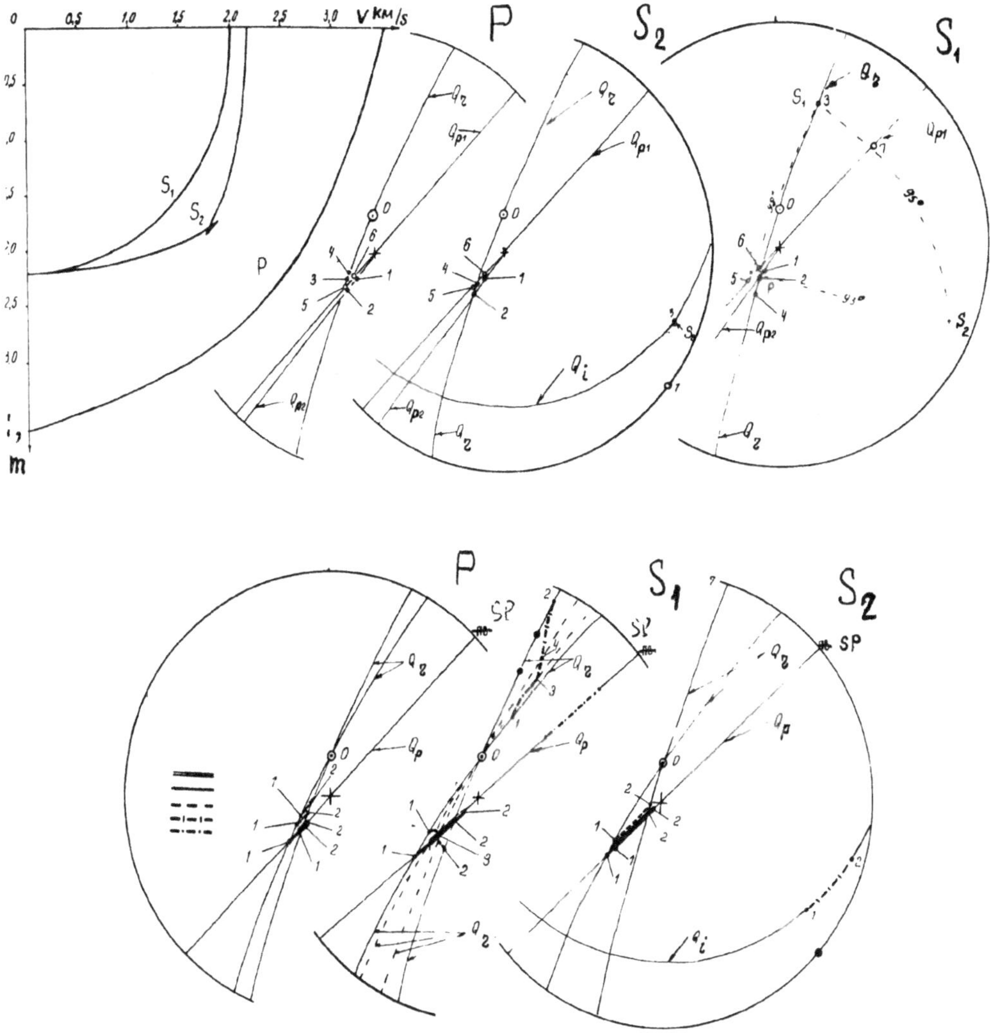

Fig. 125. Particle displacement directions in anisotropic media. *Top*: graphs of variation of velocities of *P*- and *S*-waves with depth, displacement direction in P, S_2, S_1 waves recorded at a point; the following directions are shown on stereo nets: 0 – anisotropy axis, 1 – normal and ray of incident wave, 2 – normal of refracted wave, 3 – displacement vector of refracted wave, 4 – ray of refracted wave, 5 and 6 – normal and ray of incident wave and normal and ray of refracted wave in case of an interface between two isotropic media, 7 – displacement vector of wave incident on interface between an isotropic and an anisotropic layer. *Bottom*: variations of displacement directions in refracted P-, S_2-, and S_1-waves caused by variations of directions of the incident wave: (a) interval of directions of rays and of normals of the incident wave; (b) interval of directions of normals of the refracted wave; (c) interval of directions of rays of the refracted wave; (d) displacement directions in the refracted wave; (e) region of displacement directions for the incident wave.

S_1-waves lie in the plane of rays Q_r passing through the anisotropy axis. The displacement vector of the S_1-wave (point 3) deviates from a similar vector in an isotropic medium (point 7) in the direction of the anisotropy axis. For all three waves, the refracted ray (point 4) deviates from the incidence plane, the maximum deviation being observed for the S_1-wave.

For the waves P and S_2, the vector v_n makes a greater angle with the anisotropy axis than the vector v_r; for the wave S_1, on the contrary, the deflection of the vector v_r from the anisotropy axis is greater than that of v_n. Hence, a transverse wave that in an isotropic medium has displacement directions shown by point 7, in an anisotropic medium splits into two waves S_1 and S_2. The displacement directions of all three types of waves are orthogonal.

Consider now the variations of the directions of the vectors $\mathbf{v}_r$, $\mathbf{v}_n$, and $\mathbf{u}$ for refracted P-, S_1-, and S_2-waves that take place following the variation of the direction of the incident wave (Figure 125, bottom).

Only the normals of the incident and the refracted waves lie in the vertical plane Q_P drawn through the receiver. As the result of refraction, the directions $\mathbf{v}_r$ deviate from the vertical plane, and the rays will no longer impinge on the borehole. The vectors $\mathbf{v}_n$, $\mathbf{v}_r$, and $\mathbf{u}$ for the P- and the S_1-waves lie in the same plane passing through the anisotropy axis. For the P-wave, the surface described by the rays is displaced from the vertical source-receiver plane towards the anisotropy axis. The surface described by the displacement directions is displaced even more in the same direciton. The displacement directions approach the vertical. For the S_2-wave, the rays are displaced towards the anisotropy axis, and the displacement directions lie in the isotropy plane Q_i.

The most complex surface is described by the displacement vector in the S_1-wave, in which, within the range of angles considered here, the relations between the directions of v_r and v_n, v_n and u_{S_1} as well as v_r and u_{S_1} vary as the directions change. As the direction of the normal to the refracted wave changes from point 1 to point 3, the directions of the vectors v_n and v_r draw monotonously closer, coinciding at point 3. As the direction of the normal continues to change from point 3 to point 2, the vectors v_n and v_r are observed to deflect from each other, with v_r moving away from the anisotropy axis.

The directions of the vector u_{S_1} vary correspondingly. Between the points 1 to 4 the displacement vector deflects from the normal to the ray, between the points 4 to 2 the deflection is in the direction opposite to the anisotropy axis. The examples cited demonstrate the strong influence exercised on the displacement directions by the anisotropic properties of the Earth, by the direction of the anisotropy axis and by the relations between the other parameters. Such influence may in principle be utilized to detect anisotropy and to determine its parameters.

It is worth noting that, in addition to anisotropy in the vertical direction, anisotropy in the horizontal direction is also observed in the real earth. Such anisotropy is due mainly to the stresses in the rock. This makes it possible, in particular, to utilize polarization, especially of transverse waves, to study the variation of the stresses of rock with time. However, as yet those aspects have been studied only inadequately.

As we have already seen, anisotropy exercises a strong influence on polarization, especially of transverse waves. It would be very important from the practical point of view on the one hand to use the polarization parameters to determine the anisotropic

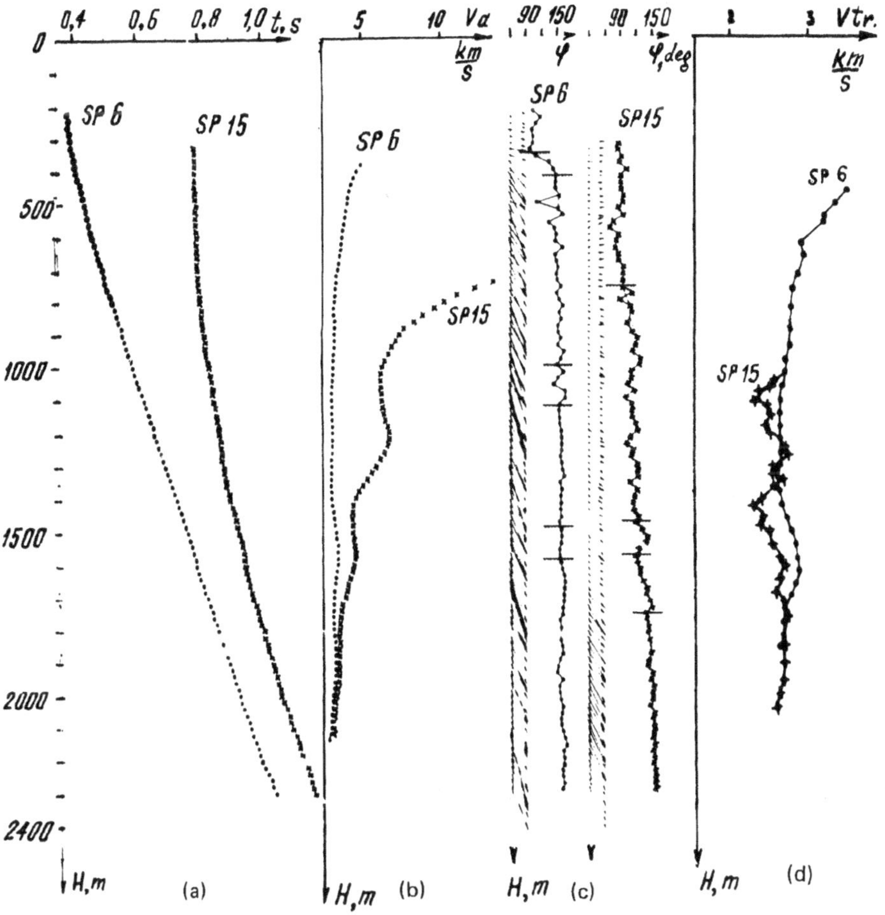

Fig. 126. Example of determination of true velocities (borehole N66 Novo–Dmitriyevskaya).

properties of the earth, and on the other to take them into account when determining other elastic properties. However, it will be necessary to continue research both in the theoretical aspect to study the relationships of displacement directions of various wave types in various media, and in the practical aspect to develop techniques for measuring trajectories of particle motion. Methods should be developed for determining the polarization of waves in conditions of complex wave fields, and the superposition of a large number of waves.

Determining velocities from observations along off-line profiles. The nature of the dependence of the displacement direction on the velocity structure within the framework of an isotropic earth model as presented above, can be used to obtain additional and independent information about the velocity structure using observations along off-line profiles.

By way of an example of the use of displacement directions, consider the determination of true velocities from off-line traveltime curves of first arrivals. When true velocities are being determined, the off-line traveltime curves are reduced to the vertical, i.e., to in-line traveltime curves, from which the true velocities are determined. However, such reduction necessitates the introduction of greatly simplifying assumptions about the earth, which make the results largely worthless. Displacement directions enable the true velocities to be determined without such reduction. Figure 126 depicts an example of the determination of the true velocity from two off-line traveltime curves obtained from shotpoints at 600 m and 1500 m offsets. The apparent velocity curves (Figure 126b) and the displacement direction curves (Figure 126c) are used for this purpose. The displacement directions are presumed to correspond to those of wave propagation, and this, as has already been pointed out, is not always correct. However, the results obtained (Figure 126d) agree on the whole with the velocity curves plotted from in-line traveltime curves.

Thus, whereas reflected waves enable the geologic section to be divided according to acoustic rigidity, and the values of changes in rigidity at the interfaces to be determined, the polarization of the direct wave provides independent data concerning the velocity change at the interfaces. By considering these data jointly, one gains a more thorough knowledge of the section.

3. Particle Displacement Directions in the Vertical Plane

The displacement directions can be represented in the form of fields, i.e., sets of directions, which permit the continuous tracking of variations of displacement directions in any plane, in particular, in the vertical. Such a field can be built for waves from one shotpoint and observed in numerous boreholes. In the absence of numerous boreholes, the field can be built from observations in a single borehole of waves from numerous shotpoints arranged on the surface along a definite line or profile. In this case, when the field is being built, all the shotpoints are imagined to be placed at one point close to the top of the borehole, and the particle displacement directions obtained are plotted at the appropriate depths and beneath the shotpoints. Such a construction is permissible in the case of an axially-symmetrical medium in the section covered by the field. The first displacement direction fields were obtained in 1961 in certain regions of Central Asia, where research on the nature of unwanted waves necessitated detailed studies of the upper portion of the section and the wave field associated with it [14].

Consider the field of particle displacement directions obtained in Southern Kirgizia (Figure 127). The upper portion of the section features terrigeneous deposits with a weak velocity differentiation. Three interfaces at depths of 18 m, 34 m, and 55 m (Figure 127a) are detectable on the velocity profile built as the result of detailed VSP studies, torpedoeing and horizontal shooting between two boreholes. At greater depths, the velocity increases linearly with the coefficient $\beta = 0.001\ m^{-1}$. It was not possible to detect the presence of a velocity gradient in comparatively thin upper layers. To obtain information about the structure of the Earth in the uppermost portion of the section, we carried out systematic observations of waves in the vertical plane at depths from 0 to 90 m from 11 shotpoints at offsets in the range $l = 0$ to 330 m. It will be seen from the displacement direction field (Figure 127b) that for $l = 30$ m, the vectors at the

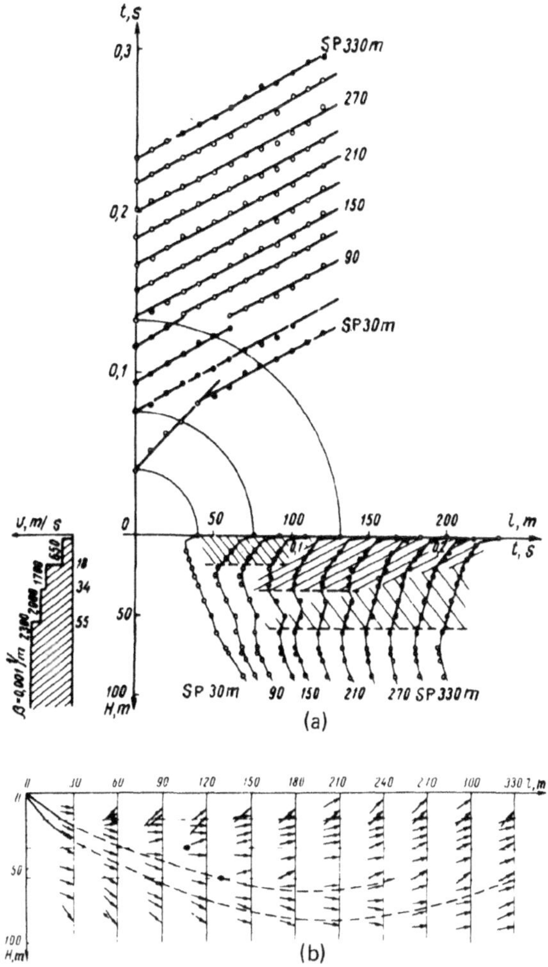

Fig. 127. Family of combined horizontal-vertical traveltime curves (a) and field of displacement directions (b). Regions of registration of refracted waves associated with different interfaces are shaded (velocity cross-section obtained from seismic logs and from horizontal well shooting).

shallowest depths (10 m and 14 m) point downwards, which corresponds to the arrival of a direct wave. As the offset increases, the first wave to be recorded at the same shallow depths is the wave refracted at the first and at deeper interfaces. At $l > 150$ m, the wave refracted at the second interface appears in the first arrivals. The particle displacement directions at a depth of 40 m for $l = 120$ m confirm its existence. This wave, having been refracted at the upper interface (the reflection is indicated by the change in the displacement direction at a depth of 20 to 14 m for $l = 150$ m), emerges at the surface. The point of emergence of the wave refracted at the third interface for l equal to about 130 m, is located at a depth of about 5.5 m. This wave interferes with those refracted at the upper interfaces, and within the field observed, appears as a first arrival at the surface for l = 330 m. At most points located close to the interfaces, a distortion of linear polarization is observable. This fact may be interesing in itself. A comparison between the displacement direction field and the observed combined traveltime curves (Figure 127a) confirms the validity of our interpretation of the wave pattern. For the sake of clarity, rays plotted from kinematic data are shown in Figure 127 by dotted lines.

Figure 128 depicts a field of displacement directions and a field of isochrons of the first arrivals plotted from data obtained in VSP studies from shotpoints located at the surface along two mutually orthogonal profiles at different offsets. A comparison between the displacement direction field and the isochron field in the vertical plane shows that the displacement directions do not coincide with those of the gradient vector of the isochron field, which is evidence of anisotropic properties of the earth. This is convincingly shown by horizontal sections of the isochron field at various levels (Figure 128b). It is most distinctly visible at the levels of 810 m and 1210 m. A detailed displacement direction field in conditions of weak velocity differentiation is presented and scrupulously analyzed in [2, 14].

The experience gained demonstrates that quantitative studies of the laws of variation of particle displacement directions enable additional information about the nature of recorded waves, and details of the velocity section to be obtained and, in particular, not only make it possible to establish the presence of a velocity gradient, but also to determine its magnitude, and the region where mode conversion of waves takes place. Of special interest are instances of the distortion of polarization or the incongruence between the direction of particle displacement in the first longitudinal wave and its direction of propagation. Such phenomena are observed in the case of downward transmitted waves in sections of the profile contiguous with the refracting interface

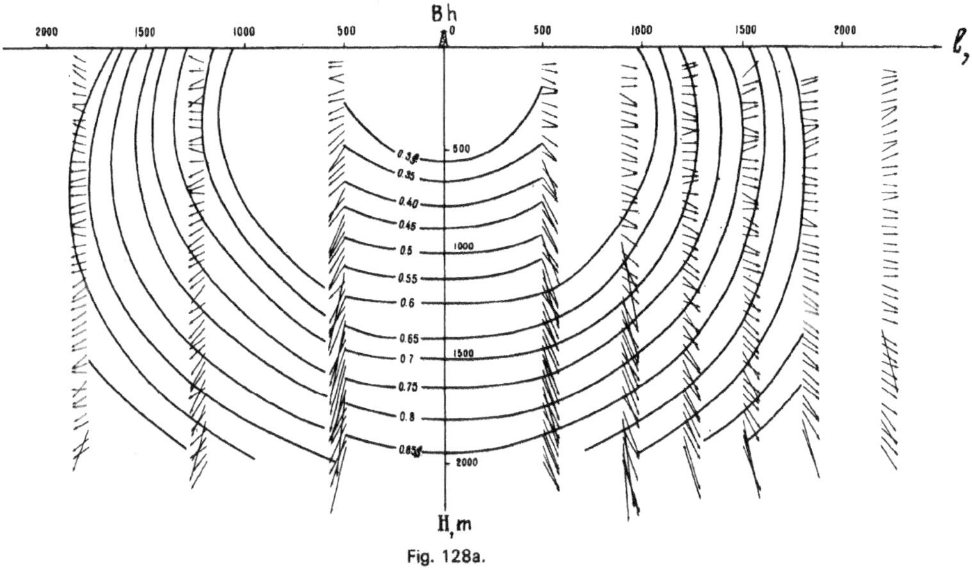

Fig. 128a.

Fig. 128a-b. Displacement direction and time fields in the vertical plane (a) and time field in horizontal planes on different profiles (b): 1 – observed times, 2 – normal fields.

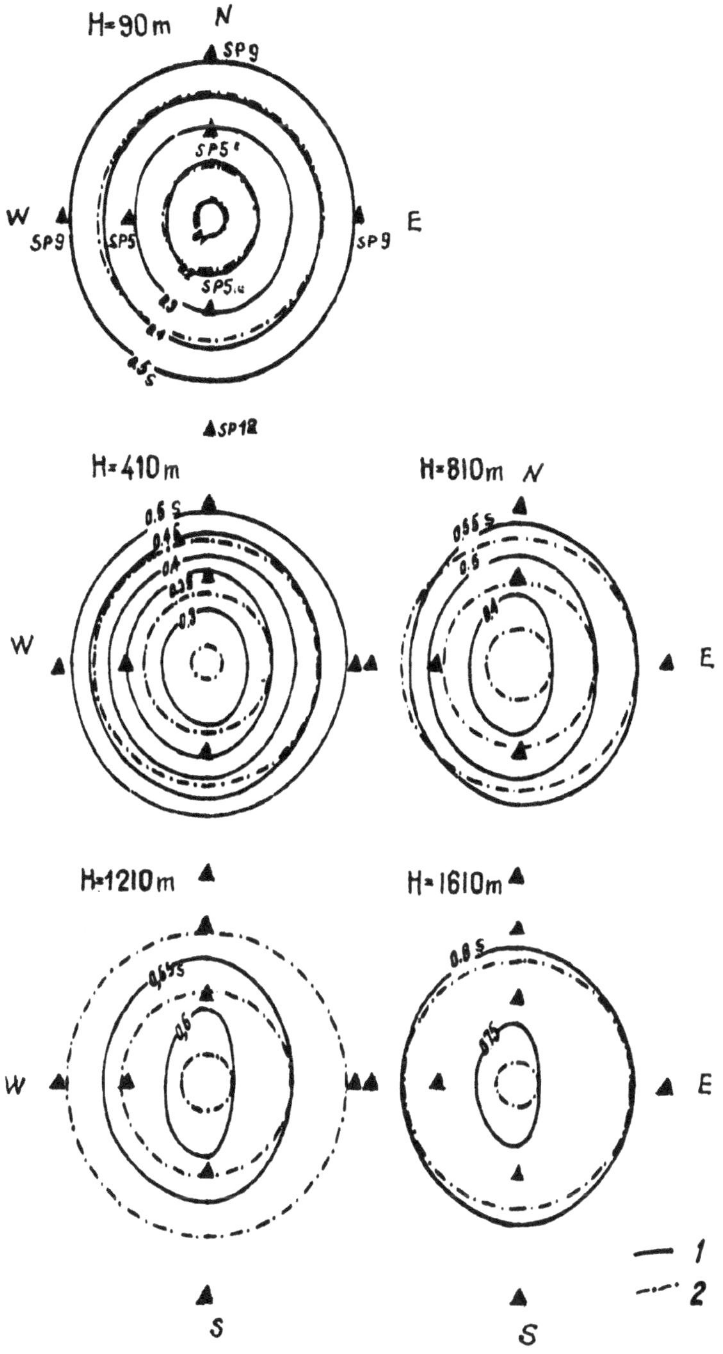
H=90m
N
SP9
W
SP9
SP5
E
SP9
0.5 s
SP12
H=410m
0.6 s
W
H=810m
N
0.65 s
0.6
0.4
E
H=1210m
W
0.6
H=1610m
0.8 s
E
S
S
1
2

Fig. 128b.

(grazing rays [2]), and of curved-path waves in the section of deepest penetration. The displacement direction of grazing rays does not coincide with that of propagation, and can be represented by two components: the radial, coinciding in direction of propagation, and a component almost coinciding in direction with that of displacement in the head wave. The latter greatly exceeds the radial at the interface, but it decays rapidly with the distance from the interface, and the radial component becomes the dominant one – the transmitted wave becomes a linearly polarized wave.

Displacement directions often enable the nature of waves to be determined when the kinematic data are inadequate. Media with a weak velocity differentiation can be regarded as such a case. It is well known that it is not always possible to identify precisely a change in the wave type from a vertical traveltime curve. This can be done much more easily from displacement directions. For example, when a direct wave converts to a head wave (the head wave emerges as a first arrival), the particle displacement direction curve crosses sharply into the section $\varphi = 0–90°$. A sharp change in the displacement directions inside the earth (away from interfaces) may serve as evidence of interference or conversion.

4. Studying Displacement Directions in Space

Generally the Earth's structure is such that the particle displacement directions vary not only in a plane, but spatially, i.e., the angles vary both in the vertical and the horizontal planes. Neglect of the azimuthal variations of the particle displacement directions may lead to substantial errors in the data interpretation. By studying the spatial variation of displacement directions, we may acquire additional independent information about the Earth's structure. For the simplest structures, methods based on the stereo net have been devised for determining the elements that describe the position of interfaces from trajectories of particle motion in sections penetrated by the borehole [11]. Progressive development of this trend and perfection of measurement techniques will enable the particle motion trajectories to be studied spatially and their rigorous quantitative processing to be performed. However, to this end one will need three-component records oriented either by means of deliberate orientation of geophones or by means of orientation sensors (see Chapter 2).

Various observation systems can be employed to study spatial displacement directions. It is technically easy to provide systems in which observations are carried out in a single hole, with the sources being arranged on the surface along lines or across an area.

Consider the results obtained in the studies of displacement directions of the first arrivals in the Zhairem polymetallic field. At first PM–VSP observations were carried out in two boreholes separated by a distance of about 600 m. Each borehole was shot from several shotpoints: borehole N3130 from four and borehole N3106 from three (Figure 129).

The vertical traveltime curves of the first arrivals are represented by smooth lines. According to plots of interval velocities and sonic logging data, the section is a high-velocity weakly differentiated one. In borehole N3106 (Figure 129a), in addition to a sharp change in the angles associated with an interface at a depth of about 80 m, an increase in the angles corresponding to a low-velocity layer, typical of nodular lime-

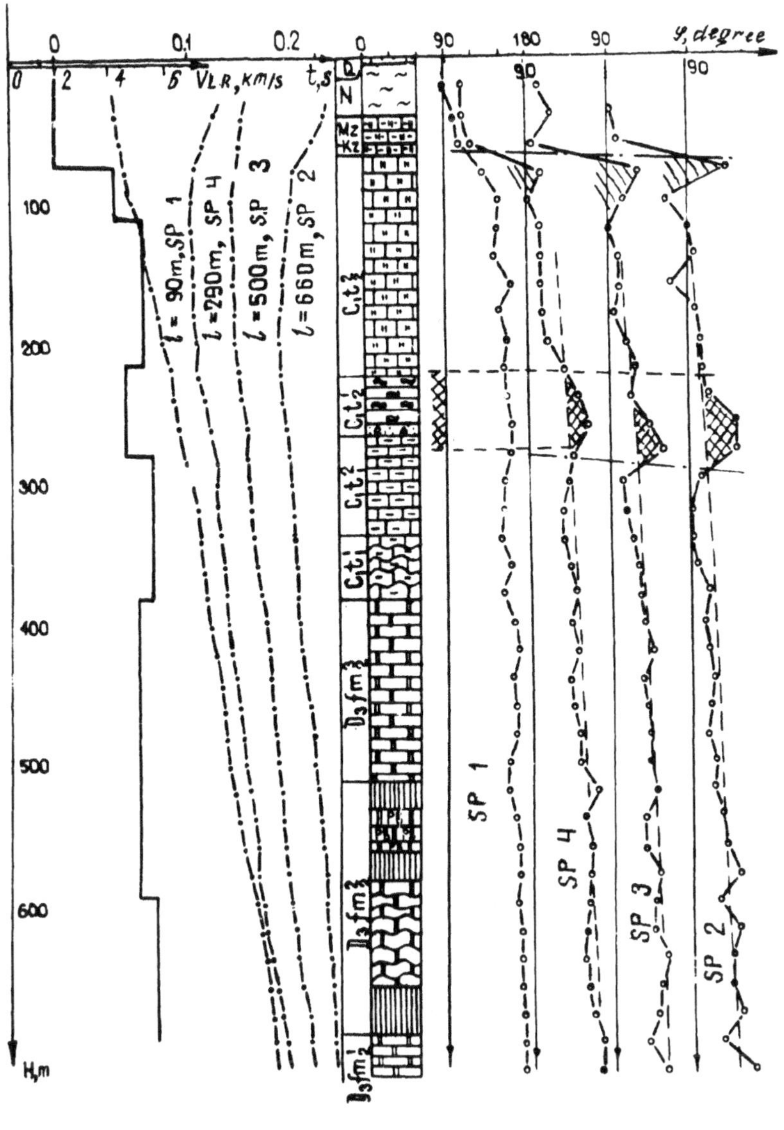

Fig. 129a.

Fig. 129a-b. Studying displacement directions along vertical profiles in the polymetallic field Zhairem (Kazakhstan): borehole N3106 (a) and borehole N3130 (b). Below-schematic representation of disposition of boreholes and shotpoints (courtesy R. N. Khairutdinov and T. G. Chastnaya): 1 – sands and clays of the Aral suit; 2 – weathering crust, 3 – quartz limestones, 4 – nodular limestones, tuffaceous sandstones and pelitic tuffs, 5 – quartz-carbonate rock, 6 – red quartz limestones, 7 – iron ore horizons, 8 – led-zinc and barite ores, 9 – quartz limestones with carbonate concretions.

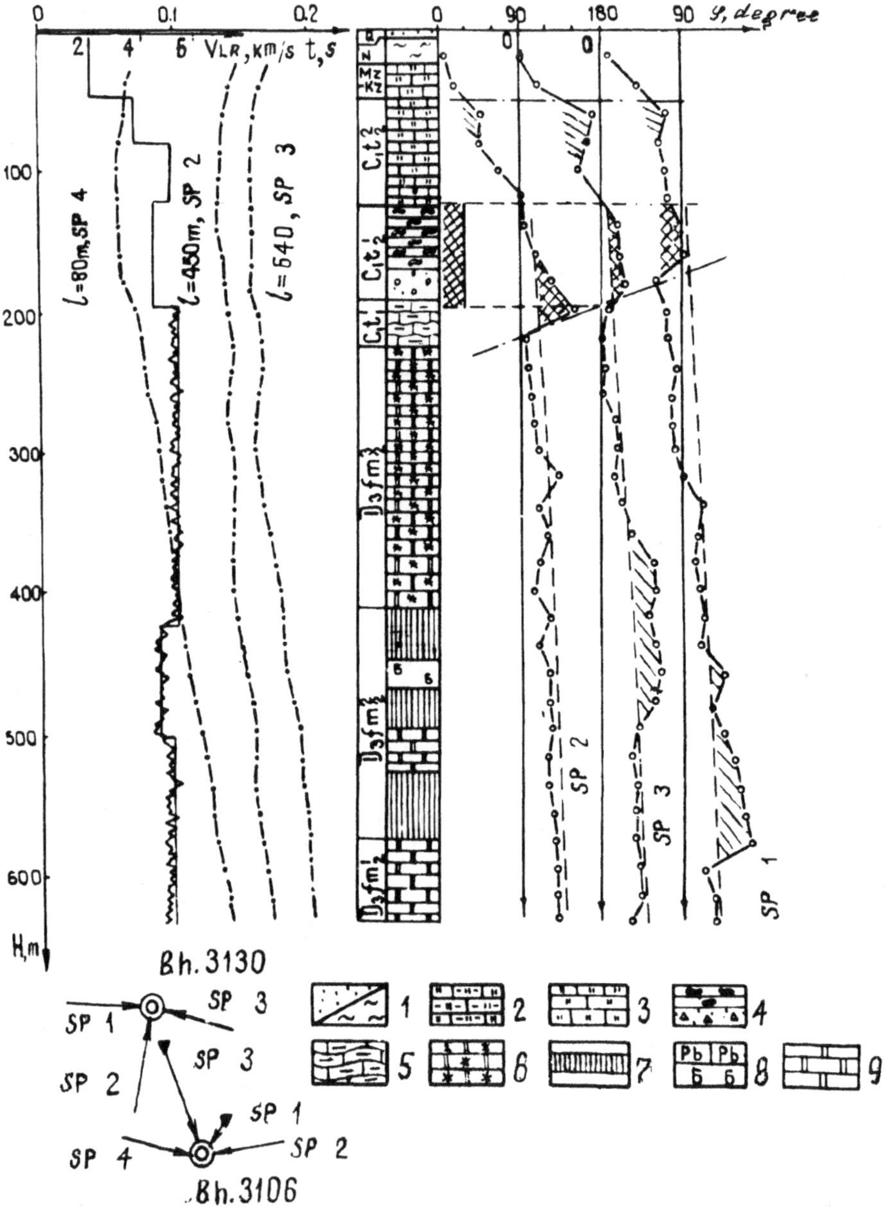

Fig. 129b.

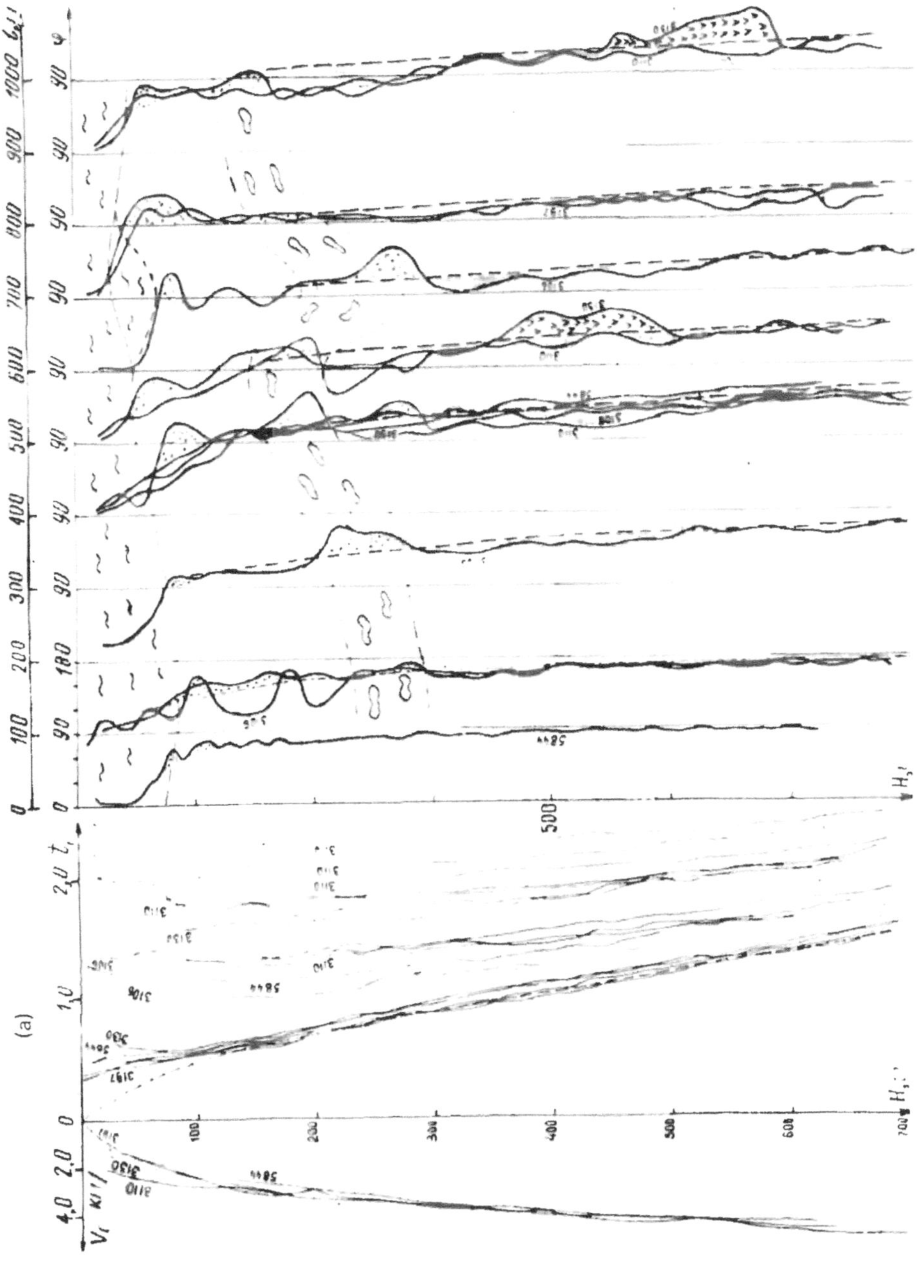

Fig. 130. Summary data on tracking heterogeneities by particle displacement directions (Zhairem): (a) plots of $v_{av}(H)$; (b) vertical traveltime curves; (c) plots of $\varphi(H)$ in the vertical plane.

stones, tufaceous sandstones and pelitic tuffs, is observed in the 220 to 280 m depth interval. The layer thickness, according to PM–VSP data, is about 40 m. In borehole N3130 (Figure 129b), the low-velocity layer appears at depths of 130 to 200 m. According to these displacement direction data, the low-velocity layer dips in the southward direction.

Angular anomalies as great as 30° to 40° are observable in borehole 3130 from shotpoint 3 in the 360 to 500 m depth range, and from shotpoint 1 in the 500 to 600 m range. As established from sonic logs, the second low-velocity layer corresponds to a seam containing lead-zinc ores with a high barite content. The depths of the anomalies and the relative disposition of the shotpoints serve as evidence of the barite body dipping in the south-east direction. The anomaly described above is not observable in borehole N3106, and the displacement direction curve on the whole coincides with theoretical curves computed for a homogeneous medium.

It should be pointed out that the relationships considered above are reliably trackable from PM VSP data in all the nine boreholes in the area of the East Zhairem syncline (Figure 130c).

It is difficult to differentiate the section using summary vertical traveltime curves plotted for the East Zhairem section. The angular plots enable the bottom of the low-velocity zone, the low-velocity layer of nodular limestones and underlying tuffaceous sandstones and pelitic tuffs to be tracked. In individual boreholes from some shotpoints, an angular anomaly associated with a barite deposit is detectable.

The second stage of the studies dealt with displacement directions in fixed geophone level observations. Recording was performed in borehole N3100 at three depths: above (at 240 m and 460 m) and below (at 760 m) the ore deposit. Vibrations from shotpoints arranged on the surface along six rectilinear profiles 1000 m long with azimuths of 0°, 45°, 135°, 180°, 225°, and 315° were recorded at each depth. On each profile shots were fired at intervals of 50 m. The disposition of shotpoints is depicted schematically in Figure 131.

Consider by way of an example the observation data obtained at the depth of 760 m. The time field is generally symmetrical (Figure 131a). It should be noted that, according to inclinometer data, at the depth of 760 m, the borehole axis is inclined at an angle of 28° to the vertical with an azimuth of 240°. The observed field of displacements, corrected for the borehole geophone tilt, is shown on the stereogram (line 1 in Figure 131b). Also shown are uncorrected displacement directions (2 in Figure 131b) and corrected directions to the sources (3 in Figure 131b) for one ray (III). Since the geophone assembly in the borehole has not been oriented, azimuthal correlation was performed during the course of processing by rotating the whole field. Prior information about the structure was also taken into account. Specifically, minimum azimuthal deviations were to be expected along the ray (V) coinciding with the line of dip. Normal values of angles with the vertical (φ) and azimuths (ω) computed with the assumption of a homogeneous medium, were subtracted from the observed data. In this way, plots of positive and negative anomalies of angles φ and ω have been obtained for every profile; for the set of profiles, they form fields of anomalies (Figure 131c, d). The region of negative anomalies φ in the south-west part corresponds to the zone of lower velocities associated with an ore body dipping in the south–east direction. The negative azimuthal anomalies associated with the general structure of the layers on the slope of an ore body correspond to the same region.

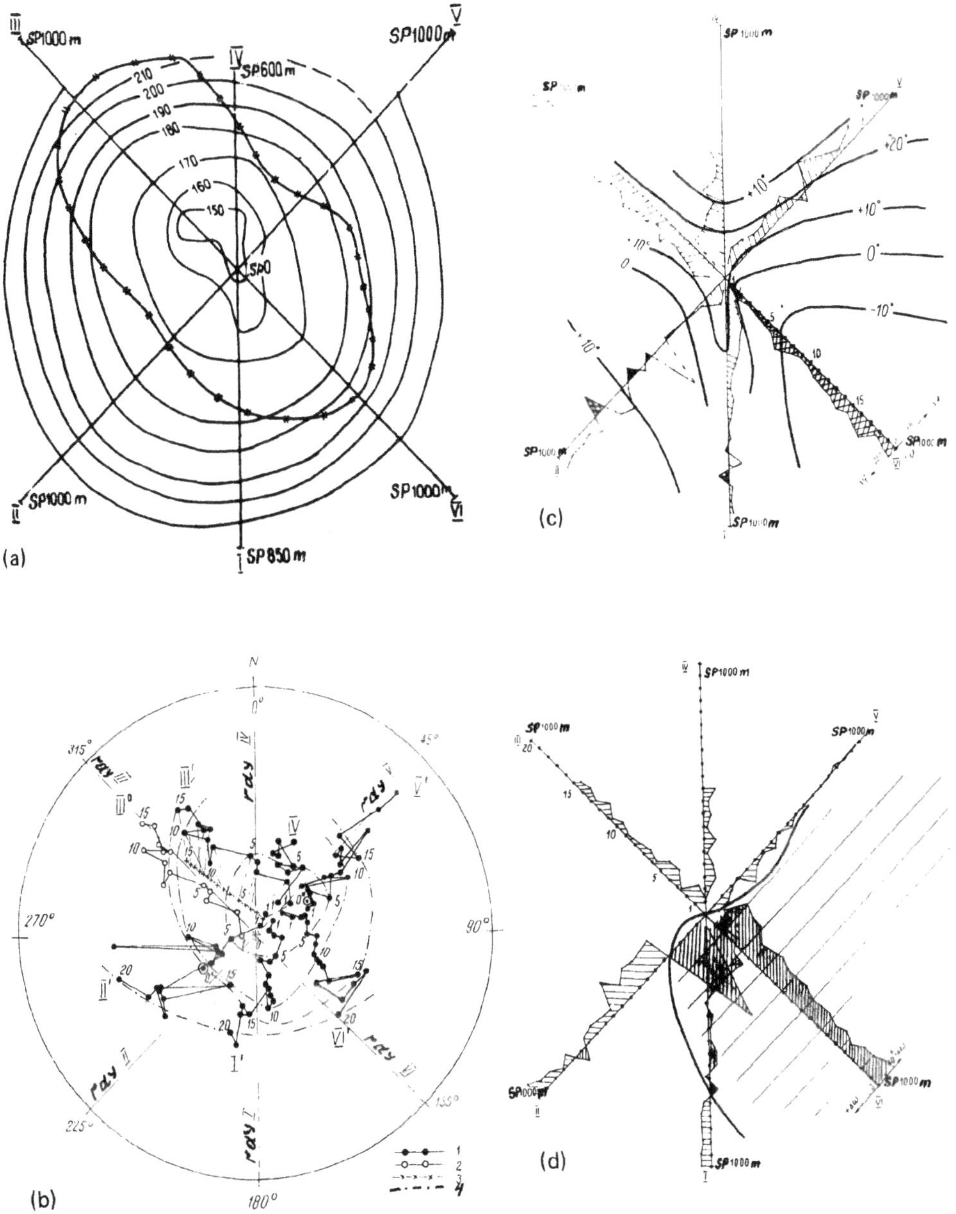

Fig. 131. Studying displacement directions in fixed geophone level observations (H = 760 m) (borehole N3100 Zhairem) (courtesy P. N. Khairutdinov and T. G. Chastnaya): (a) time field, cross mark outlines of ore contents according to geological data; (b) field of observed displacement directions (stereo net): 1 – with account taken of the borehole's tilt, 2 – no account taken of the tilt, 3 – ray III with account taken of the borehole's tilt (shotpoint line), 4 – surface described by directions from observation point to shotpoints equidistant from the borehole entrance; (c) anomalies of angles with the vertical; (d) azimuthal anomalies.

Comparison between the displacement direction and the time fields confirms that the particle displacement directions are much more sensitive to the heterogeneities of the earth than the transit times. The studies of displacement fields at various geophone levels have demonstrated that at each level such fields reflect regular variations of directions. Such regularities at different levels may be governed either by the general particulars of the structure typical of thick beds or by local characteristics. Displacement fields obtained in level observations are in good agreement with displacement directions obtained on vertical profiles (Figure 130).

5. Studying the Parameters of Elliptically-Polarized Oscillations

In the examples cited above, polarization was studied within the framework of the theory of linearly-polarized oscillations. However, as has already been pointed out, in the real earth the direct wave is, strictly speaking, not linearly polarized. Among various factors responsible for complex polarization of the direct wave, the major one is the heterogeneity of the Earth's structure. The parameters that characterize the heterogeneity should include first of all the multi-layered type of structure with which regular waves formed in the real earth are associated, as well as heterogeneities both large and small with which no regular wave is associated, but on which diffraction and scattering of body waves take place. The heterogeneities of both types exercise an influence on the polarization of the direct wave. However, the nature of such influence is not quite the same.

The effects of a layered structure. When a direct wave passes through a layered medium, its linear polarization is distorted in sections of the profile contiguous with the interfaces, due to interference with secondary waves formed at the interfaces.

To illustrate the point, we will consider a case in which linear polarization of the direct wave is distorted as it passes across an interface. Figure 132 shows a computed seismogram in a section of a profile adjoining the boundary between two half-spaces. The reflected (above the boundary) and the converted (below the boundary) waves formed in this case are not trackable up to the first arrival due to the superposition of the direct wave. The shape of the record of the direct wave changes little, and it is practically impossible to execute stratigraphic correlation of the waves. However, the existence of wave interference is already evident from the change in the nature of the polarization of the direct wave pulse. At sufficiently large distances from the boundary beyond the region of interference, the downward wave is linearly polarized. As the distance to the boundary shortens, progressively earlier phases of the direct wave begin to interfere with the reflected wave, and this destroys the linear polarization. This will be seen from the particle motion trajectories shown on the right. On trace 20 of the seismogram (numbered upwards from the interface), the downward *P* and the reflected *PP*-waves are separated in time, (the pulses are depicted at the top), and their trajectories are rectilinear. Interference of *P*- and *PP*-waves starts on trace 18. As the boundary is approached, progressively earlier phases of the direct wave begin to interfere with the reflected wave, thus resulting in the progressively greater and earlier destruction of linear polarization. On trace 10 interference spreads over the entire second period of the pulse. As the boundary is approached, linear polarization of practically the entire pulse is distorted.

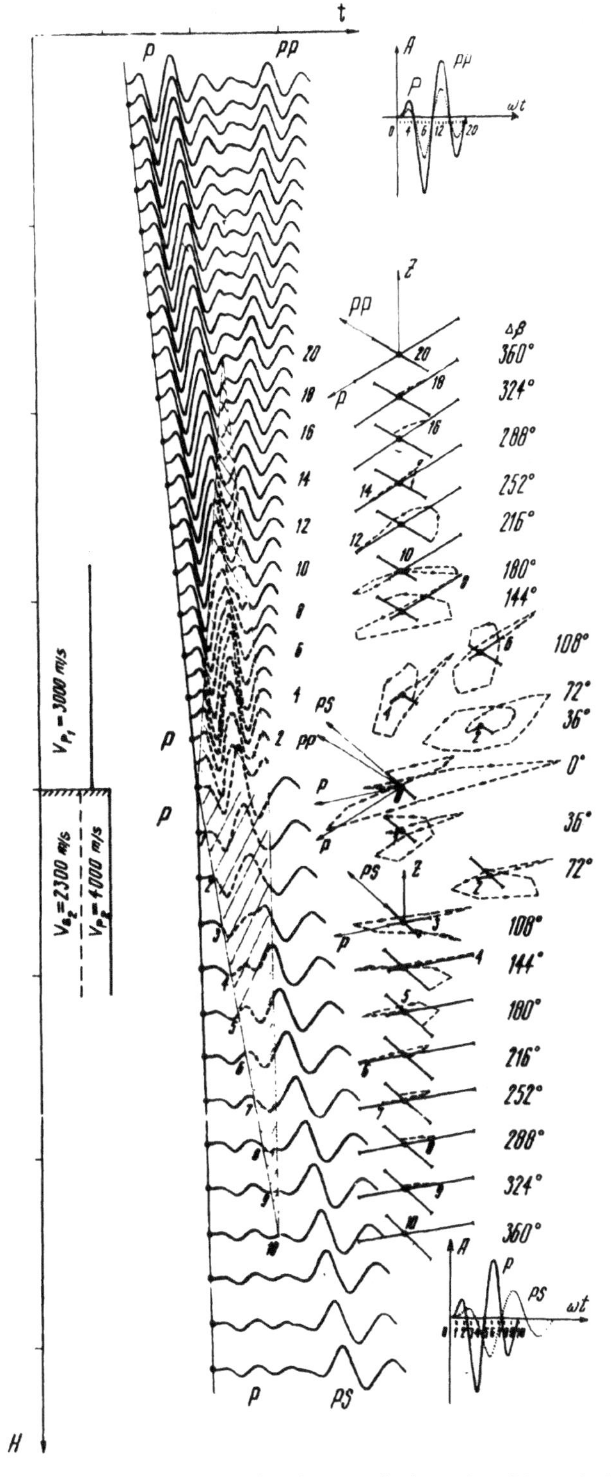

Fig. 132. Disruption of linear polarization of the first wave in the section of the vertical profile contigious with the interface. Pulses of interfering waves and trajectories of particle motion in the polarization plane are shown on the right, on the left-velocity model.

We have considered only the superposition of a longitudinal reflected wave; however, the superposition of the converted reflected wave also takes place at an interface, and the nature of distortion of polarization may be much more complex.

A similar picture is also observed below the boundary, where the direct wave interferes with the converted transmitted wave. In the computations, the period of the converted wave was assumed to be 1.5 times that of the P-wave (the pulses are shown in the lower right corner). Immediately below the boundary, the pulse as a whole is complexly polarized. The interference region diminishes with the distance from the boundary with a corresponding increase in the initial interval of the linearly-polarized pulse. The example cited demonstrates that even one interface can destroy linear polarization over a considerable interval of the profile (this interval is shaded in Figure 132). Such an effect may be observed for the pulse as a whole or for its individual phases.

In the presence of several interfaces, and especially in thin-layered media, superposition of regions with distorted polarization associated with various interfaces can occur, and the polarization of P-waves may be very complex, even along the entire vertical profile.

Observation data obtained in the Anastasiyevskaya–Troitzkaya area (Figure 133) will serve to illustrate the point. To facilitate discussion, we shall divide the geologic section into three zones, profiting from production-geophysical data. The upper very inhomogeneous zone down to 720 m consists of alternating layers of clays and sandstones. A large number of high-intensity reflected waves originate in this zone, as can be seen from the seismogram of an in-line profile shown at the left.

The middle zone in the depth interval 720 to 1400 m chiefly comprises homogeneous clays containing individual sandstone layers, some of them water-saturated. Reflected waves associated with this zone are far less numerous, and their intensities are comparatatively low. There are rather long intervals e.g. between 800 to 1100 m of the section, where reflected waves are not detectable at all.

The lower zone 1420 to 1500 m is inhomogeneous and contains gas-saturated sandstones with clay interlayers. A large field is associated with the gas deposit.

The existence of zones with drastically different structures allows favourable conditions for studying polarization parameters. The polarization parameters were determined by calculating particle trajectories on a computer from digital data obtained with a sampling rate of 0.002 s.

It will be seen from the $\varepsilon(H)$ plots (Figure 133b) that describe the ellipticity of the trajectory observed from several shotpoints arranged at various distances and in various azimuthal directions from the borehole, (the disposition of the deposit and the shotpoints is shematically represented in Figure 133a) that:

- for all shotpoints, the pattern of the curves for heterogeneous zones (the upper and the lower) is very complex, whereas it is comparatively smooth for the homogeneous (middle) zone.
- for all zones, the ellipticity increases with the separation from the shotpoint (with the increase in the wave's angle of incidence).
- minimum ellipticity is observed at the first extremum.
- for an in-line profile in a homogeneous bed, the polarization is practically linear over substantial depth intervals.
- typical of the upper heterogeneous zone, are sharp variations of the ellipticity

parameters from all shotpoints, the values of the parameters themselvesl being higher than elsewhere. In several localized depth intervals (515 to 630 m, 675 to 700 m) drastic increases in the ellipticity are observed.

– for the middle homogeneous zone, the ellipticity decreases and becomes more stable with depth. For the shotpoint at 1150 m, a more pronounced distortion of polarization is observed than for shotpoints at 1200 m and 1345 m. Quite drastic anomalies of ε are observed in the same intervals of the section with which the reflected waves are associated. In the interval of pure clays, ε drops sharply, and the direct wave is practically linearly polarized.

In the lower heterogeneous zone (1420 to 1560 m), a sharp differentiation of $\varepsilon(H)$ curves is observed for remote shotpoints. From nearby shotpoints, 600 m and 650 m, such drastic anomalies are not observable.

One cannot help noticing that the most drastic changes in the ellipticity parameters are observed in the same intervals of the section in which the reflected waves occur.

It would be very interesting to study the frequency dependence of the polarization parameters (spectral polarization analysis). In complex heterogeneous media, various monochromatic components can have a different polarization. The spectral polarization analysis of seismic waves can be executed by means of the frequency discrimination of the traces of a three-component record using band-pass filters followed by polarizational analysis of the records of each frequency component. The decomposition of the signal into individual harmonics combined with the decomposition of each monochromatic component according to its polarization is technically feasible [83]. Spectral polarization analysis demonstrates the extent of the difference in polarization between various monochromatic vector components. Such differences, being the result of the heterogeneities of the earth, should enable some idea to be gained about the nature of heterogeneities.

The spectral polarization analysis of ellipticity, carried out for an in-line profile in the frequency band from 20 Hz to 120 Hz has demonstrated that ε increases with the frequency. The results of the analysis are presented in the form of ellipticity graphs in the vertical ε_v and the horizontal ε_h planes for six spectral components 20, 40, 60, 80, 100, and 120 Hz (Figure 133).

Since the values of ε for nearby shotpoints are smaller than for remote shotpoints, the functions $\varepsilon(f)$ have been plotted on a larger scale, with the scale of ε_h being approximately twice that of ε_v. The frequency dependence of ε (its spectrum) is different for different intervals of the profile, and is chiefly determined by the parameters of secondary waves. The upper heterogeneous zone will be seen from the plots to display sharp differentiation. The anomalies in the depth range 500 to 650 m are observed chiefly on the ε_v plots, i.e., the polarization plane is close to the vertical. Drastic anomalies are observed on the ε_h plot in the depth range 450–500 m. An extensive low-frequency maximum on the ε_h plot in the upper heterogeneous zone in the range 575–650 m splits into two separate maxima at high frequencies.

The ε_v and ε_h curves for the second zone are less differentiated, their differentiation in each plane becoming more pronounced with the rise in frequency. On the whole, ε_v is smaller than ε_h, however, at different frequencies this difference does not remain the same. At frequencies 20 Hz and 40 Hz, ε_v is much smaller than ε_h, and considerably less differentiated; at 60 to 80 Hz ε_v and ε_h are almost equal. At 100 Hz and 120 Hz,

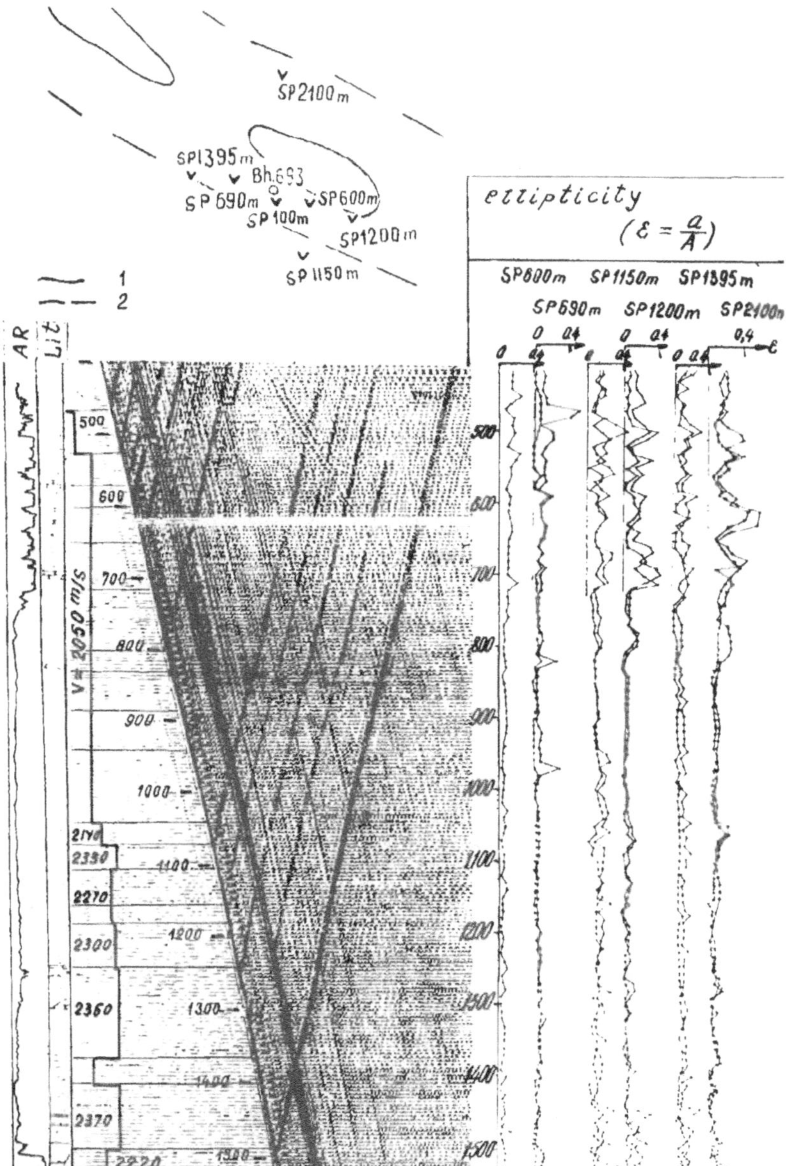
SP2100m
SP1395m
Bh.693
SP690m
SP600m
SP100m
SP1200m
SP1150m
1
2
ellipticity
(ε = a/A)
SP600m SP1150m SP1395m
SP690m SP1200m SP2100m
AR
Lit
V = 2050 m/s
V_{LR}, m/s
(a)
(b)
(c)

Fig. 133a-c.

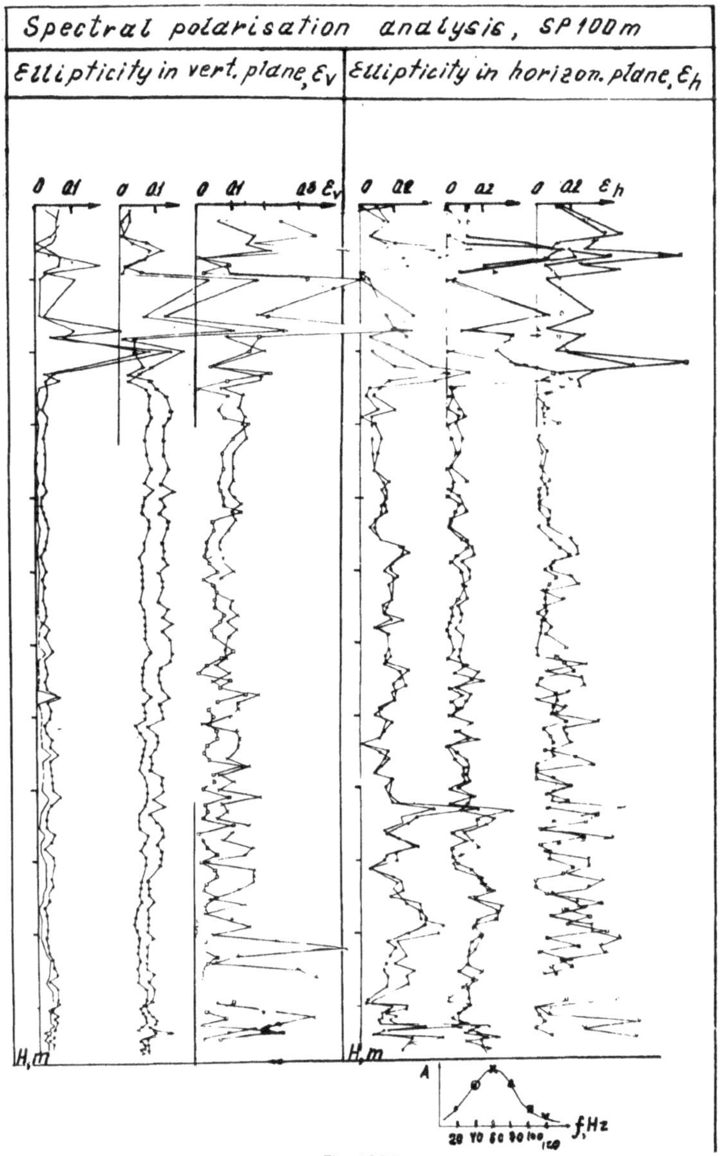

Fig. 133d.

Fig. 133a-d. Experimental studies of polarization parameters of the direct wave pulse, borehole N693 Anastasiyevsko-Troitzkaya: (a) disposition of shotpoints, production-geophysical data and VSP seismogram; (b) ellipticity of the trajectory $\varepsilon = \alpha/A$ for various shotpoints in the time interval up to the first extremum (1), up to the second (2); (c) spectral polarizational analysis of ellipticity for SP 100 m in the vertical (left) and the horizontal (right) plane 1 – 20 Hz, 2 – 40 Hz, 3 – 60 Hz, 4 – 80 Hz, 5 – 100 Hz, 6 – 120 Hz; (d) difference in displacement azimuths between adjacent extrema for SP 100 m; (e) angles in the vertical plane for various shotpoints.

ε_h is about twice as large as ε_v. With the increase in depth, the values of ε_v at frequencies of 60 to 80 Hz diminish, whereas those of ε_h increase.

In the lower heterogeneous zone, the differentiation of the $\varepsilon(f)$ curves becomes more pronounced than in the middle zone, but remains substantially less than in the upper zone.

A point worth noting is that the polarization parameters in the horizontal plane can be better correlated with the geologic section than in the vertical. The anomalies in the ε_h curves can be very well identified with the water-saturated layers of the second zone.

Some information about the relationship between the polarization parameters and the geologic section can be gained from the curves of differences in displacement azimuths of various monochromatic spectral components $\Delta\omega = \omega_{20} - \omega_i$ (Figure 133d). The $\Delta\omega$ curves are less differentiated, but in several intervals of the profile sharp anomalies are observable.

An interesting factor is the variation of the displacement direction in the vertical plane, because it is heavily dependent on velocity variations. It will be seen from Figure 133e, which depicts $\varphi(H)$ curves determined from the energy of the first extremum for different shotpoints, that the main features of the curves correlate well with one another and with the section. Most clearly defined on the curves is the low-velocity layer associated with gas-saturated sandstones (H = 1420–1560 m). This layer is also reliably detectable from a sharp rise in the attenuation coefficient (see Chapter 11). It should be pointed out that to make the study of the geologic section in conditions of non-linearly polarized vibrations more elaborate, great importance has to be attached to the very method of determining displacement directions.

Figure 134a shows for the sake of comparison, plots obtained in the course of measurements of φ in the first sample interval of 0.0025, the fourth sample interval coinciding with the first extremum, as the function of the energy of the first phase. The φ plot for the first sample differs appreciably from the other two, which practically coincide. The first sample data correlates better with the section and delineates the top of the gas-saturated sandstone layer more clearly.

The displacement angle difference $\Delta\varphi(H)$ plots (Figure 134b), present a still more impressive picture of the effect of the structure on the displacement direction from various shotpoints.

The low-velocity zones begin to manifest themselves from shotpoint 1150 m and stand out more clearly as the separation from the shotpoint is increased. Two low-velocity zones stand out very clearly on the $\Delta\varphi(H)$ plot from shotpoint 2100 m. The first is associated with water-saturated sandstones (H = 1230–1300 m), and the second, the most pronounced, is associated with the gas-saturated productive layer (H = 1450–1550 m). The $\Delta\varphi(H)$ plot agrees well with production-seismic data. It is worth special mention that the aforesaid low-velocity zones cannot be discriminated by the difference in the arrival times of the first events (shown on the left).

Thus, the results cited are convincing proof of the independence of the polarization parameters (in the local coordinate system) on the Earth's structure both in the region of the source and during propagation, and of their strong dependence on the structure of the earth in the vicinity of the point of observations. Figuratively speaking, the polarization of the direct wave in contrast to its other parameters has a 'short memory'. This is very clearly seen from the fact that there is a rapid transformation of a wave with a complex polarization into one which is practically linearly polarized which occurs as the wave enters a deeper homogeneous medium from an inhomogeneous one. This property prevents the use of polarization to 'sound out' the earth, but presents great and fundamentally novel opportunities for studying the earth in the vicinity of the point of observations.

The effects of various sized heterogeneities. The most pronounced anomalies observed in polarization parameters were associated with the reflection zones of the geologic section. Anomalies in polarization parameters not associated with reflecting interfaces are of special interest. In the example being considerd, such anomalies are observable from the curves particularly in the range 860 to 900 m. Studying polarization parameters, one is able to obtain qualitative notions about the nature of the heterogeneities. A. V. Nikolayev [147] determined the dimensions of the heterogeneities, comparing them with the diameter of the first Fresnel zone (F)

$$F = \sqrt{\frac{\lambda r_1 r_2}{r_1 + r_2}},$$

where λ is the wave length, $r_{1,2}$, are distances from the heterogeneity to the receiver and the source, respectively.

Heterogeneities with dimensions exceeding those of F are classfied as large, with dimensions comparable to F as medium, and with dimensions smaller than F as small. Spectral polarizational analysis is able to determine the period of fluctuations of polarization parameters spatially, which is almost equal to the period of fluctuations of the amplitudes and phases. More elaborate spatial observation systems are required for such observations.

Consider data from observations carried out in the Perm' province [148]. The waves were recorded in two depths intervals: above a series of thin layers in a comparatively homogeneous bed (520 to 560 m) and below that series (820 to 860 m). The lithology of the top of the series of layers is varied and is represented by clays, sandstones, carbonates, gneiss, alternating thin (4 to 6 m thick) interlayers with low and high velocities. Variations of the thickness and the extent of the thin interlayers are observable. The waves reflected from this bed have low intensities. Shots were fired along an arc profile of 600 m radius at 35 m intervals between the shotpoints. The polarization parameters of the first waves are depicted in Figure 135. It follows from the plot that:

- the polarization parameters above the thin layers in a homogeneous bed are stable and vary little from shotpoint to shotpoint (or azimuth of the source).
- below that series, the parameters vary greatly from shotpoint to shotpoint.
- it follows from the ellipticity plots that above the layers the polarization of the direct wave is chiefly linear.
- below the layers the polarization of the direct wave is not linear, and there is a great spread in the values of the polarization ellipticity.

The small dependence of the polarization parameters on the width of the time interval measured from the time of the first arrival on the one hand, and on the other a greater change in the polarization parameters when the time interval is remote from the time of the first event, suggest the presence of heterogeneities of two types: small scatters, and medium sized ones commensurate with the diameter of the first Fresnel zone, which in this case is about 85 m.

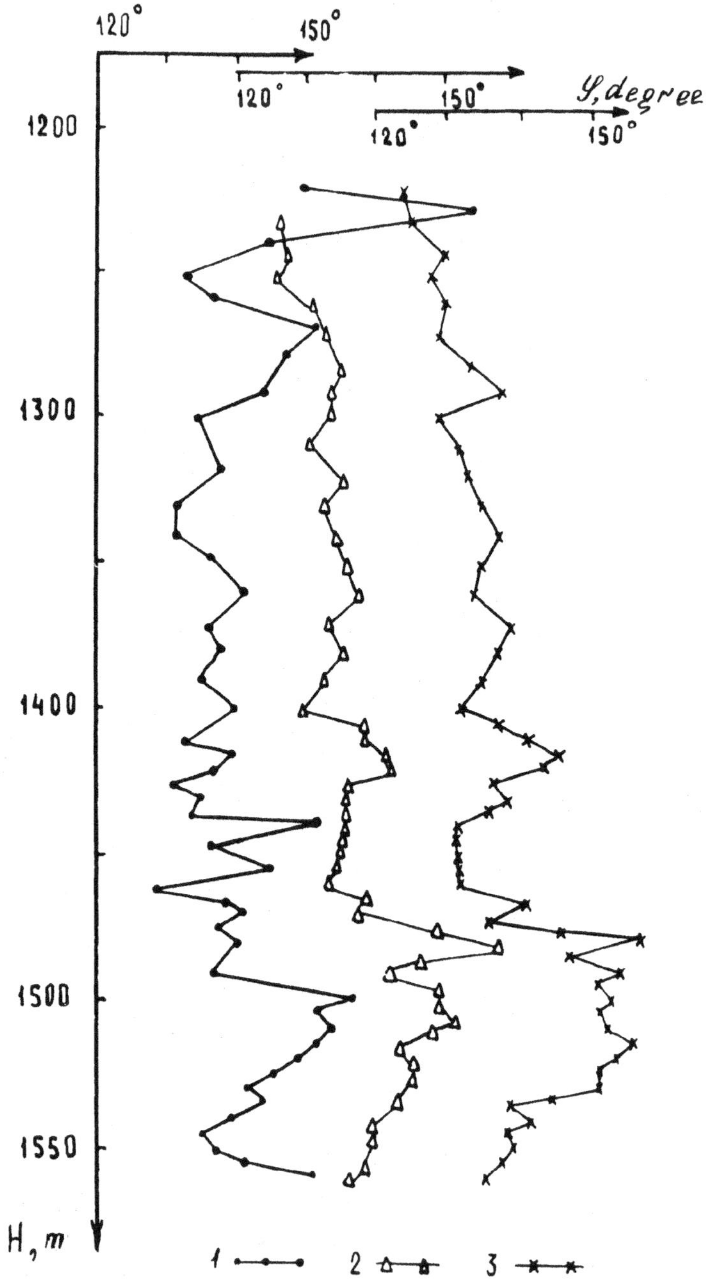

Fig. 134a.

Fig. 134a-b. Effect of structure on displacement directions (borehole N693): (a) $\varphi(H)$ plots for shotpoint 2100 m in section of the profile intersecting deposit constructed from amplitudes of the first sample (1), from amplitudes of the first maximum (2) and from the pulse's energy (3); (b) $\Delta\varphi(H)$ (right) and $\Delta t(H)$ (left) plots. Production-geophysical data and geologic column shown in the middle.

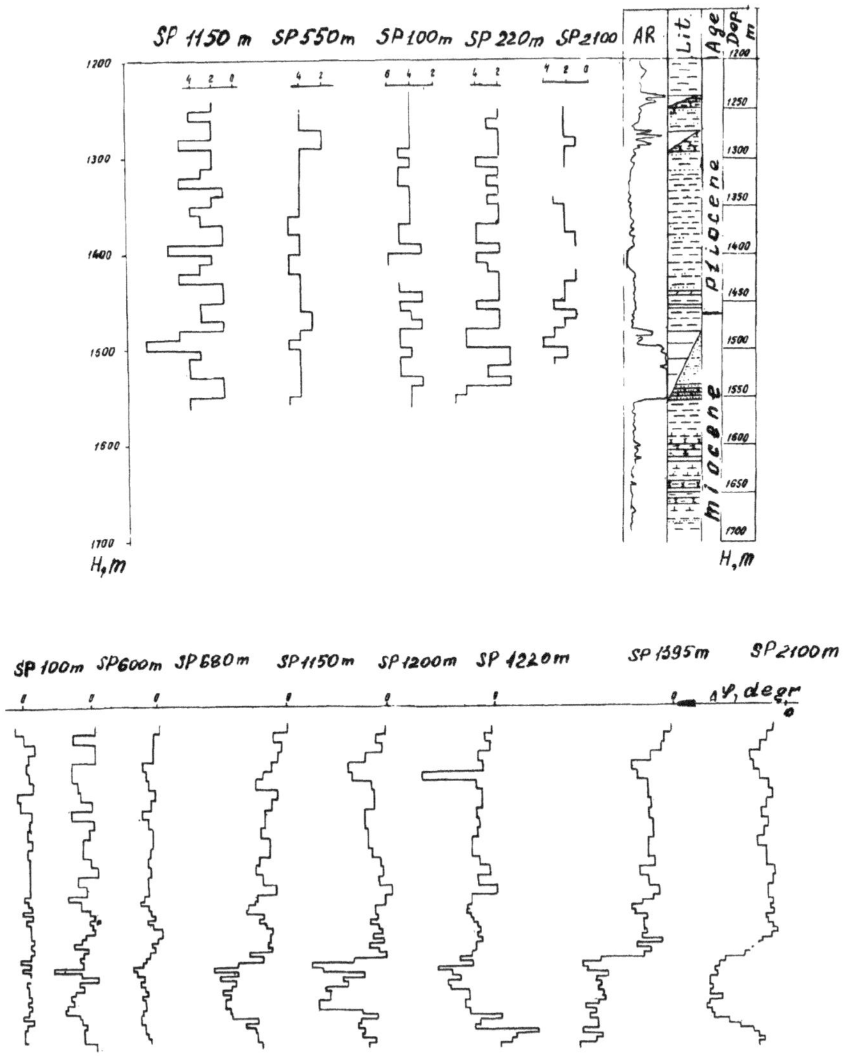

Fig. 134b.

Spectral polarizational analysis has demonstrated that (Figure 135d, e):

– above the thin series of interfaces, the maxima of the spectra of X- and Z-components of the direct wave occur at an identical frequency, and do not vary with a change of shotpoint (N);

– below the series of interfaces, the maxima of the X- and Z-spectra are displaced and have a different shape and a different dominant frequency for different shotpoints;

– above the series of layers, the displacement directions (φ and ω) depend little on the frequency;

(a)

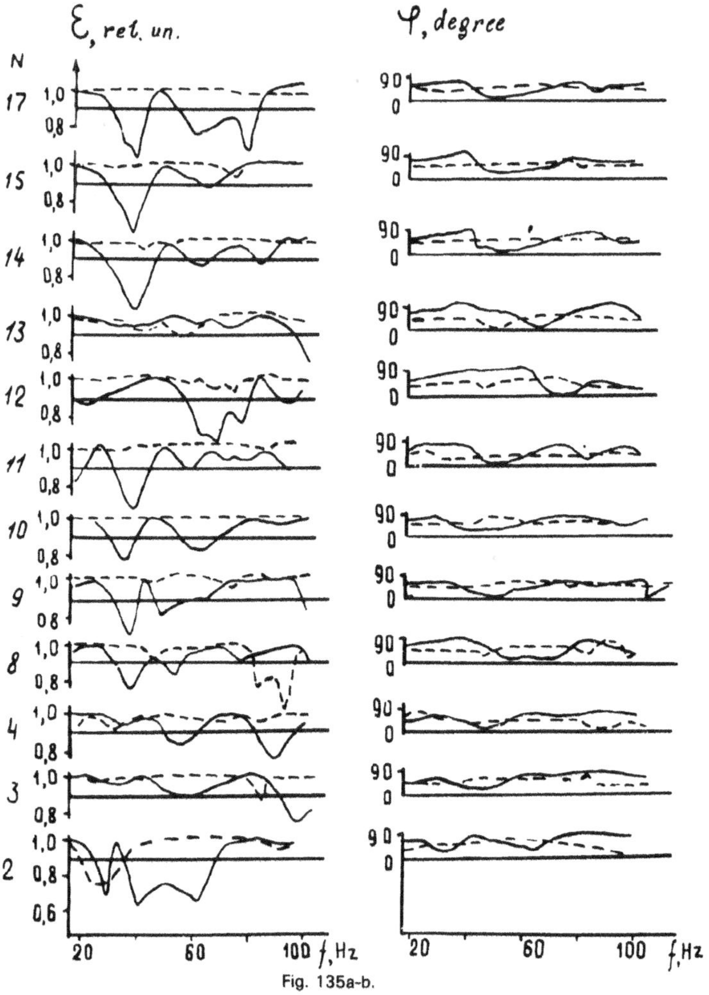

(b)

Fig. 135a-b.

Fig. 135a-e. Polarization parameters observed above (dotted line) and below (solid line) a series of layers [148]: (a) ellipticity (ε); (b) angle with the vertical (φ); (c) azimuth (ω); (d, e) angles φ for some spectral components.

– below the series there is a strong dependence of the parameters φ and ω on the frequency.

It follows from comparison of the φ graphs for several fixed frequencies that below the interfaces sharp changes in the angle take place, the amplitude of the changes increasing with the frequency. The same heterogeneities are responsible for different fluctuations of φ on different monochromatic components. The changes of the angle φ with the frequency gives some idea about the possible dimensions of the heterogeneities responsible for them. In particular, in the 51 to 70 Hz frequency band the major part

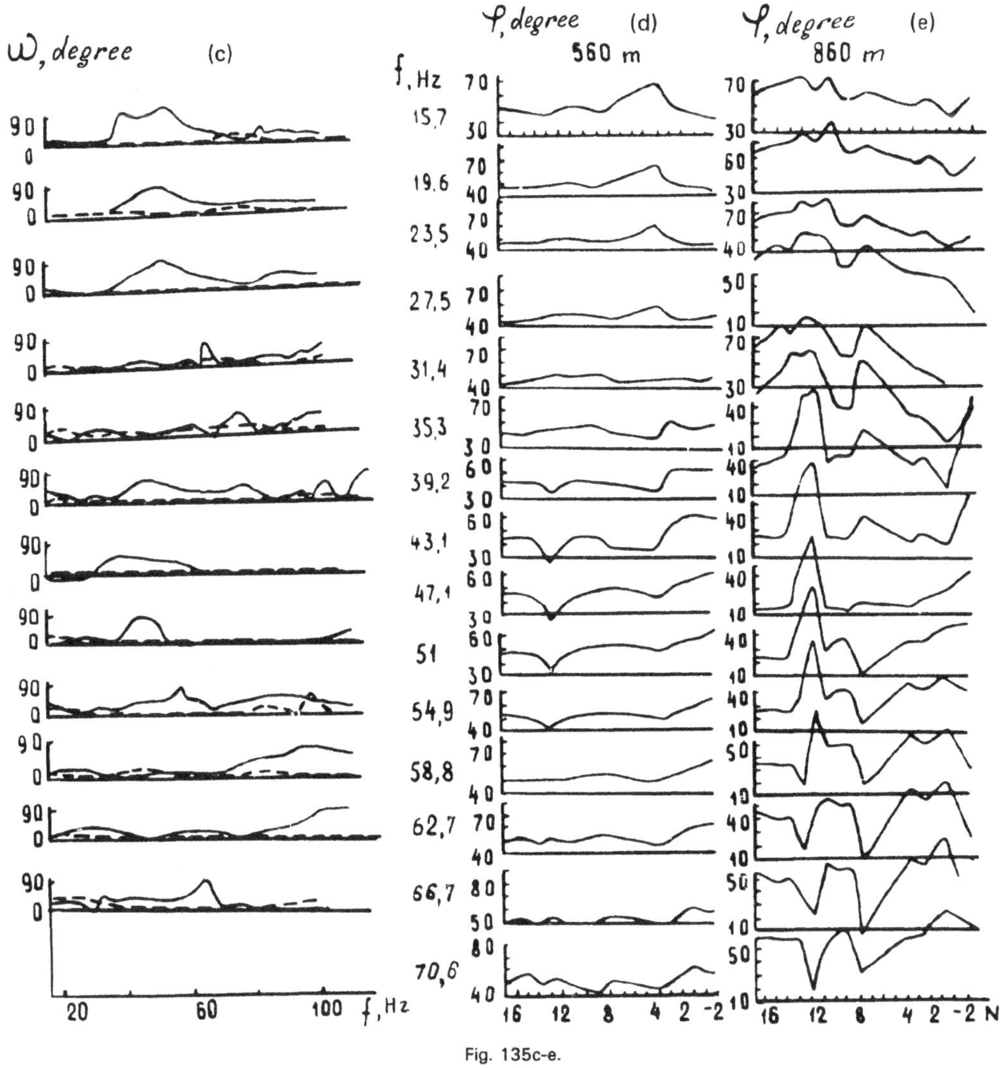

Fig. 135c-e.

is played by small scattering heterogeneities. The effect of medium-sized heterogeneities is less important.

Thus, studies of the polarization parameters in the very first portion of the direct wave as functions of the frequency, facilitate examination of the heterogeneities in the immediate vicinity of the observation point. Widening the time interval makes the effect on the polarization of the direct wave of heterogeneities at progressively greater distances from the observation point detectable.

Polarization of waves in the later part of the record. The study of the polarization of waves in the later part of the record is fraught with great difficulties, because of a complex interference field produced by the superposition of various wave types. For a quantitative analysis of the polarization of individual waves, the methods of wave discrimination will have to be perfected. However, qualitative estimates of the polarization of waves are feasible, and indicate an intimate connection existing between the polarization parameters and the Earth's structure. According to the data obtained in the Anastasiyevskaya–Troitzkaya area this connection manifests itself in azimuthal deviations of the displacement directions in the reflected *PP*- and *PS*-waves. The signs and the magnitudes of the deviations correspond to structural properties of the interfaces, so that the polarization parameters yield extra information about the shape of the structure in the region contiguous to the borehole.

The polarization of converted transmitted waves in this region displays special complexity with tangent components being present. The intensity of such components differs for different interfaces and may serve as a characteristic of the interface.

The polarization of *S* waves in the case of conventional explosive sources is determined as a rule by the conditions of superposition of S_1- and S_2-waves polarized in different directions (in an axially-symmetrical medium the corresponding waves are the *SH*- and *SV*-waves). Apart from this, an appreciable influence is exercised on the polarization of waves by the transverse converted-type *PS*-waves converted near the source. They are recorded slightly ahead of the monotype wave, interfere with it and produce a long wave train.

6. The Effects of the Low-Velocity Zone on Particle Displacement Directions

The relationships discussed above are valid only for down-hole observations beyond the low-velocity zone. The low-velocity zone greatly distorts the displacement directions, thereby limiting the potentialities of their use for quantitative interpretation.

In addition to the reflection of energy from the surface, the particle displacement direcitons observed on the surface are appreciably affected by heterogeneities of the uppermost portion of the section. The presence near the surface of one or more layers results in the appearance of interference waves and in changes in the displacement directions. Such changes depend on the relationship between the propagation velocities of elastic waves in the layer and in the underlying half-space, as well as on the ratio of the wave length λ to the layer's thickness d. The dependence of apparent angles of incidence of a wave on a free surface on the ratio λ/d computed from theoretical seismograms of interference waves demonstrates that for small λ/d, the directions of particle motion in a longitudinal wave correspond to the direction of propagation of the waves in the layer, i.e., they are governed exclusively by the upper layer. For large λ/d, the direction of particle motion corresponds to that of the waves' angle of incidence on the lower interface, and the layer has no effect on them. The transition from small λ/d to large lies somewhere in the range of values of $\lambda/d = 2.0$ to 4.5 depending on the angles of incidence. In the case of a single layer for $\lambda/d = 10$ to 15, values close to the assymptotic are obtained. On G. A. Gamburtzev's suggestion, the frequency dependence of the apparent angle of emergence of seismic radiation was used to study the law of variation of the velocity with depth in the upper portion of the section.

In conjunction with VSP development it appeared advisable to supplement computed data on the effects of the low-velocity zone on the particle displacement directions with experimental studies, and to check the correspondence between the displacement directions deep in the earth, and the waves' direction of arrival [14]. Of special interest was the appraisal of the effects of the low-velocity zone on the particle displacement directions when the low-velocity zone displays a sharp velocity contrast (3 to 5 times).

Experimental work was carried out in the Kuibyshev province, where the upper portion of the geologic section consists of beds of sand-clay deposits about 133 m thick overlying carbonate rock. The low-velocity zone, about 12 m thick, consists of loams and sands. The velocity of propagation of longitudinal waves in the zone is about 600 m s^{-1}, and below the zone 1580 m s^{-1}. Two series of experiments – VSP (with source excitaton by explosives and by impacts) and torpedoeing – enabled the particle displacement directions on the surface and under the zone to be studied.

In the surface observations, at shotpoint offsets of 240 to 680 m, two groups of waves stand out clearly on the seismograms – P_1 with $v_a = 6000$ m s^{-1}, and P_2 with $v_a = 1700$ m s^{-1} associated with the upper portion of the section (Figure 136a). The presence of two types of waves with widely different directions of arrival created favourable conditions for the experiment. The seismogram obtained, together with PM VSP seismograms, was used to determine the displacement directions in the P_1- and the P_2 waves, both on the surface and along the vertical profile. The displacement directions in the P_1-wave in the low-velocity zone make an angle of 7° with the vertical and below it 18°, the corresponding figures for the P_2-wave being 20° and 79°, respectively. The ray-path schematic depicts the wave fronts of P_1 and P_2.

Let us also illustrate the effects of the low-velocity zone on the displacement directions with the aid of PM VSP seismograms obtained with a weight freely falling from a height of 5 m and at a distance of 11 m from the top of the borehole (Figure 136b). The displacement directions determined from the seismograms are in quite satisfactory agreement with the ray-path schematic computed from the velocity section. Hence, a change of almost 180° in the particle displacement directions between the direct longitudinal and the direct transverse waves took place in a 26 m long section of the vertical profile.

In the simplest low-velocity zone structure such as just considered, it was determined by a single horizontal direction, and the particle displacement directions varied only in the vertical plane. However, even in the case of comparatively smooth relief of the base of the low-velocity zone, and great velocity contrasts across the zone, there can be large azimuthal deviations of the rays [14]. All this greatly obstructs the use of particle displacement directions observed on the surface for quantitative processing of seismic data. Attempts to take the heterogeneities of the Earth into account, and to correct the particle displacement directions failed, because of a great distortion introduced by the low-velocity zone. This refers to waves whose frequencies belong to the band used for seismic exploration. The transition to longer waves reduces the effect of the upper portion of the section, and the particle displacement directions in many cases enable the azimuth and the wave's angle of emergence to be determined.

The use of displacement directions to gain information about the Earth has not, as yet, become common practice, because some problems of wave propagation in anisotropic media still remain to be solved. At present, due to G. I. Petrashen's results [146], quantitative processing of displacement direction fields becomes a possibility, and it is to be hoped that this will produce additional and independent information about the earth.

The subjects discussed above are by no means a complete exposition of the potentialities of VSP in the study of the region around and between boreholes. Up to the present

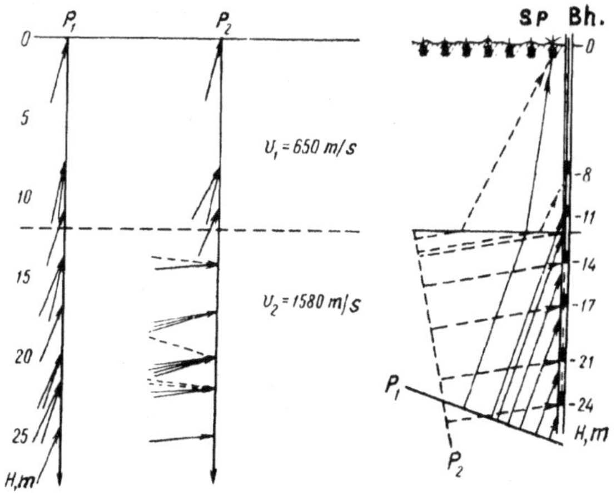

Fig. 136a.

Fig. 136a-b. Changes in particle displacement directions as the low-velocity zone is traversed: (a) for a comparatively remote explosive source (l = 280 m), seismograms: up-hole Z-component (top), PM–VSP (centre), wave fronts, rays and particle displacement directions in P_1 and P_2 waves are shown below; (b) for a nearby impact source, three-component seismograms (left), ray-path schematics and particle displacement directions (right).

time, not all the potentialities by far have been implemented; however, VSP is to become the corner-stone for the extensive use of the combination of seismic studies at the various stages of field exploration and exploitation. This will not only help to reduce the amount of drilling required for exploration, but will also speed up putting the fields into operation and improve the efficiency.

To sum up, the analysis of polarization parameters demonstrates:

– The parameters of polarization of the direct wave and the nature of their variation reflect the characteristics of the geologic section, the physical properties of rock and the section's structure. There is a clear relationship between the polarization of a direct wave and the shape and the structure of the entire seismogram. The simpler the polarization of direct waves, the simpler will be the seismogram, and the less will be the waves formed in the earth.

The polarization of seismic waves is a very sensitive parameter of the wave field, and serves as a source of additional and independent information about the earth. The study of the polarization of seismic waves enables a more detailed knowledge about the region surrounding the borehole to be obtained, than can be done by studying wave kinematics only. Joint studies of polarization in the frequency and time domains will be of vital importance in this case. It is to be expected that such studies will enable the effects of heterogeneities of various size at the point of observation, as well as the effects of large heterogeneities located at some distance from the point of observation, to be studied.

There are very good prospects for polarization to produce information about the earth in VSP, and its development should be a rapid one.

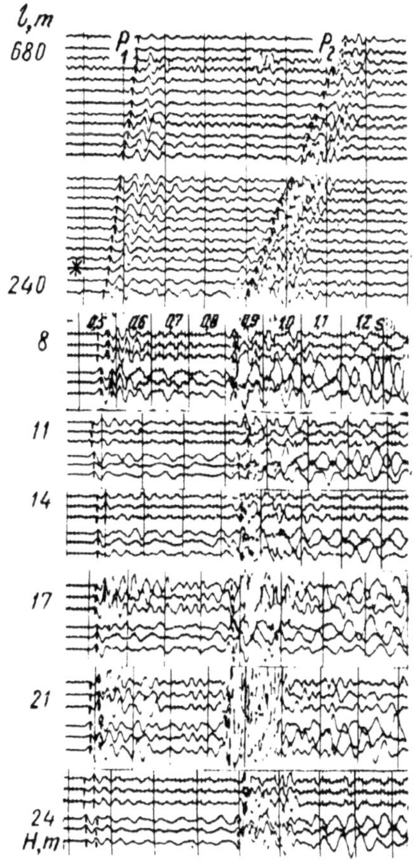
l, m
680
240
P1
P2
0.5
0.6
0.7
0.8
0.9
1.0
1.1
1.2 s
8
11
14
17
21
24
H, m

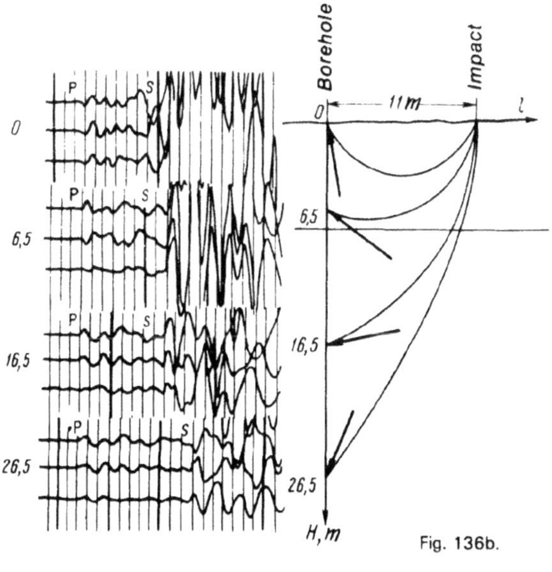
Borehole
Impact
11m
l
0
6,5
16,5
26,5
H, m
P
S

Fig. 136b.

To implement those potentialities, it is first of all necessary:

– to carry out theoretical and experimental studies of polarization of longitudinal and transverse seismic waves in actual complex media;

– to perfect measurement techniques, in particular of polarization parameters spatially oriented;

– to establish quantitative relations between the polarization parameters, the physical properties of rock and the geologic details of the section;

– to study the polarization of seismic waves in the real earth and to devise methods for quantitative interpretation of polarization parameters aimed at obtaining additional and independent information about the earth;

– to study the upper portion of the geologic section and the low-velocity zone, and their influence on the seismic signal, on the structure of seismograms and on the effectiveness of seismic studies.

The results obtained herein justify the continuation and the expansion of studies in this direction.

CHAPTER 13

THE EXPLORATION POTENTIAL OF VSP AND THE PROSPECTS FOR ITS PROGRESSIVE DEVELOPMENT

The extensive volume of VSP work performed in recent years, the experience gained in the process, and the results obtained, constitute the basis for the discussion of the exploration potential of VSP, as well as the improvements in the effectiveness of seismic exploration associated with it, and the prospects for the progressive development of the method itself. Such a discussion may reasonably start with the range of problems solved by VSP. When appraising the exploration potential of VSP, we should discuss separately the two principal applications of VSP: one aimed at improving the effectiveness of surface observations and the other aimed at studying the region around the borehole. Whereas in the early stages of the development of the method, the first application associated chiefly with the search for and the selection of structures for drilling, was the dominant one, in the course of the method's development the second application assumed progressively more importance as a self-contained method. This was because it offered the prospects of the extensive use of seismic studies not only during the stages of the search for and the selection of stuctures for deep drilling, but also at the stages of exploration and exploitation of the fields. It will also be expedient to discuss separately the potentialities of VSP for the solution of various research problems.

VSP studies have demonstrated that different wave types – both longitudinal and transverse, monotype and converted – are associated with the same interfaces in the geologic section. Explosive sources excite both longitudinal and transverse waves in the upper portion of the section. The main exploration potential is associated with the waves that propagate in directions approaching the vertical. In this respect, the reflected waves observed in the vicinity of the source, and in regional operations, the converted transmitted waves formed in the region of recording at great separations from the source, are of special interest. The polarization of oscillations is a very sensitive parameter of the wave field and may improve the effectiveness of seismic observations.

VSP results suggested that the development of methods based on the use of waves of one type should be rejected in favour of the development of a general method of seismic exploration based on the joint use of different wave types. This approach is realized in the polarization method, whose general nature follows from the use of all the parameters (including polarization) of the wave field for its analysis, and different wave types for the construction of a section, as well as from the method's applicability to all kinds of seismic studies, both surface and down-hole.

1. Problems Solved by VSP

About two hundred technical papers deal with the subject of VSP. Consider the list of problems that have been solved during the present decade by the VSP method, as reported in the literature.

Problems associated with the interpretation of surface observations

The effectiveness of VSP, as applied to surface seismic problems, is determined first of all by the possibility it presents for the sectioning of a complex wave field observed on the surface, and for the discrimination of various wave types, for the determination of their parameters as well as the elements of the section with which they are associated, and for appraising the influence of individual elements of the section on the wave field. Several groups of problems solved by the VSP method may be selected for the purpose of improving the effectiveness of surface observations.

General methodologic problems

(1) The study and selection of optimum source excitation conditions for each specific situation. (2) The study of the structure of the upper portion of the section and the low-velocity zone, and their influence on the wave field. (3) The study of the direct longitudinal and transverse waves, and their mechanism of formation and stabilization. (4) The study of the process of formation of the wave field observed on a surface seismogram. (5) The study of the relative intensities of various wave types. (6) The selection in each specific geologic situation of the method, and the optimum observation system, both surface and downhole that would guarantee an effective solution of the geologic problem.

Study of the Velocity Distribution

(1) The improvement of the resolution of sectioning of the geology according to the velocities of longitudinal and transverse waves. (2) The determination of average, interval, and ray velocities of the longitudinal and the transverse waves, with the aid of downward and upward waves in the first and later arrivals. (3) The determination of the velocity profile for *P*- and *S*-waves below the depth penetrated by the borehole. (4) The study of the laws of variation of the velocity with depth at great source-geophone separations. (5) The study of anisotropy of the real earth and its data processing consideration.

Study of Reflected Waves (Longitudinal and Transverse)

(1) The study of the unwanted wavefields. (2) The determination of the origin of waves and the discrimination of singly-reflected waves. (3) The study of multiple reflection boundaries, and the formation and relative intensities of the multiples. (4) The detection of reflecting interfaces at great depths and the study of the wave parameters associated with them. (5) The study of the means of increasing the penetration of seismic exploration. (6) Stratigraphic correlation of reflected waves. (7) The study of the connection between the nature of the wave field and the lithologic and stratigraphic features of the geologic section. (8) The computation of optimum common-depth-point systems. (9) The critical re-appraisal of the data from earlier years and their new interpretation.

Study of Refracted Waves

(1) The study of the origin of the waves constituting the intitial portion of the seismogram at great separations and, in particular, refracted waves associated with the crystal-

line basement. (2) The study of the intensity of head waves as a function of layer thickness. (3) Stratigraphic correlation of waves and the appraisal of the effectiveness of the method in a specific situation.

Study of converted transmitted waves

(1) The determination of the origin of waves recorded on horizontal components in the intitial portion of the seismogram. (2)The identification of waves recorded on the Z- and the X-components associated with the same interface. (3) The location of interfaces associated with conversion. (4) The study of the energy relationships between the converted PS upward transmitted waves and converted multiple waves in the upper part of the geologic section.

Problems involved in the study of the zone surrounding the borehole

Such problems may be subdivided into the structural and those involved in the study of the physical properties of the geologic section.

Structural Problems

(1) The study of structures away from the borehole, e.g., the study of subsalt deposits in areas with well developed multiples, or the study of a rock mass of reef origin in conditions of complex tectonics, with an extensive network of tectonic faults. The ability to study interfaces not intersected by the borehole is a specific advantage of VSP over all other methods of observation in boreholes. (2) The mapping of high-dip interfaces when the reflected waves are not trackable on the surface. This case is most frequent in seismic exploration for ores. In seismic exploration for oil this is true of the flanks of salt domes. (3) The determination of dip and strike in sections intersected by the borehole for the purpose of planning further drilling. For the solution of structural problems VSP is usually employed in combination with surface observations.

Physical Properties

(1) The study of absorption and the reflecting properties of a geologic section. (2) The sectioning of the geologic column according to elastic parameters (Young's modulus, Poisson's coefficient). (3) The study of the influence of mineral deposits on the wave field to provide a physical foundation for direct search methods for both hydrocarbons and solid minerals. (4) The detection of mineral deposits penetrated by the borehole, but not established by production trials. (5) Detection and estimates of mineral deposits not penetrated by the borehole, but located in its vicinity. (6) Establishing the contours of the deposit. (7) Detecting and predicting zones of abnormally high geo-pressure.

These problems are associated with that of predicting the geologic section, both in the vicinity of the hole and below its bottom. They may also include studies of the mechanical properties of the section for borehole design, determination of the spatial position of the bottom of the borehole and the orientation of the borehole axis used in the process of the construction of a system of interconnected boreholes.

Research Problems

(1) The study of the process of propagation of seismic waves in the real earth. (2) The

construction of an effective seismic model. (3) The study of wave dynamics in the real earth.

In addition to the problems of oil and mining geology enumerated above, VSP is also used for studying water-saturated unconsolidated deposits in coastal and shelf areas, and engineering surveys. For the solution of many such problems, VSP is the only and generally the most reliable method. VSP serves as the source of fundamental information for the qualitative appraisal of the potentialities of seismic methods.

2. VSP and the Potentialities for Improving the Effectiveness of Surface Observations at the Exploration Stage

Some of the potentialities have been discussed at length in the preceding chapters, e.g., the study of source excitation conditions, determination of the origin of waves, etc, and are considered here only in outline.

Study of Source Excitation Conditions

Of special interest in the study of energy sources for seismic exploration problems is the seismic effect of a source, the term implying its effect on the wave field.

Apart from the effect of the source itself (in the case of explosives, the factors to be considered are the charge, its shape, its chemical composition, its energy radiation directivity etc.), there is the much greater influence of the earth's heterogeneities in the vicinity of the source, which may result in the transformation of a comparatively simple and short pulse at the source into a complex train of downward waves. The source and the medium constitute a combined effect, and the influence of both factors should be considered jointly.

The conditions of excitation may exercise a substantial influence on the shape of each individual wave, and on the structure of the seismogram as a whole, and thereby on the effectiveness of seismic exploration. The most rigorous and objective appraisal of excitation conditions can be performed with PM–VSP using the direct wave and the train of downward waves immediately following it, observed at various depths.

We expect interesting results from studying the connection existing between the conditions of excitation, the structure of the seismogram and the effectiveness of seismic exploration.

VSP makes possible a quantitative evaluation of the parameters (spectrum, intensity, directivity, etc.) of sources of various types, both explosive and non-explosive, with the purpose of controlling the shape, as well as the effectiveness of various source patterns and various stacking schemes in the case of repeated shots.

In PM VSP, which is based on the simultaneous discrimination and tracking of different wave types, the excitation conditions assume a special importance. PM–VSP enables the natural directivity of an explosion to be studied, and the optimum conditions for the simultaneous excitation by a single source of both longitudinal and transverse waves to be selected. The extensive use of non-explosive sources of various types makes the problem of studying the excitation conditions a very urgent one. This is because with non-explosive sources (gas-powered, falling-weight, vibrating sources) an intense field of various wave types associated with the surface and the upper portion of the geologic section is frequently excited. Such waves seriously handicap the tracking of deep reflected waves. VSP may play a decisive role in the study of the parameters of unwanted waves.

Study of the Velocity Distribution

VSP, in contrast to conventional logging, can use not only the first arrival events, but all the waves formed in the earth and recorded in the initial and subsequent portion of the record. The velocities of the longitudinal and transverse waves can be determined from the vertical traveltime curves of longitudinal, converted and transverse waves. This makes it possible to execute more detailed sectioning of the velocity profile, as compared with normal seismic logging, by making use of both down- and upward waves passing repeatedly through the layers.

Studying the velocity structure with the aid of *S*-waves deserves a special mention. To this end, in addition to the direct transverse waves, one can use the reflected monotype *SS*, – the converted *PS* – as well as the converted downward (for nearby shotpoints) and upward (for distant shotpoints) waves. VSP enables the magnitude of the velocity to be determined not only vertically, but in any oblique direction, and this is significant in observations at great separations and especially in highly anisotropic media, for processing common-depth-point data.

When the velocity structure is built from the first arrival events, on an in-line vertical profile, the practice is to plot an interval velocity curve (as in the case of conventional well velocity surveys), which shows only layers of great thickness with great velocity contrasts. The traveltime curves of the first event recorded on off-line profiles, which are more sensitive to the fine details of the section, are used to establish velocity gradients in thick layers as well as to determine the coefficients of velocity increase with depth. The waves in the subsequent portion of the record (reflected, longitudinal, and converted) obtained from both in- and off-line profiles enable the velocity profile plotted from the first arrivals to be sectioned, and reflection interfaces, often in the form of thin layers with which waves of adequate intensity (reflected and converted) are associated, to be identified. It is advisable to carry out the combined processing of sonic logs and VSP data to determine the thickness of such individual layers and their structure. In case sonic logging is not executed, use can be made of velocity curves computed from production geophysical data.

The velocity profile thus plotted will contain finer detail than can be plotted from well-survey data in the seismic frequency band, because it takes into account all the main waves formed in it. Accordingly, such a profile to the first approximation can be regarded as a velocity model of the earth corresponding to the observed wave pattern. It should, however, be kept in mind that such a velocity model constructed for almost vertical angles of incidence, will as a rule, not agree with observations at great source separations, since with increasing distance, the effect of individual thin high-velocity layers diminishes, and the theoretical traveltime curves will overlap the observed.

VSP data is used extensively to study the velocity profile in the process of construction of an effective seismic model [47]. It should also be pointed out that VSP can provide information about the anisotropy of a geologic section. The use of traveltime curves, not only of the first arrival events, but of reflected waves as well, substantially improves the means of studying the anisotropy, which often has to be taken into account to improve the accuracy of plotting.

Detection of Reflection Horizons

With the increase in the depths being studied, it is necessary to utilize waves obtained at long recorded times, where the decrease in the intensity of the reflected wave and a substantial increase in the noise level produces a low signal to noise ratio. The increase in the charge size, as we have already seen, results also in an increase in the relative intensity of unwanted waves, and for this reason, as a rule, deep reflected waves cannot be discriminated on surface seismograms. In such conditions, the identification of reflecting horizons in zones of geologic interest located at great depths, assumes substantial importance. VSP makes it possible by means of observations at great depths to appraise and study the reflecting properties and thus to evaluate the prospects of seismic exploration. To illustrate the point, we shall cite two examples.

One major result of VSP observations carried out in the Bemkent and Karabekaul basement inflexions (Uzbekistan, Northern Kamaschi, Karakum, Gumbulak), was the detection of the reflected wave from the productive subsalt Callovian–Oxfordian limestone.

On surface seismograms, it is not possible to track this wave continuously along the profile. At the same time, this wave is the dominant one in VSP observations in the depth range from 3000 m to 2000 m (Figure 137). Work carried out in several areas has demonstrated that this wave displays complexity resulting from the heterogeneities of the Callovian–Oxfordian limestone and the overlying layers, and from the variability of the structure and lithological substitution of deposits. VSP enabled the principal characteristics of the formation of waves from productive horizons, as well as the formation of the wave field as a whole, to be studied, and a rational procedure of surface observations to be devised for the solution of the geologic problem.

Another example serves to illustrate the point: it was established from VSP data obtained in a 5500 m borehole that very weak waves that were difficult to detect and track even on seismograms recorded in the deepest part of the profile, could be identified as reflections associated with deep-lying deposits not penetrated by the borehole (Figure 16).

Determination of the Origin of Recorded Waves

The determination of wave origin is one of the main interpretation stages, which serves largely as the basis for the validity and effectiveness of seismic studies. In the majority of cases, surface observations do not provide adequate criteria for the determination of the origin of recorded waves. With the development of PM, in which various wave types are tracked simultaneously, the problem of determination of wave origin assumes special importance, because the number of waves discriminated on a seismogram rises sharply.

With VSP, the kinematic characteristics of waves of various types are more sharply expressed, and may be much more significant for the interpretation of seismic data than in the case of surface observations. The chief difficulties in VSP are associated with the complexity of the wave fields observed inside the earth. The current methods of analyzing wave fields, in particular, polar-positional correlation, substantially enhance the opportunities for the discrimination and tracking of various wave types along the

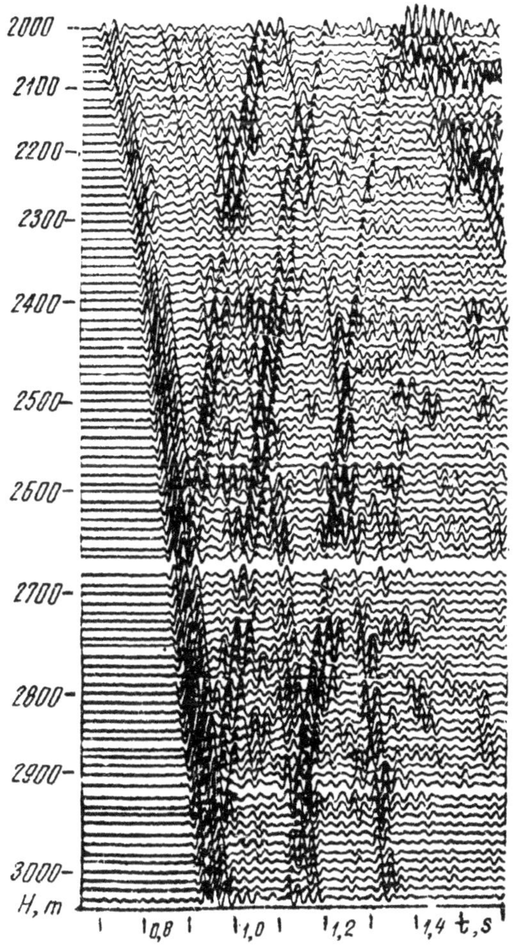

Fig. 137. VSP seismogram illustrating existence of deep reflected waves. Borehole N8 Pamuk [67].

vertical profile. It is worth pointing out that with an increase in the separation from the shotpoint, the determination of the origin of the recorded waves becomes much more difficult in VSP.

Improving Resolution

Improving resolution involves recording waves of higher frequencies up to from 100 to 130 Hz, which in addition to improving the resolution of the record, as demonstrated by numerous VSP observations, also raises the signal-to-noise ratio in many cases. In regions where the geologic section chiefly comprises terrigeneous deposits, the multiples are distinguished by lower frequencies than those of the singly-reflected waves. In such cases, the effectiveness of the reflection method can be improved by means of detailed studies of the frequency characteristics of individual elements of the section, especially in thin-layered media, and by the selection of optimum filtering.

Of special importance, is the study of the spectral characteristics of the low-velocity zone and the upper portion of the section, which as a rule displays great heterogeneity and strong absorption. The chances of raising the frequency of reflected waves recorded on the surface depend essentially on the absorption properties of the low-velocity zone and the upper portion of the section. An effective method of evaluating the filtering properties of the low-velocity zone and the upper portion of the section is the study of singly-reflected upward and multiple downward waves in VSP observations. Sometimes, it is expedient, in order to increase the frequency of the recorded waves, to carry out observations below the low-velocity zone. However, such observations are not always equally effective, because of the relative absorption properties of the earth as a whole compared with the low-velocity zone. If the filtering action of the earth as a whole is greater than that of the low-velocity zone, a wave, originally of high-frequency, arriving back at the low-velocity zone will already be so stripped of high-frequency components that the low-velocity zone will not change its spectrum, or will change it imperceptibly. This occurs in particular in low-velocity terrigeneous sections. On the contrary, in high-velocity carbonate sections, in the presence of a massive low-velocity zone, the observations beneath the zone can be very effective. This is even more true in the case of a highly absorptive zone, e.g., in peat fields and marshes (Figure 63).

Naturally, increasing the frequency of recorded waves means primarily producing a source excitation spectrum containing higher frequencies. VSP experience has proved that good results are obtained by the use of comparatively small explosive charges beneath the low-velocity zone. Among the non-explosive sources, excitation by electric pulses in a borehole may be of a special interest.

Stratigraphic Correlation of Seismic Horizons

It is not alwasy feasible to execute reliable stratigraphic correlation of seismic horizons using surface observations even in those cases when the origin of waves is beyond doubt. This is chiefly because of the inadequacy of information about the laws of variation of the velocity with depth, as well as phase shifts introduced by the field equipment. Accordingly, in the majority of cases, the correlation is rather approximate, and we can only state inside which interval the interface is located. This is true of both the reflecting and refracting interfaces. And in comparatively frequent cases, when the origin of the waves cannot be determined unambiguously, there is no sense in talking about stratigraphic correlation at all. For this reason stratigraphic correlation of waves remains one of the principal and constant goals of VSP. This goal retains its importance in all reflection-method observations (controlled-directional-reception, common-depth-point) as well as in refraction-method studies. The opportunities offered by VSP are of special interest from this point of view.

With VSP, stratigraphic correlation of *reflected waves* can be executed more reliably, since from observations in the immediate vicinity of reflecting interfaces one can obtain a more clear impression about the stratigraphic association of waves than from surface observations. However, several specific obstacles, both in principle and in practice, stand in the way of the improvement of the accuracy of stratigraphic correlation.

The principle obstacles are associated with the recognition of the wave pattern, specifically with its complexity in conditions of thin-layered media. Among practical obstacles are the neccessity for very detailed measurements (featuring small increments), the difficulty of exciting and recording high-frequency components etc. First of all it should be noted that the concept of stratigraphic correlation itself has a different meaning in different (thick- and thin-layered) media. In thick-mayered media, the seismic reflecting interfaces are associated with quite distinct geologic interfaces (lithologic differences) and can be clearly correlated with the geologic section. In thin-layered media, the concept of a single reflecting interface is an erroneous conventional one, since in this case the reflected waves are the sums of several waves, and are associated not with individual interfaces between the layers, but with some group of thin layers in the geologic section (Figure 138).

In simple thick-layered media containing comparatively few interfaces, their stratification can in principle, be established from the kinematic properties of waves reflected from them and recorded along the entire vertical profile. To this end, from observations near an interface, it should be sufficient to identify the phases of the direct and the reflected wave, and to find the depth of the common point of the traveltime curves of such phases, and to identify the phases of the waves along the horizontal and vertical profiles in the section of the profile close to the surface of the Earth. However, in practice this is fraught with difficulties, because the correlation of waves is as a rule disrupted in those sections of the profile. Near the reflecting interfaces, a broad train of high-intensity downward waves (the direct wave, the ghosts and the multiples from shallow interfaces) is superposed on a comparatively low-intensity reflected wave. Besides, in the immediate proximity of the interface the longitudinal reflected wave interferes with the converted reflected wave. In such conditions the accuracy of stratigraphic correlation is no better than several dozen meters. The improvement of the accuracy of stratigraphic correlation hangs chiefly on the possibility of suppressing the unwanted waves and tracking the reflected waves back to the reflecting interface. The discrimination of waves according to average velocity, and chiefly the method of subtraction mainly enable the reflected waves to be suppressed. However, the familiar difficulties of such methods (non-rectilinear shape of the vertical traveltime curves, disimilar oscillations being summed, both in amplitude and in shape) limit the potentialities of such discrimiantion. In this respect, the discrimination of waves according to their polarization, based on the controlled directional reception of the first kind, can be of special interest. For particular angles of incidence, polarization-positional correlation enables the component of a reflected wave to be tracked in the plane of zero displacements of the downward waves, it being possible to track separately the reflected longitudinal and the converted waves polarized in different directions.

With modern borehole instrumentation, the controlled directional reception of the first kind (CDR–1) enables the downward wave to be suppressed by about an order of magnitude, without causing the appearance of false alignments or distorting the spectral characteristics of the waves [2]. Accuracy can be improved by combining various methods of discrimination, for instance, controlled directional reception of the first kind (CDR–1) and discrimination according to the direction of propagation and according to frequency.

For stratigraphic correlation, inverse filtering is of major importance, since it reduces the duration of the seismic signal and enables a short pulse to be obtained from a

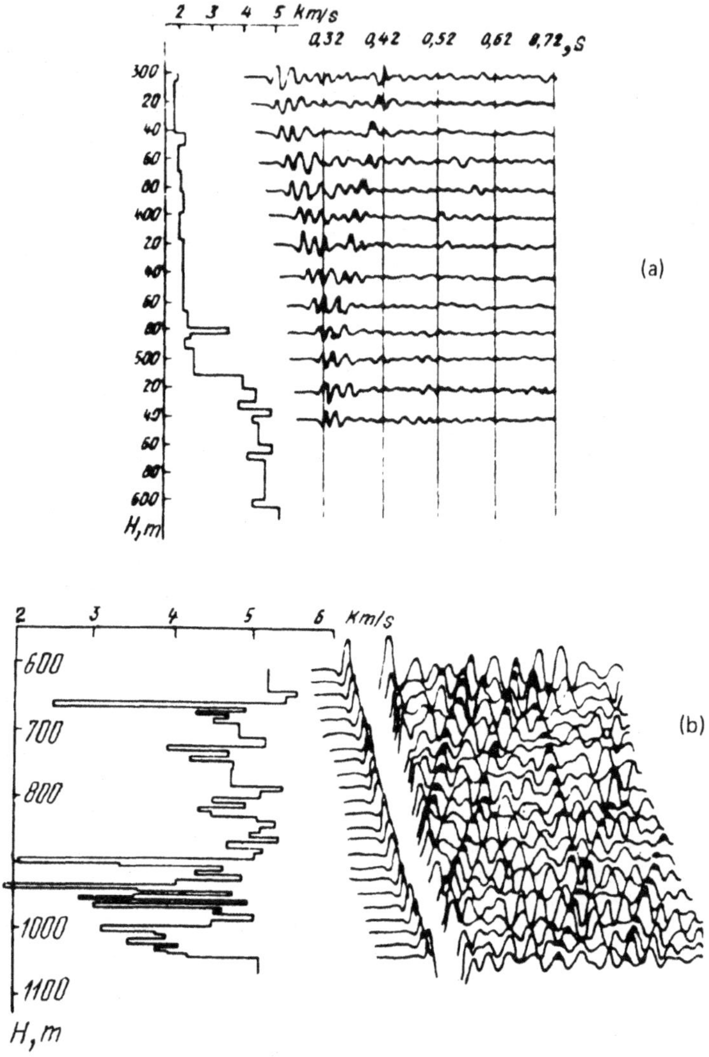

Fig. 138. Examples of stratigraphic correlation of waves from ultra-sonic logs in the presence in the cross-section of a sharp reflecting interface (a) and a packet of thin layers (b) [24].

multi-phase direct wave. This makes it possible not only to track the reflected waves in the vicinity of a reflecting interface, but also to improve the reliability of identification of identical phases of the downward and the reflected wave, and thus to improve the accuracy of stratigraphic correlation. Simultaneously, this enables waves reflected from interfaces separated by comparatively short distances to be discriminated. To improve the accuracy of stratigraphic correlation of waves, one may also make use of the maximum of the cross-correlation function of the direct and the reflected wave, which corresponds to the instant the reflected wave is formed.

Considerable difficulties are encountered in observations in the upper portion of the section where the interference of numerous waves of various types and origin makes it sometimes impossible to track the reflected waves up to the surface, and consequently to ensure correlation of phases along the vertical and the horizontal profiles. However, in such cases, the tracking of waves may be improved by discrimination according to various parameters together with the application of more complex interferential observation systems.

For the purpose of stratigraphic correlation of waves, when comparing the phases of the downward and the reflected waves, one should take into account that the time delays introduced by seismic instrumentation are different for the first distinct arrivals and the subsequent phases, and depend on the pulse shape and the dominant frequency. This may lead to an over-estimate of the depth of the point of intersection of the traveltime curves, and correspondingly the reflecting interface [24]. In addition, a change in the pulse and a displacement of the extrema associated with absorption, which increases with the distance and the extremum's ordinal number, increases the slope of the phase traveltime curves. According to VSP data, the corresponding error in the depth of the reflecting interface over an interval of 2000 m may amount to several meters. The use of inverse filtering eliminates this phenomenon.

VSP substantially extends the potentialities of stratigraphic correlation, especially when combined with production-geophysical and sonic log data, and in the majority of cases can ensure the accuracy required for the solution of exploration problems. However, all the aspects of stratigraphic correlation of waves have not been studied adequately. It is an established fact that the accuracy of stratigraphic correlation is determined chiefly by the ability to track the reflected wave right down to the first arrivals of the downward wave. This depends on the complexity and the duration of the downward wave train formed as the result of superposition of various waves associated with the upper portion of the section on the direct wave pulse. The shape of the downward wave train mainly depends on source excitaton conditions and especially on the structure of the upper portion of the section. The structure of the earth in the region of the detector exercises a much smaller influence on the shape of the downward wave. At the same time, the polarization of oscillations in the downward wave depends very little on the conditions of excitation and very much on the earth's structure. The studies of the laws governing the variation of the polarization parameters of the direct wave may bring about an appreciable improvement in the accuracy of the stratigraphic correlation of waves.

It is also worth noting that, sometimes, difficulty in identifying waves from surface and down-hole observations is occasioned by the fact that the data being analysed are not identical, having being obtained with different instrumentation parameters and different recording procedures. In addition, observations in the upper portion of the section and the low-velocity zone do not always produce high-quality data, because of a too complex mechanical design of the borehole, an improperly clamped geophone etc. With adequate high-quality data, it is usually possible to identify every phase. A point that attracts attention on the seismogram with straightened alignments of upward reflected waves, (Figure 16), is the distinct stratigraphic correlation of the reflected waves associated with the interval of the section penetrated by the borehole. The waves recorded in VSP are unambiguously identified with those recorded from surface observations, and their phases are well correlated. Note by the way that the seismogram shows, especially at great depths, how the direct wave changes from shothole to shothole.

However, in the majority of cases, sedimentary beds are composed of thin layers. In such conditions, we may speak of discriminating in the section a group of layers or a depth interval corresponding to the given combined wave (Figure 138). The situation is aggravated by the fact that the duration of the record and the distribution of intensities in a combined interference wave are liable to vary greatly, and consequently the shape of the downward and the reflected wave can be different. The correlation of waves by maximum amplitudes is not a precise method. Here we have to take account of the phase shifts introduced by the recording equipment. In such conditions a significant part can be played by synthetic seismograms.

In many cases the controlled directional reception of the first kind (CDR–1) serves as an effective instrument for stratigraphic correlation. In case the displacement directions in the direct wave do not coincide with those in the reflected wave, the latter can be discriminated in the zero displacement plane of the *P*-wave, and tracked in the region of the first arrivals (Figure 139).

Stratigraphic correlation of refracted and head waves can in principle be executed by employing the shape of the traveltime curves of the first events, and the usually observed

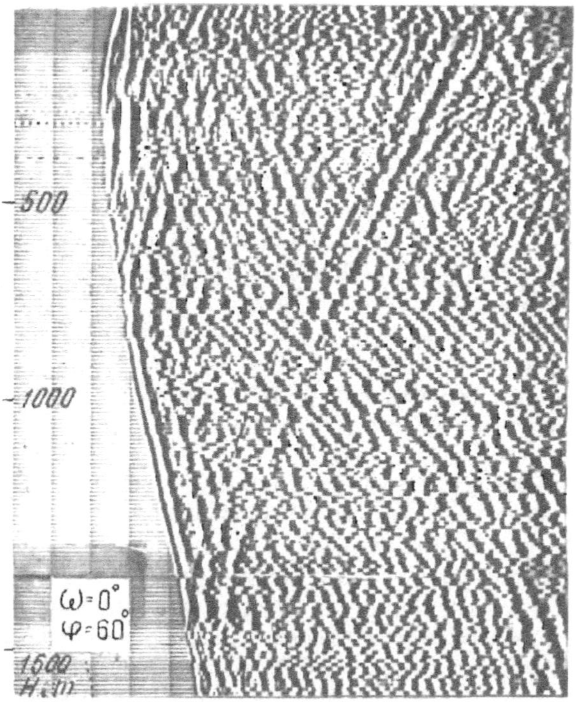

Fig. 139. Examples of stratigraphic correlation of waves with the aid of CDR–1 (borehole N693 Anastasiyevsko–Troitzkaya). Oscillation component $\omega = 300°$, $\varphi = 30°$.

distortion of linear polarization of a wave crossing an interface (this applies chiefly to head waves). The accuracy of stratigraphic correlation of refracted waves is much worse than that of reflected waves.

Synthetic Vertical Seismic Profiles

An essential part in the analysis of the complex wave field observed in VSP is played by synthetic VSP seismograms, which enable the wave field to be divided, and various waves to be studied separately. In particular, multiples associated with a thin-layered medium, when it is practically impossible to track individual waves can be analysed. By comparing the computed and the observed seismograms, it is possible to identify the main interfaces at which the multiples are formed, to study the kinematic characteristics and the regions of existence of the multiples and their effect on the wave field. It is also possible to separate those regions of the seismograms in which the singly- and the multiply-reflected waves are recorded, from those in which only the multiples are present, and to assess the influence of the multiples on the total wave field, as well as to gain a qualitative impression of the stratigraphic correlation of waves (Figure 60).

There are two variants of the Baranov–Kunetz algorithm in existence that enable both the combined wave field and the individual up- and downward wave fields to be computed [29]. The first variant may be conveniently used to compare the computed field with the observed as well as to predict the field from a known geologic section, while the second variant is advantageous for analysing the multiple reflections. Programs have been compiled for both variants. Consider as an example a synthetic seismogram computed with the second algorithm. The computation deals with plane waves at normal incidence propagating in a medium with plane and parallel interfaces and no absorption. A section typical of the Russian platform consisting mainly of terrigeneous and carbonate deposits with pronounced velocity differentiation (borehole N8 Kudinovskaya) was chosen as the model for computations. In the geologic model, one may conveniently separate a terrigeneous bed with $v_{av} = 2\ \mathrm{km\ s^{-1}}$, a carbonate bed with $v_{av} = 4$ to $4.5\ \mathrm{km\ s^{-1}}$ and a lower terrigeneous bed with $v_{av} = 3$ to $3.5\ \mathrm{km\ s^{-1}}$. The carbonate bed is characterized by a velocity gradient and by the presence of sharp interfaces associated with alternating carbonate and terrigeneous deposits.

The combined wave field is the result of the interference of numerous up- and downward waves (Figure 140a), and such a field is difficult to analyse. For the analysis of the wave field, seismograms of various wave types have been constructed separately. It is evident from the seismograms of the singly-reflected waves only (Figure 140b) that about fifteen reflected events, which do no interfere with each other, are produced by the model. They include very high-intensity waves associated with an interface in the upper portion of the low-velocity terrigeneous bed and with several interfaces in the high-velocity carbonate bed. However, a high-intensity reflected wave associated with a thin high-velocity limestone layer originates also in the lower low-velocity terrigeneous bed. The downward wave field that includes all multiple reflections from all interfaces (Figure 140c) is shown by numerous very high-intensity waves occupying a considerable time interval on the seismograms. The intensity of all the downward waves is appreciably higher than that of the singly-reflected ones. A particular characteristic is that the intensity and duration of the downward waves increase with the depth, such increases being associated with discrete elements of the geologic section. The seismogram of the

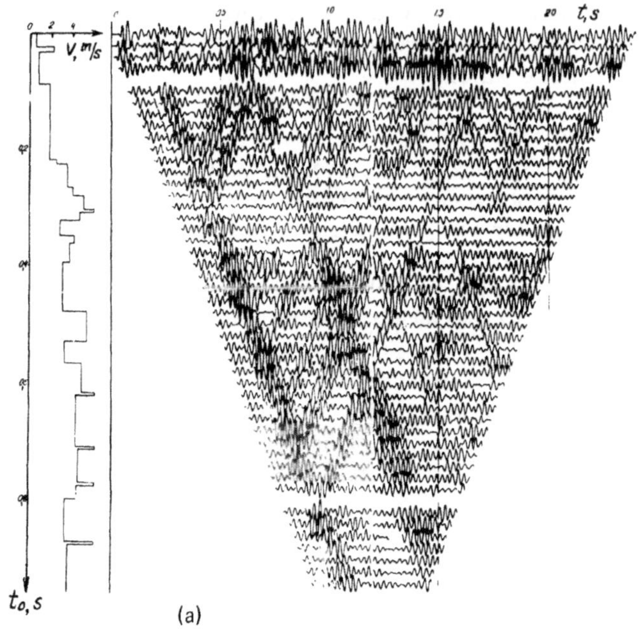

(a)

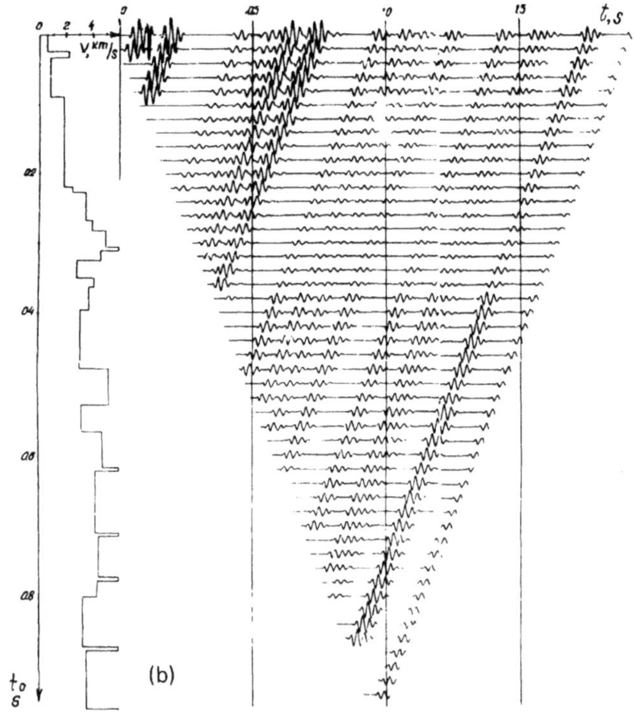

(b)

Fig. 140a-b.

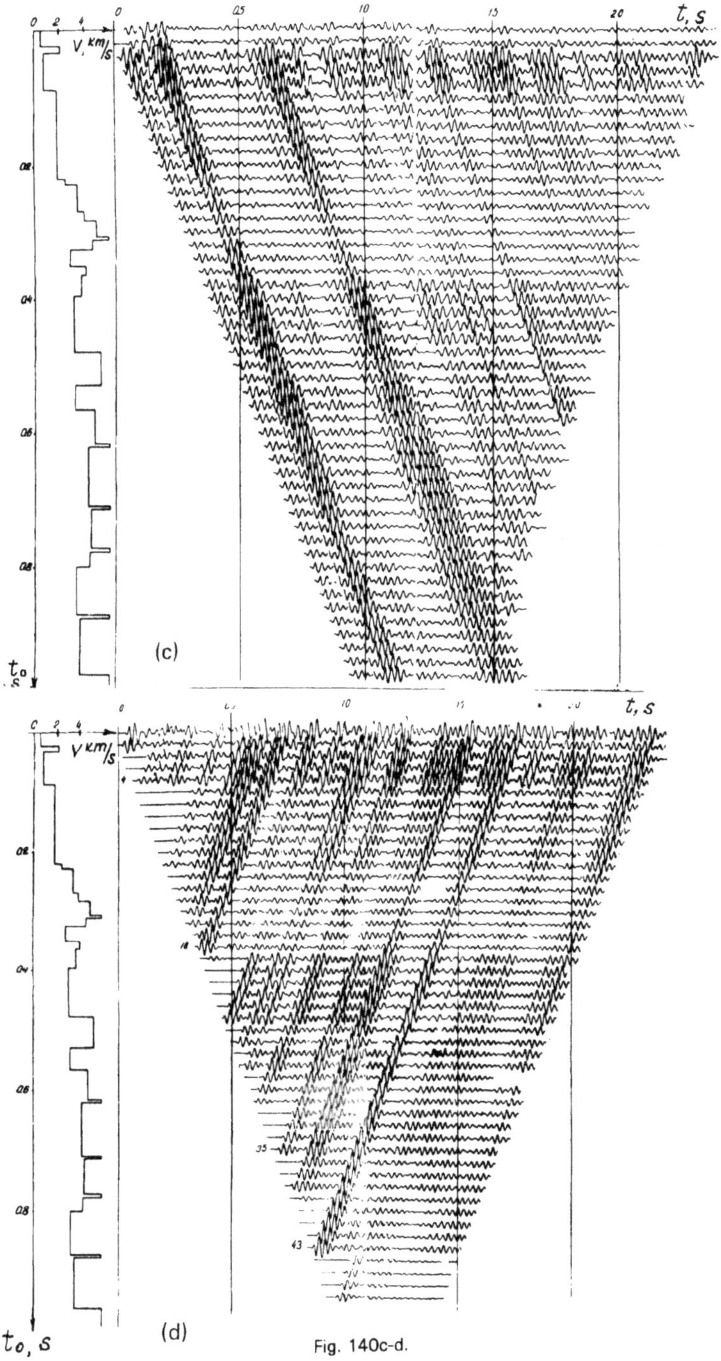

Fig. 140c-d.

Fig. 140a-d. Vertical synthetic profiling seismograms [29]: (a) combined wave field; (b) singly-reflected waves; (c) downward waves including waves of all multiplicities; (d) upward waves including waves of all multiplicities.

up-going multiples constructed with waves of all multiplicities (Figure 140d) clearly shows multiples reflected from elements within the recorded depth interval. The multiples are very abundant, their wave field is complex, and their intensity increases towards the surface. Note that, to compute synthetic seismograms, one must be in possession of a detailed model of the earth. To this end, sonic logs are normally used, but resistivity logs may be used as well. Such reconnaissance analysis carried out before field work has begun, is of great interest, since it permits the exploration potentialities to be assessed and an observation system to be selected.

Thus, synthetic seismograms present a very clear picture and provide a sound qualitative idea of the mechanism of the formation of the multiple wave field, especially in thin-layered conditions. However, the strong dependence of the synthetic seismogram on the quality of the sonic log, which is not always up to the mark, limits the potentialities of synthetic seismograms for the purpose of quantitative assessment. The situation is aggravated by the very approximate nature of data on absorption and dispersion, by the pronounced effect of a very heterogeneous upper portion of the section, and by other factors. Accordingly, synthetic seismograms should be combined with observed VSP data. This will provide more authentic information about the geologic section and enable the maximum resolution of seismic observations to be assessed.

When interpreting VSP data, especially those obtained in studies of the space around the borehole, of great importance is the use of digital models. This applies first of all to the studies of structural interfaces in sections contiguous with the borehole. Such modelling also enables the effects to be expected on VSP records to be studied, exploratory potentialities to be assessed, and an optimum observation technique for each specific geologic situation to be chosen.

Increasing the Depth Range of the Studies

The depth range of VSP studies is in principle determined by two factors: by the presence at great depths of reflecting interfaces from which waves of adequate intensity can be recorded, and by the effectiveness of their surface discrimination, which is chiefly determined by the unwanted wave field observed in the recorded depth interval of useful waves. The fact that reflected waves are formed at interfaces located at great depths, convincingly and clearly demonstrated by VSP, is the corner-stone for increasing the penetration of the seismic reflection method. In many cases when the depth of the borehole permitted observations to be made in the proximity of the crystalline basement it has been possible to discriminate waves reflected from its top or from interfaces located in its immediate vicinity (Figure 141). On the basis of VSP experience, the following plan of studies aimed at assessing the potentialities for increasing the depth range can be drawn up. To begin with, direct observations at maximum depths are carried out to identify reflected waves associated with beds of geologic interest, and their parameters determined. The next stage consists of the examination of the causes obstructing the tracking of the deep waves on the surface seismograms. Despite the fact that in the majority of cases reflecting horizons have been identified at great depths, which is of great interest from the geologic point of view, it was not possible to track useful waves up to the surface, or to discriminate and track them along a horizontal profile, because of a complex field of unwanted waves associated with the upper portion of the section. VSP is able to study the unwanted wave fields and to determine their

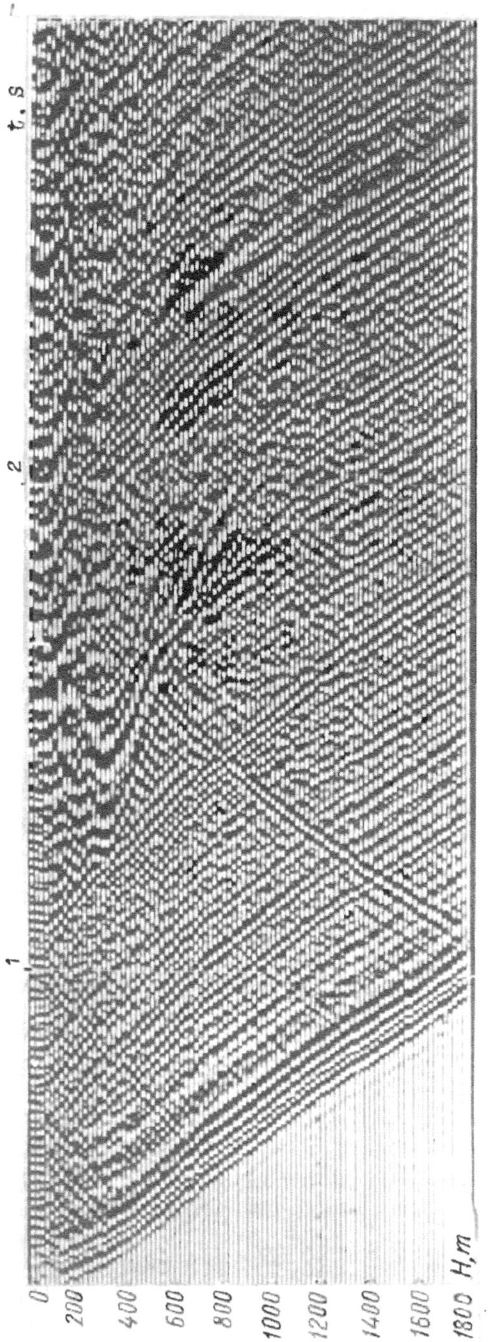

Fig. 141. **VSP** seismogram, shotpoint 400 m (borehole N318 Abinskaya, courtesy S. V. Kursin).

origin and parameters. In the reflection method, the chances of discriminating deep reflected waves on surface seismograms are chiefly determined by the level of the multiples, which in turn are mostly influenced by the base of the low-velocity zone, and by the interfaces in the upper portion of the section. Data on the origin and the parameters of useful and unwanted waves obtained in VSP, may enable the potentialities of various interference systems to be evaluated more objectively, and an optimum system for the discrimination of useful waves to be selected.

The Use of VSP Data for the Deconvolution of Surface Observations

The resolution of the seismic record in conditions of a complex interference pattern of reflected waves, the signal-to-noise ratio, the intensity of the multiples, and the correlation of seismic data with the lithology are all greatly dependent on deconvolution. It eliminates distortion of the phase and the amplitude spectra of the signals introduced by the recording system. The effectiveness of deconvolution, in turn, depends substantially on the original pulse used to compute the filter operator. The available experience of employing the pulse of the direct downward wave observed in VSP for deconvolution shows that such an approach offers good prospects for surface observations near the borehole. We shall use the example of the vertical profile shown in Figure 142 to illustrate the effectiveness of employing VSP data for the deconvolution of surface data. A clear correlation with the sonic logs is observable on the surface time cross-section plotted as the result of deconvolution employing an operator computed from VSP data. A comparison between the cross-section obtained with deconvolution employing VSP data and one plotted with standard processing shows that, despite a general similarity of the records, the deconvolution employing VSP data produced a record with better resolution. For instance, it allows suppression of sea bed reverberations recorded at about 3.0 s, and this enabled a high-intensity primary reflection (Figure 142d) to be discriminated and reliably tracked at about the same time (3.0 s). Its origin, as determined from the VSP seismogram after deconvolution, corresponds to a reflection from a powerful interface with a negative velocity contrast. Thus, VSP data appreciably enhanced the possibilities for analysing the cross-section. However, such data in principle belong only to a single subsurface point. If the wave field is stable along the line of surface observations, the deconvolution operator constructed from VSP data can be effective at great separations from the borehole, but if the wave field is changing along the profile it will be applicable chiefly in sections of the surface profile contiguous with the borehole.

VSP and the Effectiveness of the Seismic Common-Depth-Point Method

At the initial stages of development of the common-depth-point method, VSP was the chief source of information required for the selection of the optimum surface observation systems and for the details of the velocity distribution, ray velocities, the origin of waves, the relative intensity of primary- and multiple-reflected waves, the stratigraphic correlation of waves, etc.

Now, since the common-depth-point method has been developed, some of the parameters required can be obtained directly in the course of processing common-depth-point data, but even now VSP is very important for improving the effectiveness of the method. The following points are of major interest: (1) stratigraphic correlation

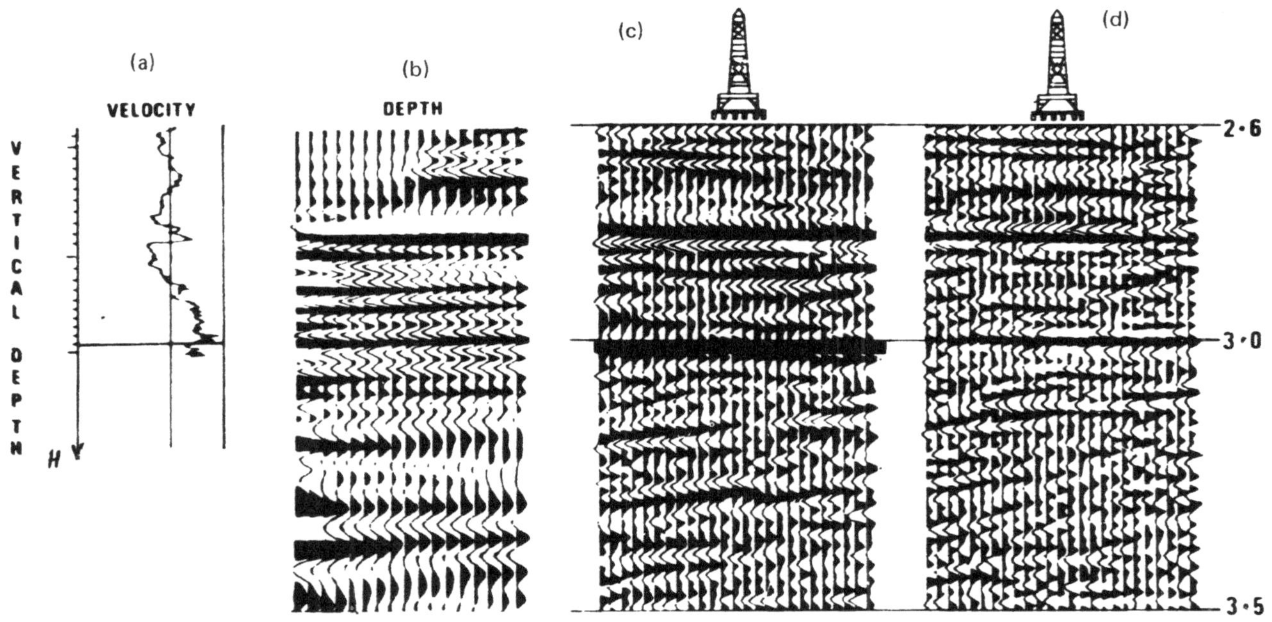

Fig. 142. Effectiveness of deconvolution on the basis of VSP observation data along a horizontal profile [142]: (a) sonic log; (b) VSP data after deconvolution; (c) up-hole seismogram after standard deconvolution; (d) up-hole seismogram after deconvolution according to VSP data.

of waves and the identification of reflecting interfaces below the depth penetrated by the borehole. The latter problem is of principal importance, since it cannot be unambiguously solved using surface observations. (2) Determination of the relative intensities of the multiples. (3) Appraisal of the effectiveness of the supression of unwanted waves, and whether or not a particular observation system is the optimum one. In the last two cases VSP yields independent results, and this may improve the effectiveness of the observations and the validity of the results.

The kinematic and dynamic parameters of the wave field of primary and multiple reflections play a significant part in the synthesis of the optimum common-depth-point observation systems. To this end, useful waves are discriminated from VSP data, and the quality of their correlation along the vertical profile is examined. By studying the causes of poor wave correlation, it is possible to discriminate unwanted waves, and to identify regions of occurence of multiple waves of different types. Kinematic characteristics of the multiples enable functions of time-delay with respect to the primary reflections versus separation from shotpoints to be calculated, and the optimum summation base, the separation between channels and the separation from the shotpoint to be determined. To select the necessary tracking multiplicity, one determines the relative intensity of the combined field of the multiples, and from it sets the required attenuation level.

Special mention should be made of the possibility of analyzing and controlling source excitation conditions along the line of the profile, with the aim of preserving the shape of the record of reflected waves. The shape of the direct wave in variable low-velocity zone conditions along the profile can change greatly, and is usually accompanied by a change in the shape of the reflected wave. The preservation of the shape of the direct wave is one of the prerequisites for improving the effectiveness of the common-depth-point method.

PM VSP and the Polarizational Common-Depth-Point Method

Of considerable importance for improving the effectiveness of seismic studies, is the joint use of different wave types – longitudinal, transverse, and converted. The polarization common-depth-point method that has been developed and is being extensively introduced into practice, enables waves to be discriminated and tracked, irrespective of their type, direction of propagation and polarization [2] by employing polarization-positional correlation. Of special interest are the converted reflected waves. VSP experience has shown the *PS*-waves to abound in the wave field and to originate at the same interface as the longitudinal waves, this enabling the geologic section to be sub-divided according to the velocities of *S*-waves and structural problems to be solved. In addition, converted waves are recorded at much later times than the longitudinal, and this in many cases facilitates the discrimination of waves against the background of interference. PM–VSP completely eliminates the difficulties resulting from ambiguous identification of longitudinal and converted waves that limits the use of converted waves in surface observations.

The extensive combination of the PM common-depth-point and the PM VSP methods is the main pre-requisite for improving the effectiveness of surface PM common-depth-point observations.

The effectiveness of application of the CDP method to VSP observations

(VSP CDP) deserves special mention. With an appropriate arrangement of sources on the surface and seismometers in a borehole, one can obtain VSP CDP systems capable of fully exploiting the advantages of a system utilizing signal stacking from a common reflection point, thereby raising the signal-to-noise ratio. Besides, such observations facilitate the studies at every depth of the dependence of the coefficient of reflection on the angle of incidence, and this is of great importance for the appreciation of changes in the properties of the layers in the vicinity of the borehole.

VSP CDP systems are more effective than surface observations, because they provide us with more precise data on wave velocities than the surface CDP observations.

Assessing the Potentialities of Surface Seismic Exploration and Selecting the Optimum Observation Conditions

To further such aims using surface observations, one would have to execute a vast volume of observations. However, in the absence of reliable criteria, in the majority of cases it is very difficult and sometimes even impossible to interpret the wave field and to single out useful waves at comparatively low intensities.

Observations deep in the earth at points in the immediate vicinity of deposits of geologic interest enable the parameters of all waves associated with them to be determined, and the most effective observation conditions to be chosen. In this respect PM VSP presents great opportunities for the discrimination of various wave types associated with the same interfaces. The discriminated waves are then tracked up to the surface, and wave propagation regions and the conditions for recording the waves are studied.

Of substantial importance in assessing the exploration potentialities is not only the intensity of the waves, but also the region in which they are trackable up to the surface. In certain cases it is expedient to use waves of lower intensity and recorded at later times instead of waves of higher intensity recorded in the initial portion of the seismogram where it may be flooded with unwanted waves. This is particularly true of converted and transverse waves.

For the solution of such problems seismic observations in boreholes may be regarded as a reconnaissance and should precede surface observations. The results obtained in such a reconnaissance are of such significance that in certain regions where extensive seismic studies are being planned, and where there are no boreholes, it might prove profitable to drill a borehole specially for VSP.

VSP data enable the results of seismic studies of earlier years to be critically re-appraised, the causes of disagreement with the results of drilling to be established, and the seismic data to be re-interpreted.

Establishing the Physical Foundations for Predictions of the Geologic Section from Surface Observations

Our experience shows the effect of various heterogeneities of the geologic section, especially of hydrocarbon deposits, on the wave field, to be much more pronounced in VSP than in surface observations. In the latter case, all such effects become more blurred. The study of the effects of such heterogeneities with the VSP method enables the viability of predicting the geologic section from surface observations to be objectively appraised and the optimum parameters for dealing with such heterogeneities in various seismologic conditions to be selected.

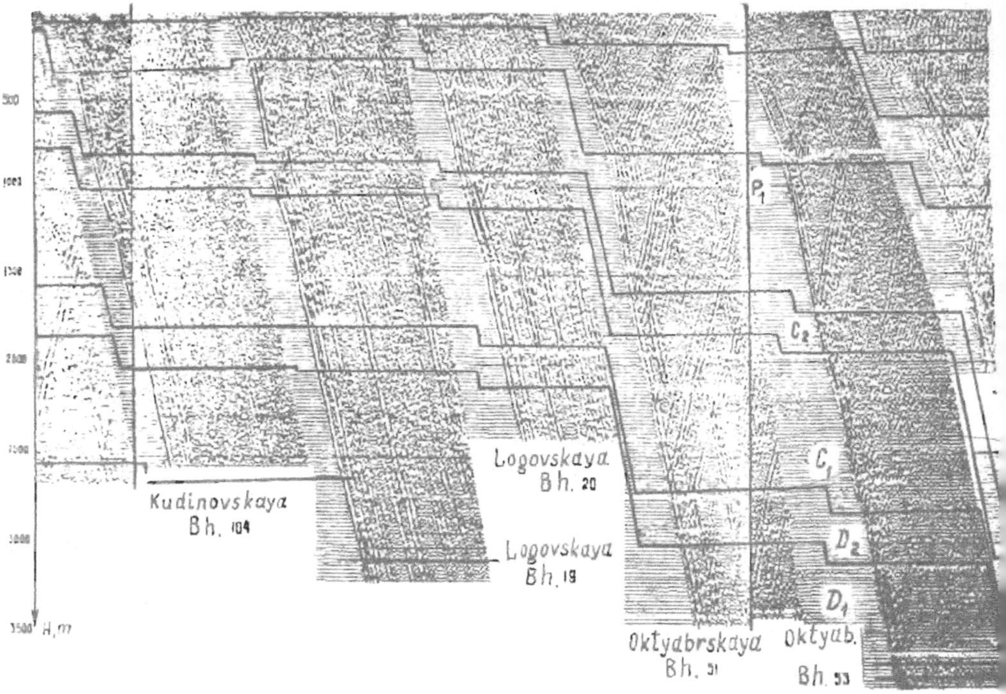

Fig. 143. VSP seismograms along a profile of boreholes in Volgograd province [103].

VSP and the Study of Regional Geologic Sections

As has been demonstrated by VSP, the wave field in many regions is so stable that it is possible to study the geologic structure by means of observations in boreholes separated by great distances. Consider two examples, in one of which the section has been mapped for the range of depths penetrated by the borehole, and in the other for depth intervals below the bottom of the borehole.

In the Volgograd province, VSP has been carried out along a profile of boreholes intersecting the Don-Medveditza ridge, the Volga monocline and emerging in the North Caspian depression (Figure 143). Because of a constant wave pattern, it has been possible to discriminate, track and identify reflected waves associated with the same stratigraphic horizons of the Permian, the Carboniferous and the Devonian age, as can clearly be seen from summary VSP seismograms recorded in a number of boreholes.

The second example deals with the study of pre-Devonian deposits and the relief of the crystalline basement in the Kama basin (Figure 144). Despite much effort spent on studies of the pre-Devonian deposits, these were ineffective until recently, due to a high level of interference on the surface, which obstructs the tracking of deep horizons. VSP was carried out in a series of boreholes arranged along a regional profile. The eastern tip of the profile (boreholes N10 Osintzevo and N42 Spas-Barda) was situated on the Kutur–Krasnoufimsk, and the western tip (borehole 232 Vavokh) on the Tatarian tip.

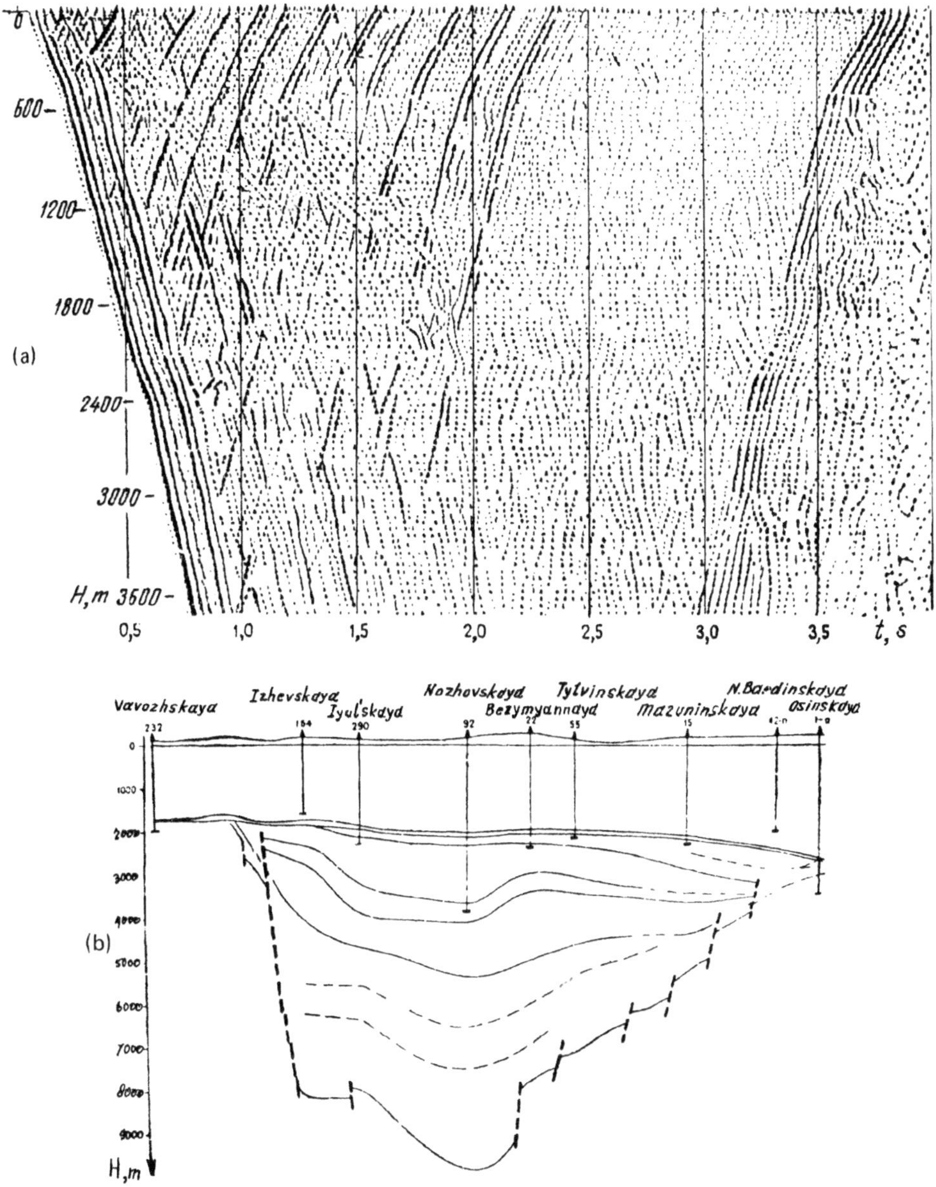

Fig. 144. Wave field from VSP data in borehole N92 Nozhovsk (a) and seismo-geologic cross-section across the region (b) [149].

The central portion of the profile (between the boreholes N164 Izhorsk and N22 Bezymyannaya), on the other hand, was situated within the bounds of a synclinorium with a sharply submerged basement.

The boreholes N232 and N10 at either end of the profile penetrated the entire sedimentary cover. The deepest borehole (N92 Nozhovskaya 4000 m deep) penetrated only the upper two kilometer-thick bed of pre-Devonian deposits. According to the VSP data, in this borehole within the time interval 1.7 s to 3.15 s (two-way transit time) seven reflected waves from interfaces beneath the borehole have been recorded. The deepest reflection, associated with the top of the crystalline basement, was recorded at a time of 3.75 to 3.85 s. By comparing the wave fields recorded in adjacent boreholes, it was possible to identify the waves and to construct a seismic cross-section from the VSP data. The constructed cross-section characterizes the structure of a region interesting from the viewpoint of its tectonics. We shall not dwell here on the geologic description of the cross-section. Note only the fact that the wave reflected from the basement was reliably recorded. The wave is represented as a multi-phase oscillation, and when the basement is deep, is recorded last, as a rule against a quiet background. Thus, VSP data enabled the structure of the Kaltasinsk anlacogen to be studied. To section the thick sedimentary bed, and trace terrigeneous and carbonate rock thinning zones on the flanks, worth exploring for oil and gas, one would do well to combine surface and down-hole observations.

Studying Unconsolidated Deposits in Marine and Continental Shelf Regions

The experience gained in the application of VSP in the sonic frequency band (vertical seismo-sonic profiling, i.e. measurements in the 1000–2000 hz band) demonstrated its high effectiveness in engineering-geology studies of coastal regions, and in the prospecting for much needed minerals occuring in the form of placers (gold, tin, etc.) VSP in the sonic frequency band improves the resolution of results of seismic observations in coastal regions [73], facilitates the appraisal of the method of continuous horizontal profiling, the stratigraphic correlation of waves and the determination of velocities, the lithological and stratigraphic sectioning of water-saturated shelf deposits, and the plotting of geologic and seismo-sonic cross-sections. At the same time, VSP does not increase the general costs. In the course of marine surveys, reflected waves from several geologic interfaces in porous beds have been identified. The depths of the interfaces established from both VSP data and the reflection method agree to within 0.3 m; the relative error in average velocities amounts to 3 to 4%. Amplitude reflection coefficients have also been obtained. The velocities in the 20 to 2000 Hz frequency band for unconsolidated ground turned out to be less than that of an infinite space of water and equal to 700 ± 100 m s^{-1}. According to theoretical calculations, they should be about 620 m s^{-1}. The frequency-dependent attenuation of hydraulic waves has also been determined.

Engineering Surveying Problems

VSP has also found application in studies of the uppermost and the most heterogeneous portion of the geologic section. On land, VSP enables the mechanical properties of rock to be determined for planning and construction purposes.

3. Potentialities of VSP during the Exploration and Exploitation Stages of a Field (Production Seismics)

The traditional stages of seismic exploration for oil are as follows: the regional, reconnaissance and the detailed. The first two deal with the general geologic structure, establish general prospects for oil and gas reserves and outline local features (folds, fractures, thinning, stratigraphic traps, etc.) that are of interest as possible oil and gas reservoirs.

The stage of detailed studies has as its main goal the mapping of structures for deep drilling and until recently was distinguished from the former two chiefly by a somewhat greater density of surface observations.

VSP has substantially expanded the potential of seismic studies, first of all for the solution of structural problems, through the amalgamation of VSP with surface observations, and so improving the effectiveness of seismic studies in the selection of locations for deep drilling [36]. However, after the structure has been mapped, and drilling has begun, seismic studies in the area usually cease.

This in our opinion explains the very great amount of unproductive drilling wasted on off-structure wells and barren rocks. Reductions in the amount of very costly unproductive drilling constitute at present one of the major reasons for improving the effectiveness of the entire complex of geologic exploration.

At the same time, appreciable progress in seismic observations, and especially PM VSP expanded the exploration potentialities of borehole surveys. In contrast to other borehole geophysical studies, concerned only with studying the geologic section along the axis of the borehole, PM VSP has revealed novel opportunities for studying the physical properties of the geologic section surrounding and between boreholes both within the depths penetrated by the borehole and ahead of the drilling bit. Such potentialities created a foundation for the application of PM VSP at such unusual, for seismic exploration, but extremely important, stages as exploration of a field, preparation for its exploitation, and exploitation itself.

The following factors determine the urgency and the importance of such studies: the current reduction in the effectiveness of geologic exploration; a rapid rise in the costs of deep drilling, and in consequence the urgency of solving the problem of the optimum configuration of boreholes to effect a substantial cut in the volume of costly drilling; the need to scientifically select the method of production; the requirement to speed up the start of production as well as controlling production from the deposit. Of great importance also is the application of VSP to the solution of technological problems associated with the drilling of the borehole itself.

The idea of carrying out seismic studies at the exploration and exploitation stages of fields is not altogether a new one. However, this trend has not only failed to find extensive use as yet, but does not even go beyond the stage of tentative experimentation, due chiefly to the predominant use of surface observations for the purpose.

The advantages of application of VSP, and especially PM VSP, at the stages of assessment and exploitation of a field are the result of, in principle, greater resolution of VSP as compared with surface observations following from a greater resolution of VSP records. This is chiefly because in VSP the distorting effect of the highly heterogeneous and absorbing low velocity zone and the upper portion of the geologic section and of other interfaces is mitigated. In addition, in VSP the direct wave pulse is known to us, and this substantially improves the effectiveness of deconvolution procedures.

PM VSP laid down the foundations for a new trend – that of production seismics or seismic studies of the rocks surrounding the borehole mainly applied at the exploration and exploitation stages of the fields. Production seismics is based on an extensive combination of PM VSP with other geophysical research methods in boreholes and in close conjunction with deep drilling. Production seismics is still in its infancy, but experience already gained justifies the discussion of its tasks [51.

Exploration of an Oilfield

At this stage, production seismics must provide corrected data concerning the structural and facies characteristics, the tectonics and corrected seismic data obtained at the reconaissance stage.

VSP problems at the stage of exploration can be subdivided into the structural, and the lithologic and the stratigraphic.

Structural Problems

Structural problems include mapping interfaces and determining their position, including those not penetrated by the borehole; determining the disposition of the layers and mapping thinning zones; establishing structural sequences, detecting and tracing tectonic faults and determining their magnitude.

Lithologic and Stratigraphic Problems

Of decisive importance for solving lithologic and stratigraphic problems are studies of the physical properties of the rocks surrounding the borehole. As seismic studies progress, research into the physical properties of rock becomes increasingly important and urgent. It is closely connected with the prediction of the geologic cross section, and includes a wide range of problems both geological and technological.

An especially noteworthy point is that VSP substantially increases the potential of stratigraphic and lithologic interpretation in studies of the space around the borehole which are based on the detailed subdivision of the geologic section. The main reason for this is that VSP enables records much less distorted by unwanted waves than in the case of surface recordings, to be obtained [58, 150–152]. The use of waves free from interference improves the effectiveness of stratigraphic and lithologic interpretation.

Several factors contribute to this, the chief being the correlation with sonic logs in the borehole and the acquisition of records less distorted by interference, especially by the multiples, than on the surface.

In seismo-stratigraphic problems, VSP plays a special role in the discovery of stratigraphic discrepancies. A scrupulous study of signals reflected from different stratigraphic horizons may make possible the mapping of stratigraphic and facies substitution zones (e.g., of sandstone by clays, etc.).

One of the main geological tasks is the study of deposits. This major problem includes discovery of the deposit penetrated by the borehole, but not detected during borehole testing; predicting a deposit not penetrated by the borehole, but located in its vicinity; studies of facies, lithologic and stratigraphic particulars of the cross-section; in the vicinity of the deposit studies of the deposit itself – the determination of its size, volume, contours in the space around the borehole; studies of the deposit's integrity and the detection of tectonic fractures; studies of reservoir properties and especially of porosity; the determination of the position of contacts; the initial appraisal of reserves; the detection of zones of abnormally high layer pressures and the determination of their magnitudes.

All such problems can be solved with the aid of VSP during the early stages of exploitation. On their solution depends the choice of the optimum method of production, which will be governed by the quality and the nature of the deposit. It should be stressed that the prediction of the geologic cross-section referred to above concerns primarily hydrocarbon deposits. However, similar problems may also be solved in the course of exploration and exploitation of fields of solid minerals.

Technological Problems include the determination of the mechanical properties of the geologic section in order to predict its suitability for drilling and to select the drilling procedure, and in particular, determination of the depths of horizons where drastically changing mechanical properties occur; the timely detection of zones of abnormally high layer pressure and the determination of pressures to select the necessary drilling technology; the determination of the spatial position of the bit (especially in emergencies) as well as the control of drilling and the orientation of borehole shafts.

Predicting the Geologic Section Below the Bottom of the Borehole

Among the problems to be solved by VSP at the exploration stage, one deserving special mention is that of predicting the section below the bottom of the borehole. Such forecasts are very important in the process of drilling, and may also be important by themselves. A forecast below the bit enables zones of abnormally high layer pressure to be detected in time, and the necessary technology for drilling to such horizons to be selected. In such conditions the overlying beds are drilled with a normal technology and subsequent drilling is carried out with the use of special technology. Such a procedure provides against accidents and reduces costly high-pressure drilling.

An important point also is the discovery of the fact that a borehole being drilled lies beyond the limits of the deposit. This makes it possible to avoid setting casing strings unneccessarily or even to stop drilling altogether.

Studies of the geologic section below the planned total depth of the borehole may also be of interest.

Moreover, predicting the section from VSP data may be valuable in itself in areas where the effectiveness of surface seismics is low due to a complex structure of the upper portion of the geologic section or to high-intensity multiples. In such conditions observations even in comparatively shallow (structural) boreholes are able to provide a general idea about the section. Specifically, such observations may be used to outline deposits. Seas and oceans are very favourable for such observations, and a vertical profile offshore enables the lithology and the nature of deposits below the sea-bed to be studied.

A method of using VSP data to predict the geologic section has been suggested and developed by A. A. Tabakov [80].

Forecasts are made on the basis of the solution of the reciprocal dynamic problem [153]. The latter reduces to plotting an acoustic impedance log along the depth axis under certain simplified assumptions about the earth (one-dimensional, isotropic, ideally elastic model).

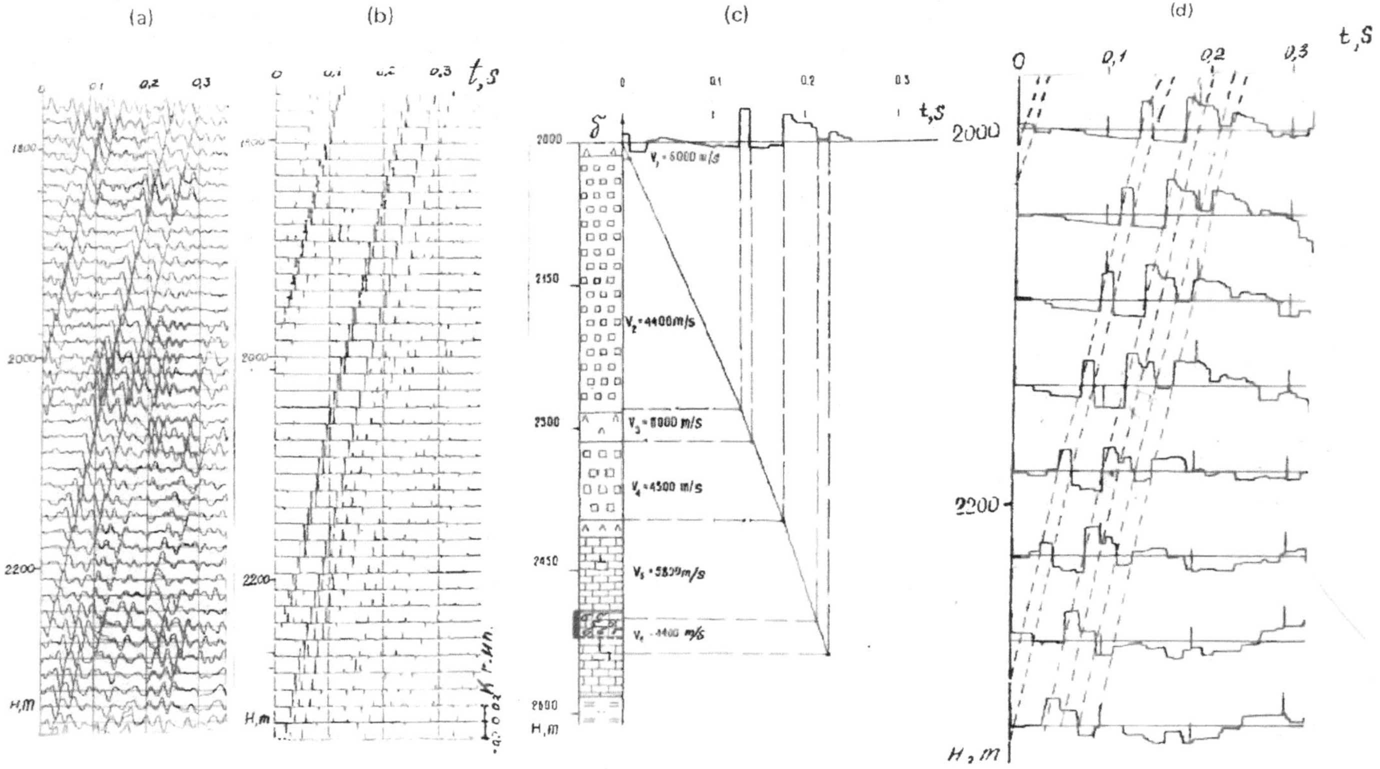

Fig. 145. Predicting cross-section below borehole bottom (borehole N7 Khauzak) [141]: (a) upward reflected waves after separation; (b) pulse seismogram; (c) comparison between predicted acoustic impedance log obtained on 2000 m level and geologic cross-section (legend: see Figure 146); (d) predicted acoustic impedance logs obtained at various prediction levels.

It should also be pointed out that methods of predicting geologic sections from surface observations data are also being extensively used at present. Such familiar methods as 'Seislog' (Canada), 'Velog' (France), pseudo-sonic logging (USSR) are being developed for the purpose. However, VSP data containing downward waves and waves reflected from interfaces located below the borehole provide substantial advantages over surface data. These advantages include, first and foremost, the ability to discriminate effectively the down- and upward waves according to their direction of propagation, the effectiveness of deconvolution of VSP data employing the direct wave, a greater band-width of recorded frequencies. All these factors combine to produce separately the downward and the upward waves less distorted by interference than is possible in surface observations. This serves as the corner-stone for the solution of the reciprocal problem.

There are various methods for the solution of the reciprocal problem: the direct solution of equations or the solution of equations after a pulse seismogram has been obtained. In the latter case the solution is based on the inverse Baranoff–Kunetz algorithm. Figure 145 shows, by way of an example, the results of prediction from observations in borehole N7 Khausak.

The data have been processed with the aid of a combination of programs, which can conventiently be subdivided into two parts: preliminary processing of the wave field, and then predicting the section. The preliminary processing programs are meant to discriminate the down- and the upward wave fields and to obtain them in the least distorted form. The results of preliminary processing have been presented in Chapter 3 (Figure 18). To begin with, we determined the shape of the down- and the upward waves, the time increments from trace to trace and the relative intensity coefficients of the waves. Subtraction was used to determine the shape. After subtraction, the correlation coefficient of the waves was 0.92 to 0.95 for the downward and 0.88 to 0.92 for the upward waves. The shape of the downward wave was used to compute the heuristic deconvolution filter, which accepts the most intense initial part of the pulse (0.070 s) and rejects the oscillations in the subsequent part of the record (up to 0.4 s). The result of the application of the filter to the direct wave is depicted in Figure 3 where, to make the record visible, the amplitudes have been increased first 16 times (up to 0.2 s) and then 32 times.

Next, the special shape of the downward wave was used for pulse shaping. The effectiveness of suppression of the downward waves is demonstrated in Figure 4. Mostly reflected waves are recorded on the seismogram. For additional compression of the signal, we used a matched shaping filter and further discrimination of the waves. This greatly improved the resolution of the recorded reflected waves, and enabled them to be tracked right up to the first arrivals (Figure 145a). Here the preliminary processing stage ends.

To predict the section, we subject the field of reflected waves (Figure 145a) to peak filtering. The trace of reflected waves, up to a specified accuracy is a convolution of the downward wave with the desired pulse seismogram. Figure 145b depicts impulse seismograms computed for various observation levels. On impulse seismograms, the top of a layer in which the high acoustic impedance increases is represented by a negative reflection coefficient, this being the result of the difference in displacement directions recorded by the geophone between a down- and an upward wave of the same sign.

The next stage in prediction is the solution of the reciprocal dynamic problem for a

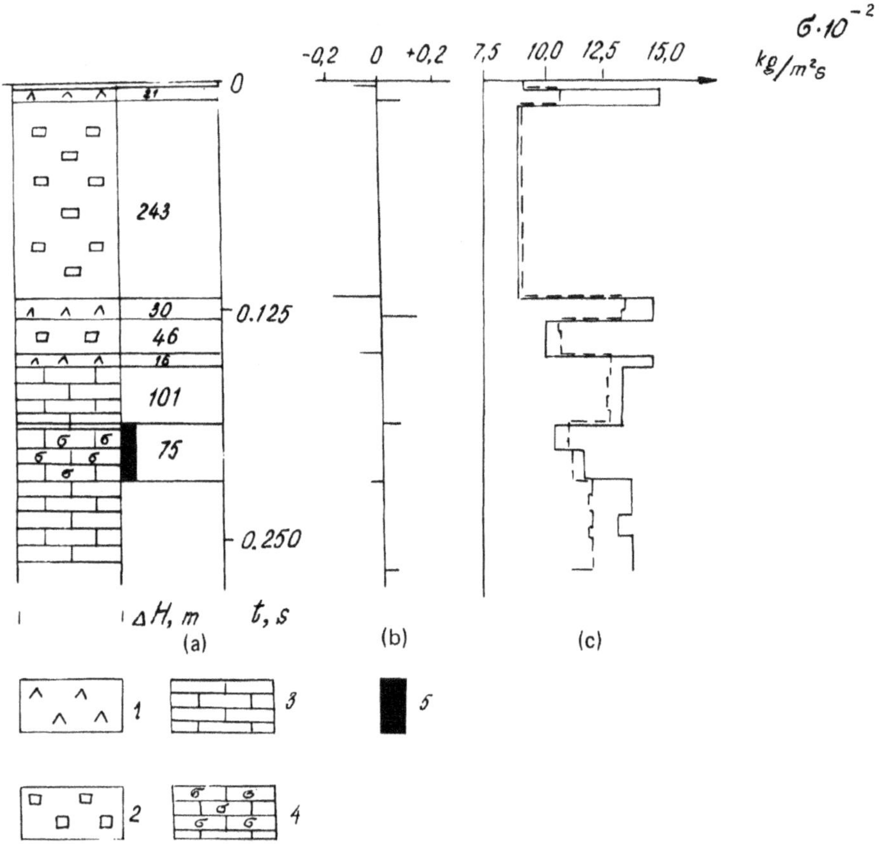

Fig. 146. Predicting cross-section below borehole bottom (borehole Dengizkul' – 1). Prediction level 1950 m: (a) geologic cross-section: 1 – anhydrides, 2 – rock salt, 3 – dense carbonates, 4 – porous reef limestones, 5 – gas-saturated section (thicknesses of layers are indicated); (b) pulse seismogram; (c) comparison between predicted (dashed) and actual acoustic impedance logs.

specified initial acoustic impedance at the level of observations, with account taken of the multiples. The solution is very sensitive to noise. When the noise level on the original records exeeds 10% of the level of reflected waves, the practice is to average the results of the solution of the reciprocal dynamic problem obtained at several points of a 150 m base. By force of this, the example depicts not the absolute values of the acoustic impedance, but their relative variations.

The predicted plot (Figure 145c) correlates well with the geologic section. Low values of the acoustic impedance correspond to the low-velocity layer associated with the gas deposit (5, Figure 146). The predicted acoustic impedance logs obtained at various levels (Figure 145d) agree well with one another and have been used as a basis for predicting the section not penetrated by the borehole.

Figure 146 depicts the forecast of the section down to a depth of 2500 m from observations executed at the 1950 m level. A comparison between the predicted (dots) and the actual acoustic impedance logs demonstrates that all the main layers are detectable on the predicted log. The interval containing the gas deposit manifests itself by a distinct drop in the acoustic impedance. In this example, in contrast to the foregoing, it has been possible to predict absolute values of the acoustic impedance.

The procedure just described has undergone extensive trials, a technological soft-ware package has been created, and specific techniques for prospecting operations have been developed guaranteeing improved resolution and better quality records. The field experience to date has proved the high effectiveness of the 'below the bit' forecast, and there is no doubt that VSP will be applied more progressively for predicting the section below the bit in close connection with the drilling process, and will substantially reduce the amount of costly deep drilling, as well as improve the effectiveness of geologic exploration.

In 1980, Kennett *et al.* published a paper on the use of VSP for predicting the section beneath the borehole in the North Sea [142]. VSP data and sonic and density logs were obtained in a borehole when the bit was at a depth of 3100 m (Figure 147a). The reflected waves in the range of depths penetrated by the borehole correlate well with the acoustic impedance log (Figure 147a, b) computed from the velocity and the density log data in the borehole. Several waves reflected from interfaces below the borehole were recorded on the VSP seismograms. After deconvolution of the VSP data employing down-going wavelets, the acoustic impedance log in the interval below 3100 m not penetrated by the borehole was computed from the recorded time, the intensity and the polarity of the waves reflected from interfaces below the borehole. The predicted impedance log is plotted below 3100 m in Figure 147a. Observations in the borehole after drilling was completed (Figure 147b) enabled the acoustic impedance log to be extended with the aid of the additional velocity and density logs with which the reflected waves recorded on the VSP seismogram are still in good agreement. Comparison of the predicted acoustic impedance changes with the measured values in the section below 3100 m are shown in Figure 147b.

Figure 147c (right), depicts the trace inversion of 18 geophone records located in the lower portion of the borehole. The reflection coefficients were computed after deconvolution and subtraction of any interbed multiples associated with the interface located below the geophone positions. The data obtained have been filtered with a 10–40 Hz band-pass filter. Figure 147c (left) depicts for comparison, the original acoustic impedance log and in the centre, the same log filtered with a zero phase 10–40 Hz band pass. The dominant reflected waves correlate well on both acoustic impedance logs. The variations of the acoustic impedance have been predicted below the final borehole bottom at 4100 m.

The depth range below the borehole in which a reliable forecast can be made depends on a number of factors, mainly, differentiation of the cross-section, data quality and total borehole depth. Experience shows that the prediction range for deep boreholes may in some cases be as great as 1.5 to 2 km.

The capacity to predict the geologic section below the borehole bottom illustrated by the examples cited may prove of great importance for the solution of numerous exploration problems. Such prediction may result in great cost savings.

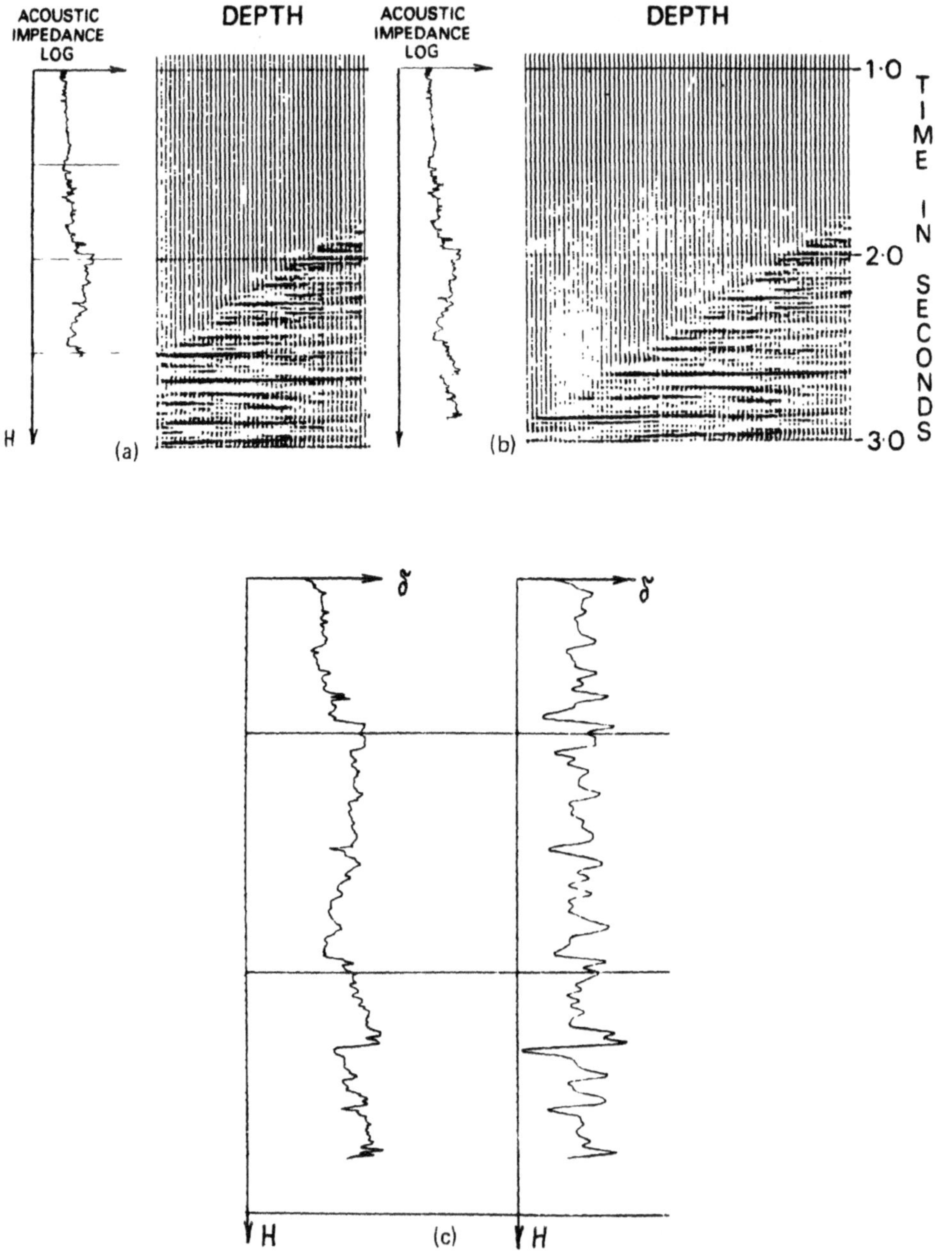

Fig. 147. Predicting cross-section below borehole bottom [142]: (a) data obtained in the course of drilling; (b) data obtained after drilling has been completed; (c) predicted acoustic impedance log prior and after filtering by a 10–40 Hz band-pass filter.

PM VSP at the Exploitation Stage

The application of VSP at the exploitation stage is based on the potential for studying the deposits listed above. Of major importance in this connection are: determination of gas-fluid contacts and their position during the process of exploitation, especially when various methods of flooding are being used; studies of the deposit contours during exploitation of the field; determination of variation of the properties of layers (deposits) in the course of drilling and exploitation; control of water pumped into the layers; appraisal of various methods of acting on the layers; control of intactness of a deposit; control of pressure variations in the deposit; planning production boosts.

The tasks enumerated above have very much in common with each other from the methodological point of view. From the viewpoint of prediction of the geologic section and exploitation of the fields, the major topics for study are the physical properties of the section, the effect of the section on the wave field; increasing the representativeness of each borehole by studying great spatial volumes around and between the boreholes.

Studying the Physical Properties of Rock

Of major interest for the solution of geologic problems are the seismic velocities of the longitudinal and transverse waves, and the absorption and reflecting properties of the section.

PM VSP enables the velocities of the longitudinal and transverse waves, and their ratio, which is one of the chief parameters of the wave field, to be studied. Experience shows that the velocities in the oil and gas zones differ from those in water-saturated parts of the reservoir. In addition, changes in velocities occur in sections of abnormally high pressure. PM VSP data may be used to study the dependence of velocity on the lithology, porosity, saturation with oil and gas and on the depth. The recording of waves in a wide frequency band also facilitates the study of velocity dispersion.

Of special interest may prove the studies of the velocity ratio of *P*- and *S*-waves in a reservoir.

Studying the Absorption and Scattering Properties of the Real Earth

The absorption of seismic energy in gas- and oil-saturated layers differs appreciably from that of layers saturated with water. Sectioning of the geologic column according to absorption properties may be of considerable interest from the geological point of view. Of special importance for the construction of an absorption model of the real earth is a study of the frequency dependence of absorption.

Studying the Reflecting Properties

Variations of the coefficient of reflection of the gas-saturated zones of a reservoir serve as one of the criteria that facilitate prediction. In VSP the reflection coefficients of interfaces are determined with a greater accuracy and over a wider frequency band than in surface observations.

Studies of the coefficients of reflection of longitudinal and transverse waves may be of great interest.

The possibility of recording reflections from gas-oil-water contacts, and hence of monitoring the position of a deposit, deserves special mention.

TABLE XIV

Parameters	Seismic data				Wellbore data												Production data					PM VSP data							
																							velocities						
	Time	Velocity	Amplitude	Character	Drill rate	Mud log	Cuttings	Cores	Drillstem test	Electric log	Acoustic log	Density log	Neutron log	Natural gamma ray log	Wireline test	Side wall samples	Tests	Pressure surveys	Water cut	Gas oil ratio	History	Time	v_P	v_S	v_P/v_S	Amplitude absorption	Spectra	Polarization	Reflection coefficient
Depth	B	B			C	C	B	A	A	A	A	A	A	A	A	A						A	B	B	B	A	B	B	A
Areal extent	B	A	C	C					D								D	B	D	D	B	A	A	A	A	A	B	C	B
Gross thickness	B	B		C	B	B	C	B		A	A	A	A	A								B	B	B	A	A	B	B	B
Net thickness		B	B	C	C	D	A	D	B	B	B	B	B		C						C	B	B	B	A	A	B	B	A
Lithology		B	B	C	C		B	A		C	B	B	B	B		A						B	B	B	A	A	B	B	B
Mechanical properties		B	C	C	C		C	A			B	B	C	B		C							B	B	A	A	B	B	B
Contacts		B	B	B		D	D	D	C	A	A	A	A		A	B	A		C	C	C	B	B	B	A	A	B	A	A
Pressure		B	C			D			B	C	C	C			B		B	A			B		A	A	A	B	B	C	B
Porosity		B	B	C	D		D	A		C	A	A	A			C		D			D		B	B	A	A	A	B	B

	Seismic data				Wellbore data												Production data					PM VSP data							
																							velocities						
Parameters	Time	Velocity	Amplitude	Character	Drill rate	Mud log	Cuttings	Cores	Drillstem test	Electric log	Acoustic log	Density log	Neutron log	Natural gamma ray log	Wireline test	Side wall samples	Tests	Pressure surveys	Water cut	Gas oil ratio	History	Time	v_P	v_S	v_P/v_S	Amplitude absorption	Spectra	Polarization	Reflection coefficient
Permeability				D			D	A	B	C	C	C	C		C	C	C	B			C		D	D	D	D	D	D	D
Relative permeability								A									D	C	C	D	D								
Fluid saturation		C	C	C		D	D	C	B	A	B	B	B		B	C	D	C	C	C	C								
Pore sizes							B	A		C	D	D	D	D		D	D	C	D			D	C	C	C	C	D	C	D
Producing mechanism	C	C	C	C					D	C									B	B	A								
Hydrocarbon properties		D	D			D	D	D	B		D	D	D		C	D	A												
Water properties								D	B	C					C		A												
Production rate								C	B		D	D	D		C	D	A	C			B								

A – one of the best sources; B – a good source; C – a useful source; D – in conjunction with other sources.

The application of PM–VSP at the exploration and exploitation stage is founded methodologically on: the use of different wave types; the development of observation systems employing not only the direct and reflected (longitudinal and transverse) waves, but also the refracted and converted transmitted waves, the development of spatial (three-dimensional) observation systems; high-resolution (high-frequency) observations; extensive use of all the parameters of the wave field; comprehensive processing of wave dynamics.

The scope of problems dealt with by production seismics include: (a) predicting the lithology and the genesis of sedimentary rock and detecting reservoirs and fluids; (b) discovery of anticlinal traps; (c) obtaining data necessary for designing deep boreholes; (d) the discovery and study of zones with abnormally high layer pressures. In this respect production seismics completely engulfs the problem of predicting the geologic section in and around the borehole. Some specific advantages of down-hole observations, in addition to the potentialities of production seismics, are volumetric observation systems, interference – free recording, correlation of seismic observation data with sonic and resistivity logs, which all enhance the reliability of forecasts of the gelogy around the borehole far beyond the limits attainable in surface observations.

To appraise the effectiveness of PM VSP as a prospecting method for hydrocarbons, compare the information supplied by the oil industry on the use of surface seismic data, various other geophysical data, and data obtained with production methods in boreholes, with data from the PM VSP technique. Table XIV (borrowed from [154]) contains surface seismic observation data, geophysical borehole and production observation data. It will be seen from the table that seismic data rank first only for the determination of the productive areal extent. For many other parameters they are rated as good. We have included in the table PM VSP data on the right-hand side. PM VSP is one of the best, or a good, source of information for many parameters. Of course, VSP should be augmented by surface observations, and only extensive combinations of observations of both types will enable the potentialities of seismic studies to be fully realised.

The studies at the exploration and exploitation stages involve a substantial increase in the volume of seismic observations to be carried out in every borehole.

This will unavoidably result in a substantial rise in the cost, but at the same time it may still be profitable, since the only alternative to seismics is more drilling whose cost rises drastically with the increase in depth of the borehole. Any supplementary data on the structure of the Earth obtained from seismic observations may result in great savings in the costs of drilling. However, because of the rising costs of seismic studies, it is necessary to obtain the most comprehensive information from a comparatively small number of boreholes, i.e. to extend the volume studied around every borehole. PM VSP makes extensive use of wave dynamics to obtain quantitative data on the properties of the reservoir required for the rational exploration and exploitation of a field.

The specific feature of production seismics is that it is employed in areas where a network of deep boreholes is being drilled, and that it is executed in close association with the drilling process. This requires first of all, a fast method of executing field surveys and a still faster system of processing the data needed for drawing up plans for continued deep drilling. To satisfy these conditions, unified optimum observation systems based on the combination of surface areal and VSP observations along both fixed level and vertical profiles should be developed. The great number of shots needed for VSP facilitates the shooting of a detailed areal system, which together with a vertical profile forms a volumetric system which is shot as a unit. In this case surface observations should provide, at many points, records as free from unwanted waves, especially from the multiples, as possible. In addition to solving structural problems, such records are needed for a detailed velocity distribution of the section with the aid of the familiar methods of pseudo-sonic logging. Of essential importance in this connection are VSP data, which enable the wave field of the multiples to be studied and shaped for subsequent subtraction. Such volumetric observation systems are conveniently operated by special production seismic teams qualified to carry out simultaneous surface- and down-hole VSP areal observations and equipped with the necessary digital instrumentation.

For fast processing of production seismics data, special computer hardware and software packages should be set up and designed for stage-by-stage processing to solve structural problems and study the physical properties of the Earth. New wave field parameters should be constantly incorporated providing progressively more detailed interpretation in conjunction with data of other geophysical methods in order to define new effective parameters that characterize the hydrocarbon deposits.

In order to develop production seismics, coordinated plans of drilling and seismic operations, which provide sufficient borehole time to shoot a unified system should simultaneously be drawn up. The results already obtained and hoped for justify a generous allocation of funds for the development and the application of production seismics.

VSP and Detailed Sectioning of the Geology in Terms of Acoustic Impedance

The opportunities that exist for detailed sectioning may prove to be of great significance for studies of the rocks around the borehole and especially their composition. A section can be detailed much more minutely, if not only wave kinematics, but more so, wave dynamics is studied. At present methods are widely in use that transform seismic traces into acoustic impedance logs [49, 50] or, as they are termed in the USSR, pseudo-sonic logs. Such transformations will be effective provided certain conditions are fulfilled:

– the seismic trace corresponds to the distribution of the reflection coefficients in time, that is to say it is free of unwanted waves;

– a sufficiently wide frequency band is recorded (3 to 4 octaves), which includes both high frequencies up to 100 Hz and low frequencies down to 5 to 10 Hz;

– effective deconvolution is an essential processing procedure.

– VSP data provides substantial advantages over surface observation data for the following reasons:

– In VSP, the discrimination of waves, according to their direction of propagation, separates the downward waves from the upward and to a considerable degree suppresses the multiples;

– VSP data are less affected by, and when both the source and the geophone are below the LVZ, completely immune from the strong filtering action of the low-velocity zone, and because of this, wide frequency band records can be obtained;

– In VSP, the same as in seismic exploration at sea, there is the opportunity to record the direct wave and use its spectrum to compute an inverse operator. In this case deconvolution enables a short practically zero-phase reflected pulse to be obtained.

– VSP provides data on the vertical variation of velocities required to compensate for the lack of low frequencies in the recorded seismic signal.

A cross-section plotted using pseudo-sonic logs reflects the lithologic characteristics, and is noteable for its resolution and high sensitivity to velocity variations. The constancy and authenticity of interval velocities obtained with pseudosonic logging form the basis for interpretation. Cross-sections plotted from pseudo-sonic logs are very important in the studies of the rocks surrounding the borehole.

4. Studying the Effect of a Deposit on Wave Field Parameters*

To illustrate the potentialities of production seismics, let us consider the results of studies into the effect of an oil- and gas deposit on the parameters of the wave field.

SEISMIC DESCRIPTION OF THE REGION OF OPERATIONS

The Anastasiyevsko–Troitzkaya area of the Krasnodar province is based on a double-arched fold that intersects the axial part of the Western Kuban inflection. The brachianticlinal fold extending in sub-latitudinal direction dips to the East and is divided into two elevations by a saddle: one is the eastern-Troitzkaya, and the other the western-Anastasiyevskaya. The Anastasiyevskaya fold is symmetrical, with its flanks rising in the Miocene sediments by angles of 10 to 14°. The Troitzkaya fold is located 60 m below the Anastasiyevskaya arch and displays angles of dip of 4 to 5°. Diapirism manifests itself in the Anastasiyvskaya fold, and it had a definite part to play in the formation of the Anastasiyevskaya–Troitzkaya elevation.

The fold is made up from Miocene to Quaternary sediments inclusively. The formations of Lower Miocene are represented by homogeneous Upper Maikop clays, those of Middle Miocene are characterized by limestone clays interspersed with marl and aleurolites; the upper Miocene includes deposits of the Sarmatian and the Meotic stages. The lithology of the Sarmatian stage is characterized chiefly by clays frequently interspersed with layers of siltstone, fine-grained sandstones and marl. Small amounts of oil have been obtained from the Lower Sarmatian and Middle Miocene deposits, but no productive reserves have been discovered. The Meotic stage is represented by clays with sand- and siltstone rock interlayers with which oil deposits are associated. The largest are associated with horizons V and IV. The deposits in other horizons with the exception of IV are not suitable for production. A gas and oil deposit is associated with horizon IV. A particular feature of its structure is the presence of a gas cap with a 156 m thick gas deposit and an oil deposit only 20 m thick. A 25 m thick non-limestone clay stack spans horizon IV terminating the Meotic stage.

THE PROCEDURE FOR OBSERVATIONS AND MEASUREMENT OF THE GEOLOGIC SECTION AND WAVE FIELD PARAMETERS

The studies in the Anastasiyevsko–Troitzkaya area have been based on the PM VSP and VSP (Z-component) procedures, and digital recording has been used. Observations

* This paragraph has been written in collaboration with Yu. D. Mirzoyan and S. E. Kamberlij.

were carried out in two deep boreholes, N693 located close to the edge of the oil/water contact, and N1641, drilled in the thickest part of the gas-saturated zone. In order to study the sediments under the deposit, BH 1641 was specially sunk by an additional 200 m to reach below the oil deposit, to a total depth of 1760 m. Observations in BH 693 have been carried out from 9 shotpoints at offset distances up to 2100 m (for different azimuths), and BH 1641 has been shot from three SP with offsets up to 900 m. An element invariably present in the general system was detailed shooting from an in-line SP utilizing a 3-point *Z*-probe moved with increments of 5 m with overlap of one geophone along the hole axis for correlation purposes.

Charge sizes of 5 to 10 electric detonators, and shot depths of 55 m for BH 1641 and 68 m for BH 693 were selected for shooting. The in-line profiles ensured a high-frequency oscillation spectrum and sufficiently high intensity of recorded waves, as well as separation of the ghost from the direct wave. On off-line profiles, dynamite charges of 0.4 to 1.2 kg shot at depths from 44 to 46 m have been employed, ensuring the simultaneous excitation of both longitudinal and transverse waves. To monitor the reproducibility of excitation conditions, in addition to surface instrumentation, a three-component borehole control instrument has been used, which was placed at a depth of 100 to 200 m close to the borehole being studied.

The PM VSP observations have been carried out with the aid of a single-point probe with a symmetrical three-component geophone cluster. The observation increments have been 10 to 20 m. A continuous seismic monitor featuring, the recording of the fourth (vertical) oscillation component has been employed.

The SDS-3 (Seismic Digital System-3) and the specialized subsystem SDS-3PM based on the ES-1040 computer have been used for data processing.

A high signal-to-noise ratio of 30 to 1 in the region of the first event enabled high-precision measurements of the wave field and the earth parameters to be accomplished.

PM VSP observations have also been employed to study the ratio of the longitudinal to transverse wave velocities ($\gamma = v_S/v_P$) and Poisson's ratio σ.

To assess the value of the geologic information obtainable from the parameters determined as the result of processing of the seismic data, they have been compared with production-geophysical data and laboratory studies of the physical properties of samples, which have been used to construct standard sections characterizing the variation of porosity, rock density, resistivity and velocities with depth.

Kinematic parameters

The values of layer and interval velocities (interval velocities are those between the tools of the probe) of longitudinal and transverse waves and the elastic parameters of the earth (γ and σ) have been studied in three boreholes: 693, 1641, and 355. The curve of interval velocities reflects well the features of the geologic section.

By carrying out a detailed study of interval times and velocities, we have been able on the one hand to check the stability and reliability of their determination and, on the other, to establish the correspondence between anomalous Δt values and the presence of an oil and gas-rich section. It is worth noting that in this case the structure of the geologic section, on account of imperceptible differences in the velocity characteristics of individual lithologic sections, is in general unsuitable for carrying out such an analysis

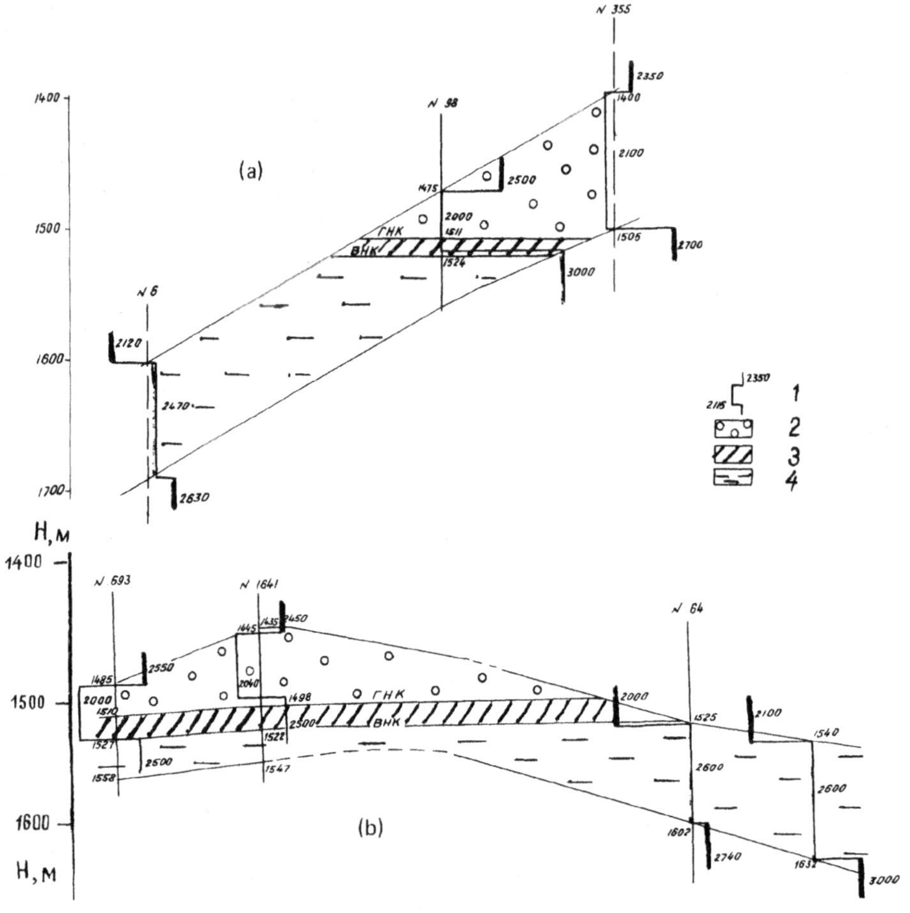

Fig. 148. Diagram of average interval velocity variation in horizon IV according to seismic logs in the Anastasiyevskaya (a) and Troitzkaya (b) area: 1 – graphs of layer velocities, 2 – gas saturated, 3 – oil saturated, 4 – water saturated.

on short time-depth intervals. The average velocities down to 1560 m do not exceed 2000 m s^{-1}, with the layer velocities changing only by 100 to 150 m s^{-1} down to the oil and gas-contact interface. Despite all these factors, in all measurements in BH 693 the oil and gas interval makes itself felt by a decrease in the interval velocities. Although the oil and gas-interval in BH 693 is not thick (20 m), and the top of the reservoir is to a considerable extent saturated with clay, such results are quite remarkable.

In the portions of the sections investigated we have also studied average interval velocities whose value in the gas-saturated portion of the sand horizon IV was 2000 m s^{-1} in BH 1641, and 2100 m s^{-1} in BH 355. On the whole, however, it should be pointed out that the velocities of longitudinal waves diminish by 20% inside the gas-saturated portion of the geologic section. This relationship is confirmed by seismic well-logging in numerous boreholes located within the boundary of the Anastasievskaya–Troitzkaya field (Figure 148). On the other hand, the wave velocity in the water-saturated portion of the layer is 2500 m s^{-1}. As to the velocity of the transverse waves, it increases from 950 m s^{-1} in Pontic clays to 1400 m s^{-1} in Meotic sandstones. Accordingly, the parameter $\gamma = v_S/v_P$ in the gas-saturated interval attains an abnormally high value of 0.65 and because of this can be regarded as one of the most significant seismic parameters. Of principal importance is the fact that this parameter can be determined with a high degree of precision from the ratios of the times of the longitudinal, the converted and the transverse waves.

However, in certain cases the lack of objective information along the total depth of the boreholes made it impossible to establish a complete correspondence between the parameters being studied. Namely, we are unable to explain the anomaly $\gamma = 0.58$ in the depth range 1050 to 1100 m in BH 1641.

Meticulous VSP observations in BH 693, 1641, and 355 have been carried out to study the variation of amplitudes, the frequency spectrum and the effective attenuation coefficient of the direct longitudinal wave.

The graphs $\lg A(H)$, the effective attentuation (α_{ef}) and the spectral band-width (Figure 149) describe their variation with the depth. The amplitude graphs for BH 693 and 1641 are in good agreement. They display sections with anomalous α_{ef}. A set of production-geophysical data for BH 1641 and standard well-logging data for BH 693 enabled the limits of individual sections to be more accurately defined and their lithologic and structural features to be established.

The first section 400 to 800 m in BH 693, and 700 to 770 m in BH 1641 (here α_{ef} has been computed from the depth of 700 m) is typical of a section of Kimmeridgean clays transformed into sand, besides a highly-porous sandy-clay saturated with fresh water. On the microsonde curve (BH 1641) these clays are represented by weakly-differentiated sections of the curve. The coefficient of heterogeneity $K = [(\rho_{max} - \rho_{min})]/\rho_{max}$ does not exceed 25%, this being an indication of uniform density. In these portions of the section the effective attenuation coefficient is 0.3×10^{-3}.

The second section 800 to 1040 m in BH 693 and 770 to 1010 m in BH 1641 manifests itself on the microsonde curve by a negligible differentiation (25%), but the caliper establishes the presence of a bottle-neck. The depth range 500 to 1040 m is characterized by a homogeneous bed with a massive texture and is distinguished from the selected range by higher porosity. Here $\alpha_{ef} = 0.6 \times 10^{-3}$ for BH 1641 and 0.8×10^{-3} for BH 693 in the upper and the middle portions.

The third portion (1040 to 1220 m) is represented by clays of the Pontic formation, which according to microsonde and caliper measurements, are constituted of a succession of variously packed micro-layers associated with the lime- and sandstone materials present in them. The lower boundary of the depth range lies in clays that separate the IInd and the IIIrd sand horizons. In BH 1641 this boundary is clearly marked on the microsonde curve at the depth of 1222 m. Here a somewhat higher attenuation $\alpha_{ef} = 2.3 \times 10^{3} - 3.7 \times 10^{-3}$ is noticeable.

The fourth interval spans 1220 to 1480 m. The main bed of containing clays with a resistivity of 1.50 Ohm/m includes the IIIrd Pontic horizon displaying high porosity and water-saturation. Here $K = 29\%$, $\alpha_{ef} = 1.2 \times 10^{-3}$ for BH 1641 and $\alpha_{ef} = 0.8 \times 10^{-3}$ for BH 693.

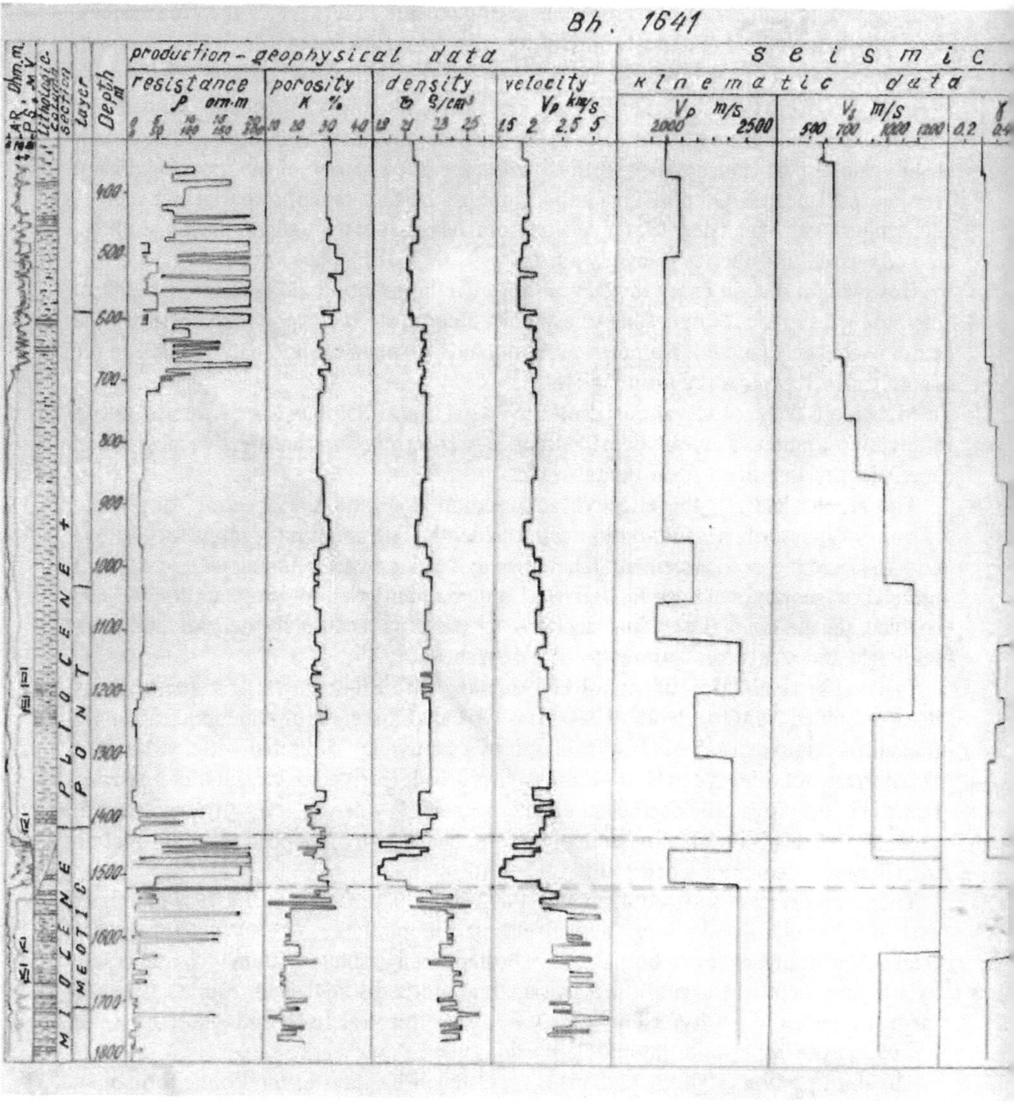

Fig. 149. Comparison of production – geophysical and seismic parameters (BH 1641 and 693 in Anastasiyevsko–Troitzkaya area).

The fifth and most interesting depth interval 1480 to 1560 m for BH 693 and 1440 to 1540 for BH 1641 corresponds to the productive Meotic horizon IV represented by a highly porous reservoir saturated with gas and oil and containing high-density centimetre-thick layers of sandstones clearly visible on the microsonde curve. This interval is distinguished from the rest by an abnormal attenuation of the *P*-wave equal to $\alpha_{ef} = 6.5 \times 10^{-3}$ for BH 1641 and $\alpha_{ef} = 3.8 \times 10^{-3}$ for BH 693. Note as well that

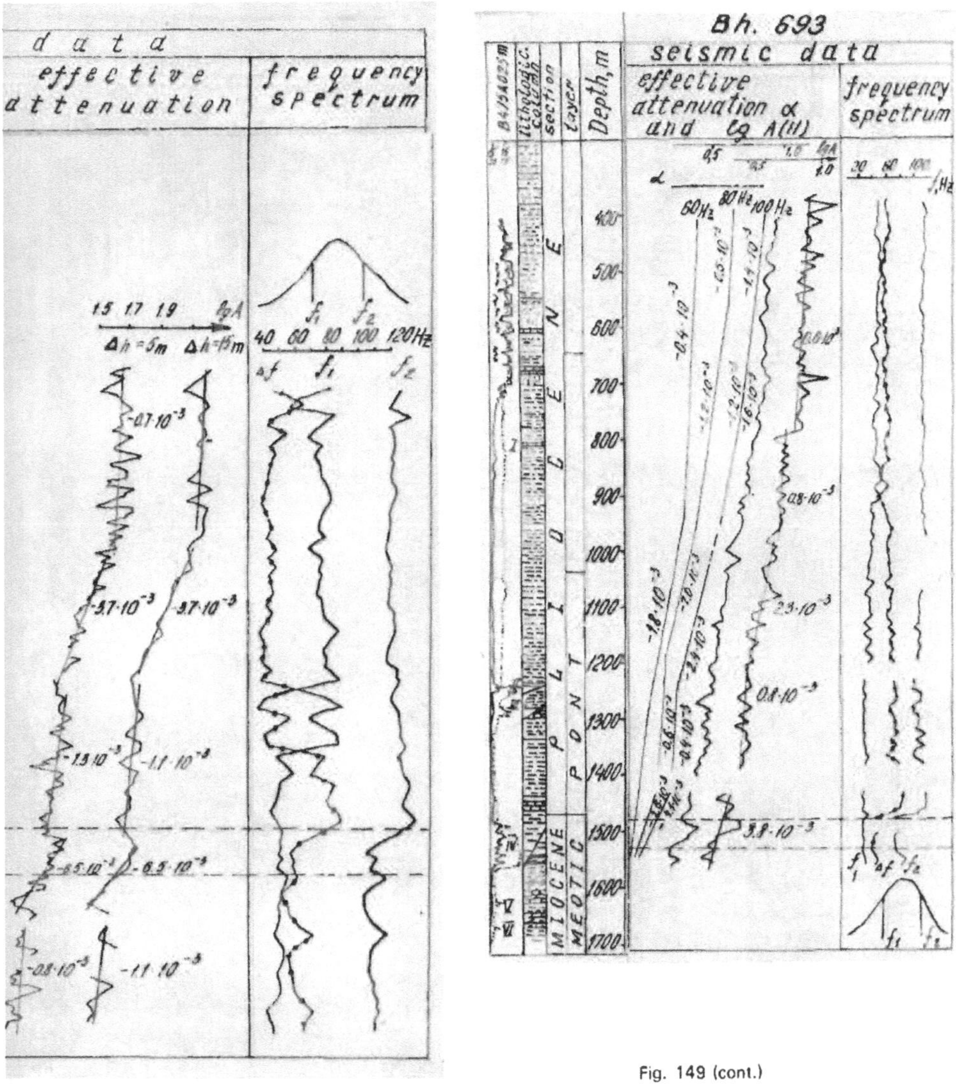

Fig. 149 (cont.)

the gas-saturated portion of the Meotic horizon IV in BH 693 is itself also distinguished by a high effective attenuation of the direct wave.

The last, sixth, interval (1540 to 1750 m) studied only in BH 1641 (Figure 149), for which purpose the borehole had been specially drilled deeper, is represented by alternating clay layers and small-capacity reservoirs. The values of the parameters being considered are appreciably lower here than in the deposit: $\alpha_{ef} = 1.1 \times 10^{-3}$; $K = 41\%$.

Hence, we may draw the following conclusions.

There is a direct correspondence between the parameter α_{ef} and the heterogeneity coefficient K as measured by the microsonde. With the increase in the amount of detail

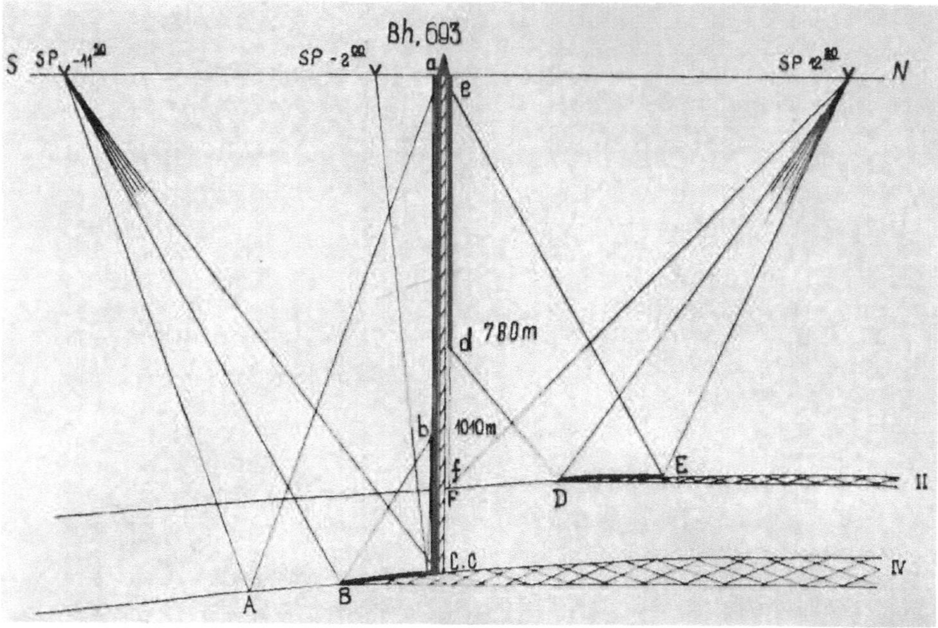

(a)

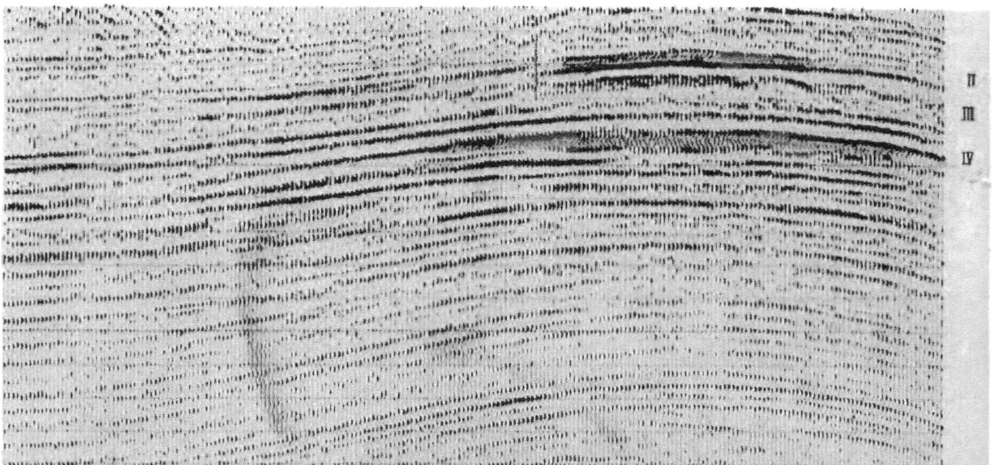

(b)

Fig. 150. Reflecting properties of a deposit: (a) schematic representation of the position of the borehole, deposits and shotpoints; (b) CDP time section.

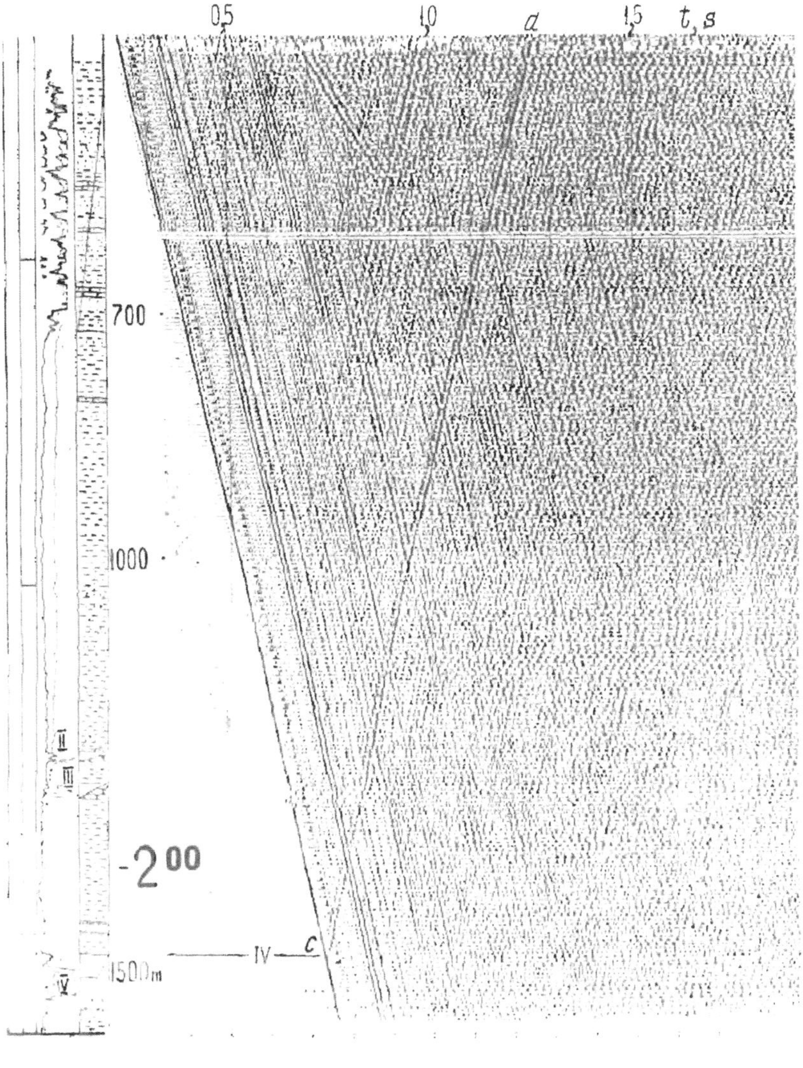

Fig 151a.

Fig. 151a-c. VSP seismograms: (a) along an in-line profile; (b, c) along off-line profiles 1150 m (b) and 1220 m (c).

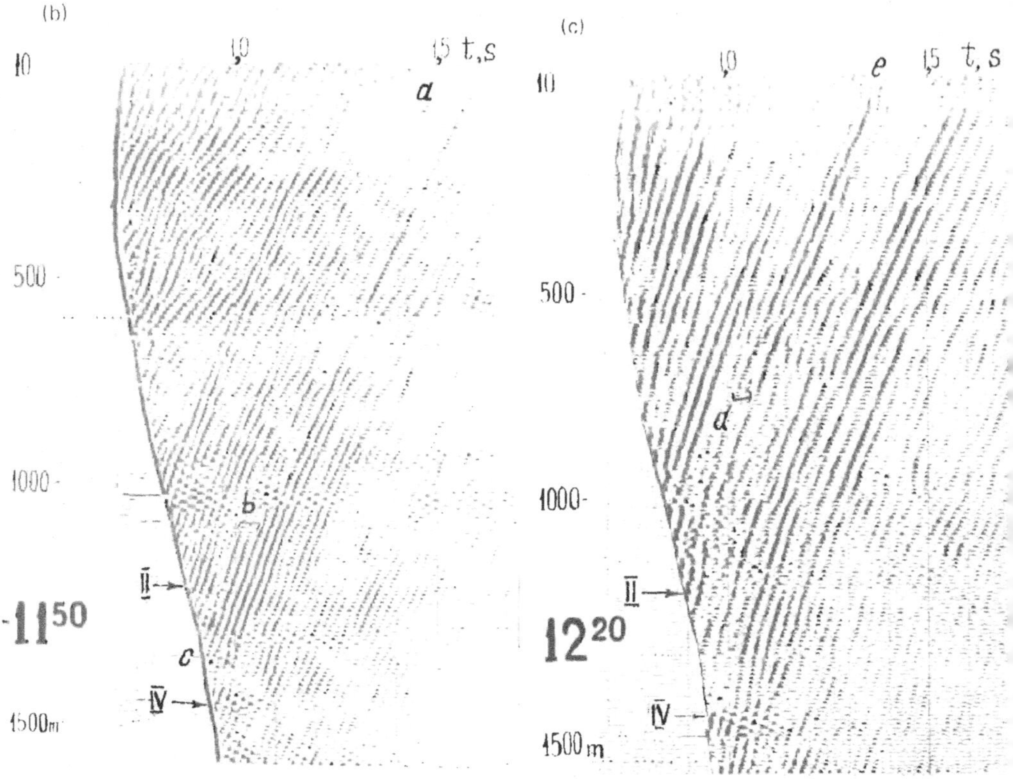

Fig 15 1b-c.

available and the improvement in the standards of measurement, the parameter α_{ef} could be used to predict the physical properties of the geologic section and its individual intervals.

A high porosity brings about an increase in α_{ef}. If the pores are all filled with water (the rock is water-saturated), minimum values will be typical of α_{ef} (interval I and in part IV). If, on the other hand, the pores are gas-saturated, the values of α_{ef} will rise abnormally (interval V). The values of the coefficient of attenuation are essentially frequency-dependent. At higher frequencies, the $\alpha_{ef}(H)$ graphs display greater differentiation and enable the geologic section to be divided according to absorption properties to reveal finer detail. The spectral band-width (Δf) of the signal decreases abnormally in the gas-saturated portion (about twice) owing to the absorption of higher frequencies.

Reflecting properties of a deposit

BH 693 penetrated the deposit in the IVth zone. The perimeter of the productive zone is located 15° to the south of the borehole. The deposit in the IInd zone has not been intersected by the borehole, with the perimeter of the gas deposit being located 400 m to the north.

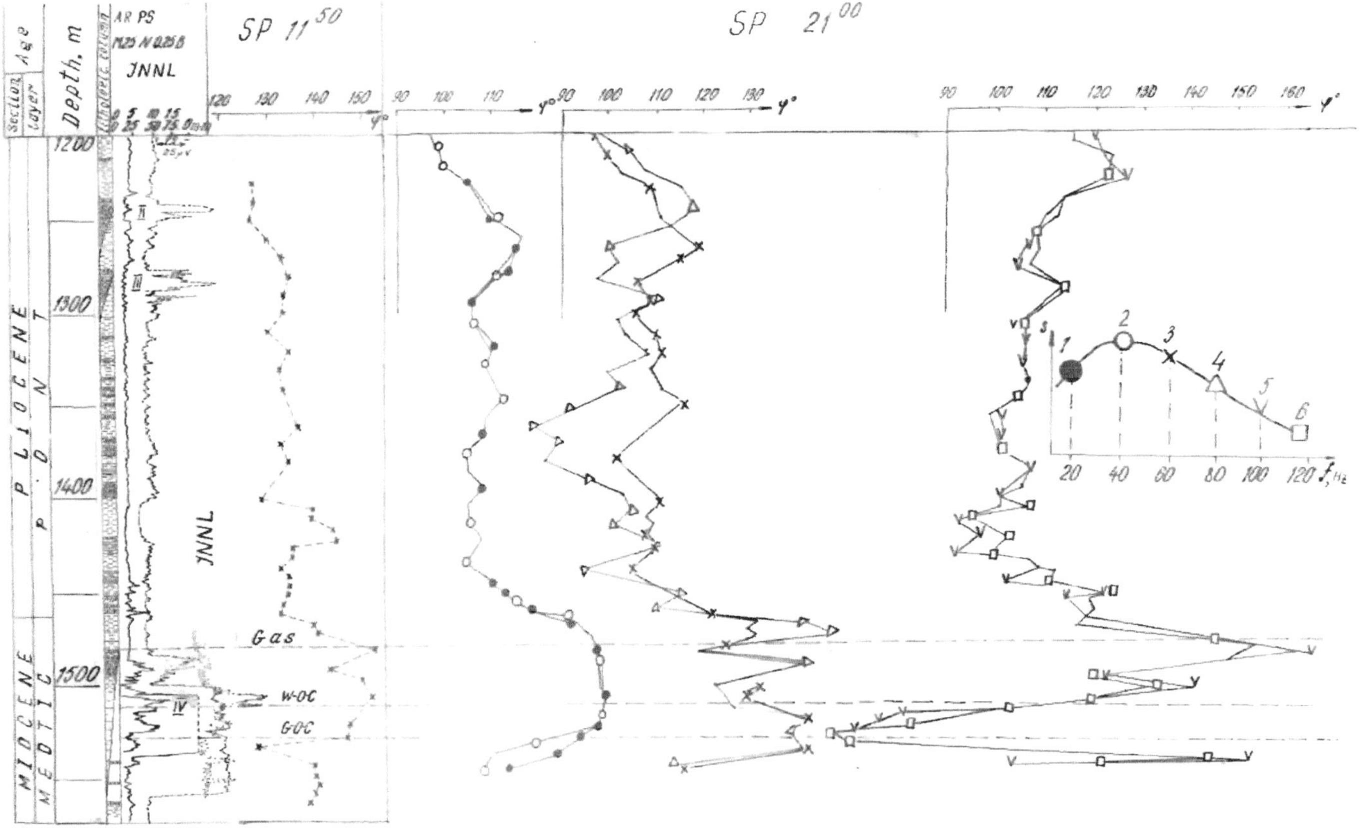

(a) (b)

Fig. 152. Comparison of production-geophysical data with the curve of displacement directions in the vertical plane (BH 693): (a) Geologic column and production – geophysical data; (b) curve of displacement directions in the vertical plane for SP 11^{50}; curves of displacement directions in the vertical plane for some monochromatic spectral components (1 – 20, 2 – 40, 3 – 60, 4 – 80, 5 – 100, 6 – 120 Hz). INNL – impulse neutron-neutron loging.

Consider how the deposit made itself manifest in the dynamic properties of the reflected waves on vertical profiles from SP -11^{50}, -2^{00}, and 12^{20} (Figures 150, 151). The reflection from the IVth Meotic zone from SP -2^{00} located 200 m to the south of the borehole is associated with its gas-saturated portion. It is distinguished from all other wave groups by its considerable intensity. The wave predominates along the entire vertical profile (Figure 151a). The dynamic properties of the IInd sand do not display any distinctive features.

The section AC of the IV horizon is covered from SP -11^{50} located to the South (offset 1150 m – Figure 150a). The seismogram from this shotpoint displays a distinctive change in the nature of the saturation (Figure 151b). The section AB belongs to the water-saturated, and the section BC – to the gas-saturated portion of the productive bed.

A similar relationship is observed up to the third shotpoint located 1220 m to the north of the borehole, which illuminates with reflected longitudinal waves the section EF of the IInd horizon. Here we notice a change in the wave intensity in the direction from the water-saturated (DF) to the gas-saturated (ED) portion of the productive zone (Figure 150a). On the VSP seismogram, a sharp rise in the amplitude is noticeable in the recorded region of the wave from the gas-saturated portion (Figure 151c).

The peculiar dynamic features of the reflected waves can be unambiguously correlated with surface PM CDP research data (Figure 150b).

The anomalous attenuation of longitudinal waves in a gas deposit established as the result of the present work is responsible for a shadow effect. This manifests itself in the attenuation of all reflections recorded below the productive portion of the sandstones.

Polarization parameters

PM VSP observations enabled, for the first time in conditions of an oil- and gas deposit, an assessment of the information value of the polarization parameters of the first longitudinal event. Within the framework of the theory of linearly-polarized oscillations, we studied the angle φ between the displacement vector and the vertical. Within the framework of the model of an elliptically-polarized oscillation, we studied the ellipticity ε of the trajectory. Consider the results for BH 693 where such studies have been most exhaustive. The nature of the variation of the angles φ is conditioned mainly by the velocity differentiation of the earth. As a wave passes through a low-velocity bed, the angles with the vertical increase, and vice-versa.

It will be seen on the $\varphi(H)$ large-scale graphs that the IVth horizon shows up from the depth of 1475 m by an increase in the angle from all shotpoints, and especially clearly for SP 11^{50} and 21^{00} (Figure 152). The nature of the $\varphi(H)$ curve corresponds to that of the apparent resistance, spontaneous polarization and neutron-neutron logs and to the variations in the physical properties.

Not only does the water-oil contact at the depth of 1527 m show up clearly, but it is even possible, although only qualitatively, to discern the intervals of the gas- and oil-saturated portion of the IVth horizon, thereby bringing out the gas-and-oil contact. Whereas gas saturation manifests itself on the $\varphi(H)$ curve by increasing angles, in the oil-saturated section the angles diminish somewhat. A greater decrease in the angles takes place in the water-saturated portion of the geologic section. The gas-saturated layer shows up still more clearly as the offsets are increased (with the increase in the angle of incidence) to 2100 m.

It is of special interest to analyse the curves of displacement directions built for various monochromatic spectral components. Whereas at the frequencies of 20 Hz and 40 Hz a layer cannot be differentiated, at 100–120 Hz it is possible to split it up, so that the top of the gas-saturated layer becomes clearly discernible. Despite the fact that in this case PM VSP observations have been carried out at intervals of 5 m, such intervals are still too large for a more reliable differentiation of a gas- and oil-saturated layer. Thus, the interpretation of the $\varphi = f(H)$ graphs demonstrates a real possibility and holds good promise for the utilization of this polarization parameter for quantitative studies of the velocity and density of thin lithologic formations and other characteristics.

It would also be interesting to consider the variation of polarization parameters within the framework of the model of an elliptically-polarized oscillation. In contrast to the displacement directions (in particular, the angle φ), which are associated chiefly with the velocity characteristics of the earth, the ellipticity of the trajectory reflects the heterogeneity of the geologic section associated with its differentiation with respect to acoustic impedance.

Figure 153a, b depicts the $\varphi(H)$ and the ellipticity $\varepsilon(H)$ curves in the vertical plane in the section of a profile that intersects a deposit obtained in BH 1641.

In the deposit interval the ellipticity rises sharply, and so does its differentiation. Emerging from the deposit, the oscillation recovers its linear polarization. It should be pointed out that polarization parameters have not been adequately studied within the framework of the model of a non-linearly-polarized oscillation. However, on account of their strong dependence on the structure of the Earth in the vicinity of the observation point, it goes without saying that their information value must be high.

Let us assess the potentialities of a complex analysis of parameters determined independently. Such an approach may appreciably improve the reliability of geologic interpretation of data and reduce its ambiguity.

With this aim in view, let us consider the graphs of drilling data, of physical properties as measured on samples and of production geophysics (including sonic logs) and seismic data obtained from measurements in BH 693 and 1641 (Figure 149).

Clearly, when studying thin layers we cannot make use of average and layer velocities that characterize mainly thick earth models. In this respect it seems more promising to use interval times, displacement angles and other parameters.

For the purpose of analysis, we have obtained $\Delta\varphi(H)$ curves determined, same as the times Δt, on short bases (10 m). Comparing the $\Delta t(H)$ and $\Delta\varphi(H)$ graphs obtained from identical shotpoints, we see them to be in good agreement, and this makes these parameters appear much more trustworthy. Obviously, the discrete nature of PM down-hole observations cannot match continuous measurements of production geophysics, however at observation points their results are in general quite close.

Comparing the parameter γ with gamma-density logs and with rock lithology, we behold considerable discrepancies between them in case of thin layers. When comparatively large intervals of the geologic section (exceeding 50 m) were used for analysis, clear relationships between γ, the density σ derived from gamma logs and the lithology were established. Thus, in the interval 300 to 855 m in BH 1641 we observe a gradual increase from 0.32 to 0.42 associated with the saturation of the sandy clay bed with fresh water. Inside this interval, the values of σ also increase from 2.1 to 2.17 g cm^{-3} this being the result of changes in the structural and lithologic properties of the geologic section

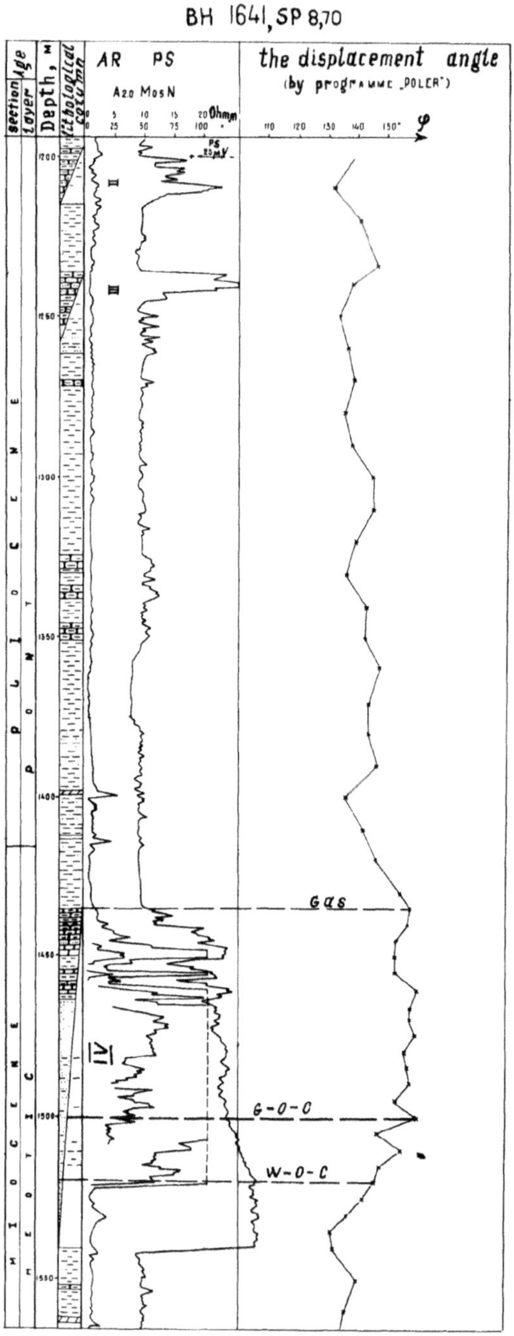
BH 1641, SP 8,70
Section
Layer
Age
Depth, m
lithological column
AR PS
A2.0 M0.5 N
Ohmm
the displacement angle
(by programme „POLER")
φ
Gas
G-O-C
W-O-C
PLIOCENE
PONT
MIOCENE
MEOTIC
II
III
IV

Fig. 153a.

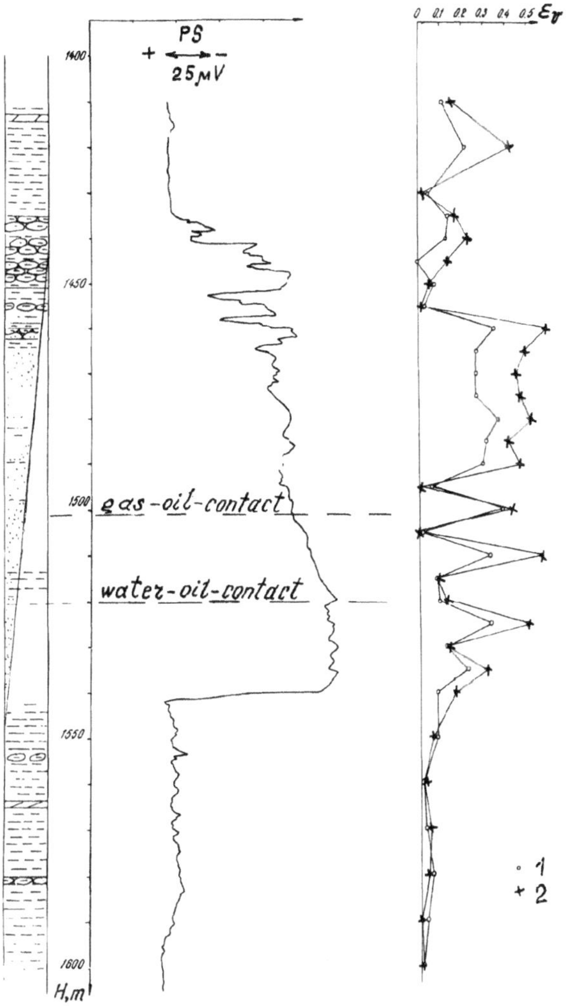

Fig. 153b.

Fig. 153a-b. Comparison of production-geophysical data with the curves of polarization directions in the vertical plane. BH 1641: (a) curve of displacement directions; (b) curve of ellipticity coefficients. 1 – 40 Hz, 2 – 60 Hz.

interval being studied. In the depth interval 855 to 1045 m represented by clays of the Kimmeridgean age there is an increase in γ to 0.57 and a decrease in σ from 2.17 to 1.95 g cm^{-3}.

Inside the oil-and-gas saturation interval, γ reaches its maximum value (in BH 693) owing to a substantial drop in the velocities of longitudinal waves (up to 20%) accompanied by a small increase in v_S. Here the density σ becomes minimum.

In general, the analysis demonstrates that the pattern of variation of the parameter

γ with the depth is in good agreement with results of other geological and geophysical studies.

Consider layer-by-layer the characteristics of two intervals of the geologic section that follow from data obtained by methods of production geophysics and by seismic wave field methods.

Interval I (1050–1120 m). A decrease in the average layer velocities v_P brings about some increase in γ at the depth 1040–1120 m. No local changes in the coefficient α_{ef} have been observed, and the spectral band-width does not change either. Here we observed only an increase in the displacement angles and a decrease in the ellipticity coefficient down to minimum values. Out of all these parameters, the angles φ display the best correlation with the geologic section. They indicate that the interval is a low-velocity one and is characterized by a low density. The resuflts obtained point to the conclusion that the interval is not a productive one.

Interval II (1485 to 1510 m in BH 693 and 1450 to 1550 m in BH 1641). Here we notice an anomalous decrease in v_P by 20%, resulting in an increase in $\gamma = 0.58$ to 0.65, as well as an anomalous increase in the attenuation $\alpha_{ef} = 6 \times 10^{-3}$, i.e. from 5 to 7 times higher than in the above lying interval. The band-width of the direct wave at the depth of 1485 m (BH 693) and 1450 m (BH 1641) is halved, and the displacement angles in the intervals 1475–1525 m (BH 693) and 1435–1535 m (BH 1641) increase abnormally. At the same time, the ellipticity coefficient assumes, at the depth of 1430 m (BH 693) and 1460 m (BH 1641), excessive values. The totality of parameter values points to the conclusion that the interval being considered is an oil- and gas-saturated one.

To sum up, the results of PM VSP and VSP studies have been used to demonstrate a strong influence of an oil- and gas-deposit on the wave field parameters and to establish a high information value of seismic parameters.

Such parameters may include the interval times (Δt), the velocities v_P and v_S, the coefficients of attenuation α_{ef} and reflection, the elastic constants γ and σ and the polarization of seismic oscillations. In conditions of the Anastasiyevsko–Troitskaya area, the oil- and gas-saturation zone is unambiguously determined by a set of wave field parameters: by a decrease in the velocities of 20%, anomalous increase in γ up to 0.58 to 0.65, over 5 to 7 times increase in the effective attenuation coefficient, decrease in the spectral band-width and a sharp increase in reflecting properties, resulting in an increase in the intensity of waves reflected from the deposit, as well as by substantial anomalies in the polarization parameters. These results may serve as a basis for the development of production seismics – seismic studies at the stages of appraisal and exploitation of a field – and present a considerable interest for laying down physical foundations for a direct method of prospecting for a deposit. They also hold out a promise of positive results in the study of oil- and gas-deposits in other regions with similar geologic conditions. At the same time the work should be continued both in the direction of extending field studies and in developing methods of data processing and interpretation with the goal of accumulating actual data and establishing relationships between the parameters of the real earth and of the wave field.

5. VSP Research Tasks

VSP may be of special significance for improving the effectiveness of various research projects. The classification of research problems is a matter of convention, since most of the problems listed above at different stages of exploration and exploitation of a field, may be regarded as problems of research. We shall consider the potentialities of VSP for the solution of only some of them, in our opinion, the most general current problems.

Studies of the Real Earth and the Wave Fields Associated with it

The propagation of seismic waves is in effect the transmission of seismic energy, and the earth can be studied as the conductor of this energy. VSP enables this most general approach to the studies of the real earth to be implemented by studying its reaction to a seismic pulse in the same way as the parameters of electric circuits are examined by studying their reaction to an electric pulse. Every surface seismogram may be regarded as the result of the earth's reaction to the seismic pulse transmitted through it by the source. But this reaction is a summary one, and this constitutes one of the stumbling blocks. VSP enables this summary reaction to be divided into separate components, and identifies those parts of the geologic section that exercise a decisive influence on the wave field. Of great interest are studies of the reflection and refraction properties of the real earth and the interconnection between the earth parameters and the wave field at various separations from the source in the region of reflected waves, as implemented in the PM reflection and refraction methods.

To utilize the polarization of oscillations to obtain direct information about the earth, it is necessary to determine its anisotropic properties and to study the process of propagation of seismic waves in anisotropic media experimentally, paying special attention to the polarization of oscillations in various wave types.

Constructing Models of the Earth

It is an established fact that every method of interpreting seismic observations is based on some model of the earth. The effectiveness of the entire process of interpretation essentially depends on the model chosen. Detailed high-frequency observations have demonstrated the real earth to be complex and in most cases made up of thin layers. As the methods of studying the earth become more detailed, the earth appears progressively more complex and heterogeneous. A layer normally regarded as homogeneous, as a rule, consists of separate thin layers or stacks of thin layers. With the development of methods of computing synthetic seismograms for thin-layered media, very simple models have been superceded by extremely complex ones containing a hundred or more layers.

VSP has demonstrated that high-intensity waves normally originate at comparatively few zones and that in the majority of cases they are associated with quite distinct geologically conspicuous elements of the section. Against their background, combined wave types associated with sections having weak velocity differentiation are as a rule of little significance. However, in conditions of thick beds of weakly differentiated deposits, i.e. in the absence of strong interfaces, the general intensity level of the reflected waves drops, and the combined waves may become dominant.

VSP experience proves that the propagation of seismic waves can be described with the aid of relatively simple models characterized by certain effective parameters. To construct such a model that would correspond to the observed wave field, rigorous principles of averaging the parameters of the real earth should be employed, so that the seismic effect could be predicted for any source excitation. At the same time the wave field itself 'averages' the geologic section and 'constructs' for itself such a model, the type of model being greatly dependent on the operating wave lengths. Since the same elements of the section affect the wave field differently at long and at short separations from the source, and for waves of different types, the seismic models of the same earth constructed for different separations and for different wave types may in principle differ. The experimental studies of wave fields are important for the construction of an earth model. VSP enables the principles for averaging the cross-section to be examined and serves as the principal source of information for the construction of a universal model of the earth. Part of such information is played by the earth's velocity parameters as well as by the dynamic characteristics of waves originating in the real earth (polarization, amplitude and shape of the waves, effective reflection and absorption coefficients). VSP stimulated the development of a wave field interpretation scheme on the basis of effective seismic models of real geologic media.

An effective seismic model [47] is based on three groups of data: the geologic and borehole production, the VSP and the surface. VSP data is used to determine the velocity model and the dynamic characteristics of the recorded waves. For such a model, the chief elements of the cross-section that exercise a decisive influence on the wave field are specified very meticulously. At the same time intermediate parts of the cross-sections are thoroughly averaged. The wave field computed for this model should correspond to the observed fields of the major waves averaged over the area. Such models have been constructed, in particular, for some regions in Uzbekistan.

The effective seismic model allows a new approach to the analysis of wave fields in VSP and surface observations to be formulated, based on a consistent utilization of progressively more detailed information about the wave fields.

Modelling Seismic Waves in the Real Earth

Knowledge of the propagation of seismic waves inside the earth in the seismic exploration frequency band can be used as a basis for modelling seismic waves in the real earth and may prove valuable for solving such difficult problems as wave guide phenomena, shadow, transitional zones, and velocity contrasts with a gradient in the underlying layer. Modelling of surface waves may also be of considerable interest and should permit the study of the mechanism of formation and propagation of surface waves and primarly the depth of penetration of the surface wave energy as a function of the wave length, this being an essential point in studies of long surface waves and the resolution of methods based on their use.

The VSP method facilitates detailed studies of the earth's structure and the construction of a dynamic model. It also provides a sufficiently comprehensive set of parameters for the model, elucidates the process of wave propagation and determines the part played by various wave types. Modelling of the real earth in its natural state in the seismic exploration frequency band, will enable the results to be applied more confidently to longer waves and greater depths than can be done by using laboratory samples in the sonic and the ultra-sonic frequency bands.

The wide variety of seismo-geologic conditions existing in the upper portion of the section greatly facilitates the choice of a model. Specifically, the wave guide that exists under the layer of permanent frost may serve as a good model of a wave guide existing at great depths (e.g. in the mantle).

Studying the Effect of the Surface and the Upper Portion of the Section on the Structure of Seismograms

As our knowledge of the mechanism of formation of the wave field grows, this problem assumes increasing importance. The upper portion of the section and the surface itself exercise a decisive influence on the pattern of seismograms. In the first instance, this influence makes itself felt in the variation of the shape of each individual wave as it travels through the upper portion of the section, or even through the low-velocity zone only. However, much greater influence on the structure of the entire seismogram is exerted by waves, which originate and propagate in the upper portion of the geologic section.

During the course of studies of the effect of the upper portion of the section, at great separations from the source (as in the correlational method of refracted waves and in deep seismic sounding), VSP observations, executed at great depths in the vicinity of the source and in the region of reception, enable the reflected-refracted and the refracted-reflected waves to be studied separately, and their combined effect to be evaluated. Numerous waves of various types originate in the upper portion of the section, and the upper portion of the section and the low-velocity zone may have a different effect on them.

It is worth pointing out that, apart from waves originating in the upper portion of the section, and propagating in the near-vertical direction, a significant part of the wave field observed at great depths may be due to waves formed as the result of the transformation of surface waves into secondary body waves. VSP has shown that a transformation does take place, and that it may be quite substantial if the high intensity of surface waves propagating in the vicinity of the source in the horizontal direction is taken into account. Such phenomena are still obscure. However, in the course of specially planned studies VSP enables the process of transformation both from body to surface waves and *vice-versa* to be examined.

CONCLUSION

(1) The transition from surface studies of seismic wave patterns already formed to studies of their process of its formation sparked off the development of the method of Vertical Seismic Profiling based on down-hole observations. Methodologically, the method is based on the correlational tracking along the vertical profile not only of the first arrivals, but of all the subsequent events on the seismogram as well.

(2) VSP is the principal method of experimental studies of seismic waves in the real earth. The method has found widespread application in the solution of a broad range of problems that exercise a decisive influence on the effectiveness of seismic exploration both in oil and in mining geology. VSP has been carried out in the overwhelming majority of areas in the USSR and is being employed on an increasing scale in foreign countries such as U.S.A., Great Britain, D.D.R., Pakistan, Cuba, etc. The range of problems and fields of application of VSP broadens from year to year. Apart from oil and mining geology, VSP is used in studies of unconsolidated deposits in coastal regions and continental shelves, in engineering surveys and in the solution of many research problems.

(3) Voluminous VSP observations have enabled the wave fields in media of variegated structure to be studied in detail. Simultaneously, it has been demonstrated that an important part is played by transverse waves excited by conventional explosive charges and formed in the Earth, and that waves of different types are associated with the same interfaces. These results served as the corner-stone for the creation of the general polarization method of seismic studies in its surface (PM reflection, PM common-depth-point methods) and down-hole (PM VSP) modifications. The general applicability and effectiveness of the polarization method stem from the fact that all the wave field parameters, including polarization, are utilized for its analysis, and that waves of different types (longitudinal, transverse, and converted) are simultaneously discriminated, tracked, and processed to obtain information about the earth.

The polarization method, as distinct from traditional methods, studies the wave field, which is essentially a vector field, with the aid of a vector and not a scalar analysis. No three-dimensional observation, no matter what observation system is employed, can be a spatial one in the full meaning of the word, if it does not employ the polarization method.

(4) Of all the variations of the polarization method, PM VSP is the most effective, due to the specific character of observations along the vertical profile, in which the directions of approach vary from point to point, even when shots are fired from the same shotpoint. In such conditions, the transition from the tracking of fixed components to tracking components performed in the course of polarization-positional correlation, enabled regular waves of various types to be discriminated and tracked irrespective of their apparent velocity and polarization. The development of VSP has established it as the chief source of information about the origin of numerous waves of various types discriminated on surface seismograms.

(5) VSP is being employed for the solution of a wide range of problems that can conveniently be subdivided into two large groups: problems associated with the improvement of the effectiveness of surface studies, and problems involving the study of the rocks surrounding the borehole. Problems in the first group include: the process of propagation of waves in the real earth; the nature of excitation conditions; the determination of the origin of recorded waves of various types; the identification of waves and their stratigraphic correlation; the appraisal of potentialities of seismic exploration, in particular, of seismo-geologic conditions (and first of all of increasing the penetration); the selection of optimum observation procedures. All these problems exercise an essential influence on the effectiveness of seismic studies and especially on the validity of the results of geologic interpretation of seismic data. Moreover, in principle, VSP enables a more general approach to the study of the real earth as the conductor of seismic energy to be implemented, and data to be acquired to construct models of actual media corresponding to the main relationships of the wave fields observed.

(6) In the studies of the region around the borehole, VSP in the initial stages of its development solved mainly structural problems. The development of PM VSP, which has provided fundamentally new opportunities for the use of wave dynamics, has enabled lithologic and structural problems also to be solved on the basis of studies of the physical properties of rocks.

The capacity of VSP to study the region around the borehole has considerably enhanced the part played by seismic exploration in the general complex of geologic exploration, and has made it possible to extend seismic studies beyond the traditional stages of the search for and identification of structures for deep drilling and to use them extensively at such important stages as the exploration and exploitation of fields. PM–VSP brought about the birth of a new concept – production seismics. Specific features typical of production seismics, or of seismics in the region surrounding the borehole, are the extensive combination with other kinds of geophysical studies in boreholes and close association with the drilling. The savings achieved with production seismics and its role in the national economy require no comment.

This effect stems from the following potential of production seismics:

– the reduction in the amount of increasingly costly deep drilling by reducing the number of deep boreholes required for the assessment of each field and through a more rational arrangement of the necessary boreholes;

– the study of a deposit and, on the basis of it, the assessment of the prospects of a field at the earliest stages;

– the choice of an optimum plan for working the field and the argumentation to support it;

– monitoring the deposit in the course of exploitation.

(7) In VSP, in contrast to surface observations, the trajectories of particle motion generally coincide with the waves' directions of arrival, and can be utilized for qualitative processing. By studying the trajectories of particle motion, one is able not only to gain a detailed picture of the wave pattern, but also to acquire additional and independent data about the earth. Studies of the polarization of waves in space may prove to be of special significance.

The experience gained confirms the existence of a clear correspondence between specific features of the geologic structure at the points of observation and the polarization parameters.

(8) VSP has substantiated the fundamental proposition of the theory of the propagation of seismic waves, about the part played by various wave types in the real earth, and enabled the exploration potentialities of the chief seismic methods to be appraised. It has been demonstrated that the chief mechanisms in the real earth by which seismic energy is returned to the surface are reflection and curved-path refraction. The reflected waves constitute the main type of waves recorded in the later portion of the seismogram. The real earth is made up of thin layers, and even a homogeneous medium (as indicated by seismic logging) contains numerous reflecting horizons. However, the dominant waves are associated with quite distinct elements of the geologic section (in most cases with individual thin layers or stacks of layers). Against their background, the combined waves have much smaller amplitudes and are observable chiefly in media with a weak velocity differentiation. Thin high-velocity layers in a low-velocity medium may exercise a strong influence on the wave field, being responsible for the formation of intense primary and multiple reflected waves. At great depths, there are, as a rule, horizons with which reflected waves are associated and this provides a fundamental opportunity for increasing the scope of seismic studies, which at present is limited chiefly by the multiples and by the effect of the upper portion of the section.

Head waves are invariably weaker than the transcritically-reflected. In cases of thin high-velocity layers, the intensity of transcritically-reflected waves may be one or two orders of magnitude higher than that of the head waves, which cannot be discriminated either in the later, or in the first arrivals. In such cases the visible first arrivals on the seismograms may be due not to head, but to transcritically-reflected waves. In the case of thick layers, the increase of the velocity with depth is of major importance and the curved waves, whose intensity depends greatly on the velocity gradient are the first to be recorded at great separations from the source. A thin-layered medium as a rule contains many reflecting horizons, and very weak head waves are associated only with some of them. The origin of waves recorded at great separations is not altogether clear. Further studies are needed to develop the physical foundations and the exploration potentialities of the method of refracted waves.

Converted waves are abundant in the real earth. Reflected converted waves have been identified almost at all depths accessible to observations, and from nearly all interfaces with which the longitudinal waves are associated, including the top of the crystalline basement. This makes it possible to increase the penetration of the method using converted reflected waves. Converted reflected waves associated with thin layers may display high intensities. Converted reflected waves have great exploration potentialities, but their implementation is still awaited.

Converted waves transmitted from below are able to improve the effectiveness of regional studies undertaken to examine the top of the crystalline basement with overlying sedimentary beds with weak velocity differentiation. In media with a sharp differentiation of the sedimentary beds, there are as yet no physical criteria for the identification on surface seismograms of longitudinal and converted waves associated with the same interface. It is necessary to increase the scope of experimentation and methodology to develop the physical foundations and to appraise the exploration potentialities of the method of converted transmitted waves.

(9) The low-velocity zone and the upper portion of the geologic section exert a decisive influence on the wave pattern observed at all separations from the source, both on the surface and down-hole, and in many cases constitute the main stumbling block that limits the exploration capabilities of all seismic observations. This influence makes itself felt first of all in the vicinity of the source, where the heterogeneities of the upper portion of the section cause a comparatively simple pulse generated by the source to be transformed into a complex train of downward waves. The shot impulse and the heterogeneities of the earth in its vicinity should be regarded as a complex mechanism that shapes the transmitted source signature. This influence manifests itself most markedly in the unwanted wave field associated with the upper portion of the section. At small separations from the shotpoint, the main classes of unwanted waves are made up of the reflected and the curved-path waves and their longitudinal, transverse and converted multiples. At great separations (as in the correlational method of reflected waves) they are the combined reflected-refracted and refracted-reflected waves in the early portion of the seismogram, and the low-velocity interference waves in the later portion of the record. The overwhelming number of waves on surface and down-hole records is in some degree or other associated with the upper portion of the geologic section.

(10) The majority of our observations have been performed in the seismic exploration frequency band; however, many of the results already obtained may be of interest for the understanding of the wave pattern observed in other methods of seismic studies – in deep seismic sounding and earthquake seismology. VSP may serve as a basis for modelling actual media in their natural state. In VSP used to study unconsolidated deposits of coastal regions and continental shelves, waves with frequencies up to 2000 Hz are recorded.

(11) The method of vertical profiling is being actively introduced industrially in addition to research applications. VSP, which includes all kinds of borehole seismic studies (logging with *P*- and *S*-waves, seismic well exploration, etc.), is the principal method of observations in boreholes, and in contrast to all other borehole geophysical methods enables the geologic section to be studied below the bottom of the borehole. VSP should be included in the mandatory suite of geophysical studies in stratigraphic test holes and other exploration boreholes. The potentialities of VSP may prove to be of such a decisive importance that in many cases in areas where seismic studies are conducted on a large scale, it might be expedient and profitable to drill special holes (if there are none) for vertical seismic profiling.

The polarization method enables VSP potentialities to be fully implemented, and it is advisable to execute a large proportion of VSP studies everywhere in their polarization version (PM VSP).

(12) The results obtained justify the continuation of large-scale research to develop the method further, so that its potentialities can be progressively implemented. The major trends in this direction are:

– The transition from conventional observations to precise quantitative measurements through the availability of an appropriate methodology and the quality of data obtained. The improvement in the accuracy of measurement of parameters of various wave types assumes a special importance. In this respect the reproducibility of polarization parameters of all waves in recurrent and independent measurements at the same point and with the same excitation is the most rigorous criterion for evaluating the data quality.

– The perfection of measurement techniques (especially in PM VSP) to obtain undistorted spatially oriented, multi-detector records in a wide frequency band (from 5 to 10 up to from 150 to 200 Hz) both in oil wells and mining boreholes.

– The development of a rational complex of down-hole (VSP, sonic logging and production geophysics) and surface production seismic observations with the aim of studying the zone surrounding the borehole, as well as the physical methods of predicting the geologic section.

– The study of the polarization of seismic waves in the real earth and the development of quantitative methods of interpretation of polarization parameters with the purpose of acquiring additional and independent data about the earth.

– The study of the upper portion of the section and the low-velocity zone, and their influence on the seismic disturbance of explosions, the structure of seismograms and the effectiveness of seismic studies generally.

All the VSP experience accumulated to date justifies the assertion that, as the method is developed, its exploration potentialities will grow, and the scope of the problems solved with its aid and its sphere of application will broaden. Accordingly, VSP studies can be extended and financed without reservations.

REFERENCES

1. Gamburtzev, G. A.: *Osnovy seismorazvedki, (Fundamentals of Seismic Exploration)*, Moscow, Gostoptekhizdat, 1959.
2. Gal'perin, E. I.: *Polyarizatzionnyi metod seismicheskikh issledovaniy, (Polarization Method of Seismic Studies)*, Moscow, Nedra, 1977. English translation: *The Polarization Method of Seismic Studies*, D. Reidel Publ. Co., Dordrecht, Holland, 1984.
3. Puzyrev, N. N.: *Measurement of Seismic Velocities in Boreholes*, Trudy VNII Geofiziki, Moscow, Gostoptekhizdat, 1957, Issue III.
4. Gurvich, I. I.: *Seismicheskaya razvedka, (Seismic Exploration)*, Moscow, Nedra, 1970.
5. Rudnitzkiy, V. P.: *Seismicheskiye issledovaniya v skvazhinakh, (Seismic Studies in Boreholes)*, Kiev, Naukova Dumka, 1968.
6. Jolly, R. N.: 'Deep-Hole Geophone Study', *Geophysics* **18,** 662–671, (1953).
7. Levin, F. K. and Lynn, R. D.: 'Deep-Hole Geophone Studies', *Geophysics* **23,** 639–664, (1958).
8. Van-Sandt, O. R. and Levin, F. K.: 'A Study of Cased and Open Holes for Deep-Hole Seismic Detection', *Geophysics* **28,** 8–14, (1963).
9. Gal'perin, E. I.: *Azimutal'nyy metod seismicheskikh nabludeniy, (Azimuthal Method of Seismic Observations)*, Moscow, Gostoptekhizdat, 1955.
10. Gal'perin, E. I.: 'Grouping of the First Kind and Method of Obtaining Multi-Component Azimuthal Seismograms', *Izv. AN SSSR, Seriya geofiziki* **9,** 1081–1098, (1957).
11. Gal'perin, E. I.: 'Determining Coordinates of a Refracting Interface with the Aid of Displacement Direction in Primary Waves Observed in Boreholes', *Izv. AN SSSR, Seriya geofiziki* **4,** 513–524, (1963).
12. Gal'perin, E. I.: 'Kinematic Characteristics of Multiples Associated with Deep Intterfaces', in *Stroyeniye zemnoi kory v oblasti perekhoda ot Aziatskogo Kontinenta k Tikhomu okeanu*, Moscow, Nauka, 1964, pp. 21–39.
13. Gal'perin, E. I.: 'On the Intensity of Head and of Transcritically-Reflected Waves', *Izv. AN SSSR* **10,** 9–14, (1966).
14. Gal'perin, E. I.: *Vertikal'noye seismicheskoye profilirovaniye, (Vertical Seismic Profiling)*, Moscow, Nedra, 1971. English translation: *Vertical Seismic Profiling*, SEG, Tulsa (U.S.A.), 1974, 2nd Ed. Moscow, Nedra, 1977.
15. Gal'perin, E. I. and Frolova, A. V.: 'Three-Component Seismic Observations in Boreholes. I: Vertical Seismic Profiling. II: Studies of Particle Displacement Directions', *Izv. AN SSSR, Seriya geofiziki* **6,** 793–809; **7,** 977–993, (1961).
16. Berdennikova, N. I., Chizhova, M. V., and Byelozerov, A. A.: 'Studying Certain Effective Parameters of the True Earth with the Aid of Down-Hole-Observation Data', in *Voprosy dinamicheskoy teorii rasprostraneniya seismicheskikh voln*, Leningrad, Nauka, 1979, pp. 130–137.
17. Berdennikova, N. I., Byelozerov, A. A., and Chizhova, M. V.: 'On the Experimental Determination of Certain Effective Parameters of the Upper Portion of the Cross-Section', in *Voprosy dinamicheskoy teorii rasprostraneniya seismicheskikh voln*, Leningrad, Nauka, issue XII, pp. 144–150.
18. Berdennikova, N. I. and Kanov, S. V.: 'Determination of Effective Absorption Coefficient from the Wave Pattern Observed in VSP', in *Voprosy dinamicheskoy teorii rasprostraneniya seismicheskikh voln*, Leningrad, 'Nauka', 1979, issue XIX, pp. 130–137.
19. Vavilova, T. I.: 'Calculating Intensities of Multiples with the Aid of the Graphoanalytical Method', in *Voprosy dinamicheskoi teorii rasprostraneniya seismicheskikh voln*, Moscow, Nauka, 1971, issue XI, pp. 90–123.
20. Pogonyaylo, G. G.: 'Theoretical and Experimental Studies of the Wave Field Observed in the Vertical Plane in One of the Areas in the Region of Lower Volga', in *Voprosy dinamicheskoy teorii rasprostroneniya seismichskikh voln*, Leningrad, Nauka, 1970, issue X, pp. 255–274.
21. Pogonyaylo, G. G., Petrashen, G. I., and Molotkov, L. A.: 'On the Effect of Thin High-Velocity Layers in the Covering Bed on the Fields of Reflected Waves at Great Separations from the Epicentre', in *Voprosy dinamicheskoy teorii rasprostraneniya seismicheskikh voln*, Leningrad, Nauka, issue XII, pp. 81–106, ill.

22. Pogonyaylo, G. G. and Gal'chinskiy, B. Yu.: 'On Peculiarities of the Effective Seismic Model at Great Separations (in the Presence of Screening Layers)', in *Voprosy dinamicheskoy teorii seismicheskikh voln*, Leningrad, Nauka, 1977, issue XVII, pp. 202–208.
23. Vasil'yev, A. A., Seval'nev, A. V., Khudzinskiy, L. L., and Shirokov, V. I.: 'Instrumentation for Studying Seismic Wave Fields at Internal Points of the Medium', *Prikladnaya geofizika*, Moscow, Nedra, 1970, issue 60, pp. 69–81.
24. Gamburtzev, A. G., Kondrat'yev, O. K., Lavrov, V. S., Polyanova, T. I., Ratz-Khizgiya, M. I., and Vasil'yev, A. V.: *Some Possibilities for Using Vertical Seismic Profiling in the Common-Depth-Point-Method*, Moscow, VIEMS, 1974.
25. Shekhtman, G. A.: *Pomekhi seismicheskogo kharaktera pri skvazhinnykh issledovaniyakh i sposoby ikh podavleniya, (Interference of Seismic Origin and Methods of Suppressing it)*, ONTI VIEMS, 1975.
26. Shekhtman, G. A. and Makarov, G. F.: 'On the Determination of Elastic Parameters of Rock in Southern Mangyshlak from VSP Data Utilizing Converted Transmitted Waves', *Neftegazovaya geologiya i geofizika* **12**, 38–40, (1970), VNII OENG.
27. Shekhtman, G. A. and Kolenkov, E. V.: 'Utilizing VSP Results in the Complex of Experimental and Methodological Studies in the Moscow Syneclise', in *VSP i uvelicheniye effektivnosti seismicheskikh issednovaniy*, Moscow, ONTI VIEMS, 1971, pp. 91–96.
28. Karus, E. V., Rudenko, G. E., and Khudzinskiy, L. L.: 'Three-Component Four-Point Borehole Probe and Some Results of Its Trials', *Izv. VUZ. Geologiya i razvedka* **3**, 147–154, (1978).
29. Rudenko, G. E. and Khudzinskiy, L. L.: 'Utilization of the Method of Synthetic Seismograms for the Analysis of the Field of Multiples', *Razvitiye metodiki skvazhinnykh seismoakusticheskikh nablyudeniy*, ONTI VNIIYa GG, 1971, pp. 148–162.
30. Rudenko, G. E. and Khudzinskiy, L. L.: 'Constructing Deep Reflecting Interfaces in Borehole Exploration with the Reflection Method Employing Computers, Appraising Accuracy of the Results', *Izv. VUZ. Geologiya i razvedka* **6**, 127–137, (1976).
31. Khudzinskiy, L. L.: 'Application of a Digital Expander in Seismic Borehole Exploration', *Izv. VUZ. Geologiya i razvedka* **2**, 116–119, (1976).
32. Khudzinskiy, L. L. and Rudenko, G. E.: 'A Complex Program for Digital Processing of Seismic Reflection-Method Borehole Exploration', *Izv. VUZ. Geologiya i razvedka* **6**, 119–123, (1975).
33. Glan, Yu. A., Korobov, A. I., and Starobinetz, A. E.: 'Determination of Interval Velocity and Thickness of a Layer not Intersected by the Borehole with the Aid of the Inverse Traveltime Curve of a Reflected Wave Using a Computer', *Express-Informatziya VIEMS. Region., razvedochnaya i promyslovaya geofizika* **1**, 35–40, (1975).
34. Teplikzkiy, V. A. and Geiman, B. M.: 'Contemporary State and Prospects of Evolution of Seismic Exploration Methods', *Trudy VNI geologo-razvedochn.* in-ta, Kamskoye otd, 1973, issue 123, pp. 269–275.
35. Teplitzkiy, V. A., Byelov, V. M., and Boichuk, N. I.: 'Seismic Observations in Boreholes with Waves Excited by a Freely-Falling Weight', in *Geofiz. issledovaniya pri izuchenii geologicheskogo stroyeniya neftegazonosnykh rayonov*, Moscow, VNIIGNI, 1974, issue 155, pp. 30–42.
36. Teplitzkiy, V. A., Artamonov, V. D. *et al.*: 'Amalgamation of Methods of Down- and Up-Hole Seismic Exploration in the Preparation of Structures for Deep Drilling', in *Geofizicheskiye metody poiskov i podgotovki neftegazoperspektivnykh struktur k glubokomu bureniyu*, Moscow, Nedra, 1978, pp. 79–87.
37. Karus, E. V., Rjabinkin, F. A., Gal'perin, E. I. *et al.*: 'Detailed Investigation of Geological Structures by Seismic Well Surveys', *Proc. World Petrol Conf.* **3**, London, 1975, pp. 247–257.
38. Wuenschel, P. C.: 'The Vertical Array in Reflection Seismology – Some Experimental Studies', *Geophysics* **41**, 219–232, (1976).
39. Gal'perin, E. I.: 'Polar-Positional Correlation of Seismic Waves', *Dokl. AN SSSR* **223**, 336–338, (1975).
40. Gal'perin, E. I.: 'Polarization of Seismic Waves and Opportunities for the Improvement of the Effectiveness of Seismic Studies', *Izv. AN SSSR, Fizika Zemli* **2**, 107–121, (1975).
41. Gal'perin, E. I. and Mirzoyan, Yu. D.: 'Polarization Method – a General Method of Seismic Exploration', *Neftegazovaya geologiya i geofizika*, Ref. Nauchno-tekhn. sbornik VNIIOE NG, 1978, No. 9, pp. 38–43.
42. Gal'perin, E. I., Frolova, A. V., Gokhblit, V. A., and Erenburg, M. S.: 'Polarization Method of Regional Studies Utilizing Remote Earthquakes', *Izv. AN SSSR. Fizika Zemli* **10**, 32–41, (1980).
43. Gal'perin, E. I., Mirzoyan, Yu. D., and Oifa, V. Ya.: 'Testing the Polarization Common-Depth-Point

Method in Different Seismo-Geologic Conditions of the Foothills of Western Caucasus', *Express-informatziya: Regional'naya, razvedochnaya i promyslovaya geofizika*, issue 16, Moscow, 1980, pp. 6–15.
44. Gal'perin, E. I., Mirzoyan, Yu. D., and Oifa, V. Ya.: 'Excitation of Seismic Waves in the Polarization Method', *Express-informatziya: Regional'naya, razvedochnaya i promyslovaya geofizika*, issue 16, Moscow, 1980, pp. 1–15.
45. Gal'perin, E. I., Mirzoyan, Yu. D., and Oifa, V. Ya.: 'Results of Employment of the Polarization Method of Vertical Seismic Profiling in the Krasnodar Province', *Express-informatziya VIEMS. Regional'naya, razvedochnaya i promyslovaya geofizika*, issue 16, 1980, pp. 27–39.
46. Berson, I. S.: *Seismorazvedka tonkosloistykh sred, (Seismic Exploration of Thin-Layered Media)*, Moscow, Nauka, 1976.
47. Gal'chinskiy, B. Yu., Byelozerov, A. A., Berdennikova, N. I. *et al.*: *Metod effektivnoy seismicheskoy modeli, (Method of an Effective Seismic Model)*, Izv. LGU, Leningrad, 1975.
48. Gogonenkov, G. N.: 'Application of Synthetic Seismograms for the Identification and Analysis of Reflected Multiples', *Izv. AN SSSR Fizika Zemli* **10**, 71–84, (1967).
49. Lindseth, R. O.: 'Seislog Process Using Seismic Reflection Traces', *Oil and Gas J.* **74**, 67–71, (1976).
50. Gogonenkov, G. N., Zakharov, E. T., and El'manovich, S. S.: 'Predicting a Detailed Velocity Cross-Section from Seismic Data', *Prikladnaya geofizika*, issue 97, Moscow, Nedra, 1980, pp. 58–72.
51. Amirov, A. N., Gal'perin, E. I., Gurvich, I. I. *et al.*: 'Production Seismics – Seismic Studies at the Stage of Exploration and Exploitation of Fields', *Izv. VUZ, seriya Razvedka nedr* **7**, 79–83, (1980).
52. Ryabinkin, L. A.: 'Controlled Directional Reception of Seismic Waves and Principles of Its Resolution', *Voprosy reguliruemogo napravlennogo priema (RNP) seismicheskikh voln Tr. MNI* **18**, 5–38, (1957).
53. Gamburtzev, G. A., Riznichenko, Yu. V., Berzon, I. S. *et al.*: *Korrelatzionnyy metod prelomlennykh voln (Correlational Method of Refracted Waves)*. Moscow, Izv. AN SSSR, 1952.
54. Puzyrev, N. N.: *Interpretatziya dannykh seismorazvedki metodom otrazhennykh voln, (Interpretation of Seismic Exploration Data Obtained with the Method of Reflected Waves)*, Moscow, Gostoptekhizdat, 1959.
55. White, J. E. *Seismic Waves. Radiation, Transmission and Attenuation*, McGraw-Hill, New York, 1965.
56. White, J. E.: 'Elastic Waves along Cylindrical Bore', *Geophysics* **27**, 327–333, (1962).
57. Gal'perin, E. I.: 'Vertical Seismic Profiling at the Stage of Exploration and Exploitation of Fields', *Dokl. AN SSSR* **253**, 1347–1349, (1980).
58. Balch, A. N., Lee, M. W., and Ryder, R. T.: 'Seismic Olefection of Porosity Development in the Madkson Limestone', presented at *47th Annual SEG International Meeting*, Calgary, 1977.
59. Kaplunov, A. I.: *Apparatura dlya seismorazvedochnykh rabot v skvazhinakh, (Instrumentation for Seismic Exploration in Boreholes)*, Moscow, Nedra, 1980.
60. Sivkov, N. R., Syrtzov, A. V., Antipin, Yu. G. *et al.*: *Metodicheskoye rukovodstvo po vertifkal'nomu seismicheskomu profilirovaniyu skvazhin, (Methodological Instruction on Vertical Seismic Profiling in Boreholes)*, Moscow, Nedra, 1978.
61. Garanin, V. A.: 'Combining Methods of Seismic Borehole Exploration and Sonic Logging for Detailed Studies of Tectonics of Coal Fields', *Express-informatziya. Regional'naya, razvedochnaya i promyslovaya geofizika*, No. 17, 1979.
62. Bondarev, V. I. and Pisetzkiy, V. B.: 'Vertical Seismic Profiling (VSP) of Engineering-Geological Boreholes for Construction Purposes', *Inzhenerno-stroitel'nyye izyskaniya*, No. 1, Stroyizdat, 1976, pp. 48–54.
63. Ryabkov, V. V. and Marifullin, S. S.: 'Effect of Conditions of Installation and of Design of the Borehole Geophone on the Qualfity of Registration of Elastic Vibrations', in *Razvedochnaya geofizika*, Moscow, Nedra, 1976, Vol. 71, pp. 49–56.
64. Clamped Detector, Gulf Research Development Co., Patent USA 166–212 (E 21 b 23/00), N 3777814.
65. Gal'perin, E. I., Vorovskiy, L. M., and Gal'perina, R. M.: 'Seismologic Observations in Boreholes and Potentialities for Improving the Useful Sensitivity of Instrumentation', *Izv. AN SSSR Fizika Zemli* **2**, 79–89, (1976).
66. Wuenschel, P. C.: *Reproducible Shot-Hole Apparatus*, Gulf Research and Development Co., Patent USA, 166–207 (G 21 V 17/02) N 3812912.
67. Barkhudariyan, A. A.: 'An Attempt to Improve Resolution of Seismic Exploration in Conditions of South-West Uzbekistan', *Express-informatziya VIEMS. Regional'naya razvedka i promysl. geofizika*, No. 6, 1975, pp. 52–57.

68. Wuenschel, P. C.: *Precision Seismology*. Gulf Res. and Develop. Co., Patent USA, cl. 340–15,5 BH (G 01 V 1/40) N 3876971.
69. Polshkov, M. K., Shneerson, M. B., Mironova, L. V. *et al.*: 'Combining Methods of *P* and *S* Waves on the Basis of Common Sources of Excitation', *Prikladnaya geofizika* **93**, 14–25, (1978).
70. Scheba, Yu. G., Lipskiy, E. A., and Kulemina, I. P.: 'Experience Gained in Applying the Vibro-Seismic Method Technology in Vertical Seismic Profiling', in *Voprosy rudnoy geologii v Kazakhstane*. Alma-Ata, No. 7, 1974, pp. 139–145.
71. Bykov, A. A., Kalinkin, G. N., Vishnyakov, V. I. *et al.*: 'Testing Electric Pulsed Excitation Source (EIVUK) in the VSP Method', in *Razvedochnaya geofizika*, Moscow, Nedra, Vol. 63, 1974, pp. 8–12.
72. Kolenkov, M. K., Shekhtman, G. A., Shneerson, M. B. *et al.*: 'Seismic Borehole Studies using a Non-Explosive Source in the Pripyat' Depression', *Prikladnaya geofizika*, Moscow, No. 96, 1979, pp. 3–12.
73. Mirandov, V. L. and Zhuravlev, A. S.: 'On the Method of Vertical Seismic Profiling in Porous Deposits', *Izv. VUZ. Geologiya i geofizika* **11**, 65–71, (1975).
74. Deubel, K.: *Die Möglichkeiten der Untersuchung dynamischer Besonderheiten seismischer Wellen mit Hilfe von Vertikalprofilierungen*, Geophys. Veroff., Karl Marx-Univ. Leipzig, Bd 1, No. 2, 1975, pp. 31–42.
75. Heimichen, W. D.: *Zur Anwendung von hochauflösender Formfiltern auf bei seismischen Vertikalprofielielierungen gewonnenes Messmaterial*, Fortschritte der Seismic, Berlin, No. 1, 1975, pp. 71–76.
76. Nakhamkin, S. A.: 'Mathematical Algorithms for Subtracting Regular Unwanted Waves in the Discrimination of Seismic Waves', *Izv. AN SSR. Fizika Zemli* **I**, No. 6, 26–35; **II**, No. 9, 23–37, (1966).
77. Potapov, O. A.: 'Analyzing VSP Data with the Optical Filtering Method', in *Vertikal'noye seismicheskoye profilirovaniye i uvelicheniye effektivnosti seismicheskikh issledovaniy*, Moscow, VIEMS, 1971, pp. 26–31.
78. Demidenko, Yu. B.: 'Some Computorized Methods of Processing VSP Results', in *Geofizicheskiye issledovaniya na Ukraine*, Kiev, Tekhnika, 1972, pp. 79–87.
79. Malovichko, A. A., Levchenko, V. K., Akhmatov, E. E. *et al.*: 'A Method of Digital Processing VSP Data with the Aim of Studying the Velocity Structure of a Cross-Section', *Tr. VNI geol. razved. neft. in-ta* **176**, 78–82, (1977).
80. Tabakov, A. A., Rakhimov, R. G., and Shashenev, M. G.: 'Method of Predicting the Cross-Section Below the Bottom of an Exploration Borehole from the VSP Data Utilizing the Method of Solution of Inverse Dynamical Problems', in *Geofiz. issledovaniya na neft' i gas v Uzbekistane*, Tashkent, SAI GIMS, Vol. 27, 1977, pp. 98–100.
81. Tabakov, A. A.: 'Inverse Filtering of VSP Data Utilizing Preliminary Wave Discrimination', in *Geofiz. issledovaniya na neft' i gas v Uzbekistane*, Tashkent. SAIGIMS, Vol. 27, 1977, pp. 105–108.
82. Bykov, I. A. and Tikhonova, I. M.: 'Algorithms for Digital Processing of Azimuthal Borehole Observations', in *Voprosy dinamicheskoy teorii rasprostraneniya seismicheskikh voln*, Leningrad, Nauka, Vol. 20, 1981, pp. 135–155.
83. Aleksandrov, S. I.: 'Methods of Polarizational Processing of Seismic Signals', in *Regional'naya, razvedochnaya i promyslovaya geofizika*, Moscow, VIEMS, Vol. 22, 1977, pp. 37–55.
84. Agayev, Kh. B., Binkin, I. G., and Baymakova, L. A. *et al.*: 'A Sub-System for Processing Data of the Polarization Version of the Common-Depth-Point Method on the EVM ES Computer in the SUS-3-System', *Express-informatziya VIEMS, Regional'naya, razvedochnaya i promyslovaya geofizika*, Vol. 23, 1960, pp. 6–18.
85. Berzon, I. S., Epinat'yeva, A. M., and Pariyskaya *et al.*: *Dinamicheskiye kharakteristiki seismicheskikh voln v real'nykh sredakh, (Dynamic Characteristics of Seismic Waves in the True Earth)*, Moscow, AN SSSR, 1962.
86. Lapin, S. I.: 'VSP Traveltime Curves Observed in Anisotropic Media', *Geologiya i geofizika* **8**, 97–104, (1978).
87. Demidenko, Yu. B.: 'Tracking and Utilizing Multiple Reflections in VSP', in *Metodika seismicheskikh issledovaniy*, Kiev, Tekhnika, 1967, pp. 14–29.
88. Teplikzkiy, V. A., Byelov, V. M., Glan, Yu. R. *et al.*: 'Method of Inverse Traveltime Curves (Procedure of Observations and of Interpretation)', *Tr. VNI geol-razved. in-ta*, Vol. 3, 1971, pp. 22–45.
89. Rudakov, A. G., Golikova, G. B., and Byelozerov, A. A.: 'On the Control of Shape of the Direct Wave in Seismic Studies', in *Voprosy dinamicheskoy teorii rasprostraneniya seismicheskikh voln*, Vol. 12, Leningrad, 'Nauka', 1973, pp. 167–175.

90. Kogan, S. Ya.: 'On the Effect of Absorption on the Shape of a Seismic Pulse', *Izv. AN SSSR. Seriya geofiziki* **90,** 1280–1289, (1961).
91. Anstey, N. A.: 'A Note on the Seismic Pulse Recorded from a Mine Explosion', *Geophysics, Prosp.* **6,** No. 4, (1958).
92. Ricker, N.: 'The Form and Laws of Propagation of Seismic Wavelets', *Geophysics* **18,** No. 1, (1953).
93. Karayev, N. A. and Prokator, O. M.: 'Field of Direct Waves Formed in the Upper Portion of the Cross-Section in Mining Regions', *Metody razvedochnoy geofiziki – Rudnaya seismorazvedka*, Leningrad, 1978, pp. 59–65.
94. Averbukh, A. G.: 'Propagation of Seismic Pulses in a Medium with an Absorption Coefficient Linearly Changing with the Frequency', *Prikladnaya geofizika*, Moscow, Nedra, 1970, Vol. 61, pp. 7–20.
95. Berdennikova, N. I. and Limbakh, Yu. I.: 'On the Spectral Peculiarities of Seismic Pulses Registered in Boreholes', in *Voprosy dinamicheskoy teorii rasprostraneniya seismicheskikh voln*, Leningrad, 'Nedra', 1966, Vol. 8, pp. 136–144.
96. McDonal, F. J. *et al.*: 'Attenuation of Shear and Compressional Waves in Pierre Shale', *Geophysics* **23,** No. 3, 421–433, (1958).
97. Bereznev, V. A. and Malovichko, A. A.: 'Some Aspects of Studies of Seismic Pulses', in *Voprosy obrabotki i interpretatzii seismicheskikh nabludeniy*, Uch. zap, Permskogo un-ta, 1974, No. 12, pp. 68–76.
98. Berlage, H. R.: 'Recherches sur le debut d'une phase', in *Publications du Bureau central seismologique international*, Ser. AL, Toulouse, 1924.
99. Pros'kin, A. D. and Udintzeva, A. N.: 'Applying VSP for Experimental Studies of the Process of Formation of Seismic Waves in Explosions near the Interfaces', in *VSP i uvelicheniye effektivnfosti seismicheskikh isslednovaniy*, Moscow, VIEMS, 1971, No. 26, pp. 84–91.
100. Tatel, H. E. and Tuve, M. A.: 'Seismic Observations at one Kilometer Depth', in *Contributions in Geophysics*, London, Pergamon Press, 1958, Vol. 1, pp. 84–91.
101. Gamburtzev, A. G.: 'Experimental Determination of Parameters of the Upper Portion of the Cross-Section by Seismic Methods', *Izv. AN SSSR. Fizika Zemli* **3,** 33–43, (1968).
102. Muzyka, I. M., Mezhbey, V. I., and Kutzenko, E. Ya.: 'VSP and Opportunities for Improving the Effectiveness of the Common-Depth-Point Method', in *Vertikal'noye seismicheskoye profilirovaniye i uvelechineniye effektovnosti seismicheskikh issledovaniy*, Moscow, VIEMS, 1971, pp. 12–17.
103. Krivonosov, V. M.: 'Results of VSP Operations in the Volgograd Province', in *VSP i uvelicheniye effektivnosti seismicheskikh issledovaniy*, Moscow, VIEMS, 1971, pp. 65–70.
104. Gal'perin, E. I., Gal'perina, R. M., Koval'skaya, I. Ya. *et al.*: 'On the Origin of Unwanted Waves Registered in Extensive Adyr Areas of Central Asia', in *Metodika i resul'taty kompleksnykh glubinnykh geofizichskikh issledovaniy*, Leningrad, Nedra, 1969, pp. 223–236.
105. Roller, A. B. and Ronin, A. L.: 'Experience Gained with VSP Used to Study the Origin of Wave Fields in Conditions of Khibin Alkaline Mass', in *Razvedochnaya geofizika*, Moscow, Nedra, 1974, Vol. 64, pp. 63–70.
106. Alekseyev, A. S., Gal'perin, E. I., and Gal'perina, R. M.: 'Computerized Correlation of Total Oscillation Vector from VSP Seismograms', in *Vertikal'noye seismicheskoye profilirovaniye i uvelicheniye effektivnosti seismicheskikh issledovaniy*, Moscow, VIEMS, 1971, pp. 39–47.
107. Levchenko, V. K. and Malovichko, A. A.: 'On the Determination of Relative Intensities of Reflected Multiples from VSP Data', in *Voprosy obrabotki i interpretatzii geofizicheskikh nabludeniy*, Uch. zap. Permskogo un-ta, 1975, No. 357, pp. 33–42.
108. Ryabkov, V. R., Sakhipov, Kh. G., and Asriyantz, L. Ya.: 'Appraising the Intensities of Reflected Multiples from Vertical Profiling Data', *Express-informatziya. Regional'naya, razvedochnaya i promyslovaya geofizika*, 1973, Vol. 18, pp. 8–29.
109. Gogonenko, G. N.: 'Appraising the Intensities of Reflected Multiples from Vertical Seismic Profiling Data', *Prikladnaya geofizika*, Moscow, Nedra, 1971, Vol. 63, pp. 6–17.
110. Skumbin, I. M. and Shirokikh, V. L.: *On the Effect of the Low-Velocity Zone on the Structure of Seismic Records*, Uch. zap. Permskogo un-ta, 1976, No. 359, pp. 17–26.
111. Gurvich, I. I., Zaidel'son, I. I., and Daddzho, Yu. R.: 'Technique of Using the Same Boreholes as Observation and Shotholes in Seismic Exploration Employing the Reflection Method', in *Razvedochnaya geofizika*, Moscow, 'Nedra', 1973, Vol. 58, pp. 24–29.
112. Bobrovnik, I. I.: 'Some Seismic Parameters of Peat Fields', in *Geofizicheskiy sbornik*, Moscow, Nedra, 1971, pp. 3–9.

113. Epinat'eva, A. M. and Yukhnin, N. F.: 'Refracted Wave from the Crystalline Basement', in *Neodnorodnost' kristallicheskogo fundamenta po seismicheskim dannym*, Moscow, Nauka, 1977, pp. 56–58.
114. Smirnova, N. O. and Ermilova, N. I.: 'On the Construction of Theoretical Seismograms in the Vicinity of the Origin', in *Voprosy dinamicheskoy teorii rasprostraneniya seismicheskikh voln*, Leningrad, Izd. LGU, 1959, sb. 3, pp. 161–213.
115. Epinat'eva, A. M. and Popova, O. G.: 'Criteria for Establishing the Origin of Waves in the Remote Zone', *Prikladnaya geofizika*, Moscow, Nedra, 1977, Vol. 86, pp. 64–71.
116. Barkhudar'yan, A. A. and Ivlev, V. D.: 'Some Results of VSP in the South–Western Part of the Bukhara and Khiva Oil- and Gas-Rich Region', in *VSP i uvelicheniye effektivnosti seismicheskikh issledovaniy*, Moscow, ONTI VIEMS, 1971, pp. 47–53.
117. Gamburtzev, G. A.: 'On the Possibility of Discriminating Transverse Reflected Waves in Seismic Exploration Utilizing the Reflection Method', *Izv. AN SSSR. Seriya geofiziki* **2,** 155–173, (1939).
118. Lash, C. C.: 'Shear Waves, Multiple Reflections and Converted Waves Found by a Deep Vertical Wave Test (Vertical Seismic Profiling), *Geophysics* **45,** No. 9, 1373–1411, (1980).
119. Molchanov, V. V.: 'Determination of Absorption and Reflection Coefficients of Transverse Waves from Observations of Transverse Waves', *Razvedochnaya geofizika*, Moscow, Nedra, 1976, Vol. 70, pp. 29–32.
120. Naidis, L. M.: 'Utilizing Transverse Waves in Seismic Studies with the Correlational Method of Refracted Waves in the Volgograd Region of the Volga Basin', *Razvedochnaya geofizika*, Moscow, 'Nedra', 1967, Vol. 21, pp. 33–44.
121. Pomerantzeva, I. V. and Mozzhenko, A. N.: *Seismicheskiye issledovaniya s apparaturoy 'Zemlya', (Seismic Studies with the 'Zemlya' Equipment)*, Moscow, 'Nedra', 1977.
122. Myl'tzin, V. N. and Kupolenko, O. V.: 'Using the Polarization Method of Seismic Studies in Regional Operations', *Express-informatziya. Regional'naya, razvedochnaya i promyslovaya geofizika*, 1980, Vol. 16, pp. 40–45.
123. Lang (Peter), D. G.: 'Technique Expands Borehole Data', *Oil and Gas Journal*, July 9 and July 16, 1979.
124. Lindsay, J. P., Dedman, E. V., and Ausburn, B. E.: 'New Seismic Techniques Define Stratigraphic, Lithologic Details', *World Oil*. May, 1978.
125. Korobov, A. I. and Oblogina, T. I.: 'Reproducing Velocity Functions $V(Z)$ from Off-Line Vertical Traveltime Curves', *Vestn. MGU. Seriya geologiya* **4,** 74–77, (1979).
126. Malovichko, P. A. and Silayev, V. A.: 'On the Possibility of Using Vertical Traveltime Curves of Reflected Waves for Computing Layer Velocities', in *Razvedochnaya geofizika*, Moscow, Nedra, 1975, Vol. 68, pp. 65–68.
127. Lipatnikov, V. A.: 'On the Combined Use of Sonic and Seismic Logs for the Solution of Seismic Exploration Problems', *Express-informatziya VIEMS, seriya: Regional'naya, razvedochnaya i promyslovaya geofizika*, Moscow, 1977, Vol. 20, pp. 32–39.
128. Petleevich, G. I. and Potul'nitzkiy, A. V.: 'Comparison of Sonic and Seismic Logs in Deep Boreholes of Carpathian Foothills', in *Geofizicheskiye issledovaniya na Ukraine*, Kiev, Tekhnika, 1971, pp. 73–77.
129. Afanasova, N. V. and Tzimbal, T. M.: 'Computing Velocity Curves from Apparent Resistance Curves for the Construction of a Seismic Model of the Earth in the Byelgorod District of the Kursk Magnetic Anomaly', in *Razvedochnaya geofizika*, Moscow, 1975, Vol. 68, 83–89.
130. Balesta, G. I. and Mokrushin, N. M.: 'Dependence of Propagation Velocity of Elastic Waves on Density of Sedimentary Rock in the Pechora and Upper Pechora Depressions', *Express-informatziya. Neftegazovaya geologiya i geofizika*, 1975, No. 12, pp. 1–13.
131. Rudenko, G. E., Khudzinskiy, L. L., Mirzoyan, Yu. D. *et al.*: 'Plotting Time Cross-Sections from Seismic Borehole Observation Data', *Izv. VUZ. Geologiya i razvedka* **6,** 110–118, (1975).
132. Shopin, Yu. G. and Litvinenko, E. R.: 'Determining the Location of Reflecting Interfaces in the Space between Boreholes from Vertical Seismic Profiling Data', *Zap. Leningradskogo Gornogo In-ta* **2,** 11–16, (1976).
133. Prokator, O. M., Rabinovich, G. L., Afanasova, N. V. *et al.*: 'Some Results of Studies in the Northern Urals', in *Seismicheskiye volnovyye polya v zonakh razlomov*, Moscow, Nedra, 1978, pp. 126–134.
134. Schoenberger, M. and Levin, P. K.: 'Apparent Attenuation Due to Intrabed Multiples', *Geophysics* **39,** No. 3, 278–291, (1974).
135. Zhadin, V. V.: 'Three-Component Measurements of Amplitudes and Propagation Velocities of

Longitudinal and of Transverse Waves in a Deep Borehole', in *Geologiya i geofizika* **10,** 129–136, (1960).
136. Wuenschel, P. C.: 'Dispersive Body Waves. An Experimental Study', *Geophysics* **30,** No. 4, 539–551, (1965).
137. Futterman, W. I.: 'Dispersive Body Waves', *Geophys. Res.* **67,** No. 13, (1962).
138. Karus, E. V. and Pasechnik, I. P.: 'Techniques of Studying Elastic and Absorbing Properties of Rock in its Natural State with Seismic and Sonic Methods', *Izv. AN SSSR. Seriya geofizika* **6,** 515–526, (1954).
139. Bokanenko, L. I.: 'Dispersion and Absorption of Transverse Waves in Vinyl Plastic and in Plexiglass', *Izv. AN SSSR. Fizika Zemli* **8,** 93–100, (1967).
140. Amirov, A. N.: 'Studies of Singly- and Multiply-Reflected Waves from Vertical Seismic Profiling Data', *Geofizicheskiy sb. AN SSSR,* Kiev, 1973, Vol. 55, pp. 86–94.
141. Tabakov, A. A. and Gul'mukhamedov, A. Yu.: 'Potentialities for Appraisal of Parameters of the Earth from Reflection Coefficients Determined from VSP Data', in *Geofizicheskiye issledovaniya na neft' i gaz v Uzbekistane,* Tashkent, SAIGIMS, 1977, Vol. 27, pp. 159–160.
142. Kennett, P., Ireson, R. L., and Conn, P. J.: 'Vertical Seismic Profiles: Their Applications in Exploration Geophysics', *Geophysical Prospecting* **28,** No. 5, 676–699, (1980).
143. Averbukh, A. G., Butznevskiy, E. M., Gel'fand, V. A. *et al.*: *Aspects of Application of Seismic Exploration for the Prediction of Oil and Gas Saturation, of Lithology, of Abnormally High Pressures and of Drillability of Rock,* VNII OENG, Tematicheskiye obzory, Moscow, 1976.
144. Pritzler, L. S., Butuzov, Yu. A., Shadkin, V. I. *et al.*: 'Inter-Borehole Sounding with an Audio Signal in Exploration of Ore Fields', in *Metody razvedochnoy geofiziki. Rudnaya seismorazvedka,* Leningrad, 1978, pp. 56–59.
145. *Method of Seismic Exploration in Boreholes.* USSR patent No. 366436. Publ. 16.I.73 in bul No. 17. Issued to G. I. Aksenovich, E. I. Gal'perin, V. L. Pokidov, A. V. Frolova, and R. N. Khairutdinov.
146. Petrashen', G. I.: *Rasprostraneniye seismichskikh voln v anisotropnykh uprugikh sredakh, (Propagation of Seismic Waves in Anisotropic Elastic Media),* Leningrad, Nauka, 1980.
147. Nikolayev, A. V.: *Seismika neodnorodnykh i mutnykh sred, (Seismics of Heterogenous and Opaque Media),* Moscow, Nauka, 1972.
148. Aleksandrov, S. I., Gamburtzev, A. G., and Starodubskaya, S. P.: 'Direct Seismic Sounding of a Heterogeneous Zone', in *Issledovaniye Zemli nevzryvnymi seismicheskimi istochnikami,* Moscow, Nauka, 1981, pp. 308–324.
149. Silayev, V. A. and Akhmatov, V. K.: 'Studies of Pre-Devonian Deposits and of the Relief of the Crystalline Basement with the VSP Method', in *Geofizicheskiye issledovaniya Uralo–Povolzh'ya,* Perm, 1974, pp. 83–90.
150. Dobrin, Milton B.: *Seismic Exploration for Stratigraphic Traps,* Seismic Stratigr. Appl. Hydrocarbon Explor., Tulsa, 1977.
151. Farr, J. B.: 'High-Resolution Seismic Methods Improve Stratigraphic Exploration', *Oil and Gas* **48,** 182–186, (1977).
152. Taner, M. T. and Sheriff, R. E.: *Application of Amplitudes, Frequency and other Attributes to Stratigraphic and Hydrocarbon Determination,* Seismic Stratigr. Appl. Hydrocarbon Explor., Tulsa, 1977.
153. Alekseyev, A. S.: 'Inverse Dynamic Problems of Seismics', in *Nekotoryye metody i algoritmy interpretatzii geofizicheskikh dannykh,* Moscow, Nauka, 1967, pp. 9–84.
154. Ausburn, B. E., Nemeth, L. K., and Reid, G. B.: 'New Seismic Techniques Aid Petroleum Engineers', *World Oil,* 52–57, (1978).

SUBJECT INDEX